T0314088

Evolution Evolving

Evolution Evolving

THE DEVELOPMENTAL ORIGINS OF ADAPTATION AND BIODIVERSITY

KEVIN N. LALA

TOBIAS ULLER, NATHALIE FEINER

MARCUS W. FELDMAN

& SCOTT F. GILBERT

ILLUSTRATIONS BY DAVID ANDREWS

PRINCETON UNIVERSITY PRESS

PRINCETON & OXFORD

Published by Princeton University Press
41 William Street, Princeton, New Jersey 08540
99 Banbury Road, Oxford OX2 6JX

press.princeton.edu

All Rights Reserved

ISBN 978-0-691-26241-3
ISBN (e-book) 978-0-691-26240-6

British Library Cataloging-in-Publication Data is available

Editorial: Alison Kalett and Hallie Schaeffer
Production Editorial: Jenny Wolkowicki
Jacket design: Karl Spurzem
Production: Danielle Amatucci
Publicity: William Pagdatoon
Copyeditor: Maia Vaswani
Jacket illustrations by David Andrews

This book has been composed in Arno Pro with Helvetica Neue

10 9 8 7 6 5 4 3 2 1

This book is dedicated to our teachers,

Conrad Waddington, Richard Lewontin, and Mary Jane West-Eberhard

CONTENTS

FOREWORD

The title of this book—*Evolution Evolving*—is designed to be read in two ways. The first reading captures the idea that the evolutionary process itself evolves over time, and to this day is still evolving. That implies that the way in which each organism evolves depends critically on how that organism works, and on the evolutionary mechanisms those characteristics afford. Not only do the traits of a given organism differ in their propensities to evolve, but organisms may themselves differ greatly in how effectively they are able to generate and find adaptive solutions. This thesis stands in marked contrast to the historically prevalent view that biologists can understand evolution without understanding "proximate mechanisms."[1] Much of the appeal of Darwin's theory of evolution by natural selection comes from the premise that the same evolutionary mechanisms account for all of life's diversity. However, without undermining the central importance of natural selection and other Darwinian foundations, a new understanding emerging within the contemporary evolutionary sciences implies that, say, yeast, oak trees, and human beings may evolve differently; indeed, that all organisms may possess a characteristic set of evolutionary mechanisms, contingent on how they develop. In this book, we set out to demonstrate that developmental mechanisms contribute centrally to an organism's capacity to evolve and may be of substantially greater evolutionary significance than was historically understood.

The second reading follows from the first. Evolutionary theory is evolving, not just through the steady accrual of new data and technologies but perhaps in a more fundamental way, with the emergence of a new way of explaining evolutionary change. That is a second key idea that we explore in this book. New data call for new ways of thinking: ways in which developmental processes are situated more centrally within evolutionary explanation than they conventionally have been. Our treatment is aligned with several academic fields, or research programs, within the evolutionary sciences that share a developmental perspective, including evo-devo (evolutionary developmental biology), eco-evo-devo (ecological evolutionary developmental biology), developmental systems theory, and the extended evolutionary synthesis. However, we also draw on important new insights from quantitative and population genetics, evolutionary ecology, the human sciences, philosophy of science, and many other fields. We wish to acknowledge a particular debt to Conrad Waddington, Richard Lewontin, and Mary Jane West Eberhard, whose inspirational writings substantially shaped our thinking, and to whom this book is dedicated. Of course, evolutionary biologists vary greatly in the extent to which they regard recent findings as demanding any reconceptualization, as well as on its scale and significance. For that reason, some of the arguments presented in this book may prove contentious. However, like most evolutionary biologists, and most scientists in general, we regard that diversity of perspective as welcome pluralism, indicative of a healthy science.

What the two readings of our title have in common—and the principal thesis that we defend in this book—is that developmental processes do more than impose constraints on selection: they also help explain adaptive evolution, and they do so in every bit as fundamental a sense as the far-better-established converse assertion that evolutionary processes explain developmental mechanisms.

In 1973, Theodosius Dobzhansky famously claimed that "nothing in biology makes sense except in the light of evolution." Yet now leading contemporary evolutionary biologists worry that "nothing in biology makes sense anymore,"[2] a quip that we borrowed for the title of the opening chapter. Bonduriansky and Day drew attention to the "monumental challenge of making sense of a rapidly growing menagerie of discoveries that violate deeply ingrained ideas."[3] No one can be surprised that in the half century since Dobzhansky made his famous claim, new data should come to light suggesting that evolution is more complex and varied than previously thought. That is the fate of all science. However, the bigger question here is whether those new findings have merely added detail and nuance to a long-standing and robust explanatory framework, or whether they might imply a new understanding of evolutionary dynamics or a different causal structure for evolutionary theory. To what extent does any "existential crisis in biology" require a shift in conceptual framework?[4] The founders of the modern evolutionary synthesis emphasized the distinction between developmental and evolutionary processes, and for most of the twentieth century, developmental biology and evolutionary biology were largely separate fields. Arguably, that separation had heuristic value. Now, however, both fields have progressed, and we believe that it may be both timely and productive for them to be brought more closely back together.

This is not a textbook, and a comprehensive account of evolutionary processes and mechanisms is beyond the scope of this book. Our focus is on understanding the developmental origins of adaptation and diversification, and we say little about genetic drift, sexual selection, inclusive fitness, and many other important evolutionary phenomena. Rather, our objective is to provide a picture of what a developmentalist take on evolution might look like, to offer a new vision of how evolution works and convey its logic and beauty. While it is impossible to address these issues without reference to some technicalities, we have tried hard to write in an accessible way, such that individuals from many backgrounds can understand and elaborate on these feedbacks. The constituency for whom we write goes beyond professional evolutionary biologists and extends to anyone with an interest in evolutionary biology. These are exciting times for the biological sciences; more than anything, we hope this book conveys something of that excitement.

The main challenge associated with writing this book was to integrate the evolutionary findings of several academic fields into a coherent and novel perspective on evolution. That has been possible only because the authors possess complementary expertise: Lala (who changed his name from Laland in 2022) has a background in evolutionary and behavioral biology, cultural evolution, and niche construction, Uller in evolutionary ecology, Feiner in evo-devo and molecular evolution, Feldman in evolutionary genetics and cultural evolution, and Gilbert in developmental biology and eco-evo-devo. The authors all share an interest in the philosophy and history of biology.

This book would have been much poorer without the generosity and wisdom of our professional colleagues. We would like to acknowledge the help of the following individuals who provided feedback on one or more chapters, or helped in other ways: Orsolya Bajer-Molnár, Guido Caniglia, Lynn Chiu, Kaleda Denton, Christian Dorninger, Flavia Fabris, Laurel Fogarty, Garrett Fundakowski, Mauricio González-Forero, Paul Griffiths, Nicole Grunstra, Benedikt Hallgrimsson, Laura Sophie Hildesheim, Quentin Horta-Lacueva, Jana Isanta-Navarro, William Jeffery, Jukka Jernvall, Elis Jones, Sven Kasser, Kathryn Kavanagh, Josie Lala-Brown, Tim Lewens, Lucas Mathieu, Ely Mermans, Armin Moczek, Gerd Müller, Deepti Negi, Mike O'Brien, John Odling-Smee, Éadin O'Mahony, Thomas Oudman, Csilla Pákozdy, Mike Palmer, Kevin Parsons, Mihaela Pavličev, Luana Poliseli, Robin Pranter, Luke Rendell, Peter Richerson, Patrick Rohner, Isabella Sarto-Jackson, Stephanie Schnorr,

Helen Spence-Jones, Hari Sridhar, Sonia Sultan, Wataru Toyokawa, Marco Treven, Masahito Tsuboi, Cristina Villegas, Gunter Wagner, Richard Watson, and Jinwen Xie.

Kevin would like to thank the Konrad Lorenz Institute for Evolution and Cognition Research at Klosterneuburg, Austria, which provided a wonderfully stimulating and supportive environment for writing through two fellowships, one of six-months' duration in 2022 and a second of six weeks' duration in 2023. In particular, he is indebted to Guido Caniglia, Eva Lackner, Gerd Müller, Isabella Sarto-Jackson, and all the KLI fellows for their friendliness, enthusiasm, and expertise, as well as to his wife and children for tolerating his absence for long periods, and all the sacrifices that entailed. Participants in two "book clubs," one based at the KLI and St. Andrews and one in the Biology Department at Lund University, provided particular detailed feedback on drafts of the entire manuscript. Kevin and Tobias greatly benefited from a grant from the John Templeton Foundation (ID: 61569, Bringing the EES to the classroom), and the authors acknowledge funding from many other sources (NIH, ERC, DSTL). We are particularly grateful to Linda Hall for administrative assistance, including in helping to compile the references, to David Andrews for producing the beautiful artwork and for his patience in working with finickity scientists who repeatedly asked him to redraw illustrations, to Lynn Chiu and Kasturi Dasgupta for bounteous ideas and help with science communication, and Anne Odling-Smee and her team at Design Science. We would also like to thank the staff at Princeton University Press, who helped so ably and professionally with the production of this book, particularly Michele Angel, Alison Kalett, Dimitri Karetnikov, Hallie Schaeffer, Maia Vaswani, and Jenny Wolkowicki. Finally, we would also like to take this opportunity to thank the many talented evolutionary scientists in diverse fields whose work we cite, on whose initiative, inspiration, and hard work we build.

Kevin, Tobias, Nathalie, Marc, and Scott
August 2023

PART I

Introduction
Why Consider
Development?

1

Nothing in Biology Makes Sense . . . Anymore

Evolutionary biologists have described many remarkable adaptations, but few as curious as the capacity of woodrats from the Mojave Desert in California to feed on creosote bushes.[1] The bushes themselves are impressive, being able to survive for years without water and to flourish by coating themselves with a highly toxic resin that deters almost all herbivores.[2] Despite this, Mojave Desert woodrats feed almost exclusively on creosote, maintaining healthy bodies while consuming quantities of toxin sufficient to kill most other animals.[3] This unusual fare allows the woodrats to exploit a novel dietary niche, largely free from competition (figure 1).

What makes the woodrats particularly fascinating is that their ability to process the creosote relies completely on the detoxifying capability of the bacteria within their guts.[4] When researchers treated the woodrats with antibiotics that wiped out that bacterial community, the woodrats dramatically lost body mass and began to deteriorate on the creosote diet.[5] Conversely, when woodrats that don't consume creosote bushes, and would normally be poisoned by it, were inoculated with the microbiota of Mojave woodrats, they thrived on creosote.[6] Here the *microbiome*—the collective noun for the array of tiny symbionts, including bacteria, archaea, protists, fungi, and viruses, that reside in organisms' bodies—is passed down through the generations by behavioral means, with each cohort acquiring the detoxifying microbes by consuming soil and feces. Experiments found that feeding feces to creosote-naïve woodrats is effective at transmitting the detoxifying bacteria.[7] Mojave woodrats have exploited this dietary niche for hundreds of years through the stable inheritance of bacteria acquired from the external environment.[8]

3

FIGURE 1. Mojave Desert woodrats feed on a toxic diet, thanks to bacteria that they reliably inherit by consuming soil and feces in their environment.

Cross to the other side of America, and just off the east coast a population of humpback whales in the Gulf of Maine has also opened up a new dietary niche. These whales prey on sand lance—an eel-shaped fish that forms large shoals—through an innovative method known as "lobtail feeding".[9] This in- volves a whale thumping the water surface with its tail, which shocks the fish below into tightening their school, and then the whale spirals around the school releasing air from its blowhole, which traps the fish in a net of bubbles, before it finally lunges up from beneath with its mouth gaping to feast on corralled fish.[10] The behavior was first observed in a single individual in 1980 and has subsequently spread to many hundreds of whales in the region. Detailed recording and analysis of the diffusion of this behavior has established that lobtail feeding is a learned trait, which individuals acquire through copying, and which has spread among close associates in a social network.[11] Young whales acquire this highly productive feeding method from older individuals, with the skill passed from one generation to the next as a cultural tradition.

Moving away from the wilds to an Emory University laboratory, researchers in 2014 were astonished by some laboratory mice that mysteriously exhibited a fear experimenters had trained into their grandparents.[12] That is not supposed to happen! For over a century, generations of students have been taught that the inheritance of acquired characteristics is a biological impossibility.[13] A mouse should not be born with knowledge that its ancestors learned during their lifetimes as this clearly violates Weismann's barrier. The "barrier," proposed in 1892 by German evolutionary biologist August Weismann, captures the hypothesis that the cell lineages that produce sperm and eggs are separated from the rest of the body early in development, and hence that whatever happens to the body cannot be inherited. Famously, Weismann severed the tails of mice, observed no reduction in tail length among their offspring, and declared Lamarckian inheritance refuted.[14]

Of course, Weismann was unaware of epigenetic inheritance. The DNA in the nuclei of cells is not naked but clothed in a variety of chemically attached molecules that affect the level of expression of nearby genes.[15] "Methylation" refers to the addition of a methyl chemical group to one of the DNA nucleotide bases.[16] When methyl groups are added to DNA they can suppress the activity of a gene, while their removal can lead to that gene's expression. The Emory University researchers showed that when mice were conditioned to be frightened of a particular smell, their offspring, and their offspring's offspring, retained this fear. That is because the odor entrainment had modified the *Olfr151* gene, which encodes the olfactory receptor specific for this odor, by removing a methyl group from it. Remarkably, this *demethylation* of the *Olfr151* gene was also seen in the sperm of these mice, and indeed their offspring's sperm.[17] The inheritance of such methylation patterns across generations is now well established in plants and some animals.[18] What is still unclear is how events in the mouse's central nervous system triggered demethylation

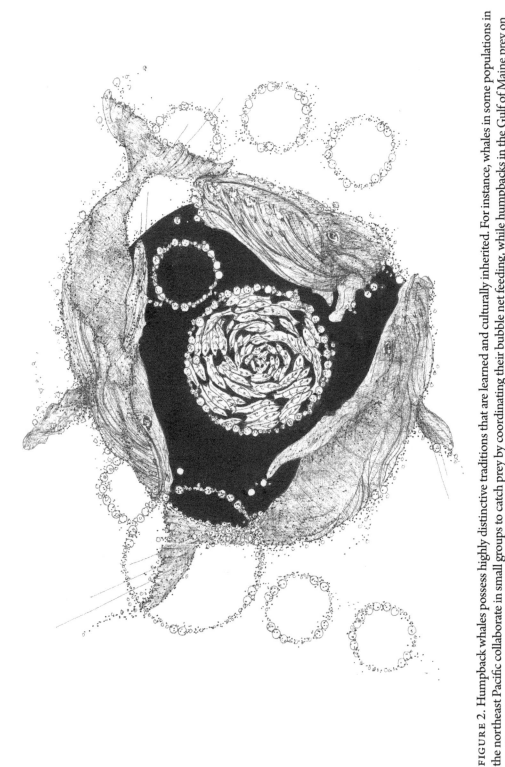

FIGURE 2. Humpback whales possess highly distinctive traditions that are learned and culturally inherited. For instance, whales in some populations in the northeast Pacific collaborate in small groups to catch prey by coordinating their bubble net feeding, while humpbacks in the Gulf of Maine prey on sand lance through a learned innovation known as "lobtail feeding." Each population has its own song, which evolves over time too rapidly to be explained by genetic evolution. These behaviors are passed across the generations by cultural transmission, independently of inherited genetic variation.

in their sperm, but one empirically demonstrated pathway involves the transfer of noncoding RNAs to the sex cells,[19] which has also been found to underlie the epigenetic inheritance of learned information in nematode worms.[20]

These three examples, in different ways, defy the classical view of heredity as mediated solely by gene transmission, and challenge our understanding of how evolution works. Generations of desert woodrats are reliant on other species' genes to exploit a novel foraging niche, and their trait of feeding on a toxic food is stable only because of resources reliably extracted from their ecological environment. The ability to exploit detoxifying enzymes confers a clear fitness advantage, and in the case of the woodrats there is evidence that natural selection has acted on this heritable phenotypic variation.[21] The example is so dramatic that it may come as a surprise that these rodents are broadly representative of many organisms that are equally dependent on their live-in microorganisms to carry out essential functions.[22] Corals rely on microalgae for energy production, legumes require bacteria for nitrogen fixation, and cows couldn't eat grass without the microbial community inside their rumen.[23] Symbiotic microorganisms can be passed from one generation to the next along diverse nongenetic pathways, including inside eggs, seeds, and embryos and through suckling, eating others' feces, or consuming regurgitated foods.[24] Even when symbionts are acquired from the external environment rather than from parents, they can still be surprisingly reliably transmitted, as the woodrat example illustrates.[25] Transgenerational microbial transmission is now recognized as a common, perhaps universal, component of animal inheritance.[26]

The humpback whales are passing foraging information that appears to enhance their biological fitness across the generations through cultural transmission, independently of inherited genetic variation. Biologists have long been aware of cultural inheritance, but have tended to regard it as a special case, germane only to humans. This belief is no longer tenable. In the last fifty years, vast evidence for "culture" has emerged through scientific investigations of a broad array of animals, both in their natural environments and in the experimental laboratory. Numerous species transmit learned knowledge through imitation and other forms of social learning, including dietary information, feeding techniques, predator recognition and avoidance methods, songs and calls, learned migratory pathways, and mate and breeding-site choices.[27] There are now—quite literally—thousands of scientific reports of learned behaviors spreading through natural animal populations by these means.[28] Familiar examples include common chimpanzees fishing for termites with sticks, birds drinking from milk bottles, and birds and whales transmitting songs.[29] Animal culture does more than contribute to inheritance, however. It allows groups of animals to adjust their behavior to match their environment. In the case of the humpback whales, the "adaptations" needed to hunt locally abundant prey did not arise through genetic mutation and natural selection, but via behavioral innovations spread through cultural transmission. Recent analysis suggests that lobtail feeding is a local refinement of the more widespread bubble net feeding, which is similar to lobtail feeding but doesn't involve the initial tail slap that stuns the fish. Intriguingly, other humpback populations in the northeast Pacific have refined bubble net feeding in a different way, creating a new cooperative strategy where groups of whales coordinate their bubble net foraging. In this North Pacific culture, individual whales take on distinctive roles, with some whales releasing bubbles, others making feeding calls, and all or most members of the group feeding (figure 2).[30] In some species—humans included—culture has become the principal means by which the animal adapts to its environment, giving rise to a new form of adaptability.[31] What is more, the spread of cultural knowledge is driving genetic evolution.[32] For instance, killer whale populations have socially learned specializations for particular prey (e.g., fish, dolphins, pinnipeds), and these specializations favored population-specific morphologies and digestive physiologies, known as *ecotypes*, among which reproductively isolated groups emerged.[33] Here, culturally learned dietary traditions have initiated and modified the natural selection of genes associated with morphologies and physiologies that match the whales' learned habits, imposing a direction on genetic evolution.[34]

And the frightened mice represent a "ripping up of the rule books" concerning how biology works and what can and cannot be inherited, as the field of epigenetics reveals hitherto inconceivable mechanisms and phenomena that defy time-honored understanding. Epigenetics is a rapidly developing field of science, and a bewildering variety of mechanisms for the regulation of gene expression have been identified.[35] While historically it was widely accepted that Weismann's barrier prevented environmentally induced changes from altering the germline (i.e., eggs and sperm), in recent years the immutability of Weismann's barrier has been undermined by experimental research demonstrating that epigenetic changes can be inherited in a wide variety of organisms.[36] There is clear evidence that epigenetic changes can strongly affect the fitness of individuals, can be subject to natural selection, and can facilitate genetic adaptation. Experiments in yeast, for instance, show that the natural selection of epigenetic changes can help populations to acquire a genetic resistance to toxins, and when experimentally deprived of this capability, populations often go extinct.[37]

Neither inherited microbiomes nor animal cultures nor epigenetic inheritance is rare in nature, as this book will make clear. A veritable cornucopia of resources other than genes are now known to be passed down the generations, including components of both egg and sperm, hormones, symbionts, epigenetic changes, antibodies, ecological resources, and learned knowledge.[38] For a century, "soft inheritance"—the view that heredity can be changed by lifetime experiences—was regarded as disreputable.[39] The doyen of evolutionary biology, Ernst Mayr, asserted that "the greatest contribution of the young science of genetics [was] to show that soft inheritance does not exist."[40] Today, soft inheritance seems to be everywhere.

Nor is it solely biologists' understanding of heredity that is being challenged. In recent years, science has revealed that there is so much exchange of genetic material across lineages that Darwin's "tree of life" now resembles a tangled network.[41] What we thought were individual organisms have turned out to be communities, which is just one of several reasons why the developing organism can no longer be parsed tidily into separate "genotype" and "phenotype" components, with the former exerting exclusive control over the latter.[42] The familiar suggestion that genes contain "instructions" is being reassessed, as the information to build bodies is distributed across numerous inherited resources and reconstructed during development.[43] Novel insights and findings like the above are pouring out of biological laboratories at a rate that leaves many researchers reeling. The challenge is to make sense of it all!

The founders of the modern evolutionary synthesis laid great emphasis on the distinction between the lifetime of an individual organism from conception to death (its "development") and the biological history of the species (its "evolution").[44] Rightly or wrongly, the view became prevalent that the processes and mechanisms underlying evolution could safely be studied without knowledge of the processes and mechanisms responsible for development.[45] For almost a century developmental biology and evolutionary biology were mostly separate fields.[46] Repeated attempts were made to bring these disciplines back together, but never with more than partial success.[47]

Now, after decades of waxing and waning, enthusiasm for developmental insights is waxing again in the evolutionary community. Invigorated by advances in experimental and theoretical methods, science is shedding new light on the developmental origins of phenotypic variation, evolutionary innovation, adaptation, speciation, and macroevolutionary patterns. This is not without controversy. The evidence that the mechanisms of cellular, molecular, and developmental biology might facilitate the generation, selection, and inheritance of adaptive phenotypic variation has been accompanied by a lively debate.[48] The authors of this book have, in various ways, been active participants in this discussion, an experience that has taught us much, and honed our perspective. We are convinced that the differences of opinion extend beyond the issue of how best to incorporate new biological knowledge, and also relate to how the history of the field is understood, as well as to philosophical issues concerning the scientific process.[49] Part of the controversy also arises from assumptions, often unstated, about how developmental processes generate phenotypic variation. That the pioneers of our perspective—notably Conrad Waddington, Richard Lewontin, and Mary Jane West-Eberhard—each emphasized in their writings that developmental processes are constructive, open-ended, and contingent, and above all not "genetically programmed," we suggest is no coincidence. Accordingly, we devote considerable attention to explaining key aspects of developmental biology, while trying to avoid too much technical detail.

Much of the debate over the role that developmental processes play in evolution relates to how researchers regard the subprocesses that underly natural selection. Harvard evolutionary biologist Richard Lewontin identified three such subprocesses: (1) there must be variation in characteristics among individuals in a population (*phenotypic variation*), (2) some variants must leave more descendants than others (*differential fitness*), and (3) offspring must resemble their parents more than they resemble unrelated individuals (*heredity*).[50] However, the historically dominant view that natural selection is the sole cause of adaptive evolution is tied to the additional, less-apparent assumption that the three subprocesses are effectively autonomous: they feed into one another, but do not modify one another's operation.[51]

Cases like the desert woodrats or the killer whales raise the possibility that developmental mechanisms do more than simply generate variation: they also modify the processes that contribute to differential fitness (e.g., when behavior learned from nonrelatives affects survival) and to heredity (e.g., when symbiotic bacteria are passed on to descendants via the ecological environment). In such instances, the three subprocesses underlying evolution by natural selection become intertwined, and understanding natural selection becomes more challenging. Current controversies concerning extragenetic inheritance (a.k.a. *nongenetic inheritance*),[52] whether developmental mechanisms constrain or facilitate evolution (i.e., *developmental bias*), whether developmental responses to environmental change can direct genetic change (i.e., *plasticity-*

led evolution), and how the activities and outputs of organisms modify selection (i.e., *niche construction*), relate to interactions between Lewontin's
subprocesses. An exciting implication of the aforementioned new data is that
the evolutionary process itself evolves, as the characteristics of evolving populations and their modes of inheritance influence how natural selection operates. To make sense of it all, evolutionary researchers may need to reconsider
both the structure of evolutionary theory and the nature of evolutionary explanations. These themes are the focus of this book.

In 1973, the influential evolutionary biologist Theodosius Dobzhansky
boldly asserted that "nothing in biology makes sense except in the light of
evolution."[53] Nearly fifty years later, leading evolutionary biologists worry that
"nothing in biology makes sense anymore" and accept the "monumental challenge of making sense of a rapidly growing menagerie of discoveries that violate deeply ingrained ideas."[54] Here, by contrast, we argue that there is a natural
order, a richness, and even an elegance to adaptive evolution implied by the
"new biology"—but its comprehension requires thinking more broadly about
evolution. Two fields of biology that historically became separated need to be
brought back together. Dobzhansky's famous assertion can be rescued if paired
with the reciprocal dictum: *nothing in evolution makes sense except in the light
of development.*[55]

2

Rodents' Teeth and Raptors' Toes

Stephen Jay Gould let the rest of us in on "an old paleontological in-joke," proclaiming that "mammalian evolution is a tale told by teeth mating to produce slightly altered descendant teeth."[1] Teeth play this central role in paleontology because they are the most durable part of our anatomy and hence are able to be well preserved as fossils, and because their highly diverse shapes and sizes reflect the dietary habits of their bearers.[2] Mastodon molars, for instance, have pointed cusps suitable for crushing branches and twigs, while mammoth molars, like those of modern elephants, are characterized by ridges that provide the surfaces needed for grinding tough grasses. The number, spacing, and size of molars are evolutionary adaptations that allow the organism to chew its particular food efficiently. These variations, in turn, reflect changes in molar development.[3] Thus, understanding tooth evolution demands the integration of insights from paleontology, developmental biology, and ecology.

Mammalian teeth provide a beautiful illustration of how knowledge of developmental biology can be used to shed light on biological evolution.[4] The developmental biology of tooth morphology is reasonably well understood. Molar teeth form in a sequence from front to back.[5] The growth of each tooth is inhibited by the preceding tooth in the sequence, but enhanced by various activating substances. Experiments with mice, conducted by evolutionary developmental biologists Kathryn Kavanagh, Jukka Jernvall, and colleagues, have established that manipulating these inhibitory and activating signals in the first molar changes the size proportions of the entire row of teeth.[6] This and other studies demonstrate that molar teeth develop as a module, in a remarkably predictable manner. For instance, knowing the size of two teeth allows the size of a third to be predicted, as the middle molar is the average of its two neighbors.[7]

Such observations led to an "inhibitory cascade" model of tooth development, which links relative levels of gene activation and inhibition in the molars to their size proportions.[8] Intriguingly, a computational analysis, based on known developmental mechanisms that were identified through experiments with laboratory mice, can accurately predict the evolutionary diversity of teeth found in a large sample of mammalian species.[9] Moreover, by manipulating key developmental interactions in the laboratory, molar teeth with cusps and ridges similar to those found in natural populations can be produced.[10] Thus evo-devo experiments link developmental mechanisms to macroevolutionary patterns.

The distribution of phenotypic variation can be visualized using a concept known as *morphospace*, which is a representation of the possible form or shape of an organism in which each axis corresponds to a variable describing some character. For instance, any human could be depicted as a point in a three-dimensional space in which one axis plots leg length, another arm length, and a third torso length. If all humans were plotted as points in the same graph, the distribution of the points would give us an indication of the range of variation in these three measures. What is remarkable about the tooth data is that virtually all mammals lie on a single path through morphospace.[11] Herbivores tend to have more even-sized teeth, and carnivores less even-sized, but all are positioned on the same line.[12] Rather than being free to realize any shape or number of teeth, robust mechanisms of development have created pathways along which natural selection may act; selection pushes evolving populations up and down these paths.[13]

Historically, such cases have often been understood as "developmental constraints"—the idea that developmental processes prevent some "impossible" morphologies from being accessible to selection. However, the existence of exceptions to the pattern—rodents, such as voles, with different ratios of teeth lie away from the line[14]—demonstrates that this way of thinking isn't quite right. The effect of development is both more subtle and more interesting. Developmental mechanisms do not screen off regions of morphospace, leaving natural selection free reign to determine morphology elsewhere. Rather, developmental processes play an important role in *adaptive* evolution by biasing the entire landscape of phenotypic variation available for selection. Developmental processes produce some combinations of traits more readily than others, thereby explaining, for instance, why some characteristics of teeth are usually coupled together.[15] Strictly, developmental processes *create* the landscape for selection, since a phenotype cannot be selected before it exists. In this manner, the mechanisms of development partly determine which features evolve, including features that are adaptive. The existence of rare forms in largely empty regions of morphospace demonstrates that such phenotypes were not impossible to produce. However, the existence of far more variability in other regions means that most natural selection is concentrated in those phenotypically variable zones. As a result, the more-probable forms of teeth are what can be found in nature. The term "developmental bias" captures the idea that developmental systems tend to generate some characters, or trait combinations, more readily than others.[16] (Developmental bias contributes to broad phylogenetic patterns of phenotypic variation, which means "phylogenetic constraints" should not be regarded as an alternative explanation to developmental bias, but rather share the same problems as constraint thinking in general).

The inhibitory cascade model not only indicates that evolutionary diversity in tooth morphology among mammals is shaped by the mechanisms by which teeth develop but also generates predictions as to what developmental and genetic changes should accompany the adaptive diversification of teeth. Genetic analysis using inbred lines of mice has identified regions of the genome associated with inhibitory signaling proteins.[17] What is more, analyses of wild and laboratory populations have shown how changing the development of mouse molars can cause tooth morphologies that are seen in other species, and have established that certain changes are much more likely than others.[18] As expected, the variance in tooth size is indeed greatest along the axis of divergence predicted by the model.

That is impressive. However, in recent years developmental biologists have been able to go further and apply the same inhibitory cascade model to predict the development and diversification of other anatomical structures that also arise through a similar process of sequentially forming segments. Finger and toe bones—known as *phalanges*—vary widely in number and size proportions across vertebrates, yet successive bones within a finger or toe exhibit highly predictable relative proportions, for essentially the same reasons as molar teeth.[19] Digits are thought to have evolved with the dispersal of amphibians onto land around four hundred million years ago.[20] As a consequence, the same rules of phalangeal variation are seen in birds, reptiles, plesiosaurs, ichthyosaurs, amphibians, and cetaceans,[21] and probably other tetrapods too. Once again, irrespective of whether they are nearly equal in size or show a

FIGURE 3. The sizes of the toe bones (phalanges) of birds, reptiles, amphibians, and cetaceans have all evolved along a highly specific pathway, predicted by a model of development. Remarkably, the model was devised based on experimental studies of molar tooth development in mice, and is equally effective in predicting tooth size and number in mammals.

skewed size gradient, phalanges have evolved along a highly specific path predicted by the inhibitory cascade model.[22] For instance, in birds, running is associated with unequal-length phalanges and perching with more equal-sized toe bones.[23] Again, there are exceptions that put to bed any suggestion that "absolute constraints" are operating to render certain ratios of toe sizes impossible. In raptors and some other birds, for instance, an elongation of the bones at the fingertips has apparently evolved convergently in multiple lineages, under selection for grasping (figure 3).[24] Thus, even the deviations from "normality" exhibit regularities.[25] Evolution is far from an "anything goes" affair.

Such studies are exciting as they help to make evolutionary biology a more predictive science.[26] They also help to explain why some adaptations exist and others do not, and why some characters are more evolvable than others. Why, then, have these ideas received comparatively little attention until recently?

The detailed answer must wait until the next chapter.[27] For the moment it suffices to point out that when biases in the generation of phenotypic variation are understood as "constraints," they can at best explain why evolution or adaptation has *not* occurred.[28] Quite naturally, evolutionary biologists focus on what they perceive to be the causes of adaptation and diversification, rather than factors that thwart such processes. A different perspective, that of evo-devo, was what motivated the previously mentioned mouse work. From an evo-devo perspective, developmental bias partially explains what evolution and adaptation *do* occur, rather than what do not, since it is focused on the variation that is commonly produced. This makes developmental bias a much more significant concept in evolutionary explanation than developmental constraints have been historically. Rodent teeth and raptor toes look like they do because the way creatures develop makes those characteristics more likely to arise.

But there is more to the distinction between "bias" and "constraint" than that—a fundamental difference that lies at the very heart of evolutionary causation. Developmental processes bias the variation that is subject to selection, but those developmental mechanisms themselves evolve through natural selection.[29] In continual interactive cycles, developmental processes bias what gets selected, but then selection modifies the developmental processes that create developmental bias. This process of *reciprocal causation* guides the evolution of morphology, and indeed all aspects of the phenotype.[30] To disregard the causal role of development in evolution on the grounds that it is a product of selection, as is common, is questionable reasoning, a point to which we return in chapters 12 and 14.[31]

A Better Explanation?

Contrast the above evo-devo perspective with how evolutionary geneticists address the integration and coevolution of anatomical features.[32] Evolutionary geneticists are actually very interested in these issues, but, rather than discussing mechanisms of development, they focus on the underlying genetic architecture of traits.[33] In quantitative genetics this entails estimating statistical variation among relatives compared with the variation across the population as a whole. For instance, evolutionary ecologist Dolph Schluter showed that the evolution of morphological diversity in the three-spined stickleback can be predicted with knowledge of the additive genetic variances and covariances of their traits, with populations tending to evolve in a direction that approximates the dimension of greatest genetic variation (known as "g_{max}" derived from the **G** matrix in quantitative genetics).[34] Schluter concludes that the available heritable variation "constrains" evolution, and asserts that "adaptive differentiation occurs principally along genetic lines of least resistance."[35] Other investigations of this "genetic channeling" hypothesis report similar conclusions.[36] That such analyses commonly state that "genetic constraints predict evolutionary divergence" might seem to imply that the above evo-devo studies contribute little that is genuinely new, and that the insights gleaned are just another means of doing what evolutionary geneticists can already do—namely, apply knowledge of covariation among traits to predict evolutionary divergence.[37]

Are developmental bias and genetic channeling just two ways of tackling the same problem, but at different levels of analysis?[38] Yes, and no. It is well established that similar genetic covariance between traits can result from vastly different underlying patterns of developmental integration of those traits. This means that in genetic channeling studies, the precise source of the morphological variation available to selection—the developmental mechanisms involved—remains unclear.[39] Covariation exists between traits, but it is not known why. At best, correlational genetic analyses go part of the way to determining the actual developmental mechanisms that are evolving, revealed by subsequent developmental analyses.[40] Minimally, a compelling argument can be made that evo-devo studies are complementary to such genetic analyses, providing otherwise missing information about such mechanisms.

But could we go further? Is there a sense in which an evolutionary approach that explicitly incorporates development might, at least sometimes, provide a *better* explanation than one that does not?[41] In an article entitled "Dissecting Explanatory Power," philosophers of science Ylikoski and Kuorikoski identified five criteria through which the relative merits of alternative explanations could be objectively evaluated. These are (1) *precision* (good explanations provide a more detailed account than poor explanations), (2) *nonsensitivity* (good explanations apply broadly across a range of background conditions and are not highly sensitive to context), (3) *factual accuracy* (while idealizations simplify reality and yet are still useful, better explanations contain fewer falsehoods than poorer explanations), (4) *degree of integration* (better explanations fit with existing knowledge, and thereby stimulate insights or answers to further questions), and (5) *cognitive salience* (good explanations are readily understood).[42]

The first of these, *precision*, we have already discussed. Evolutionary developmental biology contributes here by adding additional detail about the developmental mechanisms that generate the observed trait covariation. For instance, we learn that the sizes of successive segments, whether they are teeth, digits, or limbs, are not independent, because earlier-developing segments inhibit the growth of later-developing segments. Note, this developmental knowledge is helping to answer *why* as well as *how* questions. The inhibitory cascade is part of the explanation for why particular numbers, sizes, and shapes of teeth or phalanges evolved, when other solutions might have done the job. The approach also establishes which of countless hypothetical patterns of developmental integration of traits could be responsible for the observed variance-covariance relationships manifest in the **G** matrix, and links these processes to particular genes, or quantitative trait loci.

The evo-devo account is also relevant to the issue of *nonsensitivity*, as the inhibitory cascade model generates predictions concerning patterns of trait covariation and evolvability that, with respect to teeth, seem to hold across virtually all mammals, and for some other anatomical structures across most vertebrates. Where exceptions are found they can typically be explained by recourse to the inhibitory cascade mechanism, for instance, as extensions of the model—as in the case of the elongated toes of grasping birds.[43] In effect, this model provides explanations and predictions of phenotypic covariance and evolvability that apply, quite literally, to thousands of species of animals, across evolutionary time spans of hundreds of millions of years. In contrast, the statistical approach of quantitative genetics has thus far been restricted to making predictions concerning the evolution of only those populations in which genetic covariances have been measured, with the expectation that the "bias to the direction of evolution should be temporary and diminish with time."[44] That the **G** matrix is generally likely to be reliable only for short-term prediction is widely accepted within quantitative genetics: that is all it is expected to do.[45] We discuss this issue further in chapter 12.

The developmental perspective also provides information about the *degree of integration*. We have seen how the explanation for the mechanisms of tooth development fits with preexisting knowledge from developmental biology concerning other anatomical structures that also arise through a similar process of sequentially forming segments. This concordance has allowed predictions concerning the evolvability of molar teeth in mammals to be generalized to make novel and successful predictions concerning the evolvability of phalanges, limbs, and other structures. Conversely, in the absence of knowledge about mechanism, it is less clear how generalizations to other traits that might evolve in a similar manner could be made solely on the basis of statistical knowledge of genetic architecture and covariance.

The other criteria, *factual accuracy* and *cognitive salience*, we will assume, for the moment, are broadly equivalent across explanations. Cognitive salience is a somewhat personal criterion, with many evolutionary geneticists and evolutionary developmental biologists likely to find the detailed methods of their own field easier to comprehend than those of the other, although the core insights of both explanations are reasonably comprehensible. However, in later chapters we will draw attention to situations in which incorporating knowledge of development leads to more factually accurate explanations.

It remains to be established to what extent this comparison is fair or representative. Clearly, not all studies that include development can be expected to identify mechanisms and generate models that are both as widely applicable and as successful as the inhibitory cascade model. Nor are we particularly concerned with the relative merits of the methodologies of different fields: different approaches each have their strengths and weaknesses, and later in the book we will describe quantitative genetics studies that make effective long-term predictions about natural selection. Rather, the comparison suffices to illustrate how a consideration of developmental phenomena, at least in principle, can help evolutionary biologists to generate better explanations—not just more complete, but *really* better. That argument is central to this book. Evolutionary biology is a thriving field of science, but the need for more powerful explanations is well-documented.[46] By including development, evolutionary analyses can potentially provide more precise, more integrative, and more robust accounts for the establishment and maintenance of Earth's biodiversity.

3

How the Turtle Got Its Shell

The science of evolutionary biology has three principal goals—or, perhaps, two goals and an aspiration. One is to reconstruct the history of life on Earth. Evolutionary biologists seek to unveil the rich and often complex ancestry of living forms, tracing lineages back through time to the first appearance of life, and classifying them into groups of related forms. This encompasses the identification and dating of key events, such as the origins of eukaryotes, flowering plants, or tetrapods, and of major adaptations such as feathers, as well as documenting how those lineages and traits have changed over time. The study of fossil ammonites, the construction of a molecular phylogeny for jumping spiders, the unexpected grouping of elephants, elephant shrews, and aardvarks into a common lineage, and the use of Y-chromosome genomic variants to map human dispersal from Africa are examples of this historical reconstruction.[1]

Uncovering life's historical record is one goal of evolutionary biology. A second is to understand the processes that might have brought about the prodigious diversity and exquisite adaptations that living forms exhibit. Evolutionary biologists want to understand how organisms evolve new forms over time, including identifying the *causes* of changes in gene and phenotype frequencies, and of adaptations. How could a fish (*Phycodurus eques*) come to look so stunningly like seaweed, or a caterpillar (*Adelpha serpa*) so amazingly resemble lichen that it can hardly be seen? How is it possible that hummingbirds should have beaks that appear to be ideally shaped to extract the nectar from tubular flowers, and why in turn should the flowers on which they feed have their anthers and stigma positioned so well for the birds to transfer pollen between them? What processes led angiosperms, beetles, and cichlids to diversify into almost-inestimable varieties? Investigations of the evolution of reproductive isolation, the origins of new characters, and the diversification of lineages are also examples of this branch of biology.

Reconstructing evolutionary histories and understanding evolutionary processes have been major goals of evolutionary biology since Darwin's time. More recently, the field has begun to consider whether it might be feasible to predict how populations will evolve in the future, an increasingly pressing question in the current period of global climate change.[2] We might wonder, for instance, how rising spring temperatures will affect species of birds that depend on feeding their chicks caterpillars that are now hatching earlier.[3] Or how turtles and crocodiles might respond to climate change given that their sex is determined by temperature.[4] Prediction is a fundamental objective of many fields of science, but is it possible for evolutionary biology? Historically, researchers have been skeptical, reasoning that evolution is dependent upon too many chance events—from the randomness of mutation to the vagaries of climate or the idiosyncrasies of sexual reproduction—to display any discernible patterns.[5]

Recently, however, this attitude has begun to change as scholars recognize that evolutionary processes show some repeatable features. At least on shorter time scales, it sometimes seems possible to anticipate how evolution might proceed. Populations of guppies or *Anolis* lizards provide "natural experiments" in parallel evolution, helping researchers to understand which particular evolutionary trajectories are likely; sophisticated models fed with data from the past allow forecasting the future evolutionary dynamics of a population; controlled manipulations in laboratory populations allow alternative outcomes to be engineered and understood.[6] Prediction has emerged as a small but significant subfield that will probably become an increasingly important component of evolutionary biology in the future. We recognize it as a third objective of the field.

Reconstructing histories, understanding processes, and predicting future trajectories, then, are three goals of evolutionary biology. This book is primarily concerned with understanding evolutionary processes, and with the explanations that evolutionary biologists generate to facilitate that understanding. Here we build on the preceding chapter and illustrate how distinct subfields of evolutionary biology differ in their scientific frameworks, how the different kinds of explanation generated can be complementary, and how knowledge of development enhances explanations of evolutionary mechanisms and processes. We will also suggest that a strength of a developmental perspective is that it may sometimes enrich historical accounts and improve the accuracy of evolutionary prediction.

Pluralism in Evolutionary Explanation

Like all scientists, evolutionary biologists generate scientific explanations by framing testable hypotheses for further exploration, conducting experiments, and collecting data. Roughly speaking, to explain a phenomenon is to identify its causes. However, evolutionary explanations are far from exhaustive lists of causal influences: the world is just too complex. Whatever biologists investigate—from the intricacies of protein folding to the interacting components of a cell or the evolutionary history of butterflies—it is impossible to specify all causal influences.

Rather, what scholars hope to achieve with their explanations is to identify the "critical" or "most important" causal factors that account for the phenomena they study, and to specify how these factors interact. Scientific explanations are not intended to be complete descriptions of how the world works, which is why they rely on simplified representations, or *idealizations* (figure 4).[7] Far from trying to capture real-world complexity, scientific explanations actually endeavor to reduce that complexity, rendering the study topic manageable. Just as there would be no virtue in a map that represented the geological terrain at full size, so there is no utility to explanations that are as complex as the phenomena they explain. Good explanations are simple enough to comprehend, yet powerful enough to illuminate.

Idealizations—assumptions about the world that are false—are a universal feature of science.[8] Physicists commonly assume that objects are points in space and chemists that solutions are perfectly mixed. We all understand that these assumptions are false, but nonetheless they allow researchers to focus on some principal influences by ignoring other causal factors that, at least for the problem at hand, appear less important. What matters is that these idealizations are useful in helping scientists to build causal explanations that generate understanding and draw useful conclusions.[9] The causal factors that are neglected are not unimportant in any general sense; yet, despite this, depending on the question, their incorporation may not be necessary. After all, classical physics ignored general relativity, but was accurate enough to predict the spacecraft dynamics of moon landings.

Like other scientists, evolutionary biologists represent biological systems in ways that leave out many of the details, while retaining enough complexity that their models, observations, and experiments possess explanatory power. What is left out will very much depend on the question being addressed. A slightly disturbing implication is that all scientific representations are likely to

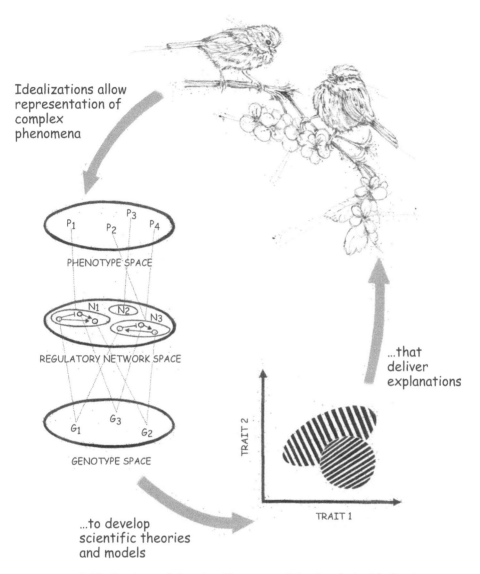

Idealizations allow representation of complex phenomena

P₁ P₂ P₃ P₄
PHENOTYPE SPACE

N1 N2 N3
REGULATORY NETWORK SPACE

G₁ G₃ G₂
GENOTYPE SPACE

...that deliver explanations

TRAIT 2

TRAIT 1

...to develop scientific theories and models

FIGURE 4. Idealization and the scientific process. Scientists devise idealizations of complex real-world phenomena, which help them to devise theories and models that, in turn, provide possible explanations.

be wrong in at least some respects. A second is that there is no single "true" explanation in evolutionary science. Rather, subfields with different goals will typically emphasize different causes, with each explanation an approximation of the underlying causal structure of the world.

Consider the question: *What caused the turtle to evolve a shell?* One explanation—the kind that a behavioral ecologist might give—is that shells protect turtles against predators and were therefore favored by natural selection. To evaluate this explanation, behavioral ecologists might seek evidence that contemporary turtles with larger or stronger shells show greater survival than those with smaller, weaker shells. They might, for instance, measure the strength of selection favoring protective shells in particular turtle populations,[10] or carry out comparative phylogenetic analyses across the 356 species in the turtle order Testudines to investigate whether variation in shell characteristics correlates with variation in predation risk. As expected, land-based turtles (tortoises), which are more vulnerable to predation, have thicker shells than aquatic turtles, which have lighter shells that allow them to swim faster.[11] This explanation is satisfying because it offers a credible cause for turtles to evolve a shell rather than staying in their original shell-less state,[12] and helps to make sense of variation in turtle morphology.

But there are other ways of framing the question that would demand a different kind of explanation. For instance, researchers could ask why turtles evolved shells rather than protecting themselves through spikes, fangs, or long legs to run away. The above explanation does not really answer this question, and from the standpoint of much of evolutionary biology it may not appear to be a very interesting question to ask. An evolutionary geneticist might reasonably assume that chance events, such as random mutations or historical contingencies, account for why phenotypic variation in shell characteristics instead of alternative defensive traits was available for selection.

Conversely, other scholars, such as evolutionary developmental biologists, are more interested in the origins of phenotypic characters. They point out that there is no obvious homologous structure to the turtle shell among its close relatives, and hence shell ancestry is an intriguing mystery. It turns out that the turtle is the only vertebrate whose shoulder is actually inside its rib skeleton.[13] So how did this remarkable innovation arise? By studying turtle embryos, evo-devo biologists have established that the upper portion of turtle shells (the carapace) develops from its ribs (figure 5). It seems that positive feedback between growth factors in the ribs and skin during early development causes the ribs to migrate laterally into the skin, rather than ventrally to form a rib cage. There, the ribs serve as a signaling center for the formation of the bony plates of the carapace. From an evo-devo perspective, despite its novelty, a carapace is surprisingly easily achieved through subtle changes in where and when key developmental genes are expressed (i.e., through changes in *heterotopy* and *heterochrony*), combined with the recruitment of existing developmental modules. Through careful experimentation, evo-devo researchers are now beginning to understand how genes contribute to shell production.[14] Such analyses help explain why shells evolved by demonstrating how they were accessible to development.

Another way of interpreting the question is to ask: *How have turtle shells changed over evolutionary time?* Did shells begin as a more localized bony plate and then get larger? Did the shell originate as a carapace and then spread around the body? Historically, in marked contrast to embryologists, paleontologists had argued that shells evolved from osteoderms—the bony scales that cover the turtles' close relatives, the crocodiles. However, once the developmental evidence showed that turtles developed their upper shells separately from their lower shells, and that the carapace arose from their ribs and not from osteoderms, paleontologists could interpret their data in a new light.[15] In addition to the developmental evidence, paleontologists unearthed an osteoderm-free 220-million-year-old turtle fossil with broad ribs and a shell that covered only its stomach (i.e., it had a *plastron*).[16] Interestingly, once paleontologists understood what early turtles must have looked like, they rapidly discovered that they had overlooked an existing fossil on museum shelves that had (paradoxically) been ruled out as a turtle because it lacked osteoderms.[17] The developmental and paleontological data eventually converged to demonstrate that the plastron, or lower half of the shell, evolved first, and that the carapace formed later from the modified ribs.

The paleontological data are fascinating because they challenge the intuitive explanation that shells evolved as protective armor. Today's turtle shells might provide an effective defense, but those of the first turtles did not cover their necks or backs, and hence offered little protection. Paleontologists now suggest that the original expansion of the ribs was an adaptation that helped stiffen the skeleton to provide a stable base from which to operate a powerful forelimb digging apparatus—a key functional requirement for burrowing animals.[18] The turtle shell began as a digging platform and only subsequently evolved into a suit of armor. Turtles didn't evolve spikes or fangs, rather than shells, because the original function of the shell was not defense, while their plastron made running away infeasible by slowing them down. However, having already

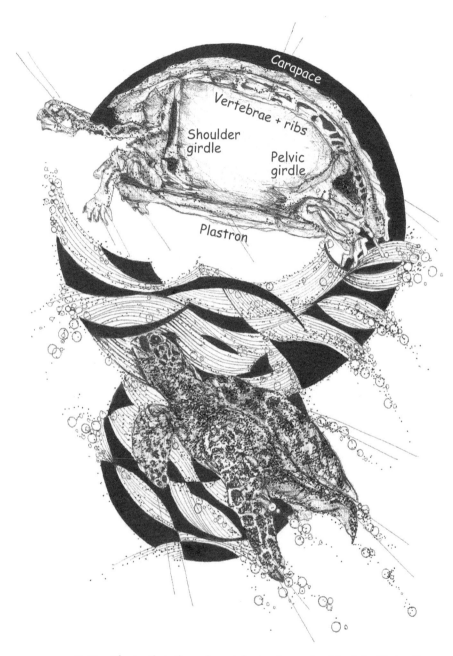

FIGURE 5. *Top.* The turtle is the only vertebrate whose shoulder is inside its rib skeleton. The evolution of this remarkable innovation makes sense only in the context of turtle development. *Bottom.* The shell is formed by expanding the ribs laterally into the dermis of the skin.

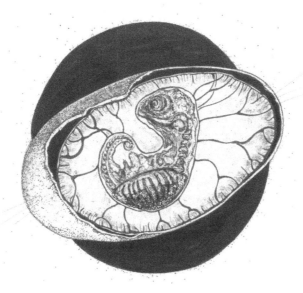

FIGURE 5 (*continued*)

evolved the developmental machinery to manufacture a lower shell, fusing all the ribs to produce a carapace was a short step to effective defense.

The turtle shell allows us to make a number of important points. First, we can see that no field alone has identified the one "true" explanation. Rather, different subfields of evolutionary biology—behavioral ecology, evo-devo, paleontology—reveal different causal patterns. Clearly, there is virtue in explanatory pluralism. Of course, evo-devo is now a well-established discipline, and hence it might seem strange to suggest, as we do, that evolutionary biology does not fully recognize the importance of development. However, evo-devo is generally considered to address different problems from other subfields, and is widely regarded as focusing on *how* rather than *why* questions. Here we emphasize that development also contributes to the *why* of adaptive evolution.

Second, while historically they may have seemed at odds, these explanations are complementary, and together provide a richer account of how and why turtles have shells than each subdiscipline could provide by itself. This richer account explains not just the original and current function of turtle shells but also the historical trajectory of their evolution, some structural and functional changes along the way, and changes in key developmental genes that may have been favored by natural selection. As we saw in the preceding chapter when discussing the evolution of vertebrate teeth and bones, knowledge of developmental mechanisms can contribute valuable evolutionary insights, ultimately leading to richer, more compelling explanations. Moreover, only through integrating the findings of molecular genetics, developmental biology, and paleontology were scientists finally able to establish how turtles are related to other animals—until recently, a major debate in evolutionary biology.[19] This ambiguity had been a serious embarrassment for vertebrate evolutionary biologists, with creationists claiming it to be a major and obvious failure of the theory of evolution. The molecular analyses of vertebrate genomes, the developmental analysis of embryonic turtle skulls, and the discovery of several early turtle skeletons eventually converged to establish turtles as related to the birds and crocodiles (known as the *archosaurians*), from which turtles branched off about 260 million years ago.[20]

Third, we can see how expectations shape explanations.[21] Fossils that showed the plastron evolving before the carapace were dismissed and largely ignored because paleontologists had assumed that shells evolved from osteoderms. Likewise, behavioral ecologists were misled into thinking that the original function of shells was defense because they currently confer protection. Scientific explanations are generally partial, tailored to the interests of the researchers, and influenced by the assumptions of the discipline,[22] which can occasionally lead to conflict. While in the long term science is typically self-correcting, in the short term scientists often find what they expect to find.[23] That is why it is good to have multiple types of explanation, which can challenge one another before converging.[24] Idealizations bias and channel research programs, for both good and bad.

The Historical Roots of Evolutionary Explanation

Given the causal complexity of biological systems, it should not come as a surprise that as broad a discipline as evolutionary biology should include multiple, coexisting research programs, each with its own explanatory aims, focused on distinct subsets of causes. At the same time, there are idealizations that pervade, and to some extent unite, the whole of evolutionary biology. Examples can be found in the dominant research agenda, which, at least since the 1950s, has endeavored to provide explanations for the adaptations of living organisms and the diversification of lineages in terms of fitness differences between individuals. Not only is the natural selection of heritable genetic variation widely regarded as the most important evolutionary mechanism, it is almost universally viewed as the sole cause of *adaptive* evolutionary change.[25] The reasons for this, which are partly historical, illustrate how the way in which evolution is represented both facilitates research and screens off—that is, encourages the neglect of—some causes of adaptation and diversification.

A major factor was the emergence in the early twentieth century of Mendelian genetics as *the* cause of heredity.[26] Prior to that time, heredity and development had not been seen as distinct processes. Yet just a few years after the rediscovery of Mendel's work, his laws had been experimentally demonstrated in numerous plants and animals, and the concept of heredity had been drastically narrowed to the transmission of discrete particles (i.e., genes) from parents to offspring.

Thomas Hunt Morgan at Columbia University played a central role in this key reconceptualization of heredity. Morgan's research group bred and crossed mutants of the fruit fly *Drosophila melanogaster* to map the genotype of this model organism onto its four pairs of chromosomes. Traits, such as eye color, were hypothesized to be determined by genes, defined as regions of the chromosome that made a difference to the character, allowing Mendelian laws to be validated by statistical tests of the offspring of specific crosses.[27]

Morgan's experiments gave rise to a theory of inheritance that focused exclusively on the transmission of *causes of differences in phenotypic traits* (i.e., genotypic differences), rather than the causes of the traits themselves. For instance, he identified twenty-five loci that affected the normal red eye color in *Drosophila*, and established that, if one locus mutated to a different form, the result was sometimes pink eyes. Even though all twenty-four nonmutated loci affected the trait, Morgan referred to the mutated locus as the *cause* of pink eyes. Trivial as it may seem today, this was an important and radical move, both because it rendered the study of heredity immensely practical, and also because to many biologists it legitimized the neglect of development, which was viewed as merely a consequence of genes. In 1910, before his work on fruit flies, Morgan had written: "We have come to look upon the problem of heredity as identical to the problem of development." However, later, in his major book of 1926, *The Theory of the Gene*, Morgan claimed that progress in biology had been inhibited by "confusing the problems of genetics with those of development" and that "the sorting out of characters in successive generations can be explained at present without reference to the way in which the gene affects the developmental process."[28] These divisions were reinforced by Wilhelm Johannsen, who defined heredity as "the presence of identical genes in ancestors and descendants," and distinguished between "genotype" and "phenotype."[29] The emergence of this new sense of causation, the separation of heredity and development, and the genotype-phenotype distinction were all vital to the development of theoretical population genetics, and had many practical advantages—for instance, in animal and crop breeding programs.[30]

The assumption that a single universal mechanism would account for inheritance, together with the genotype-phenotype distinction, enabled the transmission genetics idealization of heredity to pave the way for a focus on natural selection as the sole cause of adaptive evolution.[31] This was important, since in the early twentieth century natural selection was controversial and nonintuitive, making it a primary goal to demonstrate that fitness differences really could account for adaptation.[32] In the early part of the twentieth century, Ronald Fisher, J.B.S. Haldane, and Sewall Wright produced a body of formal theory that integrated Mendelian genetics into mathematical models describing how natural selection, mutation, and random drift alter the genetic constitution of a population.[33] Evidence that X-rays induced mutations in *Drosophila* at random locations reinforced the view that selection sufficed to account for adaptive evolution.[34] This period, which paleontologist Stephen Jay Gould characterized as the "first phase" in the coming together of the "Modern Synthesis," led to evolution being viewed as changes in allele frequencies, and adaptation as the Darwinian sorting of genes according to their effects on fitness.[35] From this perspective, phenotypes seemed largely irrelevant.[36] The approach also validated the rejection of the major alternatives to natural selection.[37] The dismissed alternatives included Lamarckism (and other forms of soft heredity, where inherited traits are influenced by the parental environment), because they were inconsistent with the known mechanisms of inheritance; saltationism (the sudden appearance of new species or traits), because mutations of large effect were assumed to be almost always deleterious; and orthogenesis (i.e., large-scale evolutionary trends), because mutation was "random" and mutation rates were thought to be too low to overcome selection.[38]

Decades later, with the benefit of hindsight and all the new knowledge that science has accumulated in the intervening period, we can see that each of these rejected hypotheses, while flawed, is actually consistent with a large amount of contemporary biological data: soft heredity is widespread (see chapters 1 and 10), evolutionary novelties can arise through mutations of large effect (see chapter 11), and macroevolutionary patterns can result from developmental biases in the generation of phenotypic variation (see chapters 2, 7, 8, and 12). Simplifying assumptions are necessary for good science, but all idealizations have a shelf life, and their deficiencies will eventually be exposed.

The period between 1930 and 1960 has traditionally been viewed as a gathering of subdisciplines into the framework devised by population genetics, sometimes referred to as a "second phase" of the Modern Synthesis.[39] However, as Gould famously proposed, and historians have independently verified, this period was also marked by a "hardening," or "constriction," of the synthesis toward selectionist accounts.[40] Theodosius Dobzhansky, George Simpson, Ernst Mayr, and other leading evolutionists all placed greater emphasis on natural selection and adaptation in their later writings compared with earlier ones, leading to less explanatory pluralism in the 1950s than there had been in the 1930s and 1940s.[41] Historians agree with Gould that empirical discoveries do not completely explain this trend and portray it as reinforcement of the above idealizations with usage, perhaps facilitated by conducive sociological factors, including the popularity of eugenics, the reformulation of human genetics in the aftermath of the Holocaust, and research funding due to Cold War anxieties about atomic bomb testing.[42] This narrowing of explanatory mechanisms also encompassed increasing belief that macroevolutionary trends and innovations also could be explained by natural selection.[43] The "hard heredity" conception was given added authority with identification of the molecular structure of DNA in 1953, and through the influence of Francis Crick's "central dogma"— the purported one-way flow of information from DNA to RNA to proteins— which allowed genes to be conceptualized as "programs" or "blueprints" for building bodies.[44] By contrast, the reputation of soft heredity was unfairly damaged by association with Lysenkoism during the 1930s and 1940s, which induced a backlash in the West against research on that topic.[45]

In the 1959 celebrations that marked the centennial anniversary of *The Origin of Species*, natural selection was presented as the explanatory centerpiece of contemporary evolutionary biology, and Darwin was championed by Ernst Mayr as the first "population thinker."[46] Mayr's 1961 distinction between "proximate" and "ultimate" causation allowed him to claim that: "the clarification of the biochemical mechanism by which the genetic program is translated into the phenotype tells us absolutely nothing about the steps by which natural selection has built up the particular genetic program."[47] No wonder that Mayr's fellow biologists were discouraged from considering how developmental processes affect evolutionary dynamics.[48]

The above historical precis is obviously incomplete and does not do justice to the diversity of views manifest within the field. Nonetheless, it suffices to establish that a number of assumptions—including that heredity arises largely through gene transmission, that evolution can be studied without consideration of development, and that natural selection is a necessary and sufficient explanation for adaptation—have played a vital role in the development of evolutionary biology. The field has unquestionably benefited from the wide-spread adoption of these assumptions. Without them, theoretical population genetics probably could not have begun, and certainly would not have been as productive as it became, and researchers perhaps would not now have the tools to infer evolutionary history and detect signals of natural selection from DNA sequences, something that is a major strength of evolutionary research today. This representation of evolution by natural selection has been successful in part because it simplifies the causal structure of evolutionary explanations. For instance, while evolutionary biologists commonly investigate the ecological sources of selection, they rarely delve further to quantify the developmental and ecological causes of fitness differences (for examples of how this can be done, see chapter 9). This allows researchers to focus on how changes in selective environments have led to changes in organisms.[49]

However, idealizations also have negative consequences; they invariably lead to the neglect of important causes—inadvertently, and sometimes advertently, discouraging their investigation. A useful illustration is provided by the history of investigations that quantify the strength of natural selection, a staple of evolutionary biology today.[50] In his classic 1986 book *Natural Selection in the Wild*, John Endler documented all published studies that had measured quantitative responses to selection in natural populations. Strikingly, the very first of these studies—conducted by Hermon Bumpus of Brown University, who demonstrated a selective advantage to large body size in the house sparrow—was published in 1899, and yet for the next seven decades only a handful of similar studies were conducted. Why was this?

According to Endler, it was partly because the prevailing thinking did not encourage such research.[51] With notable exceptions, prior to the 1970s there was a general lack of interest in ecology, or even quantitative measurement, in American population genetics, which was largely a laboratory science.[52] Endler also notes a lack of easily accessible methods, and the fact that many ecologists and evolutionary biologists believed that evolution and ecology acted on different time scales.[53] Also important were prejudices concerning which traits would affect fitness. When ecologically minded pioneers, such as Bernard Kettlewell, Edmund Ford, and Arthur Cain, were measuring selection in natural populations of moths, butterflies, and snails in the 1950s, population geneticists and many other evolutionary biologists were slow to appreciate the significance of this work, partly because they thought that traits such as color polymorphisms were unlikely to be important to fitness.[54] For instance, Ernst Mayr in 1942 wrote: "The variation in color patterns, such as bands in snails and spot patterns in lady beetles are, by themselves, obviously of very insignificant selective value."[55] As with the paleontologists' "search image" of the ancestral turtle, evolutionary biologists' expectations had clouded their judgments. Another contributing factor was the prevailing view of the strength of selection. Darwin had emphasized gradualism, and most twentieth-century evolutionary biologists followed suit.[56] There was little incentive for researchers to go out and measure selection if they believed it would take an immensely long time span to register detectable change in nature. The study of natural selection in the wild only really began in earnest in the 1970s when new perspectives led researchers to the view that evolution could be fast, and it took off in the 1980s when Lande and Arnold's classic paper on the measurement of selection provided tools that set the study of phenotypic evolution in an ecological context.[57]

Today, measuring selection in the wild is a major evolutionary industry, and there is widespread recognition that ecological and evolutionary processes operate on overlapping time frames, leading to the emergence of a vigorous subfield studying eco-evolutionary dynamics.[58] Ecology is not just the "theatre" in which evolution is "played out,"[59] but actively influences evolutionary outcomes. Both ecologists and evolutionary biologists were slow to come to this particular party.[60] The extent to which ecological factors are legitimate causes of evolution, and vice versa, that evolution over short time scales can affect ecological processes, has only recently been widely appreciated.[61]

Central to our book is the argument that the genetic idealization of evolution by natural selection also made it more difficult for evolutionary biologists to appreciate that what organisms do and how they develop might explain evolutionary episodes. John Maynard Smith is representative of many leading twentieth-century evolutionists in asserting: "One consequence of Weismann's concept of the separation of the germline and soma was to make it possible to understand genetics, and hence evolution, without understanding development."[62] Weismann's famous experiment cutting off the tails of mice was hugely influential, but somewhat misleading. The fact that, in some complex animals, the germline (i.e., eggs and sperm) is separated from the rest of the body (somatic tissue) early in development misled Weismann into concluding that environmentally caused changes in the body could not be inherited. Contemporary evolutionary biologists Russell Bonduriansky and Troy Day suggest that the "nongenetic side of heredity has been a blind spot for biology and medicine for decades."[63] They suggest that evolutionary biology might have been quite different if, instead of cutting of mouse tails, Weismann had removed "teeth" from the amoeba *Difflugia*.[64] The amoeba experiment was conducted in 1937 by Herbert Jennings, who reported stable structural inheritance of the disfigurement,[65] but by that time research into nongenetic inheritance had already been marginalized. Or perhaps Weismann might have conditioned his mice to acquire a fear of the smell of almonds, and looked to see if their offspring inherited that fear. As we described in chapter 1, Brian Dias and Kerry Ressler at Emory University published the results of that experiment in 2014, revealing that the offspring and grand-offspring of the experimental mice were spontaneously afraid of the almond smell, which appears to have been inherited through epigenetic changes in the sperm.[66] Or Weismann might have studied coral, and revealed that approximately half of somatic mutations enter the germline and are passed on to the next generation.[67]

We now know that Weismann's barrier can be remarkably porous, even in animals that sequester the germline early in embryonic development.[68] If we add to this that there is no distinct germline in single-celled organisms, that plants and many animals produce their germline from somatic tissue or stem cells,[69] and that cultural inheritance is now known to be widespread in animals, then the significance of nongenetic inheritance for evolutionary biology has been transformed, as we show in chapter 10. Idealizations shape research agendas and constrain scientific practice. There are few clearer illustrations of this fact than the observation that early geneticists should redefine heredity as the transmission of genes, subsequently infer that soft inheritance must therefore require the direct modification of gene sequences by the environment, and then go on to declare the latter "a chemical impossibility"![70]

With the benefit of hindsight, we can look back on the field and readily identify important limitations to some thinking that was prevalent in the past. Not only are some former idealizations of evolutionary biology now understood to be erroneous, but hypotheses rejected on the basis of those idealizations are now being recognized to have merit. Today, the central dogma and "hard heredity" idealizations are widely accepted to be crude simplifications.[71] Genes now look less like clearly separable units of development (and hence inheritance) and more like "complex assemblages of diverse activating and inhibiting mechanisms, acting in concert to transcribe and translate DNA templates of various makeup into functional protein products."[72] The molecular gene concept that was conceived in the early twentieth century and concretized in the 1950s to 1980s is being challenged in the twenty-first century.[73] We should not be surprised: it is the fate of all scientific idealizations that they will eventually be superseded, and that a subset of core assumptions that were useful in their day will eventually be found to be of limited validity or applicability, or even just plain wrong.

Evolution and Development

We do not dispute the fundamental role of genetics in development, heredity, and evolution. However, in this book we endeavor to make the case that certain idealizations that have served evolutionary biology well for many years need to be reevaluated. In particular, we challenge the assumption that evolutionary biologists don't need to know about development to do their job, but we will also take issue with the idea that evolutionary biologists can safely screen off the causes of selection, that development is a programmed readout of the body's nuclear genes, and that (anomalies apart) heredity is solely due to the transmission of genes.[74] Limitations of these assumptions are increasingly appreciated within the field, yet at the same time the processes that are neglected are often regarded as peripheral (e.g., as "add ons"),[75] and their explanatory roles remain contentious, in part, because idealizations encourage researchers to write off developmental processes as nonexplanatory (i.e., for the interesting aspects of evolution, such as adaptation and diversification). Reconsidering the core assumptions of evolutionary analyses, we suggest, will provide new insights into how evolution works.

In fact, other research agendas with different explanatory approaches have always coexisted with the mainstream. One of the best examples is the twentieth-century "organicist" movement. Organicists held that evolution cannot be understood solely at the level of the gene, nor be adequately described by tracking gene frequencies, but requires consideration of how genotypes develop into phenotypes, and reciprocally, how phenotypes control and regulate gene expression.[76] As might be expected, organicists adopted a whole-organism perspective, placing emphasis on how parts and whole are inextricably linked, and stressing organismal agency, as well as the reciprocal interactions between organisms and their local environments.[77] Organicism is a long-standing intellectual tradition, which reached its zenith in the 1930s and 1940s through the influential experiments, discussions, and writings of the Theoretical Biology Club, a group of intellectuals from Cambridge, Oxford, and London Universities, who took issue with the prevailing mechanistic, reductionist, and gene-centric conception of biology.[78] Perhaps the most famous member of this club was the embryologist Conrad Waddington, renowned as the father figure of epigenetics and for his classic experimental work on genetic assimilation in *Drosophila*.[79]

Waddington wanted to integrate embryology with classical genetics, and in the late 1930s he spent a year working in the *Drosophila* laboratory of Thomas

Morgan, alongside Morgan's student Theodosius Dobzhansky. However, Waddington's investigations of the genetics of *Drosophila* morphology led him to a very different perspective from that of his hosts.[80] For Waddington, gene action could be understood only by focusing on gene interactions (what we now call gene regulatory networks, or GRNs), and by recognizing extensive feedbacks with the developmental system (i.e., cell-, tissue-, and environment-specific gene expression). Waddington held the view that genes do not code for traits. Rather, genetic and epigenetic interactions in the cellular context shape the probabilities that particular traits will arise. Far from being the unrolling of a preexisting genetic program, development is a dialogue between the organism, its constituent parts, and the environment. We will expand on these themes in our description of development in chapters 5 and 6.

While Waddington's views are achieving something of a renaissance today, in the latter half of the twentieth century his work, and that of the other organicists, was dismissed by mainstream evolutionary biology.[81] The backlash in the West against research into environmental influences on heredity and development incited by the Lysenko affair led to an intellectual climate in which virtually any research on plasticity or nongenetic inheritance was equated with sympathy for Lysenkoist ideas, and even considered grounds for dismissal from American university positions.[82] While Waddington was no Lamarckian, leading evolutionists marginalized his work as too close to the Lamarckian interpretation of evolution adopted by Lysenkoists. Today, it seems extraordinary that Waddington's views on epigenetics and plasticity should be regarded as "dangerous" or "disreputable" through their tenuous association with a Lamarckian doctrine that he disavowed, when many leading mainstream evolutionists had been openly and vigorously campaigning for eugenics.[83]

Contemporary epigenetics is a broad research agenda that did not originate in the experimental laboratories of the 1990s but can be traced back to the discussions of the Theoretical Biology Club in the 1930s and 1940s. Other ideas currently receiving attention within evolutionary biology (GRNs, plasticity-led evolution, and niche construction) have their forerunners in the experiments and writings of twentieth-century organicists, as well as overlapping ideas from process philosophy, cybernetics, structuralism, and dialectical materialism.[84] While they were in no way alone in pursuing such topics, leading evolutionists, including Richard Lewontin, Stephen Jay Gould, and Mary Jane West-Eberhard have acknowledged their intellectual debt to these movements.[85]

If most of the twentieth century lacked the experimental and theoretical tools, or the supportive intellectual climate, to investigate how development contributes to evolution, the same cannot be said for the twenty-first century. Today, novel experimental approaches and advances in computational modelling create exciting new opportunities to revisit this relationship. In particular, experimental investigations of the mechanisms of trait development are shedding fresh light on the type and frequency of trait variation available to natural selection.[86] As we saw with the rodents' teeth and turtles' shells, taxonomically broad rules of development are being identified and used successfully to enrich evolutionary explanations, and reconstruct evolutionary histories.[87] Even the ability to predict evolutionary futures can also be enhanced through the study of development. Subsequent chapters will show that developmental insights do more than just add to traditional evolutionary understanding—they challenge it, portending fundamental change in the structure of evolutionary theory.

4

Understanding the Debates

For over a century, attitudes toward the role of development in biological evolution have varied greatly, occasionally generating fractious debate.[1] The debate partly revolves around differing conceptions of development, and for that reason in chapters 5 and 6 we summarize some features of developmental biology that are relevant to evolutionary science. However, much of the contention also relates to how researchers view and interpret the processes that underlie natural selection. Here we elaborate on this point and highlight how it is central to several current debates within the field, including those concerning the evolutionary significance of developmental bias, phenotypic plasticity, niche construction, and extragenetic inheritance.

Earlier we referred to Richard Lewontin's three requirements for evolution by natural selection: there must be *phenotypic variation* in the characteristics of individuals, some variants must leave more descendants than others (*differential fitness*), and offspring must resemble their parents more than other individuals (*heredity*).[2] These show that evolution by natural selection can be formulated in abstract terms—it can arise in any population of entities that fulfils these requirements, not just organisms. For instance, it has been deployed to describe technological and cultural evolution, such as the evolution of the bicycle, or of dance.[3] It follows that to explain the evolution of living organisms the principles need to be fleshed out with biological detail. Evolutionary biologists must specify *how* phenotypic variation is generated, *how* particular phenotypes come to be successful while others fail, and *how* offspring come to resemble their parents. They must also detail *how* these processes interact.

From the early decades of the twentieth century until the present, the dominant interpretation of the three requirements for evolution by natural selection assumes that evolutionarily relevant phenotypic variation arises from genetic change. It also regards differential fitness as the match between an individual's phenotype and its external environment, and treats inheritance as the passive transmission of genes from parents to offspring. A final key assumption—widely applied but rarely discussed—is that the processes generating phenotypic variation, differential fitness, and inheritance, while feeding into one another, are effectively autonomous.[4] For instance, different rates of survival among individuals with different traits determine what features are passed on to descendants, but whatever selection took place in the previous generation—it is assumed—does not typically affect the rules of inheritance in the current generation. Similarly, it is also assumed that the variation that fuels evolution arises independently of its effects on fitness, and that the acquisition of new phenotypic variants does not change how variation is transmitted from generation to generation. Finally, the selective pressures are assumed to be properties of the external environment that exist independently of the phenotypes of the focal organism and of how those phenotypes are inherited. These assumptions have the advantage that evolution can be described as an ordered set of processes: generation of phenotypic variation, differential fitness, and inheritance, each independently generating inputs for the next.

This conception of evolution by natural selection can be applied to some simple examples of adaptation in a relatively straightforward way. Consider, for illustration, the pale dorsal coloration of mice living on the sand dunes of coastal Florida.[5] The difference between the ancestral dark brown mice and the pale beach mice is partly caused by a mutation in the melanocortin-1 receptor (*Mc1r*) gene that regulates skin and fur pigmentation. Such point mutations happen randomly, not when the mice need them. Because *Mc1r* is important for the formation of brown pigment, mice with the appropriate mutation in this gene will reliably develop a paler coat than mice with a fully functional copy of the gene. Pale mice are less conspicuous on sandy soils, which gives them increased protection against predators and thus higher survival on beaches. As the beneficial mutation is passively transmitted to offspring during reproduction, the survival advantage results in an increase in frequency of the "pale" genetic variant and hence of pale mice in populations living on sand dunes.

In this explanation, all the sustained directionality in evolution—the progressive match between the phenotypes of the mice and their environment—comes from fitness differences between individuals with dark and pale coats. In the terminology popularized by Ernst Mayr, natural selection is the *ultimate* cause of the pale coloration.[6] It provides what is to most researchers a satisfactory explanation for the existence of this feature. In contrast, developmental factors are described as *proximate* causes, which explain biological systems by establishing how different components work. In this example, developmental processes would explain how mice with a mutation in the *Mc1r* gene become pale, but this is not thought to impose any directionality on the evolutionary process or its outcome. Genes are important to evolutionary explanations insofar as they influence how phenotypes are inherited, making genetics central to evolutionary biology. However, in this narrative, developmental causes do not contribute to the adaptive match between the mice and their environment. Adaptation appears solely due to the differential survival of dark and pale mice; that is, to differences in fitness.[7]

Other examples are not so straightforward, however. Recall that, in chapter 1, we described how killer whales possess culturally inherited hunting techniques for specializing on fish, seal, or dolphin prey, and that, despite often living side by side, groups of killer whales with different foraging habits are known to be reproductively isolated (figure 6).[8] In these whales the adaptive differences in diet and hunting technique are not due to genetic differences but to behavioral innovation and social learning; the different foraging behaviors are acquired and are stably transmitted across generations as cultural traditions.[9] Nonetheless, as a direct consequence of these cultural differences, which are thought to have come first, the "ecotypes" have diverged genetically because feeding on different prey elicits different patterns of natural selection.[10] As a result, killer whales exhibit morphological adaptations such as population-specific digestive enzymes, face shape, and tooth shape.[11] More and more instances of this gene-culture coevolution are being documented in animals (see chapter 10), with culture-led genetic evolution playing a particularly central role in human evolution (chapter 13).[12]

The killer whale example also violates the causal autonomy of the three processes responsible for natural selection, since the same process (social learning) not only generates phenotypic variation but also contributes to differential fitness (learned habits affect survival) and heredity (learned foraging behavior is socially transmitted to descendants). In such examples, behavioral plasticity and social learning join differential fitness in giving directionality to evolution, and thereby help to explain how and why the organisms became adapted to their environment (the transmission of maladaptive culture can also occur, but appears to be extremely rare in animals, and even in humans is the exception; see chapter 10).[13] For the whales, social learning is producing

FIGURE 6. Different populations of killer whales possess distinct learned and culturally transmitted traditions for feeding on seals, fish, or dolphins. These traditions have modified natural selection, leading to the genetic evolution of distinct "ecotypes."

a match between the behavior of the whales and the prey in their environment, which might be termed a "cultural adaptation."[14] This appears to conflict with the more-standard explanation that natural selection is the only cause of adaptation. Evolutionary biologists have recognized the existence of cultural inheritance for a long time, but only in humans, where it could be treated as a special case—a stance that is no longer tenable. Cultural adaptation and transgenerational inheritance are general properties of many animal cultures—the same complexities arise with humpback whales' lob-tail hunting of sand lance, for instance.[15] However, the killer whales' culture has gone one stage further, and triggered the natural selection of genetic variation expressed in morphological characters that complement the feeding tradition.

Related issues arise in the epigenetic inheritance of fear in mice, described in chapter 1. The example is disturbing because it implies that the same process could contribute to both differential fitness (i.e., learned habits can plausibly affect survival) and heredity (i.e., knowledge acquired in one generation is passed on to the next generation). If adaptive knowledge accrued during development can be passed on to descendants, then evolutionary directionality can arise from developmental mechanisms, and evolution would appear to work differently from how it is traditionally described. Such examples illustrate how the view that natural selection is the sole cause of adaptive evolution is tied to the assumption that Lewontin's three requirements for selection are effectively autonomous (and also to the corollary assumption that two of these processes—those responsible for generating variation and passing it on to the next generation—impose no adaptive bias on evolution).

Finally, consider a third example, identified by Darwin.[16] Through their burrowing, tunneling, mixing of organic and inorganic matter, and casting, earthworms engage in extensive soil processing. These *niche-constructing* activities dramatically affect the chemical and physical properties of the soil (e.g., generating enhanced porosity, aeration, and drainage, and more organic carbon and nitrogen) and shape the local community (e.g., affecting invertebrate abundance and diversity).[17] Earthworms' effects on the soil persist for multiple generations as an ecological legacy.[18] Moreover, there is evidence that earthworms have evolved adaptations to the modified soil environment created by their ancestors.[19]

Earthworm niche construction is typically not recognized as an evolutionary cause, but as a proximate cause that is itself the outcome of earlier natural selection.[20] Yet, ancestral earthworm activity is itself a major cause of the soil environment that favors mutations expressed in soil processing, so the causes of variation and fitness differences are not independent.[21] Unless the aim is to explain in terms of selection alone, it seems arbitrary to begin the explanation for the adaptive fit between earthworms and their soil environment with those ancestral fitness differences favoring earthworm soil processing. After all, those fitness differences arise in soil conditions that were themselves products of earlier earthworm niche construction, and so on, through endless cycles of reciprocal causation between natural selection and niche construction.[22] Here there is no clean separation between selection and variation; they are causally intertwined. Similarly, the processes underlying inheritance are also not independent of the causes of variation and fitness. Physiological analyses have established that earthworms are well adapted only when they "inherit" the ecological legacy of a modified soil environment, which is the product of ancestral earthworm niche construction,[23] just as the woodrats in chapter 1 are only well adapted because they inherit symbiotic bacteria from the soil and feces in their environment. Such ecological legacies, which are not uncommon, transform the very nature of inheritance from an exclusively genetic mechanism to a system with heritable genetic and extragenetic components. In chapters 9 and 14 we will discuss experiments with dung beetles that quantify how niche construction creates trait covariation, ecological inheritance, and fitness differences, while in chapter 14 we describe mathematical evolutionary theory for how niche construction simultaneously affects trait values, trait heritability, and fitness, and the challenge this presents to traditional views of evolution.[24]

Although niche construction is not currently regarded by mainstream biology as an evolutionary process, it highlights an important role of interactions in evolutionary causation.[25] Essentially the same issues arise with rodents' teeth, discussed in chapter 2, where biases in the generation of phenotypic variation impose direction on evolution. In chapters 7, 8, and 12 we will return to the topic of developmental bias and will present theory and data showing how the processes responsible for phenotypic variation and differential fitness are also often intertwined, again questioning the explanatory sufficiency of natural selection as the sole cause of adaptation.[26] In fact, in chapters 8–14 we will make the argument that plasticity, niche construction, and extragenetic inheritance are all forms of developmental bias, as well as ways in which the evolutionary process modifies its own mechanisms.[27]

Rescuing Strategies for Selective Explanations

Current debates over evolutionary causation would not exist if there were not ways to "rescue" the logic of ascribing adaptive change to selection alone.[28] Indeed, several strategies for doing so are seen in the evolutionary biology literature.[29] One common approach is to shift the target of explanation by "rescaling" developmental and evolutionary time; for instance, by assuming that the former is short enough to justify its neglect in considerations of the latter.[30] In the case of the killer whales, it might be argued that what requires an evolutionary explanation is not why killer whale populations acquired the ability to hunt for seals or salmon, but why they acquired a capacity for social learning. By treating only the latter as an *evolutionary* problem, it ostensibly justifies the notion that there are two fundamentally different kinds of causes in biology (proximate and ultimate) and allows the evolutionary effects of phenotypic plasticity and extragenetic inheritance to be accommodated in a genetic account of evolution by natural selection.[31]

However, this rescaling gambit has several problems.[32] One is that the explanation is seriously incomplete, as it provides only a partial historical account. In the case of the killer whales, the absence of an evolutionary explanation for why particular populations of whales possess particular feeding habits appears to leave holes in the explanation for the existence of morphological specializations. A full explanation for the morphological adaptations would require knowledge of the specific dietary traditions, and hence culture, of each population.[33] The developmental perspective, by contrast, offers a more complete evolutionary explanation. A second concern is that the rescaling hinders recognition of the role that extragenetic inheritance can play in short-term evolutionary adaptation, a topic to which we will return in chapter 10. There is substantial empirical and theoretical evidence for evolutionary adaptation arising through the natural selection of extragenetic variation, and evidence that this selection imposes directionality on genetic adaptation—not only through culturally transmitted variants but also the selection of epigenetic variation and symbionts.[34] To ignore these data would be counterproductive.

A third problem is that such rescaling is not always credible. For instance, if the cases of mammalian teeth and vertebrate digits are any guide, developmental bias can result from highly conserved mechanisms operating over time frames of tens or even hundreds of millions of years. It would seem equally reasonable to assume that whatever selection takes place in the present is, at least in part, a product of earlier developmental bias. In this example, bias and selection operate on overlapping time frames and probably always have. Not all bias is the result of selection anyway; for instance, in chapters 5–8 we describe how, for example, physical forces impose structure and biases on phenotypic variation.

Another common rescuing strategy is to treat anomalous examples, variously, as rare, trivial, nonfunctional, neutral, unstable, or as special cases.[35] For instance, if epigenetic inheritance were short-lived, or not very frequent, or if it only affected traits that contribute little to fitness, then perhaps it could be omitted from explanations for adaptive evolution.[36] However, as we shall see in chapter 10, ignoring epigenetic inheritance on such grounds is no longer tenable. There is now strong empirical and theoretical evidence that extragenetic inheritance can influence adaptive evolution even when it is short-lived, and can contribute to evolutionary dynamics even when it propagates neutral variation.

The effective autonomy of Lewontin's three requirements for evolution by natural selection can also be rescued by embracing a genetically programmed metaphor for development. Plasticity, developmental bias, and extragenetic inheritance can be problematic for the traditional view in cases that violate the assumptions that variation is blind to function, or that inheritance is the passive transmission of whatever variants persist; assumptions that, when met, leave fitness differences to be the sole source of evolutionary directionality. However, if extragenetic inheritance is considered to be under genetic control, then any adaptive directionality that extragenetic inheritance provides can—it is reasoned—ultimately be traced back to selection (we take issue with this reasoning in chapter 14).[37] Similarly, genes could conceivably "stand in" for the developmental processes that produce plasticity and bias,[38] which—notwithstanding the difficulties this approach generates—might be characterized as genetically specified products of earlier natural selection. The problem with this strategy, as we show in chapters 5 and 6, is that few if any phenotypes, including plasticity and extragenetic inheritance, can accurately be explained as arising solely from the genome.

Rescue strategies that refocus attention to genes and selection can be attractive because they appear to bypass problematic phenomena such as developmental bias, plasticity, niche construction, extragenetic inheritance, and other developmental processes, by relegating them to be only "apparent" or "superficial" causes of adaptation (i.e., solely proximate causes). For instance, it might be reasoned that similarity in selective regimes provides a sufficient and empirically well-supported explanation for the repeated evolution of similar adaptations. Consequently, as there appears to be no recognized problem in explaining the data, there is little motivation for researchers to investigate other potential causes of adaptive evolution, such as developmental biases in the production of variation, that might also contribute to observed patterns.[39] As a result, explanations reliant on idealizations that maintain the causal autonomy of the principles of evolution by natural selection will continue to support the privileged role of fitness differences, and hence natural selection, in adaptive evolution. Selection—whether past or present—is *virtually always* attributed causal priority over development.

However, rescaling may be neither possible nor desirable, and treating challenging data as anomalies or genetically determined is becoming increasingly untenable. Such explanations do not meet the *factual accuracy* criterion for evaluating explanations discussed in chapter 2. Increasing recognition of the causal entanglement of the biological processes underlying evolution by natural selection is putting pressure on the historically dominant account of adaptation and diversification. As suggested by the rodents' teeth in chapter 2, incorporation of development into evolutionary analyses promises to make evolutionary explanations more detailed and more robust, so that they fit better with existing knowledge, stimulate further insights, and can answer new or neglected evolutionary questions, including *Why did that particular trait evolve?* (chapter 9) and *How do evolutionary novelties arise?* (chapter 11).

In short, there is a framing of evolution by natural selection that makes developmental processes causally important for evolutionary adaptation and diversification; a framing that not only provides better explanations but can inspire different kinds of experiments and theory, and open up new research questions. This standpoint suggests that developmental processes are responsible for some of the creativity manifest in adaptive evolution, but while "natural selection displays the creativity of the curator," in contrast "development has the creativity of the artist."[40] These claims, together with their implications for evolutionary biology, are spelled out in the remainder of this book.

Looking Forward

Simplifications and idealizations permeate science and are necessary and desirable.[41] Yet such assumptions inevitably neglect important causal influences and may discourage their investigation. Progress in any field of science requires constant vigilance over whether representations that have served the field well in the past are productive in the present.

We have discussed several idealizations that have contributed to progress in evolutionary biology. One was the genetic conceptualization of evolution by natural selection, which became the foundation not only for a research program that aimed to provide insights about natural selection but also for evolutionary biology as a whole. Indeed, it was so successful that natural selection and adaptive evolution are sometimes treated as synonyms.[42] This success had the effect of assigning particular tasks to particular biological subdisciplines, with some fields, like genetics, becoming central to the study of evolution and others, like developmental biology, and perhaps to a lesser extent paleontology and ecology, becoming marginal to that study.[43] There were good practical reasons for these core assumptions at the time, but in the light of new insights and with the emergence of new methods, the time is now right to consider whether relaxing them might be fruitful. A wider view of evolutionary causation would allow other disciplines, not just developmental biology but also paleontology, epigenetics, ecology, zoology, botany, physiological and behavioral biology, and the human sciences, to share with genetics the responsibility for building causal explanations within evolutionary biology.[44]

Not all evolutionary biologists find this prospect alluring, and to some having natural selection as the only "true" cause of adaptive change is a fundamental feature of evolutionary theory, one that unifies evolutionary science and separates it from the rest of biology.[45] Indeed, there appears to be a genuine concern that a conceptual framework supporting broader explanations (for instance, recognizing that developmental bias and natural selection might jointly cause adaptation) would be detrimental to the field.[46] We believe this concern is misplaced; the division of labor among biological disciplines is a consequence of idealizations originally made a century ago, which enabled natural selection to be understood in the face of what appeared at the time to be the overwhelming complexity of development. It is simply unrealistic to expect that the most useful way to represent the principles of evolution by natural selection will be the same throughout nature since, as we will show in subsequent chapters, the evolutionary process, including natural selection, is itself evolving.[47]

Although limited reconciliation of development and evolution has been possible among biologists willing to regard development as an unfolding of a genetic program, no wider synthesis of evolution and development has been achieved.[48] However, there is renewed excitement that developmental and evolutionary biology are growing closer again.[49] A number of emerging fields and new methodologies represent green shoots that offer the prospect of a richer synthesis, one that integrates evolution and development, while at the same time embracing plasticity and being "eco-friendly."[50] The challenge is to demonstrate that there are evolutionary problems that can be better understood in a framework that integrates developmental biology with evolutionary genetics than through traditional genetic explanations alone. That challenge requires not only reconsidering some cherished idealizations within evolutionary biology but also incorporating changing conceptions of how development works. It is to this last topic that we turn in chapters 5 and 6.

PART II
How Development Works

5

Opening the Black Box

The almost miraculous fact that each human being develops from a single cell, too small to see with the naked eye, into a mind-bogglingly complex whole with fully functioning organs and tissues and comprising some thirty trillion cells, has inspired awe and wonder for centuries. We refer to the processes that bring phenotypes into being during an individual's life cycle as *developmental* processes, stressing that our use of the term "development" is broad, and encompasses cellular, physiological, neurological, and behavioral mechanisms that operate up to and through adulthood.[1] Specifically, by "development" we mean "the process of progressive and continuous change that generates complex phenotypes,"[2] which could start from a single cell or from an outgrowth or bud, as in yeast, hydra, or plants.

Historically, development has typically been treated as a black box by evolutionary biologists,[3] but we view matters differently. Our goal in this chapter is not to provide a comprehensive treatment of the developmental dynamics of multicellular organisms but rather to provide the reader with a taster, a brief review of the knowledge central to a developmentalist perspective on evolution.[4] Given the breadth of material that could be covered in such a summary, we have chosen to focus on those aspects of development that are particularly relevant to later chapters. Purely to illustrate the complexity of the processes, we describe some of the genes and epigenetic factors involved.[5] A short accessible summary of the general principles that emerge from this review can be found in chapter 6.

We begin by describing how genes regulate, and are regulated by, the molecular properties of the cell, and then elaborate on the roles of both genetic and nongenetic factors in the differentiation of cells into different cell types, and the formation of organs. We particularly emphasize the roles of regulatory interactions and exploratory and selective processes in development, which reveal how development is controlled by processes above the level of the gene. We will see that development is not a one-way journey from zygote to adulthood, nor is it a unidirectional translation of information encoded in DNA into phenotypes.[6] There are phenotypes present at all stages of development, and these phenotypes regulate subsequent gene expression.[7] Development proceeds by the genome and other factors, including the environment, making a phenotype that regulates the genes necessary for creating the next phenotype. There is reciprocity between genes and phenotypes in development;[8] contingency is always pervasive, and control is bidirectional.

The Genetics of Development

Traditional descriptions of the life cycle of multicellular organisms begin with the zygote (the first cell arising from the fusion of egg and sperm), but this is an arbitrary starting point. Development does not really have a beginning or an end, but rather flows continuously down the generations. Nevertheless, there is a logic to this convention, as the organism can be described as transitioning from its simplest to its most complex form.[9] The zygote begins to divide, first into two, then four, then eight cells, and so on. However, for the developing embryo to become more than just a homogeneous lump of cells, those cells need to differentiate into different types of cell. Adult humans have some two hundred different cell types in their bodies. Nerve cells, with their long projections called axons and dendrites that send and receive electrical impulses, look and behave very differently from, say, fat cells, which are bulbous spheres swollen with the fat they store, or from the flat and closely packed skin cells. To form a viable multicellular organism the cells must aggregate into tissues, such as bone, cartilage, or muscle, which interact mechanically with one another to form morphological features, such as the skull, or organs like the heart or skin.

Although there exists great cross-species diversity in developmental processes, many broad-brush features of development are remarkably conserved. For instance, in a representative vertebrate life cycle, the zygote undergoes a number of cycles of division (known as *cleavage*) to produce a ball of cells (a *blastula*). The outer surface then folds inside into a cavity (a process known as *gastrulation*), creating the three layers that form the embryo—the outer layer that will eventually form the skin, brain, and nervous system (*ectoderm*); a middle layer that gives rise to the skeleton, muscles, and vascular system (*mesoderm*); and an inner layer that becomes the gut and respiratory system (*endoderm*). Subsequently the top (*dorsal*) region folds into the embryo to create the *neural tube*, which develops into the brain and spinal cord. The latter then interacts with neighboring mesoderm cells to specify the segmented vertebrae and muscles, and these structures, in turn, interact with their neighbors to generate further tissues and organs.[10]

If all the cells in our body contain the same genes, how is it possible for such a wide diversity of cell types, tissues, and organs to arise from a single cell?[11] One answer is that the differentiation of cells into their various forms involves the acquisition and maintenance of distinctive molecular cell-type profiles that specify which genes are expressed, and hence which RNA and proteins are synthesized. There are characteristic profiles for each of the cell types, which ensure that each type performs a specific biological function, such as the synthesis of insulin in some pancreatic cells, or the accumulation of lipids that exist in fat cells.[12]

While almost every cell nucleus in our bodies contains the same genome present in the original zygote (leaving aside symbionts, like gut bacteria), only a fraction of the genome is expressed in each cell, with different genes expressed in different cell types. The unused genes in each cell are not lost or destroyed, but rather "wrapped up" in a cloak of molecules that prevents their transcription. As Barresi and Gilbert quip: "Whereas classical geneticists have likened genes to "beads on a string," molecular geneticists liken genes to "string on the beads.'"[13] The description is apt, since the long string of DNA is looped around bundles of histone proteins forming nucleosomes, which in turn are wound into tight structures called solenoids. Differential gene expression occurs in part through regulation of which regions of the DNA are unpacked and available for transcription. Many genes possess cell-type-specific *enhancers*, regions of the DNA sequence that regulate nearby structural genes that encode proteins. These enhancers can bind other proteins called *transcription factors*, which activate a gene by recruiting enzymes that break up the nucleosomes and stabilize the transcription machinery. Different transcription factors bind enhancer regions of genes to activate them, or *silencer* regions to repress transcription, and often a combination of transcription factors is necessary for a gene to be transcribed. In this manner, regulatory DNA elements, whether they reside on the same chromosome (*cis*-regulatory elements) or on different chromosomes (*trans*-regulatory elements), control whether, when, and how much a gene is transcribed.

The capacity of each cell type to express a distinctive combination of genes requires the interaction of a specific repertoire of influential transcription factors.[14] While numerous genes are expressed in multiple cell types, in many cases the combination of a small number of characteristically expressed transcription factor genes makes a big difference to the cell's morphology and behavior. Changes in the expression of sets of genes allow cells to acquire and maintain different functions, and to produce gene products that can communicate with other cells, both during embryonic development and later. Once cells acquire a particular gene expression profile, when they divide they typically pass on that profile to their daughter cells,[15] leading to stable differentiation among cell lineages.

Cell-type-specific enhancers are very important because they allow particular genes to be switched on in some regions of the body and not others. For illustration, consider the *Pax6* gene. *Pax6* has three major enhancers. One of these contains DNA sequences that bind transcription factors in the pancreas, another enhancer binds transcription factors in the nervous system, and a third binds transcription factors on the eye. This enables the *Pax6* gene to be expressed in these three types of cells, and not any others. The *Pax6* gene encodes a protein that operates as a transcription factor, also called Pax6 (to minimize confusion, throughout this book we adopt the scientific convention of italicizing the genes but not the associated proteins). In vertebrates, the Pax6 transcription factor (in combination with other transcription factors) binds to and activates various other genes, causing these genes to be expressed in the developing eye. Together, they interact to generate the cells of a functioning lens. Each "target gene" activated by Pax6 possesses an enhancer with a particular "signature" sequence, known as a DNA-binding domain: in the case of the Pax6 transcription factor, the sequence of nucleotides to which it binds is CAATTAGTCACGCTTGA.[16]

Enhancer-mediated gene expression allows mutations in major regulatory control genes to be expressed in a tissue-specific manner. An understanding of tissue-specific gene expression is itself an important contribution of developmental genetics to evolutionary theory. While a mutation in a structural gene will affect that gene's activity in every cell in which it is expressed, a mutation in an enhancer will alter the activity of that gene in one cell type and not others. This property allows for selective losses or gains of function in different parts of the body. An example is the loss of the pelvic fin in freshwater three-spined stickleback fish.[17] The *Pitx1* gene has several major enhancers. One activates the *Pitx1* gene in the spine, one activates the gene in the nose, one activates the gene in the hind limb, and another activates the gene in the thymus and sensory neurons. Three-spined sticklebacks live in both the oceans and in fresh water (i.e., rivers and lakes). In some populations of the freshwater fish there is a mutation in the hind limb enhancer that prevents the transcription factors from binding there. As a result, Pitx1 protein is not made in the hind limb cells, and the pelvic fin fails to develop. The evolutionary significance is that marine three-spines possess pelvic spines as body armor, but these have been lost in many freshwater populations because of the pelvic-spine-specific mutation.[18] The pelvic spine appears to have been selected against in freshwater environments, where predators can latch on to it.

Transcription factors, such as Pax6 or Pitx1, can have dramatic effects on phenotypes because changes in their expression affect many target genes, which then trigger a cascade of molecular signals, allowing for the coordinated expression of a large number of genes in a particular cell type, tissue, or organ. This explains, on the one hand, how a single mutation can have a major effect on the phenotype (e.g., loss of spines) and, on the other, how its impact can be regulated in an organ-specific manner (i.e., leaving other tissues unaffected).

Interconnected sets of coding genes and regulatory sequences are often referred to as a *gene regulatory network*, or GRN. The gene expression profile that characterizes a particular cell type involves many genes that are connected through the binding of transcription factors to *cis*-regulatory elements (e.g., enhancers) in a network.[19] These interactions are strongly modulated by the molecular processes within the cell, which are affected by other genes, as well as by nongenetic factors and environmental variation. However, for the moment, we will characterize a GRN as a network of interacting genes and their products, including genes that encode transcription factors or molecules expressed in intercellular signaling, and genes that function in downstream differentiation and morphogenesis.[20] GRNs can be viewed as the patterns of coexpression of genes in cells and tissues, patterns that are responsible for the properties of the cell, such as its shape, stickiness, or ability to produce products, such as insulin or hemoglobin. To illustrate, we will focus on the remarkable transformations of neural crest cells.[21]

Neural Crest Cells

The neural crest is a transient structure arising in the upper (dorsal) region of the neural tube in the vertebrate embryo. Adults do not have a neural crest, and nor do late-stage embryos. Rather, the neural crest cells migrate throughout the body, eventually differentiating into many other cell types, including cells of the skeleton and muscles of the vertebrate head, but also cells of the peripheral nervous system, adrenal glands, the skin pigment, and portions of the heart.

The fate of cells and organization of tissues is shaped by soluble factors, usually proteins, called *paracrine factors*. These factors are produced by one set of cells but influence the fates and behaviors of other cells in a local neighborhood. In some instances, the chemicals travel over longer distances and influence a larger number of cells, thereby specifying their differentiation and behavior.[22] In such instances these chemicals are called *morphogens*, and depending on their concentration different cell types are generated, with concentration thresholds that generate discrete switches in GRN activity.[23]

This is how cells become neural crest cells. Cells in the neural crest begin their existence as multipotent stem cells, their status specified by the precise timing and joint action of two paracrine factors, Wnt and Bmp.[24] If the ectodermal cells receive both Bmp and Wnt simultaneously for a prolonged period, the cells will go on to become epidermis. If the ectodermal cells experience only the Wnt signal, they become neural cells.[25] If they first receive only Wnt and then later solely Bmp they will become placodal cells that give rise to hair and teeth. Only those cells in the "Goldilocks zone" that first receive Wnt alone followed by Wnt and Bmp together will become neural crest cells.

The different genes required for *induction, specification,* and eventual *migration* of neural crest cells form coherent modules (by "modules" we mean components with their own intrinsic dynamics and integrated structure; see chapter 6), with the modules incorporated into a larger GRN.[26] The right balance of Wnt and Bmp will induce the expression of a set of transcription factors in the ectoderm (including Gbx2, Zic1, Msx1, and Tfap2) that make up the induction module. These, in combination, will activate another module in the GRN responsible for the specification of cell state. The activated transcription factors in the specification module (which include Snail1/2, Sox9, Ets1, and FoxD3) subsequently activate further genes in a migration module in the GRN (*Sox10, Lmo4, Cdh1/2/6B, Cdh7*).[27]

To move, the migrating cells literally reach out and grab the surrounding tissue and pull themselves along, by extending armlike lamellipodia that grasp the extracellular matrix, while releasing the hold behind.[28] Once migration has begun, cells begin to differentiate into their respective roles, depending on what signaling molecules are in their immediate environments. For instance, some cells that migrated toward the head proliferate and differentiate into bone and cartilage of the face, while others that migrated outward differentiate into cells containing the pigments that generate skin and eye color. In vertebrates, the heart originally forms in the neck region, and draws into it neural crest cells that generate the cardiac outflow tract (aorta and pulmonary artery). The different signals in the local tissues to which the cells migrate activate one of several distinct differentiation modules in the GRN. For instance, the presence of paracrine factor Tgf-β activates a set of transcription factors (Col2a1 and Agc1) leading to the differentiation of these cells into cartilage. Those neural crest cells experiencing further Wnt signaling activate the *Mitf* gene, which initiates their differentiation into the pigment-containing cells (melanocytes).

Gene Regulatory Networks

The above account is a highly simplified description of the neural crest GRN but highlights some important features that characterize GRNs and distinguish them both from other biological networks and from many other developmental processes.[29]

First, GRNs are constructed from the regulatory interactions of real organisms and hence can be regarded as causal representations of development. Representations of GRNs resemble complex electric circuits, sometimes involving hundreds of nodes, each usually a gene, connected by lines or edges describing the actions of gene products that activate or repress other genes (see figure 8). While typically the nodes are transcription factor genes, they could also be epigenetic elements (which modify chromatin and affect transcription), noncoding RNAs (which play diverse roles in the translation of RNA into proteins), or effector molecules (which bind to proteins and change their activity). At first, this complexity appears bewildering, and might suggest that it would be impossible to understand what the system will do, or what outputs will be generated. However, closer inspection reveals definite structure within this complexity: some particular regulatory interactions—called *circuits* or *motifs*—recur over and again within such networks.

Second, GRNs are organized in a modular fashion, with each module representing a specific regulatory state of cells during development, characterized by a unique combination of regulatory factors.[30] Modules are typically organized hierarchically, in a sequential progression. Later in this book we discuss how evolutionary innovations frequently arise through the co-option of preexisting modules into other GRNs (chapter 11), and how this is a major contributor to evolvability (chapter 12).[31]

Third, the topology of GRNs highlights how the phenotypic consequences of a mutation depend on its position in the GRN, and hence in the developmental sequence of a particular cell type, tissue, or organ.[32] Mutations in transcription factors acting early in the process likely have profound effects, often manifest across multiple tissues, while mutations in late-acting regulatory elements have smaller, localized effects.

Fourth, the production of phenotypes involves many regulatory factors that interact in complex ways, including positive feedback and feed-forward circuits. For instance, as neural crest cells migrate, they maintain their state despite exposure to a vast number of extracellular signals, in part because they express *Sox10*, a gene whose protein product is a transcription factor that is important in maintaining the cell in a multipotent state. Sox10 not only binds to the genes whose proteins promote multipotent states, it also binds to the enhancer of its own gene, thereby keeping itself active after its initial induction.[33] This is an example of positive feedback, and many such circuits across the GRN not only stabilize cell state but also vastly reduce the number of possible final outcomes. In the case of feed-forward circuits, one network node activates another, and both are required to activate another downstream target, and so forth, in a loop that contains "AND logic."[34] For instance, Tfap2a, Zic1, and Pax7 are all required for Snail expression, whilst Snail expression is critical for activating the expression of those proteins that digest the extracellular matrix, loosen cells' adhesion to other cells, and thus permit cell migration. The dynamical behavior of the circuits thus strongly influences how the GRN functions.

Fifth, GRNs can be enormously complex, and hence the effects of genes on phenotypes will often be difficult to comprehend unless researchers understand the context in which the genes are expressed. In fact, the static representation of GRNs can be misleading, both because the context of the larger network determines the behavior and function of each network motif within it, and because GRN interactions vary over time. Hence, knowing the structure of a GRN, while vital, is insufficient in itself to understand the developmental dynamics of the system. For that, computational models are typically required, and reveal that there are often a surprisingly limited number of phenotypic outcomes.[35] Which of these outcomes occur may not only depend on the topology of the network, but also on the initial conditions and strength of interactions between genes within the network.[36] The latter can depend on very many other genes, perhaps all genes expressed within a particular cell type. Thus, much of the phenotypic variation among individuals may be caused by allelic variation, which, although not part of the GRN itself, may affect how strongly functionally connected genes regulate one another.[37] Later

in the book we will show how these features of GRNs are relevant to evolutionary innovation (chapter 11) and evolvability (chapter 12).

The Epigenetics of Development

The profound challenge of understanding how a single cell could "know" how
to divide, differentiate, and form a fully functional multicellular individual has
seduced many people (both scientists and nonscientists) into assuming that
the "instructions" for doing so must be encoded in the cell's DNA. Thinking
of development in terms of GRNs can reinforce this view, as GRNs are often
represented as purely genetic. In describing the GRNs of the neural crest, for
instance, we assumed that GRNs operate at the right time, and that the acquisition of a particular gene expression profile causes directed cellular behaviors
and complex multicellular structures. However, for GRNs to specify fully, say,
the shape of a face or the color pattern of the skin, the genome would need to
exercise a level of control of individual cells that it simply cannot have. For
example, in the vertebrate nervous system, the number of connections between cells exceeds the number of structural genes in the genome by several
orders of magnitude. There is simply not enough information in the genome
to instruct each individual neuron where to move and grow and where on each
axon and dendrite to form a synapse.[38] Extragenetic sources of information
are required to build an organism. Cells therefore must respond to cues and
signals in their immediate environment to move toward the right location, to
differentiate into the appropriate cell type, and to modulate gene expression.
For example, migration of the neural crest cells is initially triggered by the
digestion of the extracellular matrix and the inability of the cells to move backward against other cells. This pushes them forward and away from one another.
For appropriate direction of movement and cessation of migration, neural
crest cells respond to both local and long-distance guidance cues. As a result,
the control of gene expression in development is to a large extent external to
the cell, and the genetic regulation of individual cell properties would not scale
up to structured tissues unless cells could respond to their surroundings and
utilize preexisting physical forces that generate order.[39]

These interactions between the GRN and other nongenetic (cellular, extracellular, or external environmental) regulatory factors are sometimes referred
to as "epigenetic," a use of the term that comes closer to the original meaning
that Conrad Waddington intended when he introduced it than how it is more
commonly used today.[40] Waddington recognized that causal factors above the

level of genotype contributed significantly to development.[41] These developmental interactions are responsible for ensuring that cells with different functions organize themselves into tissues, organs, and subsequent phenotypes. They are crucial to the relationship between genotype and phenotype, and frequently provide a level of analysis that is useful for predicting which phenotype will arise.

Physical and Chemical Forces

A snowflake is a crystal that forms when water vapor condenses into ice. Like all crystals, snowflakes self-assemble, their structure and symmetry reflecting the chemical forces that determine the angles of the molecular bonds inside the crystal, combined with the varying opportunities that the structure provides for further growth.[42] The intricate and orderly shapes of snowflakes, just like the regularities of oil droplets in water or of sand dunes, demonstrate that the appearance of complex structure is not restricted to living beings. Therefore, it is not surprising that organismal development also involves a degree of self-assembly. This was dramatically demonstrated in 1955 when researchers broke up an early amphibian embryo into its constituent cells and found, to their amazement, that the cells spontaneously re-aggregated into the original spatially segregated cell types.[43] How were the cells able to sort themselves into their proper embryonic positions? The answer is that the cells rearrange themselves into the most thermodynamically stable pattern. Cells often stick together in a ball, or tissue, because of adhesion molecules on their surface. Cell types differ in their adhesive properties, which means that the cell types with the strongest adhesion invariably sort themselves into the center of the cluster, while those with weaker adhesion remain on the periphery.[44] Much of the structure of the embryo, and the boundaries between tissues, is due to cell types having different kinds and amounts of cell adhesion molecules.[45] Tissues sometimes behave like liquids and undergo phase separation (like the oil droplets floating in water), and tissue layers in the embryo form because of this liquid-like property. Having the appropriate adhesive proteins is thought to have been necessary for the transition from a unicellular to a multicellular form.[46]

Another example of how physical forces explain aspects of development is provided by patterns of folding in tissues such as the brain. Many animal brains have a wrinkly folded property (*gyrification*) characterized by folds in the outer layers. The characteristic convoluted shape of the human brain is the result of gyrification that begins in the embryo. For the first six months of your fetal existence your brain has a smooth surface, but by the time of your birth your cerebral cortex has become covered with gyri (out-folds) and sulci (in-folds), with further folding occurring after birth. This folding brings different regions of the brain into proximity with one another and may be critically important in cognitive functions.[47] Why does the brain fold in this way? Historically, researchers have assumed that genes must specify the patterning of gyrification through spatially variable chemical signals, yet there is no evidence of gene expression patterns that match the gyral patterns.[48] The explanation for the folding is actually far simpler. The outer cortical layer (*gray matter*) expands at a faster rate than the inner tissue, generating compressive stress and leading to mechanical folding of the cortex.[49] This was superbly demonstrated when researchers used magnetic resonance images to build a 3D-printed layered gel that mimicked the developing smooth fetal brain, and then immersed it in a solvent that led to swelling of the outer layer relative to the core, mimicking cortical growth. As predicted, this put the outer layer into mechanical compression and rapidly led to sulci and gyri strikingly similar to those in fetal brains. The researchers also demonstrated that the precise pattern of folding is strongly affected by the complex curvature of the initial brain shape. Gyri and sulci that directly mimic observations in real brains develop in the absence of any prepatterning of cortical growth arising from genetic instructions; yet this packaging of the brain may be adaptive as it maximizes the amount of brain tissue that can fit inside a limited brain case.[50]

Thus, parts of developing organisms act as generic materials (like a liquid as in animal embryos, or like a deformable solid in the case of plants), while other structures can be understood through chemical processes (such as diffusion). In different organisms, homologous GRNs may utilize these physical forces in distinct ways, tweaking the amounts of activator or inhibitor substances, or manipulating the numbers of precursor cells that can differentiate. In this manner, the same set of conserved regulatory genes can generate a multiplicity of forms, by harnessing physical forces that generate order in different ways.[51]

Structure from External Cues

Self-assembly plays an important role in development but typically requires differences between cells to generate complex structure. Here, the environment can play a key role, with external forces and conditions commonly producing the axes of the developing embryo. For instance, fungal rhizomes and algae use the direction of light to establish the primary developmental axis.[52]

In the chick, the head-to-tail (anterior–posterior) axis is determined by gravity when the egg is inside the hen's uterus, prior to laying.[53] Developmental defects occur in many other organisms when they are shielded from the geomagnetic field.[54] External signals, stemming from both internal and external environments, play a key role in cell differentiation by triggering intracellular signaling pathways that affect gene expression, and these signals determine which of a number of possible cell types results.[55] In most animals, growth and cell differentiation occur in response to a combination of mechanical forces, cues from the external environment, substances secreted by other cells, and interactions with symbionts.

The first cues that enable cells to differentiate and the embryo to become an organized heterogeneous whole arise from a stockpile of maternally derived proteins and messenger RNAs (mRNAs) that have been placed into the egg. Immediately after the fusion of egg and sperm, the genome of the offspring plays no role at all in the formation of the early embryo.[56] The maternally inherited resources control basic biosynthetic processes in the early embryo, including directing the first mitotic divisions and molecular gradients that specify initial cell fate and patterning. For instance, in the sea squirt—a marine animal used in developmental studies—only those cells of early embryos that acquire high concentrations of a particular mRNA give rise to muscle cells.[57] This mRNA is stored in a particular part of the egg cytoplasm, and is inherited by any cells forming from that region of the egg. The protein encoded by this mRNA is a transcription factor that activates the muscle-producing GRN from the genome of these particular cells. When they have done their job, the maternally provided mRNAs are degraded, and zygotic gene transcription is finally activated, which carries the embryo through the rest of embryogenesis.

From the earliest stages of development through to adulthood, cell behavior, including adhesion, migration, differentiation, and division, is regulated by signaling between cells. Often, paracrine factors are produced by neighboring tissues that interact reciprocally with each other to form an organ. In the vertebrate eye, for instance, the construction of a functioning lens and retina requires each tissue repeatedly to coordinate its structure and behavior in response to signals from the other. Lens differentiation genes are induced in adjacent ectodermal cells by signals from the optical vesicle, a bulge from the brain. As the prospective lens cells differentiate, they secrete other proteins that induce the optic vesicle to form the retina. The major parts of the eye actually co-construct one another,[58] with each induction comprising a long chain of molecular events.[59]

Complicated patterns of coloration on animal skins, such as stripes or spots, can arise when two or more diffusible chemical signals interact to shape the distribution of pigment-containing cells. None of these cells "knows" what the animal should look like; rather, the pattern arises as a form of self-assembly as the cells respond to local cues and compete for space. Alan Turing (of computer fame) demonstrated mathematically that the competition between a slowly diffusing activator and a more rapidly diffusing inhibitor would result in the reorganization of the chemicals to form stable periodic patterns. For instance, the distinctive stripes of the zebra fish arise in part because the xanthophores (yellow pigment cells) and melanophores (black pigment cells) each mutually inhibit the local accumulation of the other cell type, yet long-range interactions allow xanthophores to enhance the survival of melanophores.[60] The theoretical framework of Turing patterns has been extended to recognize the critical interactions between cells that are not necessarily determined by diffusible factors.[61] Similar patterns are generated by cytoplasmic processes through which pigment cells and their precursors influence the survival of neighbors, by the physical movement of cells in response to a chemical attractant and accompanying depletion of cells in the surrounding area, and by the traction forces associated with cell migration.[62] In chapters 11 and 12 we will discuss how Turing patterns help explain the number of fingers and toes animals have, as morphogens diffuse through the developing limb bud.[63]

The Role of Mechanical Forces

The stresses to which cells are exposed can transform their biological function; for instance, triggering cell proliferation, differentiation, or migration. Physical forces emanating from neighboring cells, the extracellular matrix, or the external environment are converted into electrical or chemical signals, a process known as *mechanotransduction*. Those signals trigger further molecular signals in cascades that culminate in altered transcriptional outputs of the cell's GRNs.[64] The cells themselves, by their mere presence, or as a consequence of the chemical products and signals they generate, modify the extracellular matrix and the properties of the surrounding tissue in ways that feed back to shape their subsequent functioning, or feed forward to affect surrounding cells. For example, mechanical traction forces within the skin are thought to be responsible for establishing the initial pattern of feather bud formation in birds.[65]

An example of the importance of mechanical stress is the development and growth of the vertebrate skeleton. When loads are placed on your cartilage or bone as you walk or carry objects, individual cells experience a range of forces, including shear, compression, and tension, which they convert into biochemical signals that lead to the synthesis of transcription factors and the activation of bone-forming GRNs.[66] For instance, humans are born without bony knee-caps (patellas), which form by mechanical stress as a result of walking.[67] This process continues throughout life. Without mechanotransduction, muscles, bones, and other connective tissue would not be able to strengthen with usage. Exercise is based on mechanical stress leading to increased strength of muscles and bones. Muscles are not the only source of tension on bones, however. In the vertebrate skull, for instance, tension arises from brain tissue growth. If the brain is small it does not rattle around in the cranial cavity—rather, the brain-case is reduced to produce a good fit through reciprocal signaling. Conversely, if the brain is unusually large, because of disease or genetic mutation, the cranial cavity expands correspondingly.[68] Likewise, skin will grow as large as the body because stretching causes the epidermal stem cells to keep dividing.[69]

The fact that mechanical stress induces bone growth makes it possible for the developing organism to influence the size and shape of bones simply by exercising them. Recent experiments with cichlids provide a beautiful demonstration of how jawbone shape is affected by physical stress.[70] These fishes are known for their rapid speciation, with diversity linked to adaptations in their craniofacial skeleton associated with feeding specializations. Remarkably, it turns out that the behavior of the fish as larvae is an important source of phenotypic variation in their head and jaw shapes.[71] During vertebrate development, skeletal elements are laid down as cartilage and then later ossified into bone. Immediately after the cartilaginous lower jaw forms, but before the beginning of bone deposition, the fish start rapidly opening and closing their mouths. This "gaping" behavior is performed at incredible frequency—often exceeding two hundred times per minute—and is unlikely to be explained solely by the requirements of respiration or ion regulation. The frequency of larval gaping varies significantly among different cichlid species, but in a way that intriguingly coincides with interspecific variation in jaw shape.[72] Why should this be?

During each gaping action ligaments pull on the jaw bones, which means that gaping at higher frequency produces more mechanical load and leads to greater levels of mechanical-load-induced bone formation. Experimental manipulations that artificially increased gaping led to larger jawbones, while surgical disruption of gaping through severing the relevant muscles resulted in smaller jawbones.[73] The behavior-induced shift in mandible shape was shown to cause changes in expression of ptch1, a gene that encodes receptors in the hedgehog (Hh) signaling pathway, which participates in mechanosensing. When forced to gape at higher frequencies, the larvae showed elevated levels of *ptch1* expression, and the Hh signaling pathway is known to be able to activate the genes responsible for jaw cartilage to grow.[74] While the larger jaws of many cichlids appear to be environmentally induced by physical stress, some species activate the same molecular pathways without requiring the environmental cue.[75] This interchangeability of environmental and genetic inputs is a feature of development, to which we will return.

The larval gaping behavior does not persist into the juvenile and adult stages, but broadly similar effects of mechanical forces do continue. Juvenile development in fishes is marked by the onset of feeding from the external environment, with variation in dietary habits causing different mechanical loads to be propagated to the jaws. In this way, differences in feeding behavior lead to differential bone development, and experimental manipulation of the feeding regime during rearing (i.e., biting versus suction feeding) reliably generates differences in jaw shape.[76] Changing the mechanical environment of jaw development, either in the larval or juvenile stages—for instance, simply by switching diet—has a pronounced effect on bone development.[77]

From a traditional (i.e., "genetic program") evolutionary perspective, such "self-stimulatory" mechanisms seem quite peculiar—why not simply grow bones of the right length without having to generate additional force? In contrast, from a developmental perspective, it is expected that phenotypic evolution will capitalize on existing regulatory interactions. An advantage of the reliance on mechanical stress is that the organism's morphology can be adjusted according to its current internal and external environment, helping to ensure it remains adaptive even in the face of substantial genetic or environmental perturbation.[78] The internet is awash with videos of two-legged dogs, goats, and monkeys that run around on their rear legs like a *T. rex* or that skip like a lemur.[79] These animals are either born with severe congenital conditions or experience mutilations or crippling disease, and yet their morphology readily adjusts to demand. A quite dramatic change in one aspect of the phenotype can lead to corresponding and functionally appropriate changes in others. A compelling experimental demonstration is provided by bichir fish reared on land, whose fins develop into functional limb-like appendages surprisingly similar to those of the earliest tetrapods,[80] indicating that organisms can develop adaptive (i.e., functional) variation even under evolutionarily novel conditions. We will consider the evolutionary consequences of this in chapters 11–13, including when discussing human evolution.

The Inescapable Outside World

Temperature, humidity, light, and other properties of the environment are inescapable influences on developing organisms. Organisms can move to a warmer or cooler location, but they cannot avoid the effects of temperature on the kinetics of cellular processes. Consequently, temperature and other abiotic factors can have strong influences on gene expression and phenotypic differentiation. Familiar examples include the effect of rearing temperature on metabolic processes underlying the growth of different organs, secretion of morphogens, and accumulation of metabolites, like melanin, in cells and tissues.[81]

Organisms harness the outside world to their own ends, and many environmental factors contribute importantly to the regulatory systems that compose development. In many plants, for instance, the transition from the shoot meristem, which produces leaves, to the floral meristem, which produces a flower, is triggered by changes in day length and temperature.[82] Similarly, the butterfly *Bicyclus anynana* switches its phenotype dramatically to produce wings either with eyespots (wet-season morph) or without (dry-season morph), depending on the temperature experienced during pupation.[83] This switch in phenotype, known as a *polyphenism*, arises because higher temperatures increase the rate of production of a hormone that sustains and expands the expression of distalless, the transcription factor that determines eyespot size. The mechanism is surprisingly similar to caste determination in certain ants and bees, where more nutrition induces the formation of fertile queens through the prolonged expression of juvenile hormone.[84] In fact, hormones commonly mediate the effects of environmental stimuli on gene expression.

In many species of turtle, sex is determined by temperature. High temperature yields females while lower temperatures produce males. The higher temperatures open a thermally sensitive calcium transport protein in the cell membrane of the gonadal cells. The influx of calcium prevents the synthesis of an enzyme that would demethylate (and thereby activate) the genes for testes formation.[85] Likewise, whether or not a male dung beetle develops horns, and the size of those horns, depends not just on its genes but also on the quantity and quality of dung that it ate as a larva, which determines the titer of juvenile hormone during the larva's last molt.[86] The evolution of differences between species in the responsiveness of horn production to food has been shown to result from the amplification of ancestral nutrition-responsive gene expression, the recruitment of additional genes into nutrition-responsive pathways, and through secondary losses in the plasticity of gene expression.[87] A role for the external environment is perhaps most conspicuous in species that use environmental cues to produce distinct phenotypes, like the strikingly different diet-dependent traits of queen and worker ants; however, the environment is *always* a key determinant of development.[88]

Built by Symbionts

All multicellular organisms are hosts to symbionts, including internal bacteria, algae, protists, fungi, and viruses.[89] It has been estimated that more than half the cells in a mammalian body are those of symbionts.[90] However, symbionts are not just residents—they are workers. Signals from symbionts can be critical for normal development of the host. For instance, approximately one-third of the metabolites in human blood are derived from bacteria, and in mice, such microbially derived compounds (from the pregnant mother's gut microbiome) are crucial for normal brain and pancreas development.[91] The bacteria in the newborn gut are critical in the organogenesis of the gut capillaries and lymphoid tissues of several vertebrates, while the metamorphoses of many invertebrates from larval to adult stages also depends on symbionts.[92]

A compelling illustration of the role of symbionts in development is provided by the luminescent bacterium *Vibrio fischeri*, which constructs its own niche by scaffolding the development of the squid *Euprymna scolopes*.[93] The adult squid is equipped with a light organ composed of sacs filled with the light-emitting bacteria, but newly hatched squid have neither the bacteria nor the light organ to house them. The bacteria, which are acquired from seawater, bind to a ciliated epithelium in the squid's mantle cavity and secrete chemicals that induce hundreds of genes in the epithelium to become active. These signals lead to the differentiation of the surrounding cells into storage sacs for the bacteria. The developing storage sacs then induce the bacterial genes that cause the bioluminescence.[94] The squid reciprocates by killing other bacteria that compete to colonize the organ. In this mutualism, both organisms change the other's gene expression patterns in a beneficial way.[95]

Exploratory Processes

While physical properties of the world can contribute to complex cellular and organismal structures, the capacity for adaptive developmental change would be limited if it were not for the employment of clever—and remarkably Darwinian—processes that operate through iteratively generating variation, testing the performance of variants, and regenerating or retaining valuable functions.[96]

During development, the nervous system generates excess neurons and excess neuronal connections, sending out neural projections at random, then pruning these to retain only the most effective.[97] Neurons die if they fail to reach the tissue they will innervate, and there is competition between neurons for proteins secreted by target tissues that support neuronal survival (e.g., nerve growth factors).[98] The final neural anatomy depends very much on experience, and animals in which sensory inputs have been experimentally manipulated develop irregular innervation.[99] For example, experiments reveal that cats raised in a world of horizontal lines are "blind" to vertical objects; they have no trouble jumping up on a chair, but they bump into the chair's legs.[100]

Much of animal learning operates in a similar fashion. The primary mechanism by which animals acquire behavior is through "trial and error," in which actions followed by pleasant consequences are repeated, while those followed by disagreeable events are not.[101] Learned behavior is often the result of an exploratory search conducted over multiple trials, through which individuals hone their behavior to exploit their environment. The initial exploratory component of learning generates behavioral variability, of which the most effective behaviors are retained. Generating behavior in this way leads to extraordinary flexibility—animals can learn a wide range of functional behaviors even in novel contexts. Learning is adaptive precisely because evolution has produced the autonomy to seek out high-fitness behavioral outcomes, and to eschew activities that might negatively impact survival and reproduction.[102]

In fact, many developmental systems operate by generating variation (i.e., "exploring" possibilities), largely at random, testing variants' functionalities, and selecting good solutions for regeneration, in an iterative loop (figure 7).[103] These phenomena, known as "exploratory mechanisms," resemble adaptation by natural selection, except that they allow for information gain by the individual organism within its own lifetime, rather than the acquisition of genetic information in a population over multiple generations. For instance, the adaptive immune system generates antibodies and T cells with initially random variation, then internal selection multiplies and refines those that bind successfully to antigens, retaining the memory of effective molecules.[104] During early development, muscle precursor cells initially migrate in an undirected fashion, but only those that migrate between the tendon precursors receive the tendon-secreted signals to survive and make connections.[105] The immature

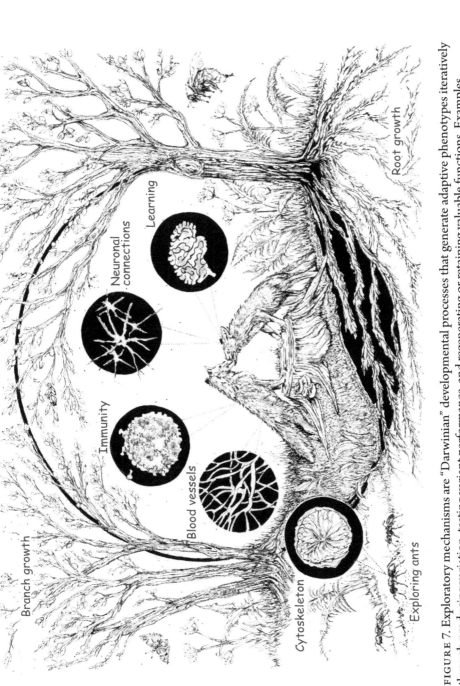

FIGURE 7. Exploratory mechanisms are "Darwinian" developmental processes that generate adaptive phenotypes iteratively through producing variation, testing variant performance, and regenerating or retaining valuable functions. Examples include adaptive immunity; brain development; the growth of blood vessels and plant roots and branches; the cytoskeleton; and learning and exploratory behavior in animals.

muscles and tendons then interact to mature each tissue. Motor neurons, too, are produced in abundance during early development but are retained only if they manage to innervate muscles. The vascular system grows on similar principles, which is why even some major blood vessels show remarkable (nongenetic) variation.[106] New vessels expand into all regions of the body, stabilizing where needed by attraction to hypoxic conditions. The size of the vessels grows in response to blood flow, which is a function of demand. Similarly, when an organ uses up oxygen, it actively promotes its own vascularization because the lack of oxygen triggers expression of a gene (*Vegf*) that produces the growth factor responsible for the differentiation of the vascular cells. The result is a well-distributed system capable of servicing virtually every cell in the body. Insect tracheal systems, much collective animal behavior (e.g., ant foraging), development of plant roots, and remodeling of bone and soft tissue (muscles, tendons) all operate on similar principles.[107] Within the cell, the dynamic behavior of microtubular systems and cytoskeletal arrangements is also the result of exploratory processes.[108] In all such cases, the genome provides the capability to explore options, "not the outcome of the exploration."[109]

Exploratory mechanisms are adaptive because rapid exploration of a large space of possibilities combined with feedback (e.g., reward/punishment) allows information to be gained from the current environment. They also help organisms to cope with internal failures in the somatic genome, epigenome, and microbiome, which are generally too numerous for genetic solutions to repair, while self-organization allows for "adaptive improvisation."[110] While the early variants generated by exploratory mechanisms may be random, later outputs (e.g., learned knowledge, antibodies, new physiological structures) will typically be adaptive because they are informed by within-lifetime selective interactions with the recent environment. At the level of biological fitness, this is distinct from both genetic mutations, which are typically neutral or maladaptive, and other forms of plasticity, many of which may not be adaptive.[111] The sheer number of variants generated means exploratory mechanisms can be costly, but these costs are offset by the organism being able to produce functional responses across a very broad range of conditions, including conditions not previously experienced by its lineage.

These properties of exploratory mechanisms confer major advantages in robustness and flexibility; they are tolerant of mutation, internal failure, environmental novelty, noise, errors, and injury.[112] Within limits, they are anatomically self-correcting in relation to functional demands. They can adapt to evolutionary changes in other parts of the organism. For example, if sensory fields grow or shrink, then the corresponding cortical areas adjust automatically,[113] while changes in tracheal and blood vessels require few, if any, mutations to accommodate to changes in morphology. Cascades of exploratory mechanisms in development can lead to coordinated change across several systems; for instance, when—without genetic change—muscle, nerve, and vascular systems respond appropriately to changes in bone growth.[114] In chapters 11–13 we discuss how these processes facilitate evolutionary change.

Control of Development

The fact that development involves so many "epigenetic" processes helps to explain why there is typically a lot of variation among individuals despite limited genetic variation in the core gene regulatory networks (GRNs). While the main properties of the development of, say, plant leaves, butterfly eyespots, or the vertebrate skull are highly robust because they are internally controlled by the wiring of GRNs, the strength of those regulatory interactions depends critically on the cellular environment. In turn, the properties of the cellular environment depend not only on the core GRNs but also on indirect effects from peripheral genes,[115] signals from surrounding cells, symbionts, abiotic factors such as light and temperature, and other environmental influences, such as nutrition, or the presence of predators, to which exploratory processes are supremely sensitive.[116] Genes are an important causal influence, but the production of a phenotype is not based on a single cause. Rather, phenotypes assemble as the outcome of contingent, reciprocal interactions among composite tissues and environmental agents.[117]

6

Five General Principles of Development

The preceding chapter provided a brief overview of some core developmental processes observed in multicellular organisms. From this material we extract five general principles of development that are relevant to the study of evolution.

First, it is apparent that development is *modular*, by which we mean its components possess their own intrinsic dynamics and integrated structure. As we saw in the previous chapter, neural crest cells are specified separately from neighboring cells that become some other tissues through the action of two key signaling pathways, Wnt and Bmp. The same Wnt and Bmp paracrine factors are used in different combinations in the construction of many other bodily organs, and the same neural crest GRNs are involved in the development of many different features. Development is modular and combinatorial.[1]

In fact, modularity occurs at all levels of biological organization. The structure of genes is modular because, by utilizing multiple separate enhancers, a protein can be expressed in several different tissues but not in others. The traits of organisms also have their own developmental dynamics; part of an organism can undergo evolutionary change without affecting other parts.[2] Sometimes a GRN can be activated in a new part of the body, transferring that character into a new place. For instance, the wings of beetles are encased in an often beautifully colored hard exoskeleton called an elytron. The evolution of this trait involved the co-option of a GRN responsible for exoskeletal development into another concerned with forewing development.[3] The fact that limbs, eyes, or eyespots can develop in parts of the organism that normally do not have these traits demonstrates that, exposed to the right signaling, cells are able to switch on the appropriate GRNs.[4] Other examples of developmental modules include parts of the organism that cannot be grafted onto other regions of the organism, but whose development nevertheless exhibits substantial autonomy, such as the separate parts of flowers.[5] Modularity is manifest even at the behavioral level, where the activities of many animals are composed of subunits. Bird and whale songs have a hierarchical organization, comprising phrases and subphrases, while numerous animals have courtship rituals comprising sequences of stereotyped actions.[6]

Of course, there must also be bounds to modularity.[7] No developmental process is entirely autonomous, and the organism is a mosaic of interacting processes, with some parts more connected than others.[8] The transcriptional regulation of signaling pathways within GRNs is highly complex, for instance, and can itself be regulated by several different transcription factors, allowing pathways to be reused in different contexts. There is "cross-talk" between these pathways, which in combination produce distinctive effects; we saw that the pattern of *Wnt* and *Bmp* expression affected the differentiation of multipotent cells into brain, skin, hair, or neural crest cells. The outputs of these circuits are also highly context dependent. Wnt can be necessary for the development of both lungs and bones only because it does not, by itself, specify which processes need to operate, nor does it interfere with the activities of a process once it gets going, but rather responds differentially to different combinations of signals. Hormones operate in a very similar manner, responding to different inputs to produce coherent effects spanning multiple tissues and behaviors, because they activate hierarchically organized modular subroutines.[9]

Such systems exhibit "weak regulatory linkage" between their individual components, leading to a mix-and-match process in which existing genes, pathways, and circuits function in different contexts;[10] this is critical to evolutionary innovation (chapter 11) and evolvability (chapter 12). For modules to be assembled into new combinations it is imperative that each building block be self-sufficient, in the sense that each one's activity depends minimally on other components.[11] In sum, that development is modular at many levels allows (1) combinatorial associations (of transcription factors or paracrine factors) to specify different tissue types, (2) the recruitment of one module into another module, and (3) the expression of genes independently in different tissues.

A second key principle is that development is "epigenetic," by which we mean it occurs through interactions between regulatory elements above the level of the gene, including interactions between cells, and interactions between tissues.[12] Development is largely about communication between cells. Paracrine factors and secreted extracellular matrices are bound by proteins on the cell membranes of neighboring cells. The binding of the secreted factors to the cell membrane activates enzymes within that cell, and these enzymes activate the transcription factors that characterize the GRN of that particular cell type. The basis of development is that genes respond to signals from outside their own cell. Genes are both active and reactive.[13] In contrast, for most of the last century, evolutionary biology has focused on identifying genetic sources of adaptive variation.[14] Genes have been attributed specific functions within development, often characterized as master regulators. Genetic change has phenotypic consequences insofar as it affects the molecular properties of cells and how cells interact with one another. However, we have described many factors beyond changes in nucleotide sequence that also play a major role in shaping developmental outcomes, and hence phenotypic variation. Physical properties of cells and tissues and a suite of environmental factors can influence gene regulation, and therefore also affect how cells interact. Cell membranes and secreted cell products influence the behavior of neighboring cells; for instance, by secreting paracrine factors that change the gene expression pattern of neighboring cells, or secreting extracellular matrices that provide "paths and guide rails" that allow the migration of other cells.[15] The composite cell types of organs and limbs engage in reciprocal induction, each group of cells instructing the other what to become. Thus, the developing lens cells of the eye instruct their adjacent cells to become retina, as the developing retina cells instruct them to become lens. Hormones respond to environmental cues by activating various GRNs, and physical forces trigger mechanosensors, which initiate a cascade of molecular changes that shape the developing skeleton. To a very large extent, it is the interplay between cells, cell products, and tissues that leads to morphogenesis and differentiation.

While it may be a convenient shorthand to think of genes as having dedicated functions, virtually all genes have multiple functions and are expressed in multiple tissues. Likewise, while it may be an appealing metaphor to characterize the genotype-to-phenotype translation as a linear "mapping," in practice there are an exceedingly large number of intervening steps in the path from DNA sequence to phenotypic character, with all kinds of feedbacks, feed-forwards, and other complex interactions, and any mapping is highly nonlinear and highly contingent on the cell type, tissue, and external environment.[16] The action of the hundreds of genes and their products expressed in trait formation are exquisitely choreographed by cellular- and tissue-level processes, resulting in cycles of gene expression and phenotype formation, until the final phenotype is generated. Mutations causing severe phenotypic change in one person might not cause that change in another. When the first human DNA sequence was determined, it was reported that Nobel laureate James Watson was homozygous at the loci responsible for Usher's syndrome and Cockayne syndrome, expected to leave him blind, deaf, and mentally retarded. He is none of these.[17]

Given the baggage carried by terms such as "epigenetic," "genotype-phenotype map," and other such labels including "genetic architecture," throughout this book we will refer to the higher-level interactions between genes and other regulatory elements as the *regulatory network*.[18] We deploy this term as a representation of the complex web of interacting regulatory elements "above" the level of the gene, including environmental cues and influences (figure 8). Our references to "regulatory networks" encompass not just interactions among genes but also among cells, tissues, and organs, recognizing that the output of these interactions depends not just on the architecture of the interacting elements but also on the initial conditions, as well as on the form and magnitude of those interactions, and on their dynamics. In chapters 11 and 12 we will describe studies showing that regulatory networks can predict phenotypes, explain innovation, and account for differences in evolvability that cannot be understood at the DNA-sequence level.

Differences in phenotypes between individuals and species depend, in no small part, on changes in developmental regulation.[19] In addition to mutations that change a protein, evolution can occur through changes in the location, timing, or amount of gene expression.[20] The bat wing, which retains its forelimb webbing by blocking the proteins that would otherwise cause the interdigit cells to die, provides an example of a change in the location of expression of a developmental module.[21] Here, a protein that is used to prevent cell death in digits is also expressed in the cells of the webbing. Changes in developmental timing are exemplified by the extended growth of the human brain, which occurs through a protracted period of cell division in the cerebral cortex, and in the enormous number of vertebrae and ribs exhibited by snakes, which results from the prolonged expression of the *Oct4* transcription factor in the rib-producing region of the embryo.[22] The evolutionary significance of changes in the amounts of key substances is illustrated by variation in beak depth among Darwin's finches; the beaks of different species arise through changes in the level of *Bmp4* expression as embryos.[23]

A third principle is that development is *constructive*, by which we mean that development is a coordinated integration of many sources of potential information, not just those arising from nuclear genes. The organism creates a

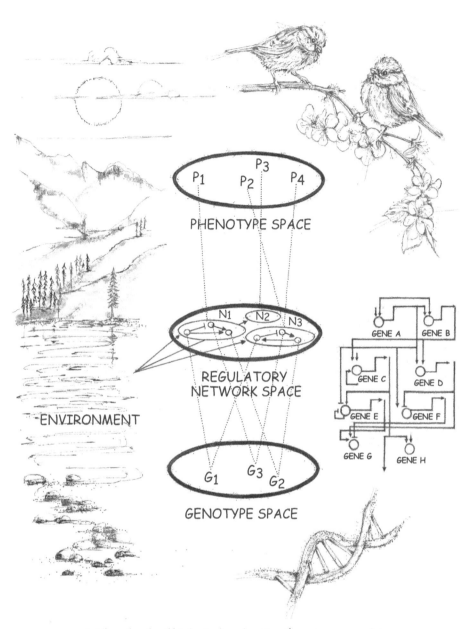

FIGURE 8. Three levels of biological explanation (genotypes, regulatory networks, and phenotypes). Regulatory networks are dynamic developmental interactions among genes and other regulatory elements, including cells, tissues, and organs. These elements, influenced by the environment, specify phenotypes, explain innovation, and account for differences in evolvability that cannot be understood only at the DNA-sequence level.

developmental trajectory by constantly responding to, and altering, internal and external states.[24] This can be contrasted with the widespread view of development as directly "programmed"—that is, unfolding according to rules and instructions specified within the genome.[25] In fact, developmental causation flows from "higher" levels of biological organization (e.g., cell-cell interaction, the immediate environment) that regulate gene expression, and back again, to generate proteins and cell behavior.[26] There is a mixture of elements of genetic specification and improvisation, and even the former is flexibly responsive to the environment.[27] For example, microbial symbionts, often inherited from the mother, are crucial in producing organs such as the rumen in cattle, and ovaries in insects such as the wasp *Asobara*. The organism has evolved to respond to these agents by developing new tissues. Other animals may encounter the same bacteria but not produce these structures. Similarly, temperature is critical for sex determination in turtles, and nutrition is critical for horn formation in male dung beetles. The response—both the manner of response and thresholds to environmental signals—can be selected. The genome is reactive as well as active. It does not directly control, but rather "listens" and responds.[28]

As shown in chapter 5, much of the inherent order and structure of the phenotype results not from the genes expressed but from natural physical forces and chemical processes, which combine to generate surprisingly rich patterns. Additionally, developmental systems respond flexibly to internal and external inputs through exploratory and selective mechanisms among microtubular, neural, muscular, vascular, tracheal, plant root, behavioral, and many other systems. Again, there are potentially significant evolutionary implications. Exploratory processes enable development of phenotypes that may not have been prescreened by earlier natural selection; yet those phenotypes may be highly adaptive and may shape the rate and direction of further evolution by determining what genetic variation is exposed to selection (chapter 8).[29]

A fourth insight is the *interchangeability* of the phenotypic consequences of a change in DNA and a change in internal or external environment. We saw this, for instance, in the case of the large cichlid jaws that could result from either physical stress or genetic change in the same developmental pathway. A gene can be activated or inactivated either genetically by mutation or epigenetically through environmental agents. This makes sense once it is recognized that much of the cause of phenotypic development does not reside completely in genetic "information," but rather at the level of the regulatory network.[30] Interacting genetic and environmental causes of development are what collectively provide the information necessary to build phenotypes, and control resides no more in one of these elements than another. Despite the shorthand language, the genome is not a code to be deciphered but a score to be interpreted.[31] Specificity in protein sequence derives from information distributed among the coding regions of the genome and all regulatory mechanisms.[32]

Nonetheless, we recognize that "information" is a tricky term with many meanings. One common usage is in the spirit of Francis Crick's claim that the sequence of nucleotide bases specifies the proteins (i.e., "Crick information").[33] However, if one maintains that there is information in DNA coding sequences that builds bodies then, by the same reasoning, one must also accept that those regulatory molecules that differentially activate, select, and alter the information in coding sequences (for instance, by determining whether, when, and where genes will be expressed or specifying how exons will be spliced together) provide additional information to build bodies, as do molecular components of some extragenetic forms of inheritance.[34] Moreover, there can be no information without a mechanism of interpreting the DNA sequence as information.[35] The zygote inherits DNA; it does not inherit "genes." Genes and gene products are constructed anew in each cell in the developing embryo by the relationships among DNA, nucleosomes, transcription factors, and RNA-splicing factors. Only certain regions of the DNA sequence are constructed into genes, and different regions of the genome can be genes in different cell types. The interpretation of *What is a gene?* is done by the cell, or a higher-order structure.[36] Development is all about the interpretation of relationships.[37]

Some readers might be tempted to infer that, as the gene regulatory machinery of cells comprises gene products, ultimately, the information necessary for development resides solely in the genome. However, there are at least three problems with that reasoning. First, information does not "reside" in anything: it is created during development in the interactions between processes.[38] Second, not all (trait, or even protein) specificity can be traced to a feature of the genome sequence, since the environment also plays an essential role in regulating gene expression.[39] Indeed, the "reactive genome" has itself evolved in conjunction with environmental inputs, which play instructive roles in the development of phenotypes. In addition, the symbionts we acquire from our mothers are critical in creating our immune systems, capillary networks, and nervous systems. Third, while many components of the regulatory machinery are gene products, the tracing of causality back to genes is an entirely arbitrary convention.[40] The regulatory machinery of cells determines gene transcription, which produces gene products that make up the regulatory machinery, which determines transcription, and so forth in cycles that flow down the generations. Tracing causality back to genes is seductive but logically flawed.[41] Development concerns relation-ships at all levels, as Susan Oyama has emphasized.[42] The following chapters illustrate how embracing this developmentalist way of thinking can aid under-standing of how genetic change leads to adaptive evolutionary change.

Finally, one of the most important implications of development for evolu-tionary biology is that, because of epigenetic developmental interactions in regulatory networks, random genetic change does not typically result in random phenotypic variation; that is, phenotypic variation is structured and *biased*. As explained in chapters 3 and 4, the historical focus on explanations in terms of fitness differences has led to the assumption that the phenotypic variation that is subject to natural selection arises randomly through genetic mutation, with mutations typically of small phenotypic effect,[43] or that the biases that arise from development are of no consequence for understanding adaptation and diversification. Key premises here are that variation is random *with respect to function* (i.e., mutations do not occur on demand), and that mutations of large effect would disrupt development and hence be highly unlikely to be adaptive. The above account of development suggests a slightly different conception. The random mutation of any gene or regulatory element will inevitably be pro-cessed by a developmental system, with the effects integrated into one or more preexisting GRNs. Phenotypic variation will be channeled by the processes of development to produce some phenotypes more readily than others, since cor-relations between traits arise through shared and often highly conserved devel-opmental regulation, epigenetic as well as genetic.[44] Moreover, the processing of mutation by preestablished developmental modules, combined with the responsiveness of developmental systems to the environment, as exemplified by exploratory processes, makes it possible for both mutation and environmen-tal challenges to produce variation that is biased with respect to function (chap-ters 7 and 8). At the same time, the modular structure of development allows major reorganization of a single system without disruption of other parts of the phenotype. As we will see in subsequent chapters, this helps to explain the rate and direction of evolution (chapter 9), the origins of evolutionary innovation (chapter 11), and variation in evolvability (chapter 12).

These then are the five general principles that we emphasize as being par-ticularly important to the field of evolutionary biology: development is modular, epigenetic, constructive, interchangeable with respect to genetic and environ-mental inputs, and prone to generate biased phenotypic variation.

PART III

The Developmental Bases of Evolutionary Processes

7

Developmental Bias

Struggling to devise a satisfactory theory of inheritance, Charles Darwin turned to domesticated animals and plants for inspiration. Breeders had generated a vast amount of data that gave clues about the inheritance of traits, through which Darwin trawled fastidiously.[1] While the resulting account of inheritance through *Pangenesis* famously flopped,[2] Darwin's investigation revealed a striking pattern, which has stood the test of time: domesticated species—particularly domesticated mammals—share a suite of behavioral, physiological, and morphological traits. That dogs, cats, horses, pigs, sheep, goats, cattle, and rabbits should all be docile and tame made sense—after all, Darwin knew they had been artificially selected for these traits. But why should domesticates have changes in their skin or coat colors? Why smaller teeth and brains, or flattened muzzles? Why floppy ears, curly tails, and frequent estrus cycles?[3] Darwin was unable to resolve this conundrum.[4]

Approximately a century later, Russian geneticist Dimitry Belyaev studied this puzzle by breeding wild foxes, generating lines in which only the tamest individuals (i.e., the calmest, most friendly) were allowed to reproduce.[5] Belyaev's breeding program, which is still going today, over sixty generations of selection later, revealed that the full suite of traits associated with domesticated animals could arise in an otherwise undomesticated species through selection for tameness alone.[6] This constellation of traits is now widely known as the "domestication syndrome" (figure 9).[7] There are, of course, many differences among domesticated mammals in the varieties of traits that have changed, leading to some disagreement over how these data should be understood, and the significance of the fox study.[8] Nonetheless, there can be no question that many of the same altered traits show up again and again in domesticated animals.[9] A compelling explanation for this had to wait until 2014, when developmental geneticist Adam Wilkins and his collaborators identified what the

FIGURE 9. Developmental bias and domestication syndrome in mammals. *Top.* Artificial selection during domestication, including for tameness and reproductive traits, favored correlated genetic variation expressed in piebald coloration, floppy ears, and other traits including smaller teeth and brains. *Bottom.* Generations of selection for tameness in wild foxes produced the full suite of traits associated with "domestication syndrome," with domesticated foxes in many ways resembling dogs.

curious assortment of traits making up this syndrome had in common: a reliance on neural crest cells.[10]

As described in chapter 5, neural crest cells are pluripotent embryonic cells that migrate from the crest of the neural tube (hence the name) and eventually differentiate into a very wide array of cell types. They are also either direct precursors to, or play vital roles in, the development of virtually all the traits associated with the domestication syndrome. For instance, neural crest cells are antecedents of bone and cartilage in the skull, jaws, ears, and tail, as well as of teeth, and of melanocytes expressed in pigmentation changes. This shared developmental origin allows virtually the entire repertoire of traits associated with domestication to be coordinated by a single mechanism: mutations in neural crest cell genes.[11] Dozens of genes have been identified as playing crucial roles in neural crest cell specification, migration, and differentiation, forming a complex GRN.[12]

The genomes of domesticated breeds contain mutations in neural crest cell genes not seen, or not common, in wild counterparts.[13] Crucially, neural crest cell genes are expressed in hormonal physiology, including in the hypothalamic-pituitary-adrenal axis, which regulates stress reactions.[14] Wilkins and colleagues hypothesized that selection for tameness had reduced reactivity by favoring mild loss-of-function mutations in genes responsible for neural crest cell formation, specification, or migration.[15] A reduction in any of these processes would inevitably produce deficits in the precursor cells of other structures and tissues that ultimately derive from neural crest cells. Different domesticated animals, and different breeds of the same species, will accumulate different mutations, which explains inter- and intraspecific trait variation. However, what the domesticates have in common—that syndrome of shared characteristics—can best be explained by developmental regulation via the neural crest GRN.[16]

Mammalian domestication illustrates how development can affect evolution. There is nothing intuitive about the suggestion that selection for tameness should lead to the evolution of curly tails or piebald coat colors. In fact, these observations baffled Darwin and numerous other researchers for over a century. That there may be genetic correlations between these traits is not really the point—the question is *why?* Only with knowledge of development can this long-standing enigma be resolved. What connects the different traits is not one or more macromutations (to date, no such mutations have been identified)[17] but the shared developmental regulation of the neural crest cell GRN. Without this GRN's regulatory interactions, and its widespread reuse in different tissues, there would be no syndrome.[18] Mammalian development effectively loads the dice to ensure that tameness, when it eventually evolves, tends to be found in animals with splotchy colored skins or small brains.

The rapid response to selection for tameness exhibited by domesticated animals is almost certainly due to selection exploiting variation in a preexisting major cluster—or axis—of naturally co-occurring traits. The same developmental axis is now biasing how wild foxes (the same species as studied by Belyaev) adapt to urban settings.[19] Kevin Parsons, evolutionary geneticist at the University of Glasgow, and colleagues compared the skull morphology of urban and rural foxes, and found that reduced muzzle length, smaller brains, and diminished sexual dimorphism have recently evolved in the urban populations. These patterns are strikingly parallel to the domestication syndrome found by Belyaev.[20] The study establishes that the developmental bias described in domesticated species cannot be dismissed as an unnatural outcome of artificial selection but also plays a part in the evolution of animals in the wild. Parsons's team examined the skull morphology of other canids, mostly from within the fox genus *Vulpes,* and found that the same developmental changes as in the urban-rural contrast were apparent across diverse *Vulpes* species, but with much greater differences among species. The most prominent interspecific variation involved a shortening of the snout accompanied by a reduced braincase, or a lengthening and narrowing of the snout accompanied by an increase in the braincase.[21] This same cluster of developmentally biased covariation has apparently been shaping the evolution of foxes for millions of years. Another study established that some variation in cichlid face shape can also be explained through mutations affecting neural crest cell migration, suggesting that this cluster of covarying traits may be found in all vertebrates.[22]

The domestication syndrome provides a particularly vivid illustration of the fact that natural and artificial selection can work with only the phenotypic variation that already exists. The point is not that all "domestication traits" are correlated responses to selection on tameness alone (that seems very unlikely), but that by causing particular traits to vary together, the developmental biology of vertebrates provides opportunities for those traits to also be consistently selected together. By generating patterns of phenotypic covariation, the mechanisms of development are responsible for determining which phenotypes evolve.

The suggestion that evolution must proceed where development leads is not merely a claim but an imperative. Yet, at the same time, developmental systems are obviously a product of evolution, which means that the direction of future evolution is dictated by past evolution, including, of course, natural selection. There is a historical explanation for the unusual biology of neural crest cells that now contributes to the morphological and behavioral evolution of urban foxes, but that too is a story of the natural selection of developmentally biased phenotypic variation. The evolutionary consequences of this causal entanglement of generative and selective processes are difficult to resolve. Historically, the whole issue has for the most part been put aside, with selection being regarded as the only creative force in evolution and development acting as a brake, capable of explaining the absence of adaptation but not its presence ("developmental constraint" rather than "developmental bias"). In this chapter and those that follow, we explain how changing this perspective can provide richer explanations for the evolutionary process.

Selection and Development Bias Are Not Alternatives

The suggestion that developmental mechanisms can bias the course of evolution has a long history, dating back at least to nineteenth-century interest in macro-evolutionary trends (e.g., "orthogenesis").[23] As described in chapter 3, such ideas were discredited in the early part of the twentieth century, with increased emphasis on natural selection as the sole explanation for sustained, adaptive change. However, they never completely went away and achieved prominence again in 1980s, following publication of Gould and Lewontin's influential "Spandrels" paper and a growth of interest in phenomena such as the sudden appearance of new forms in the fossil record ("punctuated equilibria").[24]

In such debates, the developmental mechanisms that make random genetic variation give rise to structured phenotypic variation have commonly been thought of as *constraints*: features of organisms that hinder, or even prevent, populations from evolving adaptively.[25] The term "constraint" often implies that some regions of phenotypic space that might otherwise be adaptive are not available to natural selection because they cannot be generated by the developmental system. Proponents and critics of neo-Darwinism commonly portrayed constraints as acting in opposition to natural selection, or as providing an alternative explanation to it.[26] The traditional viewpoint has been that while constraints may make certain regions of morphospace inaccessible or difficult to reach, for other regions natural selection has full reign to explain evolutionary outcomes.[27]

The evolution of the mammalian neck provides a useful example.[28] Swans, flamingos, and cranes can bend their necks into beautiful, smooth, even S-shaped curves, not achievable by any mammal. That is because, in contrast to birds and reptiles, the elongation of the mammalian neck involves lengthening each individual vertebra rather than increasing the number of neck bones. Giraffes, whose necks are commonly over two meters long, have the same number of vertebrae as mice, whose necks span a few millimeters. In fact, nearly all species of mammals, including humans, have exactly the same number of vertebrae—seven. By comparison, swans typically have twenty-three to twenty-five cervical vertebrae and flamingos eighteen to twenty, which explains how they can bend their necks so elegantly. Viewed from an engineering or design perspective, restrictions on vertebra number have *constrained* the evolution of long, slender, and maneuverable necks in mammals. Given the adaptive advantage that flexible necks seem to confer on birds, snakes, and extinct reptiles like plesiosaurs, it is difficult to envisage that such flexibility is not, and has never been, beneficial to any mammalian species, particularly when one considers the clumsy way that giraffes drink water. The absence of mammals with additional neck vertebrae is not because seven is the optimal number of vertebrae, but rather because mutations that modify the number of cervical vertebrae disrupt fundamental features of the mammalian (but not avian or reptilian) body plan.[29]

The dominant response to examples like these has been to acknowledge that developmental constraints exist but deny them any explanatory responsibility for interesting aspects of evolution.[30] For instance, in a forceful rebuttal of Gould and other critics, Charlesworth, Lande, and Slatkin first extol the potency of artificial selection as evidence that constraints play, at best, only a minor role in evolution, and then argue that "the patterns of genetic and phenotypic variation are themselves shaped by selection."[31] From this pan-selectionist perspective, selection is fully responsible for both the generation of variation and the sorting of this variation.[32] All that is left for development to account for is the absence of adaptive fit arising from idiosyncrasies that selection was not able to overcome.

More than anything, such debates highlight how conceptualizing developmental bias as constraint and interpreting constraint and selection as in opposition are unhelpful and misleading.[33] Recall, the term "developmental bias" refers to the idea that developmental systems generate some trait combinations more readily than others.[34] The distinction between *bias* and *constraint* is significant because the efficacy of artificial selection provides evidence against the ubiquity of absolute constraints but not against developmental bias—in fact, it provides evidence *for* developmental bias, as the domestication syndrome highlights (and the vertebrate neck bone example too, when properly under-stood).[35] The fact that selection *can* shape trait covariation does not imply that there must always be a functional explanation for such correlations, beyond the suggestion that in the evolution of new characters it can be expedient to reuse existing regulatory circuitry. For instance, knowledge that the neural crest GRN has evolved through natural selection would not lead any re-searcher to infer that floppy ears are an adaptation for tameness. As we will discuss in chapter 12, theoretical studies show that there can be very effective, if indirect, selection *of* the capacity to generate useful phenotypic variation from random genetic variation, and we will suggest that natural selection with-out developmental bias is actually rather a poor explanation for evolvability. Yet, despite the fact that developmental mechanisms have themselves evolved through past natural selection, there is nonetheless an important explanatory role that they can play in evolution, in providing mechanistic detail that can account for the origin of novelty, explain why variation is aligned along par-ticular axes (making some evolutionary trajectories more likely than others), and why selection *for* one trait leads to selection *of* others (in a way that goes beyond the observation that the traits are correlated). Later chapters will de-scribe how developmental processes can explain these patterns of trait covaria-tion. Developmental bias and natural selection are not alternatives: generative and selective processes jointly determine phenotypic evolution.

Phenotypic Variation versus Variability

The concept of developmental bias includes developmental constraint, but it is better suited to bring out the positive contributions of development in evolu-tion.[36] This is partly because a categorical distinction between what is biologi-cally possible and what is not neglects how development can bias outcomes within the "what-is-possible" region. Such biases can render some phenotypes highly likely to arise, irrespective of what mutations occur or which environ-mental conditions are encountered, while other phenotypes are only rarely produced. Only in extreme cases does this result in the complete inability to produce a trait.

Developmental bias refers to discrepancies in the availability of phenotypic variation. Characterizing this variation is central to the study of evolution by natural selection. Since biologists are generally concerned with complex traits that show continuous variation, such as the size, shape, and color of morpho-logical structures, they often study phenotypic evolution using what is known as "quantitative genetics." The key equation representing quantitative genetics is $\Delta z = G\beta$, which describes evolutionary change in the average values of a suite of traits (z) as the product of the G matrix, whose entries are generally thought to represent the extent to which the traits vary and covary at the genetic level and a vector of selection gradients, β, which captures each trait's contribution to fitness.[37]

At first sight, this equation seems like a suitable tool with which to investigate the evolutionary consequences of developmental bias. As discussed in chapter 2, the G matrix provides an estimate of the biasing effect of standing genetic variation and covariation on evolution,[38] and it has been suggested that the leading eigenvector of G (called g_{max}), which describes the direction with the greatest amount of additive genetic variation, predicts evolutionary trajectories because genetic covariances represent constraints on the response to selection.[39] In practice, for several reasons, G is of limited value for understanding the role of development in evolution.

First, the same pattern of genetic covariation can arise from a variety of developmental interactions.[40] G tells us how strongly a specific set of traits are statistically associated with one another, but gives us no clues as to how or why. As Mary Jane West-Eberhard emphasized, such an analysis "describes a statistical interaction, not a mechanistic developmental one. These quasi-causal terms for statistical correlations are potential traps for the innocent."[41] Second, G describes the currently existing variation but not the propensity to generate variation. Standing variation is transient, and will change through natural selection, and across environmental conditions.[42] In quantitative genetics, another matrix (M) describes how new genetic variation in each trait arises and affects other traits through pleiotropy and epistasis.[43] However, without a mechanistic representation of development, it is not possible to say very much about M either.[44] Third, genes are not the only source of bias, as chapter 8's discussion of developmental plasticity will highlight.[45] These problems might not matter if there were a simple or linear mapping of genotype to phenotype, or if any influences of developmental interactions canceled themselves out across a population, as is often assumed.[46] However, research has found the genotype-phenotype relationship to be highly nonlinear and complex, and both mathematical and computational models show that this can lead to departures from the expectations of quantitative genetic models.[47] Predicting evolution without knowledge of development is hard!

To be fair to evolutionary quantitative genetics, the approach was designed primarily to explore responses to agricultural selection rather than evolutionary consequences of development.[48] Its representation of phenotypes in terms of statistical covariance often does a good job at short-term prediction of the response to selection, and contributes in many other ways to understanding evolvability.[49] Yet, to give a complete answers to counterfactual questions such as, say, *Why did turtles evolve shells rather than long legs?* (chapter 3) or *Why do cichlids exhibit parallel evolution?* (chapter 8), researchers must also study the propensity for developmental systems to vary—their *variability*—in addition to the observed state of variation.[50]

One important aspect of variability is how random genetic variation translates into phenotypic variation, often referred to as the *genotype-phenotype map* (GP map), although here we avoid this term.[51] Given the vast number of possible genotypes, working out the correspondence between any genotype and any phenotype might seem impossible, and unlikely to lead to general conclusions.[52] However, over the past decades, theoretical and empirical studies have demonstrated that it is now feasible to study this relationship, and in doing so have revealed important insights into how evolution works.[53]

An obvious way forward has been to study phenomena with relatively simple genotype-phenotype relationships that are able to provide access to the full spectrum of possible variation. The secondary structure of RNA, for instance, is a useful system in which to study some general properties of development,[54] as the three-dimensional structure of RNA is one of the simplest, yet functionally important, phenotypes. RNA plays many different important roles in biology, from the messenger RNAs that carry information from DNA sequence to protein to various kinds of noncoding RNAs that are involved in translation, DNA replication, and gene regulation. The function of RNA is strongly dependent on its three-dimensional structure. While this is difficult to predict from the RNA sequence, the bonding pattern of a folded RNA—its secondary structure—is much easier to predict and can stand in for the three-dimensional structure.[55]

For more complex single-cell or multicellular phenotypes, much of what has been learned about developmental bias comes from experimental studies that reveal the structure of GRNs and other causal dependencies responsible for correlated changes in phenotypes. Even if developmental mechanisms are unknown, or cannot be investigated with computational models, it is sometimes possible to study genotype-phenotype relationships empirically—for example, by using mutation-accumulation lines.[56] In a mutation-accumulation experiment, inbred lines of an organism are maintained under conditions that minimize the effect of natural selection, with the aim of quantitatively estimating the rates at which spontaneous mutations occur, and examining the distribution of their effects on the phenotype. For instance, Christian Braendle and colleagues at CNRS–University of Nice in France carried out such an experiment in two species of nematode worms (*Caenorhabditis elegans* and *C. briggsae*), providing a beautiful illustration of developmental bias.[57] The authors found that spontaneous mutations produce some phenotypic forms commonly, while others were extremely rare or absent. Similar results have been observed for wing morphology in mutation-accumulation lines in the fruit fly *Drosophila*, and in a suite of morphological traits in the model plant *Arabidopsis thaliana*.[58]

These experiments reveal that the relationship between genotype and phenotype exhibits reliable and common features, patterns that theoretical models are able to reproduce.[59] One such general conclusion is that the probabilities of generating particular phenotypes through random genetic change vary by many orders of magnitude.[60] Some wing shapes or flower forms are millions of times more likely to arise than others. In other words, strong developmental bias is the norm. A second is that only a tiny fraction of the imaginable phenotypic variation is observed in nature.[61] Phenotype space is largely empty. A third, and perhaps the most provocative, finding is that it is often possible to predict which phenotypes will be found in nature, and thus to account for evolutionary change, using knowledge of which phenotypes are easily generated through mutation.[62] Fruit fly wing evolution, for instance, can be predicted with surprising accuracy with knowledge of patterns of wing variability.[63] Recent evolutionary analyses of regulatory networks help to explain why the relationship between genotype and phenotype should have these properties (see chapter 12).[64] For the moment, we focus on how these conclusions have important implications for adaptive evolution.

Developmental Bias and Adaptive Evolution

If developmental processes make some phenotypes orders of magnitude more likely to occur than others, doesn't that mean that natural selection will end up sorting between the commonly generated forms? In some respects, this suggestion is uncontentious. Various kinds of mutational biases are well established. For instance, the genetic codes of many organisms are biased toward using one of the several codons (three-nucleotide DNA sequences) that encode an amino acid.[65] Likewise, a bias toward nucleotide transitions over transversions is well established in molecular evolution, and while explanations for this have often invoked selection, a mutational cause is now considered more likely.[66] However, while mutational biases are widely recognized, they are commonly thought to be most influential when selection is absent, are rarely invoked to explain adaptation, and are often regarded as weak compared with selection in non-neutral scenarios.[67]

Consideration of phenotypes rather than DNA sequences helps to show that this reasoning is not quite correct.[68] Developmental bias is different from mutational bias, although the latter can be a source of the former. Developmental bias refers to disparities in the rate at which phenotypes arise, often assuming that all mutations are equally likely, while a mutational bias refers to differences in the rates that mutations occur. The biology of developmental processes reviewed in chapter 5 shows why developmental mechanisms do more than supply the raw material for selection. By biasing the phenotypes presented to selection, they codetermine the rates and directions of change, and equilibria approached, and thereby contribute significantly to explanations for evolution and adaptation.[69] In urban foxes, for example, selection explains why a short-muzzled, small-brained, and bold variant replaced the long-snouted, bigger-brained, wary alternative. However, developmental factors explain why snout length, brain size, and bold–shy disposition are associated, and thus why they are selected together.

That some phenotypes occur more commonly than others is, of course, well established. Over fifty years ago, paleontologist David Raup showed that only a tiny proportion of all possible snail shell shapes was realized in nature, and suggested that this was partly explained by the mechanics of growth.[70] Unfortunately, it is usually difficult to infer a developmental bias with any reliability solely from an absence of forms in nature, since that absence could also be explained by natural selection that has weeded out low-fitness variants. However, in rare cases this ambiguity can be resolved. Researchers can sample genotypes at random and compare the distribution of phenotypic variation generated in the process to that observed in nature. Correspondence would suggest that developmental bias dominates the distribution of phenotypic variation, whereas a discrepancy implies that selection is more influential.

Again, RNA secondary structure provides a powerful illustration.[71] Research in Ard Louis's lab at the University of Oxford has established that the genotype-phenotype relationship for noncoding RNA is extremely biased, with the majority of genotypes mapping onto an exceedingly small proportion of possible RNA secondary structures. Moreover, only the most commonly arising RNA secondary structures actually appear in nature. Apparently, only a tiny proportion of possible RNA structures have ever occurred for natural selection to sort among.[72] This suggests that adaptive evolution of RNA is limited to phenotypes that are easily produced by mutation—it is dominated by developmental bias, with natural selection playing the more modest role of maintaining the useful variants among this subset.[73] If the tape of life were rerun, the same spectrum of RNA secondary structures would probably evolve, because many structures that are physically possible and biochemically functional are just very unlikely to arise through mutation. For RNA, "the primary explanation" for the distribution of adaptive forms is developmental bias.[74]

The genotype-phenotype relationship for RNA secondary structure is sufficiently simple to enable phenotypes to be predicted from sequences using sophisticated theory, but the traits of many organisms are too complex for this to be feasible. Nevertheless, the biases generated by, say, neural crest cell biology, or other developmental systems, are just as real, and will inevitably channel phenotypic variation along particular pathways, thereby increasing the chances that evolution will take these trajectories.[75] In many such instances, selection will play an important role in further restricting the phenotypic variation observed in nature, sometimes perhaps even overwhelming the bias in the available phenotypic variation, and at other times being dominated by it. However, there is always developmental bias.[76] The choice between positive (bias) and restraining (constraint) metaphors is more than rhetorical.[77] The explanatory value of developmental bias is that it can help elucidate biological features that would be difficult to understand under the tacit assumption that selection acts on unbiased variation.[78] Such features include the origin and rapid adaptation of complex phenotypes, why some lineages continue to diversify while others do not, and why some features evolve repeatedly, often using the same developmental pathways.

The inhibitory cascade model of mammalian tooth development discussed in chapter 2 and the neural crest cell account of domestication described here each show, in different ways, how knowledge of development can provide richer explanations for evolutionary biology. Another example is provided by studies of the size, position, and color of eyespots in a group of approximately three hundred mycalesine butterflies that inhabit the tropical and subtropical regions of Africa and Asia.[79] The eyespots found on these butterflies' wings have concentric rings of differing colors, which have been a focus for evo-devo research for many years. Work on the model species *Bicyclus anynana* by evolutionary developmental biologist Paul Brakefield and his colleagues at Leiden and Cambridge universities has provided insights into the developmental control of the size and color composition of eyespots. A combination of artificial selection, developmental studies (e.g., gene expression analyses), and quantification of the variation observed within and among species has revealed that wing-spot characters that respond readily to selection (i.e., size of different eyespots) exhibit a predictable diversity across species for large regions of morphospace.[80] Conversely, eyespot color, which shows much more limited variability and fails to respond to antagonistic selection, shows correspondingly limited diversity across multiple genera.[81] Experimental investigations by Brakefield's team have uncovered shared developmental regulation of color across all eyespots. Here again, patterns of diversity across the wider taxonomic group can be predicted with knowledge of conserved developmental mechanisms in just a single species. The differences in the evolvability of eyespot size and coloration reflect predictable patterns of variability in the underlying developmental mechanisms.

How common is it that developmental biases are sufficiently persistent to leave a signature in adaptive evolution? This question is difficult to answer for several reasons. First, there are few examples of organisms or phenotypes where the developmental mechanisms are worked out in as much detail as the butterfly eyespots. Second, only a handful of studies have tested whether evolutionary trajectories are concordant with the phenotypic effects of random mutation. Two of the best examples come from the abovementioned nematode and fruit fly studies that quantified phenotypic variability using mutation-accumulation lines. In both cases, the evolutionary divergence between species followed closely the patterns of variability generated by mutation.[82] These studies came to the same conclusion as the example of RNA secondary structure: forms that are easily accessible in development are also those that evolve. This may be a common pattern, since studies that compare the strength of covariation in morphological characters often find that groups of organisms evolve along rather narrow morphological trajectories.[83] For example, the morphological diversity of birds' beaks—including those of Hawaiian honeycreepers, which show striking diversification—appears to have evolved along narrow paths in morphospace.[84] These conclusions also fit with recent thinking from quantitative genetics, emphasizing that there are often substantially fewer genetically independent traits than phenotypes measured, leaving strong selection concentrated along a few major axes.[85] This means that "the presence of genetic variation in a trait, and selection on that trait, is often not sufficient for the trait to respond to selection," since selectable genetic variation is aligned along a small number of axes that result from functional interactions in developmental systems.[86]

Of course, development, and hence developmental bias, evolves too.[87] For example, while the mycalesine butterflies most often fall along a developmentally favored evolutionary trajectory, the bias does not impose "constraints" that are impossible to break;[88] there is an interesting exception to the rule.[89] Unlike other mycalesine genera, the genus *Heteropsis* appears to have evolved the ability to control each eyespot's color independently, and as a result exhibits extraordinary rates of evolutionary diversification.[90] Experimental manipulations of pupal wings established that the bias had been removed through a novel regional response of the wing tissue to a conserved patterning signal.[91]

The *Heteropsis* example suggests that changes in regulatory interactions, resulting in changes in modularity and integration, provide opportunities for evolution to proceed in new directions, which often allow rapid morphological change and facilitate evolutionary diversification. At the same time, the diversity of eyespot shapes and sizes in the other mycalesine genera illustrates how the existence of developmental biases does not prevent selection from tinkering with the developmental interdependencies among genes. This is in line with theoretical models that suggest that the feedback, modular structure, and nonlinear interactions in regulatory networks allow developmental systems to exhibit both robustness (little phenotypic change even under large changes to model parameters) and innovation (i.e., large yet functionally integrated phenotypic change even under small changes to parameters).[92] Such considerations also help account for macroevolutionary trends, patterns of parallel evolution, and the relative reversibility of changes in morphology. For instance, the observation that frog species far more frequently gain and lose "thumbs" than "little fingers" over evolutionary time spans, while salamanders repeatedly experience the reverse, can be understood through studying developmental mechanisms: for both frogs and salamanders it is the last digits to form that are most often gained or lost.[93] We return to the topic of evolutionary innovations in chapter 11.

Embracing "Developmental Bias"

In biology textbooks, evolution is commonly described in terms of the spread of a mutant allele, often involving loss of function, such as a failure to produce a gene product or take up a nutrient. An example is the mutation that generates pale-colored mice on the Florida sand dunes, discussed in chapter 4.[94] Evolution does indeed often work that way, and in such instances there is no particular reason to ask why the variants that are selected are there in the first place. However, the material discussed in this chapter shows how a great deal of evolutionary adaptation is not like this, but rather involves rewiring of preexisting regulatory interactions. There is striking conservation of developmental modules and of regulatory networks across all life-forms, and the co-option of existing components into new functions is extremely common.[95] Evolutionary novelty is biased by preexisting developmental mechanisms, but that bias goes beyond the traditional conception of a phylogenetic or historical constraint. Developmental bias does not simply constrain adaptive evolution but also facilitates and directs it. The phenotypic variation widespread in nature is shaped by biases in the rate of "supply" of new variants and not just the "demands" imposed by fitness.[96] Those biases include a tendency of developmental systems to respond to genetic or environmental perturbation by generating well-integrated, functional phenotypes. Without this tendency it would be nearly impossible for organisms to adapt.[97]

The genotype-phenotype relationship is strongly affected by the regulatory network structure discussed in chapter 6, encompassing both genes and nongenetic elements, through which both plastic and nonplastic aspects of development become integrated. By making it more likely that adaptive phenotypes arise, developmental bias contributes to evolvability (chapter 12). At the same time, modular structure and nonlinear interactions in and among regulatory networks allow developmental systems to exhibit both robustness and innovation.[98] The diverse effects of genetic change, sometimes leading to changes in phenotype and other times not, arise because regulatory networks process genetic inputs, organizing them into functional phenotypic outputs.

Evolutionary biology is now starting to let go of the viewpoint that development can be understood as hindering evolution, and that developmental bias provides an alternative explanation to natural selection,[99] although constraint terminology and thinking remain prevalent. Recent evidence that phenotypic evolution commonly involves changes in regulatory interactions provides a strong motivation for evolutionary researchers to recognize and investigate how developmental systems determine variability, and how this variability in turn affects adaptation and diversification. The influence of natural selection in shaping developmental bias and, conversely, the influence of developmental bias in shaping subsequent opportunities for adaptation require mechanistic models of development to be expanded and incorporated into evolutionary theory.[100]

There are good reasons for doing this, as a focus on developmental mechanisms offers many practical benefits to evolutionary scientists. First, it helps make otherwise mysterious patterns of phenotypic covariation comprehensible across a range of organisms, leading to a richer understanding of evolutionary causation. It encourages research into which phenotypes, and which combinations of traits, tend to arise naturally over relevant time spans, helping to explain why only a limited number of phenotypic outcomes occur, why we commonly see parallel evolution in independent lineages (for instance, in blind Mexican cave fish, or freshwater sticklebacks with no pelvic spines; see chapter 8), and why traits might evolve even when they have little effect on fitness. At first sight it might appear that there is no mystery in the appearance of mottled coloration in many domesticated animals; natural selection explains its existence. Yet what is missing in this account is an explanation for why selection *for* tameness has consistently led to selection *of* mottled coloration; that is something that only developmental biology can provide. Admittedly, knowledge of patterns of genetic and phenotypic covariation affords some level of explanation, but we have seen that there are additional gains in predictive and explanatory power that come with knowledge of developmental mechanisms. With the recognition that developmental mechanisms contribute to evolutionary outcomes, explanations for evolutionary change become richer, more insightful, and more general.

Second, developmental mechanisms can be highly effective in predicting and explaining evolution. We saw this in chapter 2, where the inhibitory cascade model initially explained variation in tooth size in mice, was then generalized to account for tooth size in other rodents, and was subsequently successfully applied to explain phalange variation across vertebrates.[101] In contrast, the common conception of "constraint" as an explanation for traits that did *not* evolve or are *not* adaptive discourages consideration of the possibility that biases in the production of phenotypes may explain some natural variation, and reinforces the assumption that selection fully accounts for adaptive phenotypic variation.

We have described several instances where the prediction that divergence between lineages is shaped by the rate at which new phenotypic variants arise in development has been borne out through a combination of evolutionary analyses and detailed knowledge of developmental biology. The evolution of mammalian teeth, butterfly wing spots, fruit fly wing shapes, nematode vulva morphology, and patterns of vertebrate digit gain and loss all attest to how developmental insights can help explain evolutionary outcomes. Contrary to the traditional expectations, the effects of developmental bias are neither local nor transient but span broad taxa and persist across vast time spans.[102] Thus, a developmental bias perspective can shed new light on macroevolutionary patterns.

Third, as we saw in chapter 5, development has a number of features that enable organisms to maintain functional integration even under sometimes severe perturbation. As a result, adaptive evolutionary change begins with the origin of phenotypic variation. Developmental bias can help to explain differences in the rate of adaptive evolution, and why some adaptations evolve instead of others; its study encourages researchers to investigate what makes organisms and traits evolvable. We return to these topics in chapters 11 and 12.

Recognizing developmental bias as a *cause* of evolution, rather than a constraint, requires a change in mindset. What constitutes an evolutionary cause or process is not universally accepted by evolutionary biologists. Most textbooks emphasize that evolutionary processes are phenomena that are thought to change gene frequencies directly, such as natural selection, mutation, or random genetic drift. Developmental bias does not directly change gene frequencies, so would not qualify as a cause of adaptive evolution from this perspective. Building the contribution of developmental mechanisms into explanations for evolutionary change will therefore require a broader notion of evolutionary cause to acknowledge a steering role for variational biases in evolution.[103] An important point here is that, by recognizing developmental processes as co-determining adaptation, evolutionary biologists can encourage the study of the developmental mechanisms underlying adaptive phenotypic variation rather than viewing these as uninteresting, unknowable, or the focus of another field. We return to this issue in chapters 12 and 14.

Does the knowledge that developmental bias has evolved through natural selection undermine any causal role for developmental mechanisms in evolution? No. By comparison, the fact that mutation rates can be altered by selection does not mean that mutation should not be recognized as a cause of evolution— it merely demonstrates that the processes interact. The same reasoning applies to developmental bias: the explanatory value of a developmental perspective does not depend on whether or not bias has itself been shaped by natural selection. In practice, tracing causation back to the earlier selection that caused the bias is also not straightforward, since that selection too will have been biased by earlier developmental mechanisms. Developmental processes have been generating variational biases since the beginning of life: there is no point in time at which only selection operated. From the outset there would have been biases in forms that did not evolve, such as those generated by the physical properties of materials. Recognition of the causal role that developmental processes play in evolution therefore requires a shift in perspective.[104]

8

Plasticity-Led Evolution

In earlier chapters we described how development can be sensitive to environmental cues that trigger signaling pathways or modulate the outputs of a cell. This responsiveness of the phenotype to the environment is known as *phenotypic plasticity*, and it means that genetically similar individuals can exhibit strikingly altered traits when exposed to different conditions.[1] We also pointed out that developmental processes can direct evolutionary adaptation, help account for taxonomic diversity, and produce evolutionary novelty. Combined, these observations imply that phenotypic plasticity might also contribute to adaptive evolution, which is the focus of this chapter.

How phenotypic plasticity might play an important role in evolution becomes clearer if we regard an organism's traits as produced by regulatory interactions during development. Environmental cues are frequently processed by the same regulatory networks as genetic mutations, and, like genetic change, environmental perturbations commonly lead to biased phenotypic variation. In some cases, phenotypic variation is adaptively biased, and in others it is not, but either way plasticity can contribute to adaptive evolution by influencing which phenotypes are exposed to selection, and in which environments they appear. As this biased variation comprises multiple, often well-integrated traits, which are often tailored to the local environment, the resulting coordinated response to selection may reduce the amount of genetic change necessary to bring about adaptation.

The Blind Mexican Cave Fish

When blind Mexican cave fish (*Astyanax mexicanus*) were first discovered in 1936, they appeared so different from other known fishes that they were listed as an entirely new genus.[2] It came as a surprise when subsequent research revealed that normally sighted, surface-dwelling conspecifics of those cave fish are widely distributed across the rivers and streams of Mexico and the southern USA.[3] Moreover, the cave and surface morphs interbreed and produce fertile offspring.[4] At least thirty cave populations of this species are now recognized, mostly eyeless and with reduced pigmentation, but also sometimes exhibiting enhancements in nonvisual sensory systems, as well as changes in bone structure, fat storage, starvation resistance, metabolic rate, and foraging behavior (figure 10).[5] Analyses of their evolutionary history leave no doubt that Mexican cave fish derive from multiple colonization events,[6] and that their suite of cave-dwelling (*troglomorphic*) traits, apparently adaptations to living in the constant darkness of a cave, has evolved repeatedly.[7]

In recent years, blind Mexican cave fish have emerged as an important model system in both evo-devo and evolutionary ecology,[8] as well as an excellent vehicle to explore the interplay between developmental and evolutionary mechanisms in evolutionary adaptation.[9] This is partly for practical reasons: these fish thrive in captivity and are amenable to experimentation, including genetic manipulation. However, for evolutionary biologists, their most exciting feature is that the existence of surface-dwelling conspecifics potentially allows the evolutionary steps leading to cave fish to be retraced.[10] Their parallel evolution hints at similarities in the natural selection of different populations due to the consistent cave environment. However, that so many populations have rapidly and separately evolved the same suite of phenotypic characters has led researchers to wonder whether phenotypic plasticity might have played a role in their adaptation to the cave environment.[11] Perhaps plasticity helps to explain precisely what convergent evolution takes place?

To investigate this, William Jeffery, Helena Bilandžija, and colleagues at the University of Maryland raised surface-dwelling A. *mexicanus* in complete darkness.[12] Intriguingly, they found that these fish came to exhibit many of the traits of their blind cave fish conspecifics, including changes in the thickness of retinal layers in the eye, increased resistance to starvation, decreased metabolic rate, changes in hormone levels, downregulation of the expression of genes involved in visual perception and circadian regulation, and upregulation of genes associated with fat storage. Hence phenotypic plasticity allowed precursors of many cave-related traits to appear in surface fish within a single generation. Apparently, surface-dwelling fish already had this ability to alter their physiological and behavioral phenotypes when reared in darkness, and in cave fish these phenotypes have become exaggerated and stabilized.[13] Exposure to the stress of constant darkness seems to mobilize multiple developmental

RIVER
FISH

CAVE
FISH

FIGURE 10. Plasticity-led evolution in blind cave fish. The Mexican fish *Astyanax mexicanus* is found in both blind cave fish populations and as normally sighted populations in rivers and streams. The cave fish are typically eyeless and with reduced pigmentation, but also exhibit enhancements in nonvisual sensory systems and changes in bone structure, fat storage, and behavior. These traits do not evolve independently but are connected through interacting developmental mechanisms derived from shared regulatory networks. Experiments indicate that phenotypic plasticity has guided the evolution of cave fish adaptations, with populations evolving along trajectories created by developmental mechanisms.

mechanisms, including endocrine signaling, activation of heat-shock proteins, and other transcriptional changes, which collectively instigate major changes in the operation of the gene regulatory networks (GRNs) underlying many cave fish traits.[14] These morphological changes are accompanied by behavioral adjustments—for instance, when foraging, dark-raised surface fish increase their reliance on the lateral line (a sensory organ that allows fish to detect movement and pressure changes in water) compared with fish reared in standard lighting.[15]

Of the traits that change on exposure to the dark, some, such as increased starvation resistance or decreased metabolic rate, became more like those of the blind Mexican cave fish, with plasticity seeming to facilitate adaptively beneficial traits. Others, however, shifted in the opposite direction. Perhaps surprisingly, eye size actually increased slightly in dark-reared surface fish— but with significant increases in eye-size variability.[16] Either way, dark-rearing revealed adaptively relevant phenotypic variation upon which natural selection might act. Phenotypic plasticity may not consistently generate adaptive changes, but these and other experiments suggest that it *may* consistently increase selectable variation in relevant phenotypes.[17]

Is the emergence in surface fish of these cave fish traits just coincidence, or has plasticity played some sort of guiding role in the evolution of cave fish adaptations?[18] Obviously, plasticity alone does not explain the blind Mexican cave fish's traits, which are also known to have been subject to genetic change.[19] For plasticity to shape genetic evolution, any initial phenotypic modifications must be followed by, and partly determine, genetic change through natural selection. Evidence is provided by an experiment that subjected surface fish to a chemical that mimicked the stressful physiological effects of the cave environment. The chemical activated the same stress response genes as in surface fish reared in the low-conductivity water found in caves.[20] Surface *A. mexicanus* reared in the presence of the chemical exhibited unusually large variation in eye size, implying that the cave fish's ancestors would have responded plastically to the cave environment.[21] The researchers reasoned that if the phenotypic variation exposed by this stress had played a role in the evolution of eye loss, it would be observed in surface fish but not contemporary cave fish, since the latter will have undergone selection for reduced eye size that will have eradicated much of this variation (a process known as *genetic assimilation*).[22] Sure enough, treatment with the stress-mimicking chemical revealed extensive phenotypic variation in eye size in surface but not cave *A. mexicanus*.[23] Finally, to test whether the phenotypic variation in eye size revealed by the chemical stressor could allow the genetic evolution of eye loss, the researchers selectively bred new generations from stressed surface fish with small eyes, and found that eye size responded to selection. These studies suggest that phenotypic plasticity might have played a critical role in cave fish evolution by exposing heritable variation in eye size and in correlated traits.[24]

Why would natural selection favor small eyes? Saving energy that could be wasted on producing unnecessary organs is one possibility, and the fitness benefit of this is likely to be nontrivial.[25] Another possibility is that the selection of eye loss has occurred indirectly, as a consequence of selection for other aspects of the phenotype.[26] Regulatory elements expressed in eye formation are enmeshed in GRNs, and, as a result, they play multiple roles in the formation of many aspects of the phenotype. These include traits that are directly advantageous in the dark, and are credible as targets of selection on other organs (such as taste buds, lateral line, brain structure, and jaw shape).[27] There may have been selection *of* smaller eyes, but not necessarily selection *for* smaller eyes.

The development of the vertebrate eye is relatively well understood.[28] A network of key regulatory genes includes the retinal homeobox (Rx) gene, which is required to specify the retina and activates *Pax6*, the major gene responsible for forming the eye field in the neural plate. The Pax6 protein initiates a cascade of other transcription factors that collectively generate a single eye-forming field at the front of the forebrain. The sonic hedgehog gene (*Shh*) represses expression of the *Pax6* gene in the center of the eye-forming field, separating it into two fields, and hence two eyes. If the Shh protein is inhibited there, the result is a single eye in the center of the head, a condition known as "cyclopia" after the Cyclopes in Greek mythology. However, if *Shh* is overexpressed in the eye region, then the *Pax6* domain shrinks and small defective eyes are formed. (This also enlarges the jaws, and increases the number of taste receptors).[29] That is what happens during blind Mexican cave fish development, and is one of the key molecular changes associated with the blind cave fish phenotype.[30] Overexpression of Shh in the developing eyes of surface fish induces the controlled death of lens cells in the eye (apoptosis), and the result is a fish that strongly resembles the blind cave fish (a "copy" of a genetic change by an environmental effect is known as a *phenocopy*).[31]

Transplantation of a surface fish lens into a blind cave fish optic cup is able to restore the eye, while lens transplantation from cave fish to surface fish results in eye degeneration.[32] These findings not only provide a beautiful illustration of the mutual induction of the vertebrate lens and retina, discussed in chapter 5, but also suggest that the genes and regulatory networks involved in cave fish eye development remain fully intact but have merely been rerouted toward a different developmental outcome.[33] Interestingly, lens transplantation and rescue of eye development also changes the orbital and nasal bone structures of cave fish to resemble those of surface fish.[34] There are substantial differences between surface fish and cave fish in their head and face bones, partly because facial bones develop differently in the presence of eyes. Further evidence that the same underlying developmental mechanisms and molecular pathways are involved comes from the observation that, while different genetic loci are associated with eye loss in different cave fish populations, the populations nonetheless mostly share the same morphology.[35] In other words, different mutations may have been fixed in different populations, but selection has consistently acted on the same regulatory mechanisms.

The parallel evolution of cave fish populations makes sense once it is recognized that multiple cave fish traits do not evolve independently but are connected through interacting developmental mechanisms derived from underlying regulatory networks, as we saw for the domestication syndrome. Here again, there is a package of developmentally linked traits for which the underlying mechanisms generate some patterns of trait covariation more readily than others (i.e., developmental biases), and as a result some characters are favored through selection on others.[36]

Some cave fish traits are most likely analogous to the floppy ears or curly tails in domesticated animals—neutral or mildly deleterious traits indirectly favored by selection because they are produced by the same mechanisms as characters that were selected directly. One curious anatomical feature of many cave fish populations is the fragmentation of facial bones.[37] Bones that in surface fish are a single structure have literally been broken up into little pieces in cave fish. A team of puzzled researchers remarked: "It is difficult to imagine bone fragmentation as a beneficial trait, yet it has been observed across multiple cavefish populations."[38]

Once again, consideration of developmental mechanisms allows us to make sense of this oddity by suggesting that bone fragmentation has been favored indirectly. Experimental analysis of three cave fish populations revealed that fragmentation of facial bones was consistently associated with a significant increase in the numbers and density of neuromasts (clusters of sensory cells that enable fish to detect water currents) on these bones. These neuromasts are almost certainly responsible for the enhanced mechanosensory capabilities of cave fish.[39] Surface fish reared in the dark develop more neuromasts than those reared in the light and show increased dependency on the lateral line when foraging.[40] This helps to explain why in darkness cave fish forage more efficiently than surface fish and are better able to locate food.[41] Other behavioral responses are tied into this package of advantageous traits. Surface fish feed inefficiently from the bottom, with their bodies at a 90° angle to the substrate, while cave fish forage at a 45° angle, which makes better use of their shovel-like lower jaws.[42] Apparently, the cave fish phenotype was selected not so much because smaller eyes were advantageous but because refinements in other senses, including lateral line, olfaction, and taste, as well as morphological features such as larger jaws and physiological traits such as reduced metabolic rate, are beneficial in the dark.[43]

We have seen that being reared in conditions that mimic the cave environment (constant darkness, low conductivity water) triggers rapid morphological, metabolic, and molecular changes in surface fish, many of which appear necessary for the survival of early colonizers. However, once the initial challenge has been survived, caves are thought to be a relatively benign, predator-free environment.[44] Thus, phenotypic plasticity may enable the population to persist, allowing natural selection to incorporate more robust adaptations.[45] However, plasticity plays a far more active role than merely buying time: plastic responses can uncover preexisting cryptic genetic variation, which facilitates the rapid evolution of cave-related traits.[46] Because of this plasticity, either adaptive phenotypes or increased phenotypic variability or both have arisen within a single generation. The plastic responses were sometimes adaptive and sometimes maladaptive, yet the increased phenotypic variability did not arise at random: the evolved response to the cave environment matched the increased variability in the suites of traits that responded plastically to the dark environment in surface fish. This variability can be interpreted, and to some extent predicted, with knowledge of developmental coregulation. If a new blind Mexican cave fish population were discovered tomorrow, researchers could anticipate aspects of its phenotype, including traits like fragmented bones that are not obviously adaptive. A recent origin of similar traits from surface populations has been proposed for several cave-dwelling animals, both invertebrates and vertebrates, raising the possibility that plasticity may be a general mechanism that facilitates colonization of cave environments.[47]

There are fascinating parallels between the cave fish and domestication examples. In both, the patterns of phenotypic variation are determined by shared developmental dependencies, and in both there is evidence that evolutionary outcomes are aligned with the main axis of variation, consistent with the suggestion that natural selection directs evolving populations along trajectories created by developmental mechanisms.[48] Irrespective of whether the input is genetic or environmental change, the phenotypic outcome is the result of the processing of novel inputs by the same regulatory networks. This is what allows environmental perturbation to mimic mutation (phenocopies), and vice-versa (genetic assimilation).[49] At any point in time, genetically and environmentally induced biases jointly determine the phenotype distribution, shaping the correlated variation that is exposed to selection and thereby influencing the rate and direction of evolution.[50] We explore the consequences of this for evolvability in chapter 12.

Plasticity Leads, Genes Follow

The hypothesis that phenotypic plasticity can precede, facilitate, and perhaps even direct adaptive evolution has been around for a long time.[51] An experimental proof of principle was provided by Conrad Waddington in 1953, and as far back as the late nineteenth century several scholars—most famously James Mark Baldwin—promoted the idea.[52] Yet for decades the hypothesis was viewed skeptically, and has received detailed consideration only in recent years, following Mary Jane West-Eberhard's influential writings and her claim that "genes are followers, not leaders, in evolution."[53]

The central idea is that novel traits may start out as environmentally induced variants that later become fixed or stabilized (i.e., more robustly elicited in development) through the natural selection of genetic variation. While the induced phenotype may be novel, the developmental pathway to produce the variation was already there, which makes it legitimate to consider "development, not selection, [as] the first order cause of design."[54] Plastic responses may or may not be reversible or adaptive. However, the novel environmental inputs are always processed by existing developmental regulatory mechanisms, which makes plasticity a significant source of developmental bias that influences the distribution of phenotypes upon which selection can act. If there is heritable phenotypic variation within the population, adaptive change will follow. The effect of plasticity on subsequent evolution may be transient or long-lasting. In the latter case, plasticity may enable populations to adapt, and plastic responses may direct evolution toward particular adaptations, as apparently occurred in the cave fish example. As populations adapt, the environmental responsiveness itself may evolve, since selection can favor enhanced plasticity (e.g., leading to the evolution of a *polyphenism*), reduced plasticity (*canalization* or *genetic assimilation*), or no evolutionary change.[55] However, it is not the extent of plasticity that matters—the key point is that the evolutionary trajectory the population will follow, the rate at which it evolves, and any equilibrium approached may be guided by the characteristics of the plastic response.[56]

In previous chapters, we stressed the similarity between mutational and environmentally induced effects on the phenotype, which are typically processed and channeled by the same regulatory networks and can produce similar outcomes. Yet, the traditional assumption has been that mutations are likely to have greater evolutionary potential, on the grounds that genetic differences, but not environmental effects, are heritable. However, there are several reasons for thinking that environmentally induced variation may often have evolutionary consequences.[57]

First, while most mutations are either neutral or deleterious, plastic responses can be tailored to the inducing environment, and hence generate adaptive phenotypic variation. For instance, in the cave fish example, dark-reared surface fish have a reduced metabolic rate and increased starvation resistance.[58] If environmental conditions persist for multiple generations, adaptive environmentally induced phenotypes will be favored by natural selection, thus influencing selection on underlying genetic variation.

Second, in contrast to beneficial mutations, which arise in just a single individual, environmental stress will typically affect multiple individuals simultaneously, reducing the likelihood that adaptive phenotypes are lost by chance.[59] Moreover, the fact that environmental change can affect many traits in many individuals means that the effects of the environment on evolution are likely to be ubiquitous.[60] This is probably one reason why there have been so many instances of parallel evolution among the Mexican cave fish: in plasticity-led scenarios the evolutionary response reflects both the similarity of the selective environments and of the plastic responses.

Third, the evolution of adaptive plasticity can allow cryptic genetic variation that might otherwise be eliminated by selection to accumulate, at least in instances where adaptive plasticity reduces phenotypic differences between genotypes and thereby weakens selection on such genotypes.[61] This cryptic genetic variation can subsequently be expressed when novel or extreme environments are encountered. For instance, increased variability in eye size was seen in surface fish reared in dark and low-conductivity conditions.[62] By increasing the variability of environmentally sensitive plastic traits, the accumulation of cryptic genetic variation potentially increases the evolvability of those characters (see chapter 12).

Fourth, many of the developmental mechanisms that we discussed in chapter 5 enable individuals to accommodate environments that are radically different from those they usually experience. As a result, even in novel environments, the distribution of phenotypes that are exposed to selection is highly non-random, and may be biased along fitness-relevant dimensions. Again, the convergent evolution of cave fish provides an example, since multiple traits of dark-reared surface fish show plastic adjustments that generate patterns of covariation that mimic the patterns of covariation seen in derived cave fish phenotypes.[63] Here, the axes of maximum phenotypic variability in the surface-fish plastic response appear to be aligned with the evolutionary response to selection in the derived cave fish populations. Why this should be is discussed later in this chapter.

The above line of reasoning—often labelled "plasticity-led" or "plasticity-first" evolution[64]—entails that the appearance of adaptive novelty does not typically require mutation, but results from developmental reorganization and the incorporation of environmental inputs.[65] The key point is not that plasticity comes first (after all, evolutionary responses may capitalize on the prior existence of relevant genetic variation), but that plasticity directs, or "leads," adaptive evolution.[66] By biasing the phenotypic variation that is exposed to selection, and by orchestrating coordinated morphological, physiological, and behavioral responses to selection, plasticity can both accelerate and, perhaps less commonly, slow down adaptive evolution. More importantly, plasticity is a major source of developmental bias, and thereby helps determine the *direction* and *dynamics* of evolution by natural selection.[67]

The blind Mexican cave fish provide a valuable proof of principle that plasticity-led evolution can happen. However, whether it is an important, or even the dominant, mode of adaptive evolution remains contentious.[68] In recent years the hypothesis has received considerable attention, and supporting evidence has been found in flowering plants, nematodes, crustaceans, insects, and birds, as well as other fishes and amphibians.[69] Spadefoot toads (genus *Spea*) provide one of the best-studied examples.[70] These amphibians have evolved a fascinating sensitivity to diet: tadpoles fed detritus develop into small omnivores, while those fed large animal prey such as shrimp become giant carnivores, with the huge jaw muscles and short gut that are typical of carnivores.[71] Such polyphenisms may be critical for survival in harsh environments such as the Sonoran Desert. The tadpoles exist in ephemeral ponds produced by early rains. If the rains come again, the pond remains, and the tadpoles develop at a rate that allows them to metamorphose into robust toads. However, if rains do not come, the pond starts to dry up, and the tadpoles undergo a stress response that allows cryptic variation to be expressed. Some tadpoles develop large carnivorous jaws and guts, allowing them to eat other tadpoles.[72]

This remarkable polyphenism is found only in *Spea*: other frogs and toads develop into omnivores. This makes it possible to test if "genes are followers" in adaptive evolution by investigating how tadpoles of the most closely related genus (*Scaphiopus* species) respond when fed shrimp—a novel diet for these species. Consistent with plasticity-led evolution, those tadpoles also developed a shorter gut, and exhibited greater variability in both gut length and gene expression, compared with tadpoles reared on their normal diets.[73] However, *Spea* tadpoles were able to assimilate this diet more efficiently than *Scaphiopus* tadpoles, suggesting that natural selection had enhanced *Spea*'s ability to digest flesh.[74]

Polyphenisms, like the tadpole morphs in spadefoot toads, are common in nature—for example, genetically identical ants can develop into workers, soldiers, or queens, and butterflies can have different wing patterns depending on whether they emerge in the wet or dry season[75]—and they are commonly regulated by hormones. The same holds for a temperature-dependent polyphenism exhibited by the hornworm moth (*Manduca quinquemaculata*), whose caterpillars emerge black during cold temperatures (helping them to absorb sunlight) and green at warmer temperatures (which affords better camouflage). In contrast, the closely related tobacco hornworm moth (*M. sexta*) has green caterpillars irrespective of the temperature, although a mutant black form is also found. The black mutant has reduced levels of juvenile hormone, which affects many traits including skin color. An artificial selection experiment in the hornworm moth (*M. quinquemaculata*), conducted by developmental biologists Yuichiro Suzuki and Frederik Nijhout, sheds light on how such threshold shifts can occur.[76]

Suzuki and Nijhout subjected the black *M. sexta* mutant to a burst of high temperature early in development—a heat shock—and found that some of the caterpillars turned green. Thirteen generations of breeding from those caterpillars that changed color after heat shock, and subjecting the offspring to the same burst of high temperature, produced a polyphenic line of caterpillars that, without the heat shock, were black at low temperatures and green at higher temperatures, just like their naturally polyphenic cousin *M. quinquemaculata*. Conversely, breeding from those individuals that failed to change color created a *monophenic* line that was always black.

Experimental manipulations implicated juvenile hormone levels as an important target of selection, with higher levels in the polyphenic line than in monophenic or control lines. The results of this study suggest that natural selection can readily adjust how organisms respond to environmental cues by modifying endocrinal regulation of development. The black *M. sexta* caterpillar appears to result from a "sensitizing" mutation that shifts the juvenile hormone titer down to just below the switch threshold.[77] In insects, higher temperatures often lead to increases in levels of hormone production, and juvenile hormone responds to the heat shock in this way. In some individuals with the right hormone levels, the resulting shift is sufficient for them to cross the threshold and produce a green skin. Selection that adjusts the levels and timing of hormone secretion is postulated to have played important roles in the evolution of several seasonal polyphenisms.[78] Likewise, Alexander Badyaev showed that coordinated environmentally sensitive hormonal regulation of diverse morphological, physiological, and behavioral traits in both mothers and chicks may have facilitated the adaptive divergence of the house finch during colonization of North America.[79] Hormones are not the only mechanism underlying environmental influences on development, but a coordinated response of multiple traits to selection is commonly mediated by a shared regulatory machinery sensitive to environmental inputs.[80]

Diversification and Adaptive Radiation

Phenotypic plasticity may also play a role in promoting speciation and adaptive radiation.[81] By producing different phenotypes under different conditions, potentially exposing genetic differences to selection, plasticity provides the potential to diversify. A large body of theory supports this conclusion, establishing that plasticity can lead to the evolution of phenotypic differences between populations.[82] Likewise, a substantial number of empirical studies report that plasticity in an ancestor produces intraspecific variation that mirrors interspecific variation within the same clade.[83] For example, young marine three-spined sticklebacks reared on typical freshwater benthic (bottom-dwelling) and limnetic (mid-water) diets develop a morphology that strikingly resembles the derived freshwater forms.[84] Fish feeding on benthic prey tend to have shorter jaws, which are better for grazing algae or crushing snails, while those feeding in the water column have longer jaws that provide increased suction of prey, such as zooplankton.[85] This implies that environmentally induced, intraspecific variation might underlie phenotypic differences between species.[86] Several studies of insects, fishes, amphibians, and birds report that populations of the same species exposed to different foods or host plants, and that differ in their expression of environmentally induced morphologies, appear to be evolving reproductive isolation, and thus may represent incipient speciation.[87] Comparative analyses are consistent with this conclusion, since fish and amphibian clades that exhibit resource polyphenisms are more species-rich than sister clades lacking this form of plasticity.[88]

Thus, the diversity of a clade is not determined solely by the diversity of external environments but may also depend on the plasticity of the ancestors.[89] West-Eberhard's "flexible stem" hypothesis posits that if an ancestral stem group repeatedly colonizes similar environments, developmental plasticity should consistently give rise to similar phenotypes.[90] As the cave fish example illustrates, plasticity can consistently bias the phenotypic variability of a particular suite of responsive traits, which helps to explain patterns of parallel evolution. The same logic applies to adaptive radiations. Repeated evolution has historically been interpreted as resulting from convergent natural selection in similar environments; however, developmental bias arising from plasticity may also be an important contributing factor. For instance, the dramatic parallel evolution of cichlid fishes of the African Rift Valley lakes seems to have been influenced by the capacity of the relevant developmental mechanisms to generate variation biased along particular axes of morphospace, guided by plastic responses to different diets and the mechanical stresses they elicit, as well as egg and hatchling environments.[91] The same molecular mechanisms most likely underlie this plastic and evolutionary response.[92]

How typical are such cases? To answer this question, a representative sample of studies must be reviewed to assess whether the general pattern supports the hypothesis. A recent meta-analysis of plant adaptation did just that.[93] The authors analyzed data from 34 studies in which plants of two (or more) populations of the same species were both grown in their own environment and transplanted to the other population's environment (these studies are called "reciprocal transplant experiments"). The authors reasoned that, if plasticity were directing the course of adaptive evolution, the change in phenotype when plants were grown in the new environment should be similar to the divergence between the two locally adapted populations when grown in their own environment. The logic is directly equivalent to interpreting the resemblance between surface *A. mexicanus* reared in the dark and cave fish as being consistent with plasticity-led evolution. The authors found that it was indeed possible to predict the phenotypes of locally adapted populations on the basis of plastic responses. In fact, locally adapted plants pretty much resemble what their ancestors would look like if thrust into the local environment, but with some fine-tuning by natural selection (*genetic accommodation*).[94] Plasticity itself did not evolve but, as predicted, phenotypic variation was higher in novel environments.[95] The findings fit the expectation that plasticity is an important source of developmental bias that leaves a signature during local adaptation.[96]

While the evidence for plasticity-led evolution is rapidly accumulating, not all cases fit.[97] Nor should they. While it is plausible that many cases of evolution would be plasticity-led, it also makes sense that some will not be. This appears to be the case in an adaptive radiation of *Anolis* lizards, which colonized the

Greater Antillean islands approximately fifty million years ago. A recent comparative analysis found the most plastic features of the locomotor skeleton were neither the features that contributed most to divergence, nor the traits that improved the lizards' performance in navigating their microhabitat.[98]

If a role for plasticity in evolution is neither universal nor inevitable, when will plasticity take the lead in adaptive evolution? One suggestion is that environmentally induced phenotypes are most likely to lead when environmental challenges are severe.[99] Under such circumstances, plasticity may cause dramatic changes in the distribution of phenotypes compared with the effect of standing genetic variation. Moreover, how organisms respond becomes crucial, since selection in novel environments is likely to be strong and favor those individuals that are able to maintain some level of functionality.[100] The individuals may not be very fit in absolute terms, but, if the perturbation is recurrent or affects a large proportion of the population, natural selection is likely to allow at least some of the induced variants to be retained, and the resulting adaptations will tend to resemble the environmentally induced phenotypes.[101] The dramatic change in conditions experienced by surface fish entering a pitch-black cave seems to fit these criteria, but perhaps not tree-trunk-adapted lizards placed on a bush. Arguably, the most extreme example of a severe environmental challenge is the transition from water onto land. That the bone structure and musculature of bichir fish forced to walk on their pectoral fins rather than swim resemble those of stem tetrapods supports this reasoning.[102]

Niche Construction

Organisms can also create developmental biases by behaving in ways that systematically modify their external environments, a process known as *niche construction*.[103] These modifications include the construction of nests, burrows, pupation chambers, and social environments by animals, and the alteration of soil properties through the secretion of exudates and allelopaths by plants.[104] Often parents construct developmental environments for their offspring, and thereby regulate the external conditions the offspring experience; for instance, buffering them from predators or thermal fluctuations. Niche-constructing activities are significant in the context of this chapter as they generate two routes to correlated phenotypic variation: organisms can construct environments suited to their phenotypes (e.g., when birds build nests appropriate in size and strength for their chicks), but also organisms develop in response to self- or parentally constructed environmental components (e.g., chick development responds plastically to nest conditions).[105]

Dung beetles again provide an example. Females of *Onthophagus* dig tunnels underneath cowpats, within which they construct brood balls of dung containing a single egg. These balls constitute a safe, thermally buffered home for their offspring, prestocked with all the food (i.e., dung) that the emerging larvae will need to complete their development and metamorphosis.[106] Mothers also endow each brood ball with a fecal pellet (or "pedestal"), which the larva consumes on hatching, and through which the larva acquires its mother's gut biota.[107] Pedestal microbiota are enriched for genes that are critical for cellulose degradation and nitrogen fixation. Deprived of the pedestal microbiota, larvae show drastically higher mortality, and those beetles that do survive develop more slowly and emerge as smaller adults.[108]

The larvae themselves also engage in diverse forms of niche construction, including defecating throughout the brood ball and mixing the dung (which distributes the microbiota), repairing the brood ball, and constructing a pupation chamber. The spreading of larval feces throughout the brood ball appears to generate a symbiont-mediated external rumen that promotes the breakdown of chitin, lignin, and cellulose and thereby transforms the brood-ball carbohydrates into a more digestible form. As a consequence, just like maternal niche construction, larval activities enhance larval growth and fitness.[109]

Maternal and larval niche construction generate covariation between diverse niche-constructing traits and other phenotypic characters—including growth, horn size and shape, duration of development, and degree of sexual dimorphism—that respond plastically to the constructed environment.[110] Niche construction biases the nature of the phenotypic variation exposed to natural selection, generating an axis of trait covariation encompassing both niche-constructing traits and plastically expressed morphological traits. Selection can favor more or less brood-ball niche construction and bigger or smaller beetles will result. Likewise, it can favor shorter or longer development, or more or less sexual dimorphism, with a likely impact on the extent of niche construction. These patterns of covariation constitute a form of developmental bias, as some trait combinations occur more frequently than others.[111] Here, biased covariation arises not because of pleiotropic action of genes expressed in niche-construction but because of feedback from niche construction to the developing organism via the external environment.[112] Interestingly, there is experimental evidence that maternal and larval niche construction can partly compensate for each other, and that each dung beetle species has a specialist microbiome,[113] raising the possibility that trade-offs in niche construction (stronger vs. weaker, parental vs. offspring, host-microbiome interactions) have contributed to the prolific divergence of *Onthophagus*, of which there are more than two thousand species.[114]

Dung beetles are far from unusual in being reliant on their microbiome to break down cellulose: most animals cannot digest plant cell walls. Nor are dung beetles unique in passing on crucial microbes by eating feces. Coprophagous transmission is common—that is how termites acquire the symbionts that allow them to digest wood.[115] In chapter 5 we described the bobtail squid whose luminescent bacteria construct homes in the squid's light organ, which illustrates how symbionts can engage in niche construction inside their hosts' bodies, and in the process play critical roles in the hosts' development.[116]

The establishment of symbioses between host animals and the plant-digesting microorganisms in their guts is now thought to have been a necessary precondition to the evolution of herbivory.[117] Feeding on plant material is almost certainly a derived trait, with herbivores evolving from carnivorous ancestors. Herbivory is particularly widespread in mammals: cattle, sheep, antelopes, deer, and giraffes, for instance, all chew the cud regurgitated from their rumen, where bacteria break down plant materials. It has been estimated that over 70 percent of a cow's energy derives from this microbial digestion of plant material.[118] The rumen itself is a gut adaptation that evolved specifically to house symbionts, which ferment nondigestible plant tissue into fatty acids that the host can digest. These fatty acids, in turn, cause the rumen and its musculature to grow and differentiate, and in this manner the gut bacteria help construct their rumen niche.[119] Symbionts can be "agents of developmental plasticity" that through their niche construction "remodel host anatomy."[120]

Herbivory is known to have evolved many times, with each instance tightly coupled to a particular group of bacterial symbionts.[121] Each independent origin is also associated with a rapid and diverse adaptive radiation, with diversification contingent on the herbivorous diets and the mutualisms that underpin them.[122] Once herbivores acquired plant-wall digesting and plant-toxin neutralizing symbionts, they could exploit an array of behaviors (e.g., grazing, browsing, sap consumption) and morphologies (e.g., tooth shape, jaw musculature, gut length). The evolution of herbivory appears to have been critically dependent on symbiont niche construction, which in turn may be a major source of convergence in herbivores.[123]

Learning and Behavior

While behavior is perhaps the most plastic of traits, and learning is an important source of behavioral plasticity, learning is rarely brought up in discussions of plasticity-led evolution. Yet some of the most spectacular examples of plastic traits that precede and determine the nature of genetic change arise from human learning and culture. For example, our ancestors learned to incorporate novel foods into their diets (e.g., dairy products, excess starch, alcohol) which triggered the selection of genetic variants expressed in more efficient metabolism of those foods (i.e., *LCT, AMY1, ADH*).[124] Similar phenomena are reported in other animals. British great tits, for instance, have learned to exploit bird feeders, and as a consequence have evolved stronger jaws that are more efficient at processing feeder food than those of great tits in the relatively feeder-free Netherlands.[125] Likewise, through greater cultural learning, Sumatran orangutans appear to have evolved enhanced cognition relative to Bornean orangutans.[126]

By allowing animals to tune their behavior to environments, including to novel conditions, learning becomes an important aspect of adaptive plasticity, and another source of developmental bias.[127] Through learning, animals can introduce new behaviors into their population's repertoire (e.g., when birds learn to evade a novel predator) and generate adaptive phenotypic change in the absence of any immediate environmental stressor (e.g., when orangutans proactively devise new food-processing techniques, such as eating palm hearts).[128] Such "behavioral innovations" are extensively documented in animals, and social learning allows them to be propagated to other individuals, including nonrelatives and across generations.[129] Learned behaviors also typically spread in a biased way, with fitness-enhancing learned behavior more likely to be propagated than bad information.[130] For instance, red-winged blackbirds copy feeding conspecifics *except* when they exhibit an aversive reaction to a specific food, nine-spined sticklebacks monitor the foraging success of other fish and subsequently select the richer of alternative food patches, and bats that are unsuccessful at foraging follow previously successful bats to feeding sites.[131] These empirical findings are supported by mathematical theory, which has suggested that organisms can evolve plastic responses that are adaptive even in environments the lineage has never encountered.[132] Learning can both buffer genetic variation and expose it to selection, but either way plasticity affects genetic evolution.[133] More generally, forms of phenotypic plasticity that are reliant on exploratory and selective processes (e.g., adaptive immune system, vascular system, nervous system)[134] allow organisms to respond to evolutionarily novel environments with phenotypic variation aligned to functional demands.

Studies of animal learning also show that whether plasticity is lost during genetic accommodation (where genetic adaptation follows plastic changes) depends critically on the mechanism of plasticity. For instance, experiments have uncovered an adaptive specialization in the learning of stickleback fishes, with nine-spined but not three-spined sticklebacks able to discern which are the richest food patches by observing the feeding rates of other fish.[135] Interestingly, when male nine-spines are in a reproductive state they switch to behaving like three-spines and ignore this "public information," while, in contrast, reproductive nine-spine females increase their reliance on social information.[136] These shifts in stickleback learning with reproductive condition imply that the adaptive specialization probably evolved through endocrinal changes, without compromising the general learning ability of the animals.[137] Adaptive specialization in learning can also occur through upregulating animals' perceptual systems, rather than constraining learning capabilities. This appears to be the case for monkeys that are predisposed to acquire a fear of snakes, for whom snake-shaped objects have become particularly salient.[138] This refinement of perceptual and motivational systems makes sense: individuals will experience multiple learned challenges during their lifetime, and overadjusting to a single threat could compromise future flexibility.

Behavior is also likely to generate developmental biases when it is influenced by endocrinal pathways, which commonly also affect other aspects of the phenotype, such as physiology and morphology, and are highly sensitive to environmental inputs. There is considerable intra- and interspecific variation in behavioral traits such as dispersal rates, exploration rates, activity levels, aggression, boldness, and sociality, across diverse animal taxa, and extensive covariation among these has been noted.[139] For instance, dispersive animals are generally more active, bolder, and more exploratory than philopatric or less dispersive individuals.[140] These personality traits are known to be sensitive to stress and maternal effects, suggesting that variation in dispersal may be an outcome of stress-induced maternal effects. In western bluebirds, for instance, females breeding in territories with few nest cavities experience heightened aggression from nest-site competitors at a time when they are producing eggs, and as a consequence allocate more testosterone to the clutch, which in turn leads to more-aggressive sons, compared with those of females in low competition territories.[141] The organizational effects of androgens on sons' aggression predisposes them to particular breeding and dispersal strategies, as more-aggressive individuals are better able to compete for territories, while less aggressive ones are better able to act as helpers at the parental nest.[142] In this manner, stress-induced maternal effects help to explain behavioral strategies in western bluebirds that are not seen in other members of the genus, including remaining on the natal territory to breed nonaggressively, and cooperative breeding by helping at the parental nest.[143] The key point is that, by eliciting coordinated behavioral, physiological, and morphological changes, including across generations, hormonal changes arising during development greatly reduce the number of evolutionary steps needed for the evolution of complex novel behavioral phenotypes.[144]

The Evolution of Plasticity

We have offered numerous examples of plastic responses to environmental challenges that are aligned with the local adaptation of closely related populations or species. While that correspondence has generated a great deal of academic interest, and has frequently been taken as evidence for plasticity-led evolution, this interpretation remains contentious.[145] The controversy arises in part because plastic responses, and the developmental mechanisms that produce them, are themselves sculpted by earlier natural selection. However, the logic of reciprocal causation applies just as surely to plastic traits as to other sources of developmental bias. Just as the fact that bias can evolve doesn't preclude it from being causally significant in evolution, so the same holds for plasticity. Since its inception, life has always exhibited phenotypic plasticity, not least because organisms are built from plastic materials. We return to this issue in chapter 12, where we discuss how the evolution of plasticity can affect genetic evolvability.[146]

9

The Causes of Selection

Deserts are among the most extreme environments on Earth. Soil surfaces in the Sahara or Mohave can reach 75°C–80°C, yet fall well below freezing on winter nights.[1] What makes a desert, however, is not the unforgiving temperatures but an acute lack of precipitation.[2] Some Antarctic deserts have not seen rain for fourteen million years![3] Yet, in spite of the hostile conditions, a host of desert organisms valiantly eke out a living. Scientists have long been fascinated by the adaptations that allow plants and animals to flourish in this stark wilderness. Biologists, in particular, have been attracted to deserts as natural laboratories where the punishing setting would seem to accentuate the potency of natural selection.

Camels are among the most celebrated of desert creatures.[4] In scorching temperatures, these animals can go for days without drinking, and for months in milder conditions where green food is available. Camels cope with levels of water stress that would kill most other animals, largely because their tissues are highly resistant to osmotic pressure. The camel's famous hump is primarily a food store that can be drawn on during long journeys when forage is scarce.[5] While cattle lose gallons of fluids every day in urine and feces, camels' exceptionally long intestines and hyperefficient kidneys allow them to recycle virtually all their fluids.[6] A unique vascular system uses blood cooled in the nasal passages through respiratory evaporation to keep camels' brains several degrees lower than their bodies. Two layers of long eyelashes protects camels' eyes from sand, and an additional transparent eyelid, which opens from side to side, works like a windshield wiper to clean the eyeball surface, while still allowing good vision in a sandstorm.[7] Each of these traits evolved through the natural selection of heritable genetic variation.[8] Camel adaptations exemplify the formidable logic of Darwinian evolution: through generations of differential survival and reproduction, the properties of environments inexorably honed the properties of organisms.[9]

The desert rhubarb, a plant endemic to the Negev Desert in Israel, takes a different approach to survival in an arid climate.[10] Unlike many desert plants, which have evolved small leaves or spines to reduce water loss through evaporation, the desert rhubarb has enormous, highly ridged leaves that act like an upside-down umbrella, collecting rainwater and channeling it down to its single dominant taproot. This central root grows and then shrinks again, creating a cavity lined with root exudates in which water can accumulate. Through self-irrigation, the plant generates moisture levels in the vicinity of its root system over six times those of the surrounding soil: it literally constructs a "mini oasis" in the desert.[11]

While camels are archetypical "endurers," their morphological adaptations allowing them to withstand the punishing conditions, the desert rhubarb is an "evader," able to dodge climatic extremities through effective marshalling of resources. Rather than simply withstanding the desert environment, it engineers benign soil conditions more typical of a Mediterranean climate.[12] The plant, no less than the camel, is remarkably well suited for desert life, but in the plant's case the complementarity of organism and environment arises through the organism changing the environment.

The "hand in glove" idiom is often deployed to describe evolutionary adaptation. It captures the complementarity of organism and environment arising through natural selection. Ironically, unlike, say, the coevolution of hummingbirds' long beaks and the tubular flowers from which they feed, there has been no selection over aeons for human hands to fit a world replete with gloves. Instead, throughout human history, gloves were deftly manufactured to suit hands of all sizes. Environments can be constructed to suit organisms, just as organisms can evolve to fit environments.

Termites provide an example of this alternative pathway to adaptive complementarity. These insects occupy a variety of habitats, but the fungus-cultivating termites are able to inhabit dryer and hotter habitats than most other termites.[13] Each colony constructs a massive mound, several meters tall, permeated with an elaborate network of tunnels, above a subterranean nest. Through erosion, the mound constantly loses dry soil, which is continually replaced with wet soil deposited by the termites.[14] In this manner, the structure itself evolves over the lifetime of the colony, with the termites determining the patterns and rates of soil deposition.[15] Mounds are known to persist for hundreds, even thousands of years, far longer than an individual termite.[16] Constant remodeling of these structures allows the colonies to manage the nest environment, and ensure that suitable temperature, humidity, and concentrations of respiratory gases are maintained. The workers bring to the nest macerated plant forage, which is then composted by a fungus cultured on "combs" that they construct.[17] The fungus combs not only comprise a hoard of farmed food but are also a water store that regulates humidity.[18] Nest moisture is enhanced by active transport, with the termites "mining" water from deep below the surface and carrying droplets or dollops of wet soil to the nest.[19] As long as subterranean water is available, these animals can create a humid and stable sanctuary in the desert.[20] In doing so, they change the selective regime: selective advantage that might have accrued to evolving internal organs of physiology better able to deal with water scarcity, or thick cuticles that limit desiccation, as seen in other termites, now accrues to the ability to build environments where water is abundant.[21]

There are more prosaic means of evading an arid climate than devising irrigation systems. Countless small animals avoid overheating through behavioral means, seeking out microhabitats such as rock crevices, or the shade of plants, or hiding away in underground burrows. A burrow is an extremely effective means to avoid both the heat and the cold. Fluctuations in burrow temperatures are much smaller than those in the external environment, while humidity too is more constant.[22] Desert insects, spiders, scorpions, frogs, and rodents all escape high temperatures by retreating into burrows, especially in the summer. Several species of desert beetles dig meter-long trenches in the sand dunes, which during night-time fogs collect moisture from the air that the beetles then extract.[23] Desert snails hide under rocks, emerging at nightfall, climbing to the top of tall structures where a cooling breeze may be found, burying themselves for long periods of dormancy, or sealing their shells with a thick wall of mucus.[24] Desert frogs too can spend much of the year safely aestivating in underground burrows.[25]

Many species forage at night, thus avoiding high temperatures.[26] Arabian oryx[27] dig into the sand to expose cooler layers, and then sit in these pits where they lose body heat through conduction.[28] Desert vultures urinate on their legs to cool themselves by evaporation.[29] Cape ground squirrels use their bushy tails as a parasol, doubling them back over their heads in a manner that creates shade and reduces body temperature.[30] Similarly, ostriches hold out their wings like an umbrella to shade their torsos.[31] The desert rhubarb might seem exotic, but its strategy is logically equivalent to those of many organisms that evade environmental extremities by constructing more-benign conditions for themselves.[32]

Humans are among these. A billion people are thought to live in deserts today, forging a difficult living alongside the snakes and the scorpions. Humans cool themselves by standing upright and sweating,[33] but they also rely on a suite of behaviors—seeking shade; traveling at night; wearing loose-fitting clothing—that are not biological adaptations but rather forms of phenotypic plasticity, mediated by learning, culture, and traditional ecological knowledge. Yet there can be no doubt that these behaviors are adaptive, as people not suitably prepared for desert existence are likely to lose their lives.[34] Thousands die each year under the desert sun: hikers who lost their way, travelers whose cars broke down on a lonely road, or those migrating in search of a new home. In 2020, the bodies of 225 migrants crossing the desert from Mexico to the USA were found, making it the deadliest year on record,[35] tragedies that underscore the necessity of effective survival strategies in the desert.

We can now see that there are at least three logically distinct, if not mutually exclusive, means of evading the desert extremes: First, organisms can inherit morphological *adaptations* that facilitate their survival, such as humps or sweat glands.[36] Second, organisms can build a more tolerable local microenvironment; for instance, by digging burrows, manufacturing trenches, or sealing openings, all of which are examples of *niche construction*.[37] And third, organisms can adjust themselves and their behavior to the conditions—for example, by foraging at night, creating shade, or wearing a headscarf, which are forms of *phenotypic plasticity*.[38] Whether they ameliorate the local environment or make adjustments to themselves, and whether nor not these responses are adaptations,[39] organisms partly determine the conditions they experience, and hence the intensity and character of the selection that ensues. The activities of living creatures shape the relationship between themselves and their environments, and thereby determine which phenotypes are fit and which are not. Selection depends on what individual organisms do.

Understanding Selection

Any population whose members exhibit *phenotypic variation* in their morphology, physiology, or behavior that is *heritable*, and with the variants differing in their ability to leave descendants (*differential fitness*), is expected to undergo evolution by natural selection.[40] Those individuals possessing advantageous traits will be more likely to leave descendants, and be more likely to pass on genes that underlie their beneficial characteristics, than individuals lacking such phenotypes. As a consequence, species change over time, as favorable traits spread through populations, generating adaptations to their environments.[41]

Leaving chance events to one side, the fate of any phenotypic variant—be it an unusually long camel intestine or a more heavily ridged desert rhubarb leaf—will depend on the relative fitness of those individuals that possess the variant. If, on average, the trait enhances the likelihood that its carriers will survive and reproduce, over time the trait should increase in frequency in the population through natural selection. In this manner, fitness differences are perceived to *cause* trait frequency changes in populations, and natural selection is thought to *explain* both the evolution of adaptations and the diversification of life-forms. However, the causes of evolution are more slippery, and explanations for evolutionary change more challenging, than they appear at first.

Perhaps surprisingly, whether fitness differences can be construed as *causes* of trait frequency changes remains a long-standing point of contention, with some biologists and philosophers viewing such changes as inherently entailed by fitness differences rather than causes of them.[42] Consider the assertion that dark-colored moths have higher fitness than light-colored moths in polluted environments. Intrinsic to this claim is the expectation that dark-colored moths will be more likely to survive and reproduce, and hence increase in frequency. More generally, an evolving population may simultaneously undergo (1) *changes in its genetic composition,* as some genotypes die out and others become more frequent; (2) *changes in adaptedness,* if the population becomes better suited to its environment; and (3) *changes in trait frequency,* as some traits increase and others decrease in representation.[43] Fitness is a specification of the expected relative rates of birth or death that particular genotypes experience, and wrapped up in that description is an implicit expectation of concomitant changes in adaptedness and trait frequency.

Population genetics quantifies these associations by specifying how differences in fitness will be reflected in changes in trait frequencies.[44] Treating fitness differences as a cause is thus a scientific convention, one that follows directly from the population genetic representation of evolution by natural selection.[45] Richard Lewontin famously described population genetics as the "auto mechanics of evolutionary biology": researchers "plug in" parameter values and "crank out" predictions.[46] Accordingly, explanations for evolutionary change are commonly provided in terms of the existence of fitness differences, and this approach has proven productive. Yet, however common or intuitive it may be to think that way, these are not the only possible explanations. It is equally valid, and we suggest often more satisfactory, to explain adaptations by reference to the causes of fitness differences. For instance, the evolution of a camel's extra-long intestine can be understood in terms of the causes of the births and deaths of camels with guts of different lengths.[47] In Lewontin's terms, inherent in any explanation of adaptive evolution through natural selection are the developmental *causes of trait characteristics* (and hence of phenotypic variation), the *causes of organisms' suitability to their environment* (and hence fitness differences), and the *causes of parent-offspring similarity* (i.e., heredity).[48]

Source Laws for Evolutionary Biology

In a landmark analysis of natural selection, philosopher Elliott Sober distinguished between what he called "source laws" and "consequence laws": "The former describe the circumstances that produce forces; the latter describe how forces, once they exist, produce changes in the systems they impinge upon."[49] Sober shows how both are required to provide a complete understanding of a system.

Sober characterizes evolutionary theory as rich in consequence laws, which are "the province of population genetics," but "impoverished" with respect to source laws (we interpret "laws" as meaning "regularities" in this context).[50] Indeed, the latter are often regarded as the subject matter of other fields, including physiology and ecology. Sober's point is similar to the claim that fitness differences are not causes of adaptation,[51] and to Lewontin's reference to "auto mechanics." Population genetics provides valuable tools with which to track the genetic consequences of fitness differences (as well as mutation and random drift), and to comprehend their ramifications. The effects of differences in fitness can be understood by statistical aggregation, but they tell us little about the causes of variation in fitness.

A complete understanding of the causes of evolution requires the specification of source laws, as Sober emphasized. Since the causes of fitness differences are found in the relationship between individuals and their surroundings, understanding selection requires attention to both ecological and developmental processes. Models of evolution that treat fitness as dependent on individual interactions, population structure, and demography demonstrate that such details matter to evolutionary trajectories and end points.[52] Nevertheless, the traditional assumptions that the complexities of development make the origins of phenotypic variation unpredictable, and that ecological processes are no less fickle and idiosyncratic, have hindered the formulation of such source laws for selection.[53] Consequence laws are what is left.[54]

Evolutionary biology's concentration on consequence laws, while taking the existence of fitness differences for granted, explains why fitness differences are the starting point for most explanations of adaptive evolution. When Lynch (responding to claims of neglected causal influences emerging from evolutionary developmental biology) asks the rhetorical question, "Have evolutionary biologists developed a giant blind spot?," it is hard to imagine that the answer (at least, from a population geneticist) could be anything other than "No."[55] However, explanations that jump straight to fitness differences mask particular kinds of causal patterns that could be important in evolution, including regularities in the *origin* of natural selection. Such regularities in the sources of selection lie hidden within the concept of fitness differences. Strikingly, many claims made in recent years for "neglected" causal factors in evolution can be interpreted as putative regularities in the causes of natural selection.[56]

In chapters 7 and 8, we described how developmental mechanisms, including the plastic responses of organisms to novel conditions, can bias the pool of phenotypic variation and introduce evolutionary novelties;[57] in this chapter (and chapter 8), we described how the niche-constructing activities of individual organisms can create or modify fitness differences; and in chapter 10, we will discuss how the experiences of organisms can shape the inheritance of characters.[58] Developmental bias, plasticity, niche construction, epigenetic inheritance, developmental symbiosis, and animal culture share the property that they represent aggregated individual-level processes that occur within and to individuals and impose direction on evolutionary change.[59] These individual-level processes do not just constitute population-level effects at another level of analysis: individual-level processes are *causes* of those population-level effects, and may therefore impose regularities that, at least potentially, help determine the strength and direction of evolution by natural selection. If such regularities exist, they will not, as some commentators appear to fear, overturn the principles of population genetics, since source laws and consequence laws are complementary.[60] Rather, the identification of such regularities will enrich genetic accounts. The evolutionary credentials of candidate developmental processes depend on their ability to uncover such patterns and thus contribute to the scientific understanding of evolution.[61]

In chapter 8, we described how river fish raised in darkness exhibited similar traits to blind Mexican cave fish, including changes in the retina, modifications to skull morphology, increased resistance to starvation, and decreased metabolic rate.[62] The plastic response of dark-reared fish involves multiple developmental mechanisms, including endocrine signaling, activation of heat-shock proteins, and other transcriptional changes, which together instigate major changes in the operation of the regulatory networks underlying the codeveloping, and hence coevolving, traits. Critically, the same pattern of correlated variation in developmentally codependent morphological, physiological, and behavioral traits, in response to dark rearing, has appeared in multiple experimental populations. The fact that different populations respond similarly explains why a virtually identical response to natural selection was manifest in over twenty independent cave fish populations.[63] There is an impressive regularity to the plastic response—an anticipated pattern to the trait variation, fitness differences, and trait heritability that enables the evolution of the blind cave fish phenotype to be confidently predicted. Such examples highlight how plastic adjustments of organisms to environmental conditions do more than simply fine-tune the phenotype. Plasticity contributes to evolution by natural selection by specifying the developmentally biased distribution of phenotypic variation, as well as the viability and reproductive potential of each variant.

The well-studied cave fish case has more general implications for understanding the role of plasticity in evolution. For instance, it implies that when desert organisms attune themselves to extreme conditions—for example, by foraging at night, or seeking shade[64]—their plastic responses are also likely to generate regularities in the phenotypic variation exposed to selection, regularities in the fitness of those variants, and regularities in the response to selection. When developmental biology was in its infancy, claims along these lines would have appeared fanciful speculation, but as the field matures they are increasingly looking like plausible and empirically verifiable hypotheses. For instance, we have seen how knowledge of the mechanisms of mammalian tooth development, described in chapter 2, led to successful predictions about patterns of covariation in tooth morphology across widespread taxa.[65] One of the reasons development is important in evolution is because taxonomically shared developmental mechanisms cause replicable patterns in phenotypic variation spanning multiple coregulated traits. These patterns, in turn, can impose regularity on what phenotypes become selected, and when they become selected, resulting in evolutionary trends, parallel evolution, and differences in evolvability.

The same points apply to niche construction. In chapter 8, we also described how the activities of dung beetles generate covariation between diverse niche-constructing traits and other phenotypic characters, including growth, duration of development, and degree of sexual dimorphism.[66] Like developmental plasticity, niche construction biases the nature of phenotypic variation exposed to natural selection, generating a major axis of covariation encompassing both niche-constructing traits and plastically expressed morphological traits.[67] Once again, the well-studied dung beetle case has wider implications. It suggests, for instance, that when desert animals dig burrows, mounds, or trenches or seal up openings, their niche construction is also likely to generate regularities in the phenotypic variants that are exposed to selection, regularities in the fitness of those variants, and regularities in the response to selection.[68] Ecology is important in evolution, not only because survival and reproductive success depend on the biotic and abiotic environment, part of which is constructed by organisms, but also because taxonomically shared developmental mechanisms generate consistent patterns in that niche construction, again generating biased phenotypic variation in multiple coregulated traits, as well as biasing the frequency of certain environmental conditions. Undoubtedly, there will be morphological and physiological traits that are coregulated with burrow or mound digging, just as there are for dung beetle brood-ball manufacture and processing. These developmentally coregulated traits are causes of covariation and causes of fitness differences. Regularities of development may impose regularity on ecology, and hence selection.[69]

This reasoning applies to *all* organisms in *all* environments. Being too big to seek shade or bury themselves away in burrows, camels evolved physiological adaptations for coping with desert conditions. Had there been giant cacti in whose shade camels could shelter, they might have evolved a very different physiology. "Endurers," like "evaders," play a major role in determining their own fate. Creating microenvironments, taking up minerals, storing food, timing activities, or simply changing position or orientation are among the many ways by which living organisms modify themselves or their conditions, transduce the physical signals of the world, or create statistical patterns of resources that differ from the patterns in the macroenvironment.[70] For instance, by developing narrow thick leaves that minimize water loss in the sun, and broad thin leaves that increase the quantity of photosynthetically active light experienced in the shade, and by orientating those leaves optimally, the developmental plasticity of plants allows them partly to control their light and water environments.[71] In doing so, plants weaken selection due to extreme light and water conditions.

The same holds for emperor penguins. Winter in the Antarctic is an extreme environment where the icy landscape can be as cold as −50°C. Yet by huddling together to keep warm, emperor penguins raise the temperature in the middle of the huddle to as high as +37°C.[72] Their actions reduce selection on physiological adaptations for coping with the cold, and create selection for traits such as the competitive ability to stay in the middle of the huddle, or the capability to breed in Antarctic conditions. Migrating animals also change their experienced environment, commonly leaving behind an austere habitat with little food and entering a location rich in nutritious fare. This movement can hide from selection phenotypic variation in characters affecting the ability to cope with the winter months, such as foraging ability, and expose to selection phenotypic variation in traits related to long-distance migration, such as flight muscles. In this way, fitness differences can be constructed by organisms.

In sum, plasticity and niche construction are underpinned by taxonomically shared developmental mechanisms,[73] and their effects often aggregate across a population to generate statistical regularities. Those patterns are easy to miss because the developmental causes of fitness differences are not currently central to the study of evolution. Instead, multiple distinct causes are subsumed under the label "natural selection" in a manner that obscures regularities whose analysis could generate richer explanations and more powerful predictions.[74]

Life's Inductive Gamble

Any system exhibiting *variation, differential fitness,* and *heredity* is expected to undergo natural selection.[75] In the process, it will accumulate information in the structure of the DNA coding sequence and associated regulatory molecules, as well as the cellular and external environmental context, which results in the manufacture of the proteins necessary to build and maintain bodies.[76] As all organisms with descendants must have survived at least long enough to reproduce, that inherited information specifies the pattern of protein assembly necessary to produce viable phenotypes in the ancestral environment. Over time, when subject to natural selection, genes and their regulatory molecules accumulate information about what led to adaptive phenotypes in the past.[77]

Building bodies in a similar manner to one's forebears is a reliable way of producing an adaptive phenotype, provided the environment has not changed. The catch is that there is no guarantee that present conditions will resemble those in the past, so the reuse of inherited information in the present amounts to a gamble that the selective environments of contemporary organisms will be sufficiently similar to the selective environments of their ancestors to make whatever was adaptive before adaptive again.[78] Organisms cannot know what they will encounter in time to come, and are forced to wager that the future will be sufficiently like the past to make inherited information valuable.

All of life is contingent on this fundamental inductive gamble.[79] The fact that species go extinct shows that generalizing from past to present is far from foolproof; yet, in spite of this, many lineages persist for millions of years. How is survival in the face of constant change possible? The predictive power of the genome is only part of the story. Organisms must do more than gamble on the past: they must seek to control the future.

Organisms act on their environment to keep it within tolerable bounds, and thereby ensure their inherited information remains relevant. Life persists in part because organisms are active agents that construct their environments, transforming them in reliable, predictable, and often homeostatic ways.[80] The external world may be erratic, but the experienced environment is what matters, and if the conditions an organism encounters retain some constancy across generations *because* they have been regulated by the organism, the inductive gamble will be more likely to pay off. To the extent that organisms control environmental conditions, thereby enhancing consistency across generations—like the desert rhubarb collecting water near its taproot, or the termites mining water deep below the surface—they increase the chance that their inductive gamble will pay off. Far from being only a matter of chance, the inductive lottery is rigged by sly incumbents.[81]

Bird's nests, termite mounds, and mammal burrows all experience reduced variation in temperature and humidity, and more moderate conditions, than the external environment.[82] The point holds, more generally, for virtually all developmental environments constructed by parents for their offspring, as well as for the animals that dodge in and out of the shade or avoid other extremes through their choices rather than their constructions.[83] The predictability of the experienced environment is contingent on the capacity of organisms to maintain conditions within manageable ranges.[84] Niche construction evolves rapidly,[85] and is generally adaptive, because it often increases the chances that organisms and their descendants will remain within tolerable conditions.[86]

Organisms have another trick up their sleeves to help ensure their inductive gamble pays off: they can let the environment specify their phenotype. Like environmental regulation, adaptive plasticity allows organisms to maintain complementarity between themselves and their environments, helping to ensure that inherited information remains relevant.[87] Adaptive plasticity also readily evolves.[88] Many forms of plasticity also rely on inherited information to shape how individuals should develop in response to conditions encountered, and hence are no less of an inductive gamble. However, as described in chapter 5, developmental systems, including the vertebrate immune and vascular systems, the nervous system, and animal learning, sometimes operate through *exploratory mechanisms*—that is, by iteratively generating variation, testing the variants' effectiveness, and selecting good solutions.[89] Exploratory mechanisms are distinctive because they allow for supplementary information gain by individual organisms, information that is particularly valuable because it concerns the *current*, rather than past, environment. These information-gaining configurations greatly enhance the predictive power of life's inductive reasoning, helping organisms to function effectively in an only weakly predictable world.[90]

Moreover, niche construction and adaptive plasticity reinforce each other. Organisms, from nest-building ants to brood-ball-burying dung beetles, reliably construct safe and benign rearing environments for their young, helping to ensure that the developmental responses of their offspring are more likely to be adaptive. At the same time, the adaptive plasticity of organisms helps them to upgrade their environments—as, for instance, where birds, insects, and spiders improve their nest building or nest site selection with experience.[91] Plastic responses allow parents to update the diverse developmental resources, from epigenetic marks to hormones in the egg to learned dietary preferences, thereby enhancing the developmental environment that they create for their offspring.[92]

Prediction and regulatory control go hand in hand. An unpredictable, capricious, and capacious external environment can be better managed by constructing a physical boundary, such as a nest, hive, or burrow, to create a part of the environment that is more amenable to regulation. The adaptive logic is identical to those animal organs that constitute "internalized" components of the external world. Your lungs and gastrointestinal tract, for instance, are topologically an "external" environment, but enfolded "internally." Why did these structures evolve? The answer is internalization brings down the costs of homeostasis, by rendering the part of the environment with which the organism interacts easier to control, allowing more effective exploitation of the activities of symbionts.[93] The elaborate infrastructure of the lung internalizes a small piece of the atmosphere in a way that regulates conditions on both sides of the epithelial boundary of the lung where respiratory gas exchange occurs.[94] The internalization of lungs in terrestrial vertebrates, like the building of mounds by termites, allows for effective management of temperature, humidity, and gas exchange in the experienced environment, enhancing the predictive power of the genome.

If environments change too much or too quickly, the evolving population may lag further and further behind the environment, in which case natural selection will not improve the fit of organism to environment, and chance events, such as genetic drift, will dominate. Conversely, by damping out variability in the experienced environment, and by flexibly adjusting the organism to experienced conditions, niche construction and adaptive plasticity not only enhance fitness, but also help to ensure that the evolving population can track the environment.[95] How organisms develop, and what they do, determines not only what is selected but also how strong the selection is.

Selection Of and Selection For

In his analysis of natural selection, Sober also distinguished between the *selection of* objects and *selection for* properties. To illustrate this, Sober used "a toy that my niece once enjoyed playing with before it was confiscated to serve the higher purposes of philosophy."[96] The toy is a "selection machine": a transparent container, structured into levels by dividing partitions, and containing balls of differing sizes. Each partition is perforated with holes of the same size, with the holes at each level larger than those on the level below. Balls of a particular size have the same color—the smallest balls, for instance, are green, while larger balls have different colors. If the balls are initially at the top, shaking the toy eventually redistributes them throughout the toy, with the smallest balls traveling furthest through the holes to the lower levels, and larger balls trapped at higher levels. The *smallest* balls are the objects that are selected, but *green* balls have been selected at the same time.

Darwin's analysis of animal domestication revealed that *selection of* floppy ears and curly tails was associated with *selection for* tameness, while more recent investigations of blind cave fish link *selection of* fragmented bones to *selection for* more sensory neuromast cells that detect water currents. By highlighting the distinction between selection of green balls and selection for smallness, Sober's selection toy provides a helpful analogy illuminating the causes and effects of selection. Balls have been selected for their smallness, not for being green. The toy's partitions represent the filtering role of the environment, with organisms like balls, selected according to their size, and other nonfunctional traits hitching a ride.[97]

Helpful though this analogy may be, there are important respects in which the selection toy gives a misleading impression of the action of natural selection—a characterization that portrays organisms in overly passive terms. For instance, the structural features of the toy—the balls and the perforated partitions—were factory made. The balls played no part in building the partitions. The partitions simply exist, and the balls are selected according to the partition's properties (i.e., the size and shape of the holes). However, termites construct mounds and regulate temperature and humidity within them, which redirects selection away from internal organs that deal with water scarcity, such as thick cuticles, toward selection for effective behavioral strategies for humidity control, such as digging down to the water table to retrieve moisture. And desert rhubarb plants collect water next to their primary root, generating selection for huge, heavily ridged leaves, as opposed to selection for small leaves or spines. The actions of living organisms determine what is selected for. It is as if the balls in the selection toy themselves build the partitions and cut out the holes. Perhaps we can imagine that the partitions are made of rolled cookie dough, with each ball cutting away a shape, like a cookie cutter. Round balls cut round holes, cubed balls cut squares, and elephant-shaped balls cut elephant-shaped holes. The analogy is not overstretched, given that we know organisms manufacture microenvironments, such as burrows, nests, and soil conditions, to suit their own characteristics.[98] Living creatures modify the environment in ways that partly determine which traits constitute the major determinants of fitness. The balls themselves govern whether there is selection for smallness, roundness, or greenness.

Furthermore, the balls in the selection toy are solid, inflexible objects that move only when acted upon by external forces. They are not able to choose which level of the toy they will occupy, and nor are they able to change their own size or shape. Yet we have also seen that, in the real world, desert animals can evade the extreme conditions through behavioral plasticity; for instance, hiding under rocks like snails, urinating on their legs like vultures, or creating their own shade like ostriches. Rather than transforming the environment, these organisms have adjusted their phenotype to match the world. It is as if the balls in the selection toy are made of a squidgy and malleable jelly, and are able to squeeze through tiny holes irrespective of their circumference. When needed, the balls can have a smaller diameter, just as the desert creatures can become heat tolerant. Perhaps we can imagine two selection toys, one with solid balls and the other with squidgy balls. The properties of the balls govern whether there will be selection for smallness or "squidginess." (There are other pathways to adaptive fit that could also be mentioned here. For instance, in chapter 1 we discussed Mohave Desert woodrats that team up with a bacterial symbiont to digest creosote. The analogous situation with the toy might involve the balls releasing an acid to eat away the partition and make the holes bigger.)

Finally, in the selection toy, the association between ball size and color is entirely arbitrary. There is no reason why small balls need to be green—we are implicitly conscious that they could just as easily be red. This gives the impression that, in selecting for smallness, greenness has been randomly chosen. But here again the analogy is misleading. Floppy ears are not linked to the tameness of domesticated animals by chance: the components of domestication syndrome are connected by joint developmental regulation through the neural crest gene regulatory network (GRN).[99] Nor are the fragmented bones of cave fish coupled with an improved ability to sense vibrations in the water through historical contingency. Fragmentation is a by-product of the increased production of sensory cells on these bones, cells that form the lateral line and are responsible for this enhanced mechanosensory capability.[100] In the real world, developmental mechanisms connect the *selection of* traits to the *selection for* traits, with a powerful implication. Through investigating developmental bias, evolutionary biologists may understand and predict which traits there would be *selection of*, alongside the character selected *for*.

The selection toy starts with a preexisting partition and round balls. Philosophers and children alike take as given smallness to be the trait selected for, and greenness as a coincident character. Likewise, as pointed out above, evolutionary biology often begins with the identification of fitness differences, and works through the ramifications of these for character evolution, adaptation, and speciation, detecting correlated change in other characters in the process. In this respect, the selection toy provides an excellent analogy for how natural selection is viewed by biologists. The focus on consequence laws means that it is rarely asked why character X is a major determinant of fitness differences, as opposed to characters Y or Z, over and above a tacit assumption that the environment has made it so. That camels should relay blood from their nasal passages to cool their brains makes sense for an animal exposed to the desert heat, and confirmation of a putative adaptation comes with evidence that camels with more effective vascular cooling systems exhibit enhanced survival or reproduction relative to those with less efficient systems. What hasn't been asked is why camels don't have fluffy tails that they can use as parasols like ground squirrels, or why they don't dig cooling pits like oryx, or why they don't possess any of the other tissue-cooling strategies seen in the natural world. We assume that chance, or historical contingency, explains why camels rely on a unique brain-cooling mechanism, while other organisms deploy alternative strategies, and we are content that selection fully explains camel adaptations.

This focus, while productive, comes at a cost, in the form of limitations on the power of evolutionary explanations. The nonchalant attribution to "selection" of all evolution arising from fitness differences masks hidden determinants of the *sources* of fitness differences. This would not matter if fitness differences arose by chance, or if those characters selected *for* were packaged with the selection *of* other characters in a coincidental or unknowable manner, but that is not the case. In the real world, balls are green for a reason.

Development brings regularity, or order, to the production of fitness differences, which derives from potentially knowable, and increasingly known, mechanisms that determine which characters will reliably be associated with fitness. Well-studied developmental mechanisms in Mexican cave fish explain why at least twenty-two separate populations that wandered into caves should each generate a remarkably similar distribution of fitness differences and as a consequence independently evolve a virtually identical constellation of traits. Likewise, well-understood shared developmental mechanisms explain why over thirty domesticated animals should all have evolved a domestication syndrome. Subsumed within the conventional explanation for these instances of parallel evolution are hidden causes of evolution that operate by biasing selection in consistent ways. Identifying the causes of these patterns enhances evolutionary explanations. These examples illustrate how knowledge of developmental processes allows evolutionary biologists to go beyond explaining why a character evolved through natural selection to explain why alternative adaptations did not evolve, and why other traits were favored in the process.

Dynamic Adaptive Landscapes

The idea that organisms might set their evolutionary agenda by determining the selection they experience is not new. In 1985, Richard Lewontin wrote: "The organism influences its own evolution, by being both the object of natural selection and the creator of the conditions of that selection."[101] "Adaptive landscapes" can be a helpful visual metaphor for understanding the role that organisms play in shaping natural selection. An adaptive landscape is a graph that plots mean fitness on the vertical axis, and one or more of the population's traits or genes on the horizontal axes.[102] The resulting landscape resembles a rugged surface, in which peaks represent high-fitness trait combinations (i.e., local optima), and valleys represent phenotypes with lower fitness. Genetic (and phenotypic) change is represented by a shift in the position of the population on the surface. Through natural selection, the population will move toward exhibiting the traits suited to the local habitat, which is often represented on the landscape as climbing a local fitness peak (figure 11, top).

Adaptive landscapes are more than just a visualization tool, and play an important role in quantitative genetics theory, where formal models describe a multidimensional surface and a population's movements on it.[103] Evolutionary processes move populations around the hills and valleys of the landscape. Selection, for instance, is typically portrayed as populations climbing a nearby hill, while genetic drift is characterized as a random walk across the surface. We can readily visualize the evolution of burrow building or shade seeking through natural selection as a population moving up a local fitness peak representing these traits. However, that image would be misleadingly static.

What niche construction and developmental plasticity actually do is modify the shape of the landscape.[104] They are part of the nexus of feedback and causation, in which the developmental adjustments of individual organisms constantly remodel the fitness landscape, while organism-modified environments and experienced conditions continually select for changes in organisms.[105] A recent theoretical analysis by Mauricio González-Forero illustrated how developmental processes determine the admissible evolutionary pathways across the landscape, with evolutionary outcomes occurring "at path peaks rather than landscape peaks" (figure 11, middle). His analysis also shows how niche construction modifies the fitness landscape to create new peaks, and thereby affects which equilibria are approached (figure 11, bottom).[106]

González-Forero's findings are in accord with other recent theory.[107] For instance, recognizing the deficiencies of traditional landscape models for dealing with such phenomena, Mark Tanaka of the University of New South Wales and his colleagues devised a novel mathematical framework to model adaptive evolution on two linked landscapes.[108] In their formulation, the evolution of the organism occurs by hill climbing on an *adaptive* landscape, while simultaneously the evolution of the environment occurs via hill climbing on a *constructive* landscape. The rate at which selection drives populations up the adaptive landscapes is described by Fisher's fundamental theorem: the increase in fitness is equal to the additive genetic variance in fitness.[109] Fisher

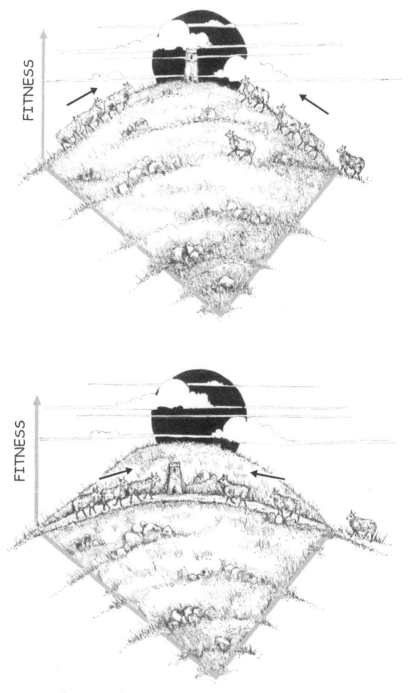

FIGURE 11 (*top, middle*)

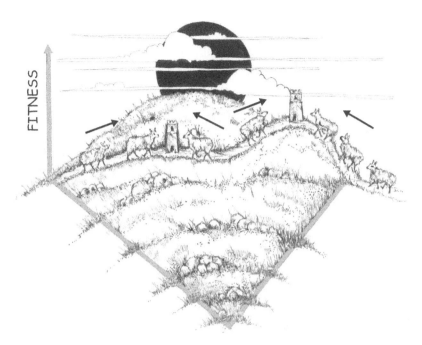

FIGURE 11. How developmental mechanisms affect evolution on adaptive landscapes. An adaptive landscape plots mean fitness against one or more of the population's traits or genes. In the resulting surface, peaks represent high-fitness trait combinations and valleys are phenotypes with lower fitness. *Top.* Conventional analyses commonly assume that through natural selection the population will converge on the traits suited to the local habitat, represented as climbing a local fitness peak. *Middle.* However, recent evolutionary theory shows that developmental processes determine the admissible pathways across the landscape, with natural selection pushing populations to converge on path peaks rather than landscape peaks. *Bottom.* Further theory shows that niche construction modifies the fitness landscape, creating new fitness peaks and affecting which equilibria are approached.

assumed that since populations can't continue to get fitter indefinitely, the process must be balanced by a corresponding deterioration of the environment. Tanaka and colleagues' model of coupled landscapes shows mathematically that when factors like niche construction are considered, this conservation of fitness need not hold. Organisms can alter their environment in such a way as to improve their absolute fitness (like building mounds on top of the hills), or they can spoil their environment and lower their fitness (like digging troughs into the landscape).[110]

Tanaka and colleagues' approach is logically equivalent to modelling sexual selection. Just as we can't adequately understand the evolution of secondary sexual characters, such as peacocks' tails, without tracking the coevolution of peahens' mating preferences, so the evolution of traits whose fitness depends on features of the environment modified by organisms can't be described without incorporating niche construction. The same point holds for developmental plasticity.[111] Mating preferences have been shaped by earlier natural and sexual selection, but this does not undermine the legitimacy of researchers treating mating preferences as an important aspect of the cause of sexual selection. In such instances, causation is seen to be reciprocal.[112] However, reciprocal causation is no less relevant in situations where niche construction and plasticity modify patterns of selection and have themselves been shaped by earlier natural selection. If we accept that mating preferences are causes of sexual selection, the same reasoning implies that niche construction and plasticity are causes of selection.[113] We return to this issue in chapter 14.

Practical Implications

We have stressed that how organisms develop and what organisms do partly determine the strength and direction of natural selection. In doing so, we regard developmental bias, plasticity, and niche construction as causes of natural selection. The term "cause of selection" is, however, understood in different ways, so here we elaborate on what we mean by this.

Evolutionary biology sets out to provide satisfactory explanations for the adaptations of living organisms. For instance, the conventional answer to the question *Why does a camel possess a hump?* is that the existence of this trait in current populations is explained by the hump on average increasing the fitness of camels, and hence being favored by selection, in the context of the hot, dry, sandy environments that ancestral camels inhabited. In formal terms, we can state that there is a causal relationship between the camel hump phenotype and a camel's fitness, which can be quantified as a selection gradient that describes how selection changes the mean of the phenotypic distribution.[114]

There are now many hundreds of studies demonstrating that quantitative traits are under selection in natural populations.[115] Many use a framework devised by Russell Lande and Stevan Arnold, which quantifies selection on a focal trait in terms of a selection gradient that controls for selection on correlated traits.[116] These studies have generated thousands of estimates of the direction and strength of natural selection in the wild, allowing for the comparison of standardized coefficients across traits and study systems and for their use in meta-analyses to test hypotheses concerning the impact of selection.[117] This approach is one of the key methodologies of contemporary evolutionary biology.

Yet, it has long been recognized that quantifying selection in this manner does not provide information about why the character is associated with fitness; it does not demonstrate a causal connection between the trait and fitness, but rather isolates correlates of fitness differences, and correlation is not causation.[118] For example, positive selection gradients could be detected for floppy ears in domesticated foxes and fragmented bones in cave fish, even though few biologists would argue these traits caused the selection.[119]

Nor do such studies generally identify why there is an association between the character and fitness. We want to be able to do more than merely state that camels evolved a hump—we also want to explain that they did so because the hump was able to act as a store of fat, which is adaptive in an environment in which food is scarce. In other words, good evolutionary explanations would go beyond specifying what was selected (fat storage in a hump) to stipulate the adaptive function (it acts as a food store) and source of selection (an environment in which food is scarce).[120]

Methods exist for addressing these concerns.[121] Experimenters can manipulate a putative environmental agent, predicting that this will change the relationship between the trait and fitness.[122] More recently, statistical tools for inferring causal relations on the bases of patterns of covariation have been deployed to quantify the causal pathways contributing to natural selection.[123] These analyses provide valuable tools to determine whether there has been selection *for* a trait, and hence identify the *causes of fitness differences*.[124]

Yet even with such experimental and statistical methods, this chapter has emphasized that one foundational question would remain unanswered. Returning to the camel example, the analyses would not explain why camels evolved humps rather than other ways of coping with food shortage. What is it that caused humps to be the solution, as opposed to the numerous other functionally equivalent adaptations that desert organisms deploy, ranging from storing food in burrows to hibernating to migrating?[125] Nor would they explain what change in developmental mechanisms caused camels to accumulate fat cells above the spine.

Here developmental biology can contribute—although it is easier to investigate these issues in animals that are more tractable to laboratory study than camels. Let's return to experiments by Daniel Schwab, Patrick Rohner, Armin

Moczek, and others at Indiana University, who established that dung beetle niche construction directly causes and strengthens the relations between key dung beetle traits and fitness.[126] Recall (chapter 8) that the larvae of these beetles are provisioned by their mothers with a brood ball, which they process in various ways, including defecating and mixing (which redistributes maternally inherited microbes and allows for more efficient use of carbon substrates) and constructing a pupation chamber. The authors conducted experimental manipulations that either permitted (NC+) or cancelled (NC−) the environmental effects of this niche construction by regularly relocating the larvae into fresh dung. They showed that the modification of environments by larvae was essential for the maintenance of sexual dimorphism. Larval niche construction not only substantially enhanced larval growth and adult body size—two of the most important life-history traits that evolutionary ecologists study—but also determined the relationship between key secondary sexual traits (including foreleg size, eye size, and horn size) and fitness. Beetle niche construction consistently shaped phenotypic variation, in the process dramatically altering the scaling relationship between some key traits under selection and fitness. The answer to the question of why adult beetle horn size is under sexual selection, as opposed to some other beetle character, is that this trait is unusually sensitive to the manner in which larval niche construction makes carbon nutrients available to the growing beetle. In this case, it appears that the nutrition-dependent growth pathway has been integrated with doublesex, a gene involved in constructing the secondary sex characters.[127] Such studies provide *explanations for why those particular traits have been favored by selection*.[128] Here the selection is determined by beetle niche construction, and it is in this sense that we can regard niche construction as a cause of selection.[129]

A second major finding was that closely related beetle species had diverged in their reliance on this niche construction, and associated traits including host-microbiome relationships.[130] This is also important because it shows how selection on the reaction norms created by niche construction can allow populations to evolve along developmentally biased pathways that represent clusters of mechanistically associated phenotypes.[131] In other words, larval, like maternal, niche construction generates developmental bias, creating corridors in multidimensional phenotype space along which populations can diverge in their body and trait size, life history, and behavior in a coordinated manner.[132] Beetle niche construction, quite literally, created distinctive population-specific reaction norms, triggering bouts of divergent selection that generated distinct beetle species. The fact that beetle niche construction is an evolved character does not change this fact. This point was made by Lewontin in the 1950s, but remains poorly appreciated.[133]

To date there have been relatively few attempts to manipulate the mechanisms of niche construction experimentally to examine the fitness consequences, but this and other dung beetle work provide a clear illustration of how that can be done.[134] Not all selection arises from organism-modified sources. The desert rhubarb's physiology may be adapted to its self-manufactured microclimate, but the camel's hump is more an adaptation to autonomous features of the environment. Nonetheless, extensive evidence suggests that niche construction and developmental plasticity may often create or transform the relationship between phenotypes and fitness.[135]

Deserts can be so vast that it is tempting to regard them as beyond organismal control. Yet organisms almost always have a say in the conditions of life that they experience. In countless different ways, including moving, hiding, burying, digging, timing, collecting, defecating, and sweating, living organisms cause natural selection.[136] Butterflies are known to respond plastically to low food availability as caterpillars by developing stronger thoracic muscles, which enables them to fly to better habitats; in doing so, they switch selection away from coping with low food availability to favoring traits influencing dispersal.[137] Western bluebirds respond to heightened aggression from nest-site competitors by allocating more testosterone to the clutch, modifying the reaction norms of several offspring traits, including dispersal, boldness, and helping at the nest.[138] Multiple plant species produce light and shade leaves depending on conditions, which damps fitness differences between individuals due to the light regime.[139] In competition to attract females, male house finches alter the fitness landscape for sexual selection by choosing social environments in which they are more attractive.[140] These and other examples suggest that the selection that occurs is often determined by how plasticity and niche construction are expressed in that population at that time.[141]

There is much untapped potential for experiments that cancel or enhance niche construction, habitat choice, or developmental plasticity to determine how they influence which characters are selected.[142] Such studies are likely to produce richer explanations that go beyond identifying correlates or causes of fitness differences to explain why those particular traits, and not others, evolved. Building on long-standing interest in the roles of plasticity, choice, and behavior in evolution,[143] such studies would stimulate a relatively underinvestigated line of enquiry for evolutionary biology, focused on identifying the causes of natural selection.

Empirical studies show that environmental features constructed by organisms have distinct properties and generate different patterns of selection from aspects of environments that change independently of the organism.[144] For instance, a recent meta-analysis of selection gradients found that selection in response to aspects of the environment regulated by organisms is consistently weaker and less variable than selection arising from autonomous aspects of environments. Organisms modify environments in characteristic, nonrandom ways, thereby imposing systematic biases on natural selection.[145]

A case can be made that a deeper understanding of evolutionary causation should be viewed as a priority for the field. That is because, for all the rich insights generated by studies of phenotypic evolution in the wild, only a tiny proportion of the variance in reproductive success—typically less than 5 percent—has been explained, leading to calls to broaden methodologies.[146] Equally important are the insights that knowledge of development can provide to the relationships between traits under selection. Most investigations of characters under selection quantify readily measurable traits. Combining and integrating developmental and selection studies can enhance the predictive power of evolutionary investigations by helping researchers to predict a priori which traits will exhibit a correlated response to selection and which will not, and to explain—at a mechanistic rather than statistical level—why selection for one trait has generated correlated selection of others. There is no conflict between developmental and quantitative genetic approaches here: they are complementary, and together they can lead to deeper causal understanding.

While researchers may not need to know what causes fitness differences in order to generate evolutionary explanations, they can often do a better job at both prediction and explanation if they do.[147] As explained in chapter 7, strong selection is frequently concentrated along a few major multivariate axes, each representing a cluster of correlated traits.[148] The mechanistic causes of these axes are diverse. Sometimes trait covariation arises through pleiotropy, as, for instance, where hormonal changes in house finches and western bluebirds brought about coordinated changes in morphology, physiology, and behavior.[149] In other systems, traits are products of the same GRNs, comprising a number of functionally interacting genes and other regulatory factors, as for domestication syndrome and blind Mexican cave fish.[150] Sometimes axes of covariation are found in multiple species, as we saw when discussing the role of symbionts in the evolution of herbivory.[151] Coordinated change can even occur without causally relevant genetic covariation, as for instance where the cultural practice of dairy farming created selection on genetic variation in adult lactose absorption.[152] Investing in understanding how those axes are created by functional interactions in developmental systems will likely pay dividends for evolutionary researchers. It will shed light on the causes of natural selection, bringing a mechanistic precision to how selection operates in a particular population or species, and leading to more satisfactory explanations for why particular traits evolved.[153] Such studies also potentially reveal regularities in natural selection that produce general patterns of evolutionary adaptation and diversification. It's the difference between casually remarking that you'll be driving east, and specifying that you'll be taking Interstate 80 then Route 95 to visit grandma in New York City.

10

Inheritance beyond the Gene

During the twentieth century, scholars of evolution focused on the slowly changing aspects of evolutionary adaptation and the most stable components of inheritance—namely, genetic processes. We predict that in the twenty-first century a new consensus will emerge, in which adaptation is understood to involve multiple complementary mechanisms operating on different time scales, supported by diverse forms of inheritance beyond genes, and where faster-acting (that is, extragenetic) processes are recognized to be important components of the adaptive system.[1]

Opening this chapter with a prediction is appropriate, since its focus is the nature of biological inheritance, and our principal claim is that inheritance is really all about prediction.[2] In 1969, Conrad Waddington presciently claimed: "The main issue in evolution is how populations deal with unknown futures." Organisms face "an unknown, but usually not wholly unforecastable, future," which they can survive only through successfully anticipating which traits will be required by themselves and by their descendants.[3]

Inheritance comprises more than a single discrete package of genes and other cytoplasmic resources; it is a time-distributed developmental process by which diverse developmental resources become available to the next generation.[4] Many extragenetic inheritance processes are best regarded—not as noise, fine-tuning, or baroque "add ons"[5]—but as essential tools for short-term, rapid-response adaptation.[6] Parent-offspring similarity may be just a side-effect of a process whose ultimate function is not stable transmission but accurate prediction.[7]

This view of inheritance contrasts starkly with the narrative that has dominated biology for over a century and according to which biological inheritance has been almost synonymous with genetic transmission.[8] Look up "inheritance" or "heredity" in an authoritative textbook, dictionary, or website, and you will find claims such as "an inherited trait is one that is genetically deter-mined," or "the study of heredity in biology is genetics," or definitions like "the process by which genetic information is passed on from parent to child."[9] The current market-leading evolution textbook states that "hereditary variations are based on . . . genes," which is in line with most contemporary thinking.[10]

Biologists are, of course, aware of the explosion of interest in epigenetics over recent decades, and evolutionists have paid particular attention to in-stances where epigenetic marks are transmitted across generations, dodging Weismann's barrier.[11] Supplementing extensive research on epigenetics is an avalanche of investigations into domains concerned with extragenetic inheri-tance, the most prominent of which are studies of parental effects, the inherited microbiome, and animal cultures.[12] This literature has generated excitement and debate, but little consensus. As Eva Jablonka and Marion Lamb's pioneer-ing writings demonstrate, "there is more to heredity than genes," but how important extragenetic inheritance is to evolutionary adaptation remains a matter of dispute.[13]

Extragenetic inheritance has been buried in murky unintelligibility for over a century. This abstruseness results from a combination of its great complexity and mechanistic diversity, relative to which Mendelian genetics appears posi-tively simple. "Soft inheritance"—the view that heredity could be changed by developmental experiences—has been shunned and neglected for decades.[14] Even today, fundamental questions go largely unanswered: *Why are these ad-ditional forms of inheritance present? What do they do? Do they work independently or in concert with genetic transmission?* The molecular detail can be overwhelm-ing, but we suggest that beneath the complexity lies a relatively straightforward logic. Extragenetic inheritance does an important job in evolution, but that job is, in the main, distinct from that of genetic inheritance.[15] Understanding its role requires thinking differently about the processes of adaptation.

The Breadth of Extragenetic Inheritance

We assume that the reader is familiar with genetic inheritance, which remains central to heredity. Here our focus is on the less-familiar extragenetic forms of inheritance. Diverse resources other than genes are known to be passed from parents to offspring, including components of both egg and sperm, hormones, symbionts, epigenetic marks, small RNAs, antibodies, ecological resources, and learned knowledge.[16] Traditionally considered proximate causes of devel-opment, it is now evident that some of these factors can lead to both short- and

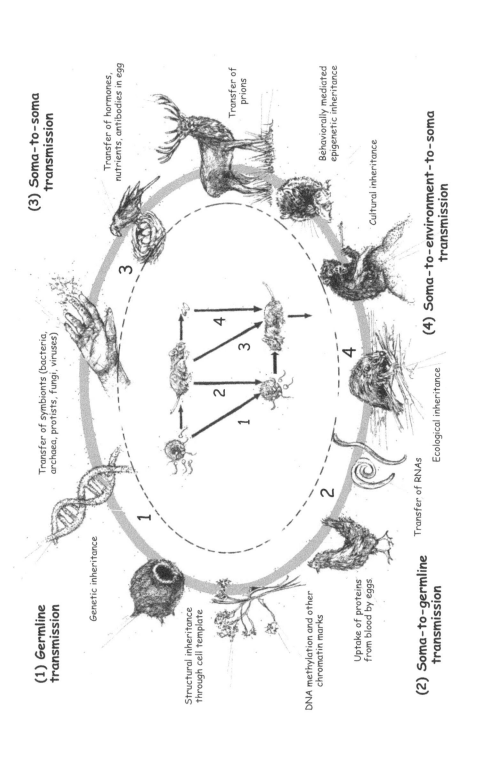

(1) Germline
transmission

Genetic inheritance

Structural inheritance
through cell template

DNA methylation and other
chromatin marks

Uptake of proteins
from blood by eggs.

(2) Soma-to-germline
transmission

Transfer of RNAs

Ecological inheritance

(4) Soma-to-environment-to-soma
transmission

Cultural inheritance

Behaviorally mediated
epigenetic inheritance

Transfer of
prions

Transfer of hormones,
nutrients, antibodies in egg

(3) Soma-to-soma
transmission

Transfer of symbionts (bacteria,
archaea, protists, fungi, viruses)

FIGURE 12. There is more to inheritance than genes. For a century, "soft" forms of inheritance were disputed, but today nongenetic inheritance is well established. The data point to four inheritance pathways: (1) Germline transmission includes genetic inheritance but also inherited epigenetic marks such as DNA methylation and histone modifications. In single-celled organisms, modifications of cell structure can be faithfully passed on for many generations. (2) Soma-to-germline transmission encompasses the uptake of proteins from the parent's blood into the egg in birds, and the transfer of noncoding RNAs to the sex cells, found to underlie the inheritance of learned information in nematode worms. (3) Soma-to-soma transmission occurs through the direct transfer from parent's to offspring's bodies of symbionts (microorganisms such as bacteria and fungi), hormones, nutrients, antibodies, and prion- and pathogen-borne diseases. Epigenetic inheritance is behaviorally mediated along this pathway, as exemplified by the inheritance of maternal care in rodents. Cultural inheritance is not unique to humans, with extensive evidence that birds, primates, whales, fish, and invertebrates like ants and bees all have cultural traditions. Finally, (4) soma-to-environment-to-soma transmission occurs when traits are passed down the generations through the external environments, as an "ecological inheritance." For instance, young beavers inherit the dams, lakes, and constructed lodges of their parents.

long-term inheritance of phenotypes. Many researchers are now attempting to integrate these factors into an extended concept of heredity.[17]

A comprehensive approach might be organized around inheritance pathways (e.g., "germline," "soma-to-soma," as in figure 12)[18] but for simplicity we discuss extragenetic inheritance in terms of four broad domains: *parental effects, epigenetic inheritance, the inherited microbiome,* and *animal culture.*[19]

Parental Effects

In single-celled organisms (e.g., bacteria, protozoans, many algae, and some fungi), whether reproduction is asexual or sexual, cellular components other than DNA that guide development are made available to the offspring by their parents. These include forms of structural inheritance observed in single-cell eukaryotes in which components of the cell, or the cell structure itself, act as a template.[20] The shape and arrangement of the cell cortex have been experimentally demonstrated to be inherited in a wide variety of single-cell eukaryotes via such templates.[21] In *Difflugia, Paramecium,* and other single-celled organisms, for instance, experimental modifications of cell structure are faithfully passed on for many generations.

In multicellular organisms, everything in the egg and sperm apart from DNA is potentially a component of extragenetic inheritance.[22] The first cues that enable the embryo's cells to divide and differentiate arise from maternally derived proteins and mRNAs placed in the egg, with the offspring's genes initially playing no role at all.[23] In live-bearing organisms, like mammals, the mother can transfer resources to her offspring throughout embryonic development. For example, a mother's diet can influence the development of the embryo's olfactory bulbs, which later shapes the offspring's attraction to smells.[24] Many resources, including hormones, nutrients, antibodies, symbionts, and regulatory molecules such as RNAs and transcription factors, as well as toxins and parasites, are transmitted from parents to offspring via the cytoplasm, in egg yolk, through the placenta, in mother's milk, or via some other pathway.[25] Where, when, and how these are dispersed, as well as egg and seed size, are maternally determined.[26] Parental effects play important roles in plants and in other organisms that experience minimal parental care,[27] where resources in the egg help to establish embryonic axes and shape pattern formation (see chapter 5).[28] In other species, parents have important impacts on their offspring's development well beyond birth—for instance, through food provisioning, grooming, and defense.[29]

All causal effects of the parental phenotype on offspring phenotype can be regarded as *parental effects*.[30] Although a strict separation between parental effects and, for example, epigenetic inheritance or animal culture is therefore not possible, we discuss them separately below. Many parental effects are thought to have evolved because they allow parents to build and control a reliable developmental environment for their offspring: Components of the egg help ensure that the energetic and molecular resources necessary for the embryo's early development are reliably present.[31] Generation after generation, dogs resemble dogs and cats look like cats, not just because of inherited genes but also because of gene expression whose stability is due to presence of extragenetic resources in the egg and surrounding milieu.[32]

Our definition of parental effects also includes the choice and modification by parents of external environmental conditions, which can influence the environmental resources experienced by their descendants (*ecological inheritance*).[33] Numerous animals are born in or on nests, burrows, holes, webs, pupal cases, or plants and other foods, manufactured or chosen by their parents.[34] Ovipositing insects choose host plants on which to lay their eggs, and thereby guarantee that nutritious food is available for their larvae, and many plants select specific locations for their seedlings.[35] Ecological inheritance is a means by which developmental factors of environmental origin become as dependable as genomic factors.[36] Theory predicts that parental effects, including ecological inheritance, evolve under a broad range of conditions, and can strongly affect evolutionary rates and dynamics.[37]

However, the role that parental effects play in evolution is not restricted to promoting intergenerational stability in development. Many such effects are now believed to be means by which parents, most frequently mothers, can adjust the development and phenotype of their offspring to match environmental conditions.[38] Parents may be better placed than their offspring to detect and interpret an environmental cue, with the parental environment guiding offspring development.[39] In this way, parents fine-tune their offspring's phenotypes to anticipated conditions. Experiments have shown that factors such as yolk testosterone, symbionts, or antibodies in the egg can exert strong influences on offspring growth, mortality, life-history traits, and patterns of dispersal.[40] In earlier chapters we saw how female bluebirds' exposure to aggressive encounters from nest-site competitors influenced how much testosterone they allocated to their clutches, which in turn affected the aggressiveness and dispersal of their offspring.[41]

Not all parental effects are adaptive; environments can change, rendering parental contributions outdated.[42] Theory suggests that if the environment is unpredictable across generations, predictive parental effects should be weak or absent.[43] In such conditions "bet-hedging" maternal effects may be favored, since mothers with a randomized mix of offspring phenotypes would leave at least some survivors.[44] Support for this is found in experimental studies in nematodes, where populations experiencing similar temperatures across generations evolved a positive maternal effect, while populations in environments with no correlation in temperature across generations did not.[45] Interestingly, if parents and offspring experienced different conditions, with the environment switching between oxygen-deprived and normal conditions each generation, the nematodes evolved a negative parental effect.[46] This shows that it is the predictability of the environment across generations, rather than its intergenerational similarity, that is crucial to the evolution of parental effects. When statistical models were fitted to the experimental data, it was found that a combination of phenotypic plasticity, anticipatory maternal effects, and epigenetic inheritance provided the best fit,[47] implying that organisms may draw on multiple sources of information to predict the upcoming environments.

There are other reasons why parental effects may be maladaptive. Once an inheritance channel is established, other factors may be inadvertently passed along, and the channel becomes vulnerable to exploitation by parasites. Rogue elements that disrupt offspring development, such as transmissible diseases and toxins, can sneak into the goody bag of parentally transferred resources. For instance, nematodes are passed from one generation of lizards to the next by infecting their eggs before they acquire a defensive shell, and ecological inheritance can include factors that negatively affect fitness, such as prion diseases contracted through exposure to feces, infected soil, or corpses.[48]

Epigenetic Inheritance

Cellular DNA is not naked but clothed in a variety of chemically attached molecules that affect whether genes are expressed.[49] The term "epigenetics" often refers to such effects, which don't involve changes in the DNA sequence but nonetheless influence gene activity. Some epigenetic attachments and detachments are specified by the DNA sequence, others occur at random like mutations, and others occur in response to environmental cues, such as foods eaten, pollutants, or social interactions.[50]

Epigenetics is a rapidly developing field of science, and a bewildering variety of mechanisms have been identified.[51] To prevent our account from being swamped by molecular detail, we characterize epigenetic inheritance mechanisms as falling into three broad categories,[52] although it is important to bear in mind that this is a simplification. The first is *DNA methylation*, which usually refers to the addition of a methyl group to one of the DNA nucleotide bases (cytosine) and can block transcription factors from binding to a gene and thereby suppress its expression.[53] The second is *histone modification*. DNA is usually wound around spool-like proteins called histones, chemical modifications of which can affect how tightly the DNA is wound, and in this and other ways upregulate, downregulate, or silence transcription of genes.[54] The third is *noncoding RNAs*, including both small and long noncoding RNAs, which regulate gene expression posttranscriptionally, often by binding to, and thereby silencing, RNA molecules.[55]

Each of these epigenetic mechanisms can survive across cell division, allowing groups of cells to maintain different patterns of gene expression even though they share the same DNA. This stability is fundamental to cell differentiation in multicellular organisms. However, it is also now well established that some epigenetic modifications can persist through the production of sex cells (meiosis) and hence be transmitted across generations.[56] Most mammals undergo two rounds of erasure of epigenetic marks, the first at the time of gamete production and the second in the embryo, which removes most—but crucially not all—epigenetic attachments.[57] In other animals and plants, however, where segregation of the germline occurs considerably later, there are ample opportunities for environmentally induced epigenetic changes to be inherited.[58]

Some between-generation effects involving epigenetic mechanisms have become famous because they appear to affect our own species. One widely discussed example concerns the long-term consequences of prenatal exposure to a famine known as the Dutch Hunger Winter. At the end of World War II, the Nazis blocked food supplies to the Netherlands, and before it was liberated more than thirty thousand people had died of starvation. A large proportion of the surviving offspring of nearly starving pregnant women developed health disorders. Not only was parental malnutrition linked to offspring low birth weight, diabetes, obesity, coronary heart disease, and cancer, but some of these effects were passed on to grandchildren. A pregnant mother's diet affected not only her children's health but also that of her grandchildren.[59]

How can the environment that an individual's parent was exposed to when it was in the womb be "remembered"? Epigenetics provides testable answers. There is a critical developmental window of time during which malnutrition can promote disease development through the modification of epigenetic factors. The Dutch Hunger Winter silenced certain genes in unborn children through methylation.[60] For instance, famine led to higher body mass index (BMI) by triggering the methylation of a gene called *PIM3*, which is involved in burning the body's fuel, leading to health problems in both offspring and grand-offspring. For other genes, such as insulin-like growth factor II (*IGF2*), which is critical to human growth and development, there was reduced methylation.[61] Much current interest in epigenetics is driven by medical science, with inherited gene regulation implicated in several hereditary diseases, and potentially accounting for some "missing heritability."[62] While the extent to which epigenetic effects in humans persist across several generations remains poorly understood, similar results have been reported in animal models, where some epigenetic changes have been found to be transmitted through the germline for five successive generations.[63]

Cases such as these can no longer be regarded as exotic.[64] Many hundreds of studies now demonstrate that the transmission of epigenetic variants can lead to stable inheritance of phenotypes in the absence of differences in structural DNA.[65] The epigenetic inheritance of fears learned by their ancestors in mice, described in chapter 1, is just one dramatic illustration.[66] Other familiar examples include the epigenetic inheritance of flower shape in toadflax, and the inheritance of obesity and coat color in agouti mice, both of which are mediated by DNA methylation.[67] Genetically identical mice can be sleek and gray or obese and yellow depending on the methylation pattern produced by the mother's diet. A recent review of five hundred studies reported that multigenerational inheritance of epigenetic marks is common in both plants and animals; indeed, there are few studies in plants in which epigenetic marks are *not* transmitted to offspring and grand-offspring generations.[68] In some taxa (e.g., nematode worms) epigenetic inheritance can last over twenty generations, but in most it generally peters out in less than three to five generations (but that does not mean it is not evolutionarily significant, as we will see).[69] One implication is that part of the heritable diversity seen in nature likely results from differences in inherited gene regulation.

Environmental challenges—for instance, maternal smoking, exposure of animals to insecticides, and of crops to heavy metals[70]—can generate an inherited epigenetic "memory" that lasts generations, showing that epigenetic inheritance need not be beneficial.[71] Recent, though still controversial, evidence suggests that pregnant holocaust victims and 9/11 survivors may have transmitted epigenetic marks associated with their trauma to their children.[72] Nonetheless, an intriguing feature of the Dutch Hunger Winter studies is the suggestion that some epigenetically mediated parental effects might be benefi-cial.[73] For instance, reprogramming offspring to upregulate the storage of food as body fat through the methylation of *PIM3* makes sense if an organism is starving. Such cases are variously known as *predictive adaptive responses, adap-tive parental effects,* or *anticipatory parental effects,* and there is some evidence that such responses may indeed be adaptive.[74] Of course, once the famine ended and food became available, such reprogramming may have become dis-advantageous, since it resulted in obesity, but it would probably have been adaptive had the famine continued.[75] Such "mismatches" where parentally inherited methylation patterns don't align with the actual environment are thought to be responsible for some cardiovascular disease and the prevalence of human obesity.[76]

Like parental effects in general, modification of offspring gene expression through epigenetic mechanisms can be adaptive to the extent that parents are able to detect accurately the current environment and use this cue to predict the phenotypes best suited to future offspring environments.[77] Novel pheno-types can be rapidly induced in response to environmental change through epigenetic mechanisms, which, when inherited, leave the offspring primed for the environment predicted by parental experience.[78] Consistent with this, ex-periments with inbred genotypes of the annual plant *Polygonum* conducted by Sonia Sultan and colleagues at Wesleyan University found that parental plants raised in shade produced offspring with increased fitness in shade but reduced fitness in sun.[79] Likewise, pathogen infections trigger immune responses at both challenged and remote DNA sites in many plants and animals, with chro-matin modifications and DNA demethylation at defense-related genes under-lying this *defense priming,* and with the "primed" state epigenetically inherited by subsequent generations.[80]

Moreover, at least in some species, there appear to be mechanisms that adjust the duration and impact of epigenetic inheritance to match the period and scale of environmental challenge.[81] Leah Houri-Ze'evi and her colleagues at Tel Aviv University found that repeated activation of the small RNAs responsible for epigenetic inheritance extended the duration of epigenetic inheritance in nematodes.[82] The more the lineage was exposed to the environ-mental trigger, the more small RNAs were transmitted across generations, establishing a memory trace of ancestral environmental conditions. Con-versely, exposure to stress, including starvation or high temperature, induced resetting of ancestral small RNA responses and a genome-wide reduction in levels of heritable small RNA.[83] This stress-dependent termination of small

RNA inheritance protects organisms from inheriting obsolete epigenetic information. In nematodes, a regulatory feedback loop appears to tune the duration of transgenerational gene silencing, with consistent environmental triggers extending epigenetic inheritance by the production of additional small RNAs, but with exposure to stress or change in state leading to the "memory" being forgotten. Similar findings are reported in other taxa: for instance, Sonia Sultan's group also found that *Polygonum* plants grown in drought conditions that have drought-stressed parents have higher fitness than those with no such ancestral experience, and those whose parents and grand-parents were both drought-stressed do better still.[84] That a "tunable mecha-nism" determines the duration of epigenetic inheritance renders it far more plausible that it can be adaptive.[85]

By contrast, in yeasts some genes appear to be switched on and off largely at random through histone modifications, with active and inactive gene ex-pression states heritable for up to twenty generations.[86] The process is restricted to specific regions of the genome, with the rate of phenotype switching through gene silencing dependent on environmental factors, such as tempera-ture or food availability.[87] This does not imply that such inherited epigenetic factors do not contribute to adaptation, however, as there is also good evi-dence that epigenetic variation can be subject to natural selection.[88] For ex-ample, in *Arabidopsis thaliana* (thale-cress), a small flowering plant widely used in genetic studies, epigenetic variation was found to alter many ecologi-cally important traits, including flowering time and root length.[89] When Marc Schmid and colleagues at the University of Zurich conducted selection experi-ments on *Arabidopsis* they found that, after five generations of selection, stably inherited differences, for instance in flowering time and morphological fea-tures, were linked to epigenetic variation.[90] Populations under selection showed a reduction in epigenetic diversity, including changes in methylation state that were associated with the phenotypic changes, but no evidence for genetic changes.[91] The study not only provides an experimental demonstra-tion that epigenetic variation can be subject to selection but raises the possibil-ity that some responses to selection in the wild are based solely on epigenetic variation.[92]

If environments are changeable, predicting conditions that descendants will experience can be something of a gamble. Even so, epigenetic inheritance can still be adaptive by facilitating a heritable form of bet hedging.[93] Epigene-tic variation allows phenotypes to be variable in the absence of genetic diver-sity, so that at least a fraction of the population can persist in stressful, rapidly changing, or unpredictable environments, thereby buffering the lineage against extinction.[94] Bacterial "persister" cells may provide an example; some bacterial populations are able to tolerate antibiotics by becoming dormant, yet they can spring back to life again once the stress is removed, with epigenetic mechanisms implicated as playing an important role.[95] Mathematical modelling suggests that the fitness disadvantage accrued through producing a suboptimal phenotype in the current generation can be more than offset by the advantages that phenotype would confer if the environment changed, with the rate of phenotype switching evolving to match the periodicity of environmental fluctuation.[96] Experiments in yeast and nematodes reinforce these conclusions: yeast populations with high rates of epigenetic switching grow faster in environments with short periodicity, while lower rates of switching are beneficial in more slowly changing conditions.[97] By responding to rapid environmental change through the selection of heritable phenotypic variation, genetic variation can be preserved.[98] But even in stable environments some low level of epigenetic switching may be adaptive.[99]

Thus, while transgenerational epigenetic inheritance may often involve reduced offspring fitness,[100] this does not preclude epigenetic inheritance from contributing to evolutionary adaptation through selection or bet hedging. After all, probably an even larger fraction of genetic mutations are maladaptive. Rare adaptive variants, genetic or epigenetic, may increase in frequency through selection. The fact that much epigenetic variation is environmentally induced means that, unlike mutation, it can appear in multiple individuals simultaneously, leaving potentially adaptive epigenetic variants less vulnerable than potentially favorable mutations to loss through drift or population extinction. Nor is the selection of epigenetic variation reliant on associated genetic variation. A landmark study, again in *Arabidopsis*, found that DNA methylation is gained and lost at rates high enough to decouple it from genetic variation, yet low enough to sustain "long-term" responses to selection.[101] This quantification is important, as the lack of stability of "epimutations" has been claimed—erroneously—to imply that epigenetic inheritance will play no role in evolution.[102]

In addition to predicting adaptive responses through *detection* of environmental cues, and by generating phenotypic variation that can be subject to *selection*,[103] there is a third experimentally demonstrated way in which epigenetic inheritance can affect evolutionary dynamics, namely genetic assimilation.[104] Dragan Stajic and colleagues investigated the adaptation of genetically uniform populations of yeast to a toxic chemical, in the process manipulating levels of heritable silencing of a gene that in silenced form conferred resistance.[105] They found that multigenerational heritable silencing through epigenetic inheritance was a powerful contributor to evolutionary adaptation, without which the majority of populations went extinct. Silencing aided adaptation by helping the population to survive and grow, which increased the supply of adaptive mutations. Among such mutations were those that enhanced the effectiveness of epigenetic silencing mechanisms, leading to genetic assimilation of the silent (i.e., resistant) phenotype.[106] This study illustrates how even transient forms of inheritance can facilitate adaptation to novel environments in which evolutionary rescue by mutation may be difficult.[107]

Moreover, the DNA nucleotide base that is subject to methylation (cytosine, or C), is known to mutate spontaneously to an alternative nucleotide base (thymine, or T) at a higher rate when methylated.[108] This means that, other factors being equal, methylated regions of DNA exhibit a higher propensity for mutation than unmethylated regions—a rate estimated to be ten to fifty times higher.[109] Other mutation rates are also affected by epigenetic processes.[110] The impact on mutation can be nontrivial—for instance, in human cancer cell lines, variation in epigenetic marks accounts for 40 percent of variation in the mutation rate.[111] The evolutionary significance of epigenetic marks stems not from their mutagenesis per se, but rather from the potential this creates for directing genetic evolution. By enhancing the supply of mutations, either by increasing population size or by providing a targeted site for mutation, epigenetic modifications provide a means for functionally equivalent adaptive genetic variation to supplant epigenetic variation, providing a pathway for genetic assimilation.[112] Genetic mutations may occur and subsequently become assimilated into the genome, through DNA methylation, histone modifications, and noncoding RNAs.[113] For instance, an analysis of genetic differences between domestic chickens and red junglefowl established that these differences are disproportionately associated with methylated genes, which "seem to be hotspots of mutations related to speciation."[114]

Whether epigenetic inheritance is adaptive through accurate detection of environmental cues, through a combination of bet hedging and selection, or through triggering genetic change, there is strong evidence that its disruption reduces adaptability. In green algae, for instance, chemical interference with histone acetylation or DNA methylation resulted in impaired adaptation to salt, nitrogen, and carbon dioxide stresses, with adaptation arising through new patterns of epigenetic marking.[115] Populations of yeast were found to evolve higher levels of drug resistance through epigenetic inheritance.[116] In another experiment, yeast were able to adapt to a caffeine-rich environment both by evolving the epigenetic silencing of caffeine-sensitive genes and through genetic mutation.[117] Here adaptation through epigenetic change preceded genetic mutation, providing clear evidence that heritable epigenetic silencing can act as a buffer against environmental challenges.[118] Such experiments are important because they establish a causal link between epigenetic inheritance and adaptation, and allow researchers to control for, or quantify, the effects of genetic induction of epigenetic variation.[119] Field studies reinforce these findings; for instance, epigenetic differences are correlated with adaptive phenotypic differences (e.g., between river and lake snails), even in populations that lack genetic diversity (e.g., Japanese knotweed).[120] Epigenetic variation is sometimes more strongly correlated with habitat and adaptive phenotypes than genetic variation (e.g., in salt marsh perennials).[121]

In sum, a growing literature provides strong evidence that epigenetic inheritance plays at least three important roles in evolutionary adaptation—environment *detection*, *selection* of phenotypic variation, and the *redirection* of genetic evolution. That these roles can be important for evolution has also been confirmed using mathematical models.[122]

The Inherited Microbiome

In earlier chapters we described how desert woodrats exploit an unusual niche through the ecological inheritance of symbiotic bacteria that detoxify creosote.[123] We also introduced other organisms, including corals, termites, bobtail squid, legumes, and cattle, that are similarly dependent on their resident microorganisms to carry out essential functions.[124] The "microbiome," which includes symbiotic bacteria, archaea, protists, fungi, and viruses, can be passed from one generation to the next through a broad variety of mechanisms, including asexual reproduction, via eggs and seeds, internally through the transfer of symbionts to developing oocytes or embryos, or through "intimate neighborhood transmission," such as passage through the birth canal, drinking mother's milk, eating mother's feces, or provisioning offspring with regurgitated foods.[125] In invertebrates, transmission is often accomplished through "egg smearing," where mothers coat their eggs with microbes as they lay them, which is somewhat similar to the way in which human infants are "smeared" with maternal vaginal microbes as they leave the birth canal.[126] Many other species, such as dung beetles, provision offspring with a microbe-rich fecal pellet, or lay eggs in their feces, which are consumed when the larvae hatch.[127]

Even when symbionts are acquired from the external environment, they may still be reliably inherited, as the woodrat and bobtail squid examples illustrate.[128] Intergenerational microbial transmission is now recognized to be a common, perhaps universal, component of animal inheritance.[129]

The desert woodrats also illustrate how the inherited microbiome can play similar roles in evolution to epigenetic inheritance. Here, gut microbes enhance tolerance to plant toxins, allowing the woodrats to expand their dietary niche, while buffering their survival in the challenging desert conditions.[130] Plant secondary compounds, which are frequently toxic, are a major determinant of the dietary niche breadth of numerous herbivores, many of which rely on their microbiome for detoxification.[131] The ability to exploit detoxified enzymes confers a clear fitness advantage, and, like epigenetic inheritance, selection can act on this heritable trait. Interestingly, while woodrats, from other regions, that don't consume creosote were able to consume this toxic food when inoculated with the microbiota of Mojave woodrats, they did not achieve the same level of consumption or growth.[132] Apparently, Mojave woodrats have evolved detoxification adaptations that bolster their specialization on this diet, a case of genetic accommodation.[133]

The adaptation of dogs to a starch-rich agricultural diet provides another example. Simone Rampelli and colleagues extracted and sequenced ancient DNA from 3,500-year-old fossil dog feces found in Solarolo, Italy.[134] They analyzed this DNA and that of the dogs' gut microbiome and compared them to those of wolves and modern dogs. They found the Solarolo dogs had not yet evolved the characteristic high copy number of starch-digesting genes found in both humans and modern dogs, yet were consuming a starch-rich agricultural diet.[135] This was possible only because the Solarolo dogs' gut microbiome was enriched for the amylase enzymes that break down starch, compensating for their host's deficiency.[136] Neolithic dogs apparently responded to the starch-rich diet by expanding microbial functionalities devoted to breaking down starch, allowing the dogs to extract energy from new agricultural foods. In European dogs, selection for increased copy number of starch-digesting (amylase) genes actually occurred thousands of years after they adopted this diet, illustrating both the adaptive buffering role of the gut microbiome and the eventual genetic assimilation of this functionality. A similar mechanism is thought to have been important to the dietary evolution of several domesticated animals.[137]

There may, however, be a faster route for hosts to acquire the useful genes of symbionts—horizontal gene transfer. While most plant-eating insects need symbiotic microbes to digest energy-rich plant material such as pectins and cellulose, there are some beetles whose genomes include this function. Sequence analyses reveal that the genes did not come from the standard evolutionary pathway but rather from the horizontal transfer of genetic material from microbes to hosts.[138]

Other studies illustrate the substantial impact of locally adapted microbes on host phenotypes and fitness.[139] Plants that grow on geothermal soils are associated with a heat-tolerant mold called *Curvularia*, and exposure to this mold increases the survival of other plants when subjected to heat stress.[140] Salt-tolerant fungi produce salt tolerance in nonadapted plant populations.[141] Bean bugs and soy bugs each acquire pesticide resistance from a pesticide-degrading bacterium, coral microbiomes protect against pathogens and extreme heat, and seasonal shifts in the storage of fat in hibernating brown bears are mediated by their microbiome.[142] Adaptation through the acquisition of symbionts can have important ecological ramifications; for instance, the red turpentine beetle recently became a pest when it acquired a fungal symbiont that allowed it to infest live trees, and millions of trees in China have perished as a consequence.[143]

Relying on the microbiome may facilitate survival and underpin phenotypic change in the short term, but in the longer term we might expect that hosts would evolve their own functionally equivalent adaptations, or alternatively evolve adaptations that can guarantee the presence of locally adapted microbial populations. For example, in some, but not most, coral species the symbiotic algae have evolved to be contained within their host's gametes.[144] Further examples of host taxa that have taken the second route and specialized in microbe-constructed niches include bark beetles, pine weevils, and coffee berry borers, all of which rely on their microbiomes to detoxify plant secondary compounds.[145]

The woodrat and dog studies, in particular, are compelling illustrations of how the gut microbiome can be a crucial "fast-adaptive partner,"[146] enabling microbially acquired adaptive responses to rapid dietary changes, and complementing the slower-acting response to selection observed in their hosts. Analogous to plasticity-led evolution, there is growing empirical evidence of microbial-led adaptation in animal evolution, providing a pathway to accelerated niche expansion, evolution of parasitoid resistance, adaptation to heat stress, and speciation.[147] In fruit flies, a particularly striking example involves commensal bacteria that play a surprising role in speciation.[148] When a population of flies was divided into two and the subpopulations reared on two distinct food media (molasses and starch), after just a single generation flies exhibited a strong preference for mating with individuals reared on the same food, with this almost instantly acquired preference stably inherited for many subsequent generations.[149] The authors were able to demonstrate that the fly microbiome was responsible for the rapid appearance of this mating preference: treatment with antibiotics eradicated diet-induced mating preferences, and inoculating antibiotic-treated flies with bacteria from the original medium restored them.[150] While fruit flies have a short generation time, their symbiotic bacteria reproduce at least one order of magnitude faster, allowing multiple generations of bacterial adaptation to the new food to occur in a single fly generation.[151] Further analyses suggested that the bacteria altered the flies' mating preferences by shifting the levels of cuticular sex pheromones.[152] Symbiotic bacteria are known to contribute to the odor of many animals, raising the possibility that they play a general role in animal pre-mating isolation.[153]

Some bacterial species share a parallel evolutionary history with their hosts. Such intimate symbiosis might be expected for hosts that transmit their symbionts directly through their eggs or embryos,[154] but is also found in other species. For instance, when placed in an aquarium with free-floating bacteria, three species of *Hydra* will recruit the precise set of microbes characteristic of their species.[155] This coevolution of microbes and hosts has been seen in many organisms, including humans.[156] Over fifty microbial species co-diversified along with human populations, and they appear to have independently evolved traits that lead to host dependency. That said, comparative evidence suggests that only a fraction of microbial lineages co-speciate with their hosts, and the overall distribution of microbial functions in the gut does not always match host phylogeny.[157] There is some concordance across animal taxa between a host's diet and the composition of its microbiome, but only sometimes is this due to the microbiome being vertically inherited with its hosts.[158] More commonly, the microbiome is evolutionarily labile, with shifts in diet and the horizontal transfer of microorganisms between host taxa also contributing to the correspondence.[159] Behaviors, such as the desert woodrats' tendency to collect and eat the feces of other animals, such as jackrabbits, facilitate a sharing of the microbiome that can parallel shifts in diet and aid expansion of dietary niches.[160] This sharing of the microbiome across a community was demonstrated in fruit bats, where an entire bat colony's microbiome changed collectively over time.[161] Group-level microbiome similarities have been reported in several vertebrates, including humans.[162] Indeed, the sharing of beneficial microbes has been suggested to be one of the major benefits of group living.[163]

Bumblebees, for instance, require either direct contact with nest mates or feeding on their feces to acquire their normal gut microbiota, without which they are far more susceptible to parasites.[164] Such findings highlight discrepancies between the inheritance pathways of the host's genes and its microbiome.

In sum, evidence is emerging that inherited microbiomes play important roles in evolutionary adaptation, allowing rapid evolutionary responses to changing or challenging environmental conditions, responses that can subsequently be stabilized through selection on the host. Once again, theoretical modelling supports these empirical findings. For instance, Dino Osmanovic and colleagues developed a formal model that shows how Darwinian selection operating simultaneously at two levels—hosts and their microbiomes—can generate rapid adaptation of the system as a whole.[165] In the model, exposure to a toxin led in a single host generation to selection of resistant bacteria, which in turn increased the tolerance in the host's offspring. The findings are strikingly evocative of the fruit fly experiment above, where adaptation of the bacterial community generated pre-mating isolation within a single generation. Other theoretical work by Joan Roughgarden and colleagues has demonstrated that a mix of horizontal and vertical symbiont transfer does not prevent *holobionts* (hosts plus their microbiome) from operating as functional wholes, evolutionary interactors, or exhibitors of adaptation.[166] The holobiont concept remains contentious, partly because some researchers assert that vertical transmission and high partner fidelity are required, and it is not clear in what proportion of cases the microbiome is vertically inherited with its hosts.[167] However, even skeptics accept that selection at the level of the symbiotic community occurs in some cases.[168] The "genome" of a holobiont probably requires relatively high partner fidelity if it is to evolve as a unit,[169] but as the desert woodrat example illustrates, and theory confirms, even when symbionts are primarily acquired from the environment, the microbiome can still be reliably inherited and can still play vital roles in evolutionary adaptation.[170]

Animal Culture

Evolutionary researchers have traditionally viewed cultural inheritance to be of marginal significance for species other than humans.[171] Yet, as we described in chapter 1, animal culture is now well documented in a wide range of animals, including invertebrates,[172] and observations of natural populations are backed up by hundreds of experimental demonstrations of animal social learning.[173] Among the most compelling are cross-fostering studies in great tits and blue tits, in which birds raised by the other species shifted numerous aspects of their behavior toward that of their foster parent, including the height in trees at which they forage, their choice of prey, their calls and songs, and even their choice of mate.[174] Studies that translocate entire populations of fishes provide similar evidence for culture.[175] Even in insects, the evidence for culture is surprisingly strong.[176] Charles Darwin speculated that bees learned by copying to cut holes in flowers to rob them of nectar, and his intuition has since been confirmed.[177] Social learning can no longer be regarded as the sole province of humans, apes, or large-brained mammals; it is widespread in animals.[178]

Animal culture may play similar evolutionary roles to epigenetic inheritance. Most obviously, it can allow individual animals to adjust their behavior to match their environments; for instance, to locate new foods or identify novel threats (equivalent to "environment detection"). By copying others, animals can acquire high-payoff, up-to-date and potentially adaptive knowledge, and do so quicker and with less energy expenditure than learning for themselves.[179] Behavior patterns that are repeatedly performed tend to be the most productive, whereas unproductive behavior is generally quickly discarded (i.e., "selection" of phenotypic variation).[180] Groups often make more effective decisions than individuals and are better at identifying difficult-to-find solutions, as investigations of "collective," "swarm," or "crowd" intelligence attest.[181] Among thousands of reports of social learning in natural populations of animals, almost none (apart from humans) involve the spread of "bad" information.[182]

Cultural information can be highly diverse. For instance, chimpanzee populations not only vary in their dietary traditions, such as fishing for termites, ants, or honey or cracking open nuts with stone hammers, they also exhibit other learned traditions such as grooming with particular postures, dancing in the rain, and using plants as medicines.[183] Likewise, in guppies there have been experimental demonstrations of social learning of routes to food sites, food patch preferences, novel feeding behaviors, mating preferences, predator evasion behavior, predator recognition, and predator inspection tactics.[184] Animals do not pick up only an occasional trait socially: often a sizeable proportion of their behavior is acquired that way.[185] Experiments reveal a mechanistic plurality to cultural transmission. Rats, for instance, can acquire a preference for a novel food by observing other rats feeding on that food, through picking up dietary cues on other rats' breath, through following other rats to food sites, through following scent trails, through detecting excretory marks at food sites, through consuming other rats' feces, and from dietary cues in their mother's milk.[186] These different information channels often reinforce one another, to stabilize information transmission, but, interestingly, when they impart conflicting messages, transmission rapidly breaks down, preventing the spread of misinformation.[187] That is one reason why rats prove so difficult to exterminate.[188] Together, these mechanisms are so effective that they support colony-wide dietary traditions for efficiently exploiting safe and nutritious foods, while leaving poisonous foods largely untouched.[189] This "belt and braces" quality may be a general characteristic of extragenetic inheritance.

Considerable effort has been devoted to establishing that candidate "cultural" differences between animal populations are not explained by genetic differences: experiments and developmental observations leave little doubt that many of these behaviors are indeed socially acquired.[190] Primates, whales, birds, fishes, and some other animals maintain stable differences between populations as behavioral traditions.[191] While the capacity to learn from others is underpinned by genetic transmission, the content of culture—what food is eaten or which migration routes are taken—is not genetically determined.[192] Evolved biases exist but appear not to constrain learning very much. For instance, the learning of fears through observation is an ancient mechanism extremely widespread in animals, and both monkeys and birds can acquire novel fears socially, including of entirely arbitrary objects.[193] Primate brains contain neurons that respond rapidly and selectively to images of snakes.[194] Possibly, because snakes were a reliably present threat in ancestral environments, selection has tuned monkey perceptual systems to make snake-shaped objects particularly salient, but without undermining general learning capabilities.

Culturally transmitted behaviors can become locally adapted through a combination of natural and cultural selection.[195] We saw that killer whale populations exhibit learned specializations on particular prey (e.g., fish, dolphins, pinnipeds) that are associated with population-specific morphology and digestion; these specializations are known as *ecotypes*.[196] Here, culturally learned dietary traditions have apparently initiated and modified the selection of genes for morphologies and physiologies that match the whales' learned habits, thereby redirecting genetic evolution.[197] However, animal culture can be evolutionarily important even when short-lived.[198] In a single mating season, for example, in insects, fishes, birds and mammals, socially learned "fads" can develop in the qualities that individuals find attractive in their partners, and mathematical models have shown that such "mate choice copying" can strongly affect the intensity of sexual selection.[199] Likewise, socially learned mobbing of cuckoos by reed warblers is thought to have influenced the evolution of brood parasitism, while the social learning of traits involved in mating decisions, such as song, can bias the evolution of signals and affect population divergence, hybridization, and speciation.[200]

Cultural transmission differs from genetic inheritance in important ways. For instance, it allows behavioral traits to spread among unrelated individuals in a population, as we saw for lobtail feeding by humpback whales.[201] Such horizontal transmission is a common feature of animal culture.[202] One consequence is that socially acquired information is generally up to date and relevant. Another key difference is that behavioral innovations, while analogous to genetic mutations in that they introduce novel variation, are usually not random but are adaptively biased.[203] Through learning, for instance, animals can discover and exploit new foods, or devise novel means to avoid a threat. Learning can even generate seemingly adaptive phenotypic change in the absence of any immediate environmental change or stressor (e.g., orangutans exploiting palm heart).[204] Social learning too is typically nonrandom and adaptive.[205] In chapter 8 we described how some birds copy the feeding behavior of conspecifics except when they observe an aversive reaction to the food, how some fish monitor the foraging success of other fish to select the richer of alternative food patches, and how insects and birds copy the nest-site decisions of successful individuals.[206] However, cultural interaction can inadvertently provide a pathway for disease transmission too.[207]

Culture can mimic adaptation by natural selection by generating incremental increases in fitness over time, as the foraging of bighorn sheep demonstrates.[208] Here, in gradually adjusting to novel habitats, learning caused offspring fitness to be higher than that of their parents, with more offspring in each generation exploiting high-quality forage than their parents.[209] This is equivalent to adaptation without natural selection. It should not come as a surprise, given that in our own species cultural evolution has long been known to generate increments in trait fitness over time, ranging from improvements in the design of artifacts such as axes, bows, and canoes to increases in agricultural yields through irrigation, fertilizers, and selective breeding (although in humans there is also compelling evidence that culture can also propagate maladaptive behavior). We will discuss the impact of culture on human evolution in chapter 13. Unquestionably, in our species, the use of symbols and invention of language and writing substantially enhanced the volume and accuracy of culturally transmitted knowledge.[210] Here it suffices to note that humans have clearly redirected their own and other species' evolution through their culture; for instance, by cultivating crops, domesticating livestock, and causing extinctions.[211]

We have described how dairy farming created the selection pressure leading to the spread of alleles for human adult lactase persistence.[212] Likewise, agricultural practices led to greater consumption of starch, protein, and alcohol, generating selection for alleles related to their metabolism,[213] a point supported by gene-culture coevolutionary theory.[214] Theoretical work has explored how learning and culture affect genetic evolution, finding that culture can strongly affect evolutionary dynamics.[215]

Predicting the Future

In the previous chapter we described the reliance of organisms on inherited genetic information as equivalent to a gamble that current environments will be sufficiently similar to ancestral conditions to render past-adapted traits adaptive again. Although genetic information may predict which phenotype will thrive in an anticipated world, genes provide only a long-term forecast, a composite "memory" of what worked in the past, averaged across the entire population and over aeons of time.[216] The stability that adaptive genetic information confers is important; without it populations could change radically with each environmental alteration. Yet, except for microorganisms, genetic evolution is not generally nimble enough to allow organisms to adjust to sudden or unanticipated changes in conditions: like a juggernaut, it has too much momentum and too little acceleration or maneuverability.

To some extent organisms can compensate for limitations in the predictive power of their genes through developmental plasticity, which allows organisms to update their prediction of a well-functioning phenotype in the light of cues from their immediate local environment. Individuals often have access to reliable information early in development that allows them to accommodate to the environmental conditions that they will experience at a later stage. Plants, for instance, can as seedlings be primed by their environments for stresses that they will experience later in life.[217] However, the detection of environmental cues by a developing organism is not always possible or optimal, either because offspring may still be in the maternal environment, where information can be accessed only indirectly via their mother, or because they are not yet sufficiently mature to detect environmental cues reliably, or because there are fitness benefits to making key developmental decisions early. Moreover, while genetic information can be too general, self-reliance risks being too specific and generating only a local, idiosyncratic forecast. This is where extragenetic inheritance comes in.[218]

The preceding sections illustrate the roles that extragenetic inheritance can play in evolution. Extragenetic inheritance can contribute to long-term adaptation by helping to ensure that descendants' genes are expressed in a cellular environment with the resources necessary for robust development. However, we have seen that extragenetic inheritance can also operate as a fast-response capability, allowing survival and adaptation to novel, challenging, or changeable conditions. Recall how heritable epigenetic gene silencing in yeast rapidly conferred resistance to a toxin and without this silencing most populations went extinct, or how in dogs the gut microbiome enabled a fast adaptive response to an agricultural diet centuries before genetic adaptation occurred, or how social learning allowed rats to avoid extermination prior to the evolution of genetic resistance.[219] Extragenetic inheritance can establish an evolutionary beachhead, upon which genetic adaptation can build.

We propose three principal means by which extragenetic inheritance enhances evolutionary adaptation: *detection*, *selection*, and *redirection*. Parental effects including epigenetic inheritance can be adaptive if they allow parents to transmit information to their offspring about the environment that they have detected. This information can help to predict offspring phenotypes that will flourish in anticipated environments, with parents contributing to these phenotypes by passing on relevant resources. Epigenetically inherited adaptations to light and shade in *Polygonum*, defense priming in plants, and temperature priming in nematodes provide examples.[220] However, epigenetic variation plays more than one role in evolution, and the *selection* of heritable epigenetic variants adds to the predictive power of plasticity and maternal effects, as studies of selection in nematode worms established.[221] For all the benefits of drawing on parental effects to guide offspring development, a parent is still just a single individual, localized in time and space, and with a very specific assessment of the state of the conditions. Supplementing the developmental decision-making apparatus with epigenetic information, accrued through a handful of recent generations of selection on a local population, improves prediction. Formal theory confirms this conclusion, showing that the natural selection of epigenetic variation complements detection-based mechanisms, and supplements predictive information arising from genetic variation;[222] those organisms that integrate detection- and selection- based cues are better able to predict future environmental states and generate adaptive phenotypes.[223] Culturally acquired knowledge is generally adaptive for similar reasons—it too has been subject to a few generations of selection, which helps to guarantee its utility. Like the inherited microbiome, culturally transmitted variants can undergo multiple generations of (cultural) selection within a single "host" generation, and this selection on two time scales can generate "Lamarckian-like adaptation" in the system.[224]

The selection of extragenetic variation can generate information about traits that have recently proved adaptive. There is always a risk that environmental change will render that information obsolete, but this risk is reduced by extragenetic inheritance that samples recent generations, and the evolution of safeguards, such as stress-dependent termination of small RNA inheritance and abandonment of outdated cultural knowledge.[225] Epigenetic switching, cultural learning, and hosting microbiomes can be advantageous in changing or unpredictable conditions. Unlike genetic mutations, environmentally responsive, and potentially adaptive, phenotypes can arise in multiple individuals and be altered or reversed.[226]

Some prominent biologists have argued that extragenetic inheritance is insufficiently stable to play an important role in adaptive evolution,[227] but we suggest that it is significant precisely because of that instability. We envisage an evolutionary process in which genomes and robust parental effects (over hundreds to thousands of generations), cultural knowledge (over tens of generations), epigenetic modifications (over a handful of generations), other parental effects (over a single generation), and phenotypic plasticity (in the current generation) collectively contribute to adaptive evolution.[228] Extragenetic inheritance is what allows populations to adjust to rapidly changing aspects of their world.[229] Many of us have been conditioned by our knowledge of genetic transmission to assume that extragenetic inheritance must operate in a similar way, but it probably plays a quite different role in evolution.[230] Organisms require a capability to cope with environments that change on all temporal and spatial scales, since appropriate genetic mutations cannot be guaranteed.[231]

We have also seen that extragenetic inheritance can *redirect* genetic evolution, including by triggering genetic assimilation. Through various means—improving population survival, increasing population size, providing a targeted site for mutation, and modifying selection—epigenetic modifications can open a path by which functionally equivalent adaptive genetic variation can replace or supplement epigenetic variation.[232] The inherited microbiome and animal culture operate in similar ways, giving time for the host to evolve equivalent and complementary adaptations (i.e., the more efficient metabolism of creosote in woodrats, or the population-specific morphology and digestion of killer whales).[233]

Explicit recognition of these three distinct roles that extragenetic inheritance plays helps to make sense of taxonomic diversity in adaptive mechanisms. All organisms require some fast-response adaptive machinery, but which kind they rely on will depend on each organism's characteristics. Complex animals can respond to novel challenges through learning, including socially guided movement,[234] and for them adaptation through the selection of epigenetic variation may be less important. Animals with simple nervous systems and short generations, such as nematodes, are probably more reliant on epigenetic inheritance than animals with complex brains and long generation times.[235] Plants, which are sessile and lack brains, will be more influenced by epigenetic inheritance than animals.[236] Other factors being equal, the longer the lifespan of the organism, the fewer the generations of epigenetic inheritance that are likely to prove adaptive, as longer generation times mean more time for the environment to change.[237] Plasticity and heritable bet hedging are likely to be particularly important for asexual organisms as mechanisms to increase phenotypic diversity.[238]

The various inheritance pathways operate in a "functionally interdependent" manner to reinforce and complement each other.[239] We have already seen this in cases of genetic assimilation, for instance in yeast, woodrats, and monkeys, where genetic reinforcement of extragenetic mechanisms appears to bolster the stability of trait inheritance.[240] Such genetic reinforcement is far from inevitable. Only a subset of epigenetic variation has been found to be tightly associated with genetic variation,[241] and the same holds for cultural variation.[242] Tight genetic regulation of extragenetic inheritance would be maladaptive in rapidly changing conditions because genetic evolution is too slow.[243] For extragenetic inheritance to function effectively as a rapid-response capability it must maintain some degree of autonomy from genetic variation. The downside is that once an organism's genes do not control other inheritance pathways, the organism becomes vulnerable to the transfer of maladaptive elements, including toxins, parasites, and parasitic information.

A Developmentalist View of Inheritance

Heredity is more than a package of genes and cellular resources handed over at conception like the baton in a relay race: it is a continuous process of developmental reconstruction that spans the life cycle.[244] All forms of inheritance collectively guide offspring development by contributing to the production of a phenotype predicted to match the expected environment, where that "prediction" is based on transmitted genes and updated by inherited extragenetic information accrued through both detection and selection mechanisms.

Parents construct offspring developmental environments by transferring resources internally and externally, choosing and building structures, and regulating conditions, in ways that enable reliable implementation of genetically inherited predispositions.[245] Extragenetic inheritance mechanisms operate in concert with genes, and with one another. For instance, ecological inheritance, in the form of enduring physical traces that accompany foraging or tool use, is vital to stable cultural transmission.[246] Young chimpanzees learn appropriate wand selection for termite fishing by utilizing their mothers' discarded wands.[247] Epigenetic inheritance too can be behaviorally mediated, as the inheritance of maternal care in rodents illustrates.[248] Highly nurtured rat pups tend to grow up to be calm adults that nurture their own offspring, while rat pups that receive little licking and grooming grow up to become anxious mothers that don't lick or groom their pups. Cross-fostered offspring develop a stress response resembling that of their foster mothers. The stress response and nurturing of a mother shapes her offspring's epigenomes and physiology.[249] The stability of traits across generations is a team effort.

The foundational concept of heredity via genetic transmission has been very productive in evolutionary biology, and population and quantitative genetics would not have developed without it. However, this does not mean that transmission genetics constitutes a complete description of the inheritance of biological features, nor that its implementation in evolutionary theory constitutes the best description of evolutionary dynamics. Biology has moved on, and progress in understanding both trait inheritance and evolutionary dynamics will be enhanced by incorporating relevant components of extragenetic inheritance into evolutionary models and experimentation.[250]

What, then, is transmitted across generations? We suggest it is the developmental means to construct phenotypes that are predicted to match anticipated environmental conditions.[251] Those "means" include not only genes but also many other resources that parents pass on to descendants, as well as activities parents engage in to construct the environmental context in which their offspring develop. If there are similarities between the traits of parents and offspring it is because, within lineages, phenotypes are reliably reconstructed across generations.[252] That reconstruction is consistent because it is informed by processes of environmental detection and selection, operating at a range of temporal and spatial scales, including, but not restricted to, the selection of genetic variation. As Waddington stressed, all organisms face "an unknown, but usually not wholly unforecastable, future."[253] The true function of heredity is to make an informed forecast.[254]

PART IV

Implications of the Developmental Perspective

11

Novelty and Innovation

Thomas Huxley was an indefatigable champion of Charles Darwin's ideas, but there was one important respect in which the two men profoundly disagreed. Darwin emphasized that new characters arose through gradual change, and ascribed apparent discontinuities in the fossil record to its imperfection. Huxley, by contrast, favored a saltationist view of evolution and regarded morphological gaps in the fossil record as genuinely reflective of evolutionary patterns.[1] From its first enunciation, the adequacy of Darwinian evolutionary theory to account for trait novelty and innovation was a major point of contention. From early religious apologists such as George Mivart, through paleontologists' advocacy of punctuated equilibria in the 1970s and 1980s, up to the contemporary challenge of some evolutionary developmental biologists, evolutionary explanations for phenotypic novelty have elicited controversy.[2]

That there should be truly foundational questions in evolutionary biology—questions that have motivated scientific interest since the mid-nineteenth century but remain to be adequately resolved in the twenty-first—is a sobering thought. *How do novel traits originate?* is such a question.[3] Many leading twentieth-century evolutionary biologists—notably Mayr, Dobzhansky, and Simpson—followed Darwin in assuming that evolutionary innovations arose through the continuous accumulation of small changes in structure, sometimes leading to novel functions. Discontinuities in form were often interpreted as reflecting the patchiness of the ecological environment, with the associated implication that only certain forms would be favored by selection.[4] Even today, evolutionary biology focuses primarily on the origin, accumulation, and potential fixation of variants of preexisting traits that differ in modest ways from ancestral forms. Together with the concept of natural selection, evolutionary biology inherited Darwin's intuition that all change arises through "descent with modification."

This way of thinking has served evolutionary biology well, but it becomes problematic when attempting to explain truly novel characters. In biology, homology commonly refers to the similarity of structure among different species based upon their descent from a common ancestor. Novel characters, almost by definition, lack obvious homology to preexisting structures, so where do they come from? We encountered this challenge before, in chapter 3, when discussing the turtle's shell, which is so dramatically different from anything else in nature that it was hard to explain how it might have arisen. At first glance, this remarkable evolutionary innovation cannot be resolved through gradual changes of shape and size, while maintaining the basic architecture of the turtle body, because a viable intermediate state seems to be impossible.[5] The same holds for the carnivorous leaves of the pitcher plant, or the light-producing organ of fireflies. Evolutionary biologists sometimes have to think out of the standard box to propose how such characters arose. A developmental perspective is necessary.[6]

The Challenge of Discontinuous Variation

The terms "novelty" and "innovation" are used differently in genetics, evolutionary biology, evolutionary developmental biology, and paleontology, sometimes referring to new structures, new functions, or new structure-function combinations, but also to traits that are characteristic of new taxa, or that have an unprecedented evolutionary or ecological impact.[7] While many commentators distinguish between "novelty" and "innovation," we see little consensus and will use these terms interchangeably. For us, the most relevant properties of evolutionary novelties and innovations are that they are distinct from, or discontinuous with, existing phenotypic variation, and that, while they certainly can evolve through the slow accumulation of small-effect mutations, they may, at least sometimes, arise through radical developmental reorganization (although these alternatives are not mutually exclusive).[8] Examples of evolutionary novelties include a new body plan; a new constructional element, such as teeth or feathers; a new character, such as a trunk; or even a new gene or cell type.[9] Our focus is generally at the trait level, but innovation can arise at all levels of biological organization; for instance, the origin of histones allowed histone tails to become available for chemical modification, which subsequently became a core feature of gene regulation, critical to both cell differentiation and epigenetic inheritance.[10]

The insect wing constitutes a classic example of an evolutionary novelty. At the phenotypic level, insect wings lack obvious homology to other insect appendages and cannot easily be understood as a refinement of preexisting structures. To make matters more difficult, because most insects have wings (the exceptions being those that have secondarily lost them, and some very primitive forms) any comparisons with wingless or partially winged forms must involve highly divergent lineages. Whatever variation once existed has been lost deep in time and is unavailable for present-day comparative evolutionary analyses.[11] Lacking both obvious correspondence to other traits and significant phenotypic variation within or across populations of closely related species, constructing an adequate framework in which to address the issue of where insect wings come from has proven challenging.[12] As a consequence, evolutionary biology has accumulated a great deal of information about the quantitative and population genetic architecture of insect wing size and shape, but comparatively little on the origin of the insect wing.[13]

Evolutionary developmental biology provides a different way of thinking about evolutionary novelty.[14] Evo-devo focuses on how traits are made during development, and how the process of building a particular trait compares to that of other traits, regardless of whether they do or do not share obvious homology. This makes a difference, because new complex traits often arise through the co-option and reuse of existing developmental circuitry.[15] Analyzing the mechanisms of development, and identifying common pathways by which characters are built, provides clues as to origins of evolutionary novelties. By opening up the "black box" and filling it with the nitty-gritty of developmental motifs, signaling pathways, and the diffusion of morphogens, evo-devo has revealed "deep homologies" that can explain the origin of novel characters.[16]

Innovation through Co-option of Developmental Circuitry

Innovation commonly arises from the recruitment of preexisting developmental modules. The insect wing provides an excellent example. The origin of the insect wing had been debated by evolutionary biologists for over 130 years. One hypothesis proposed that wings evolved as novel structures on the body wall, and another that wings evolved from preexisting outgrowths (*exites*) of the leg. These hypotheses appeared to be in conflict, but evolutionary developmental biologists have shown how both can be correct.[17] Experiments by Heather Bruce and Nipam Patel suggest that leg segments in the common ancestor of contemporary insects and crustaceans were incorporated into the body wall of the newly evolved insects, and only later were co-opted to form wings.[18] The authors conclude that: "both the leg exite and body wall theories are correct, but each is relevant to different phylogenetic time points: crustacean leg exites evolved into body wall lobes, then subsequently into wings. While wings are an outgrowth of what is now the insect body wall, they owe their origin to the leg segment of an ancestral arthropod."[19] Insect wings may exhibit a novel function, but they evolved from preexisting ancestral structures.

Moreover, once these wing-forming modules were constructed, they too could be modified. Consider again the head horns of dung beetles. These horns, which are secondary sexual traits used in competition for mates, are not modified versions of ancestral appendages, but genuine evolutionary novelties.[20] Their origin also illustrates how evolution can repurpose existing regulatory networks to scaffold innovation, and how the characteristics of such innovations are inevitably biased by the properties of those ancient networks. A recent experiment, which inhibited the expression of genes associated with wing development in dung beetle larvae and pupae, found that the central (medial prothoracic) horn was also reduced in size and split into two bilateral projections.[21] It turns out that horns arise from tissues that would have formed the top (dorsal) part of the wing, and that the wings' gene regulatory network is used to commence their development, before recruiting another set of genes unrelated to wing formation during their subsequent growth and development.[22] That wing-related genes are required for the initiation of horn development might seem remarkable, but such reuse is a general feature of development.

The head-pattern mechanisms that functioned ancestrally in the dung beetle embryo have also been co-opted to integrate novel horns into the adult head.[23] For example, the transcription factor orthodenticle (otd) plays a central role in the embryonic head development of a variety of animals. Experimental investigations have established a major role for otd in the formation and positioning of horns in dung beetles.[24] Downregulating *otd* removed horns from typically horn-bearing regions, while otd is not expressed in the corresponding regions of hornless species such as *Tribolium*.[25] These and other data suggest that components of an ancient GRN—of which *otd* is just one part—already involved in embryonic head development have been co-opted in adult head development, where they integrate horns in a manner that does not compromise overall head patterning.[26] The reuse of this preexisting circuitry provides a built-in mechanism for spatial specification, and as a result biases the position of the horns to certain locations.[27] That there is a disproportionate abundance of species in this group with paired posterior horns in these precise locations is almost certainly no coincidence.[28] Such reuse may not only have facilitated the initial positioning of horns during early *Onthophagus* evolution but also have biased subsequent diversification of horn positions across the entire clade.[29] The beetle horn exemplifies how many novel structures can evolve by reuse of preexisting parts, and how this co-option can produce developmental bias that shapes the diversification of a clade.[30]

Dung beetles also show how disrupting the self-inhibition of developmental modules can lead to striking evolutionary innovations. When Eduardo Zattara and colleagues experimentally downregulated the expression of *otd*, an entirely novel compound eye was produced in the middle of the beetle forehead.[31] Even more impressive was the finding, from behavioral tests, that the new eye was at least partially functional and integrated with the central nervous system.[32] Perturbing the level of one transcription factor cannot possibly explain all aspects of the development of this highly complex organ. Rather, the finding illustrates how simple developmental signals can switch ancient circuitry on and off, with the effect of incorporating core processes into development, or removing other processes, resulting in novel structures. If a novel compound eye were to be favored by selection, there would be no requirement for all the individual components of the new eye to coevolve: in one fell swoop an existing eye-making package can be taken "off the shelf" and utilized in a novel location.

Evolutionary innovation in the construction of analogous structures and organs in unrelated organisms can involve the repeated co-option of the same developmental pathways.[33] Some of the most spectacular cases include the independent evolution of eyes across diverse animal phyla, the evolution of contractile hearts in both vertebrates and invertebrates, and the formation of outgrowths, from the legs of insects to the tube feet of echinoderms or the siphons of sea squirts.[34] In each instance, reuse of preexisting genes, pathways, and morphogenetic processes has generated functionally similar outcomes (although subsequent lineage-specific modifications can be so drastic that the structures cannot be diagnosed as homologous in the classical sense). Rather than implying constraint, such cases illustrate how developmental systems can actually create evolutionary trajectories, biasing adaptive evolution toward specific complex structures independently of taxonomic context.[35] The result is that the number of genetic changes needed to evolve an eye, heart, or appendage is significantly fewer than what one might imagine without knowledge of the developmental circuitry.

Innovation through Developmental Plasticity

Phenotypic plasticity is a ubiquitous aspect of development, not least because living organisms are made up of materials that change their shapes and properties in response to environmental conditions. Temperature changes, for instance, can cause proteins to fold differently, thus determining whether or not a particular gene is expressed, while fluctuations in pressure can affect the assembly and dynamic properties of the molecules of a cell.[36] In this way, environmental triggers can elicit changes in phenotypes.

The scale and sensitivity of plastic responses is also subject to natural selection, which harnesses and stabilizes these responses and the physical properties of molecules and cells.[37] For instance, cells become stickier at higher temperatures, but selection appears to have favored specific cell mechanics and membrane-cytoskeleton interaction that modulate and stabilize cell adhesion and ensure robust patterns of cellular aggregation during development at a range of temperatures.[38] Environmentally induced cellular stress plausibly induced an alternative regulatory state, which elicited new cell types that have subsequently been fine-tuned and stabilized by selection.[39] Likewise, the simplest way that cells in a tissue mass can be induced to differentiate is by responding differentially to different concentrations of diffusible proteins.[40] Subsequent natural selection could have reinforced the reliability of those phenotypic changes that enhanced survival, so the transport of morphogens would not be solely due to diffusion but also dependent on active cellular processes.[41]

Many innovations in animal body plans are thought to have arisen through selection that mobilized physical forces and preexisting plastic responses, which stabilized their use in development.[42] Plasticity-led evolution is almost certainly not a recent development, but an inherent feature of evolution since the origin of life.[43] The evolution of form has probably always been biased by the physical, chemical, and material properties of its constituent substances.[44] For instance, early metazoan forms were probably highly environment-dependent multicellular assemblies that acquired stable patterns of development and evolutionary robustness only subsequently, through stabilizing selection.[45]

Once environment-sensitive developmental pathways and circuits evolved, they were able to be repeatedly reused. Again, the horned dung beetles provide an example. The insulin/insulin-like signaling pathway (IIS) is a highly conserved pathway well-known for regulating growth in response to nutrition.[46] In insects, rich nutritional environments cause insulin-producing cells in the brain to secrete insulin-like peptides that bind to receptors in target tissues and induce cell growth and proliferation. Many tissues respond plastically to diet in this manner. The evolution of head horns involved the co-option of this preexisting pathway, rendering horn formation responsive to nutrition—a characteristic of many secondary sexual characters.[47] Furthermore, with the doublesex gene (which is spliced in a sex-specific manner, producing sexually dimorphic trait development) recruited into the horn regulatory network and linked to IIS, horns evolved nutrition-responsive growth in males but not females. This innovation is thought to be central to the dramatic radiation in sexual dimorphisms and polyphenisms that characterizes these beetles.[48] Any evolutionary innovation in some cells or tissues that reuses a preexisting developmental mechanism inevitably generates developmental bias, since reliance on the same circuitry generates covariation (i.e., horn size covaries with other nutritionally sensitive tissues reliant on the IIS pathway).

Predicting all the functions that a biological structure might have, and therefore what might be selected, is virtually impossible,[49] and changes in development can yield unexpected new phenotypes. Consider the development of eusociality in ants. Evidence from developmental biology shows that wing polyphenisms can be generated at many levels, with environmental signals interacting with wing-forming GRNs.[50] In those larvae destined to be queens, the insulin-signaling pathway activates the wing imaginal discs to proliferate and differentiate, while those larvae destined to become workers have their wing-forming GRN silenced by having less nutrition, so that their wing discs degenerate. Silencing the wing GRN prevents workers from participating in mating flights, thereby potentially creating a reproductive division of labor. That wing development uses nutrition-sensitive circuitry may not only have led to the evolution of wing polyphenisms but also have contributed to the origin or maintenance of their eusocial condition.[51] Following the origin of wing polyphenism and eusociality in ants, a diverse array of mating strategies subsequently evolved.[52] There is considerable current excitement over the hypotheses that the evolutionary origins of eusociality may be rooted in the developmental plasticity of ancestral groups, that polyphenisms in social behavior can become encoded in the genome through the reuse of regulatory regions, and that the evolution of various aspects of worker social behavior could have been biased by this reuse of key developmental genes and pathways.[53]

Conversely, when formerly adaptive developmental pathways are silenced by mutation they are no longer the direct target of selection and can remain latent for a period of time until once-again triggered by genetic or environmental change. For example, in some species of ants there are large-headed

FIGURE 13. Plasticity and innovation in bichir fish. Bichir fish raised on land begin to walk faster and more efficiently than aquatic-reared fish. Their bones and muscles change shape to allow these movements, mimicking changes seen in the fossil record. Experimental studies imply that developmental plasticity may have facilitated the transition to walking on land.

"supersoldiers," in addition to the usual soldier and worker castes. Supersoldiers use their heads to block the entrances to nests during army ant raids. The queens are known to induce different castes through differential emission of juvenile hormone, and administration of a chemical mimic of juvenile hormone to related species that do not naturally produce supersoldiers, but whose ancestors did, elicited the supersoldier caste in one-third of these species.[54] This illustrates how cryptic developmental pathways may retain the potential for expression that can be rereleased through an environmental challenge (or mutation).[55] Many adaptations that appear to have arisen rapidly have effectively had a hidden head start.[56]

Coordinated plastic adjustments of the skeleton, musculature, and behavior may have been critical for animals to negotiate severe environmental challenges, such as the transition from water to land.[57] Bichirs are African eel-like fish with functional lungs and pectoral fins that they can use to drag themselves across land (figure 13)[58]—traits they share with fishes that moved onto land four hundred million years ago. An experiment that raised bichirs in ter-restrial conditions found that, compared with aquatic-reared fishes, they rap-idly began to walk faster and more efficiently, with their heads raised and their fins close to their bodies.[59] The neck and shoulder bones and muscles of land-reared fish changed shape to allow these movements, in a way that mirrored changes seen in the fossil record of fishes that evolved into amphibians. The authors conclude that "environmental induced developmental plasticity facili-tated the origin of the terrestrial traits that led to tetrapods."[60] As we described in chapter 8, such examples demonstrate that plastic responses can generate patterns of variation that are functional and hence enable initial adaptive di-vergence. Were the focus solely on structure, bichirs might be regarded as compatible with a gradualist view of novelty. However, the key point here is that environmentally induced changes in bichir fins facilitated the use of these fins for a novel function. The importance of developmental plasticity here is not that it generates radically new structures but rather that it enables flexible and integrated changes in structure that support a radically new function—walking on land.

Innovation through Symbionts

Symbionts play a key role in evolutionary innovation, often stimulating nov-elty by making new environmental resources accessible to the host, but trap-ping their host into a particular way of life in the process. For instance, for over 150 million years the pea aphid has had an obligate symbiont, the bacterium *Buchnera aphidocola*. The aphid needs the symbiont to produce essential amino acids from its nutrient-poor diet of phloem and xylem fluids. The bac-terium has allowed the aphid to become a specialist in drinking sap. Aphid mouthparts evolved into a needlelike structure called a *stylet*, which they insert into the sap-filled plant tissues to suck up fluids. While acquiring symbionts allowed these insects to open up a new ecological niche, their specialist mouthparts now lock them into this lifestyle.[61] Similarly, cattle are dependent on a diet of plants that can be digested and fermented by the symbionts in their rumen. The ability to digest plant material allowed cattle to graze pastures that other animals could not feed on. This promoted the evolution of teeth and intestines suited to the grinding and digestion of plant material. Now the feet and mouths of cattle are no longer capable of catching and killing animals, and they are obligate herbivores.[62] In this way, microbial symbionts can bias the possible evolutionary trajectories of their hosts.

Viral symbionts are also an important source of evolutionary innovation. Endogenous retroviruses are DNA sequences of viral origin—transposable elements that are now part of the host genome. Viruses bring new DNA into the genome, which can serve as enhancers, promoters, or protein-encoding genes. About 5–8 percent of the human genome consists of retroviruses, and about 50% of the human genome consists of sequences originating from viruses.[63] In eutherian (placental) mammals, the uterus has evolved a new cell type, the uterine decidual cell, which is capable of responding to progesterone and thereby facilitating the embedding of the embryo and coordinating its maintenance throughout pregnancy. The uterine decidual cell type evolved through the insertion of transposable elements carrying the binding sites for specific transcription factors into regions of the genome that could be regulated by progesterone and cyclic AMP molecules. The result was the construction of a new GRN that coordinated gene expression by progesterone and cyclic AMP signaling. This network is what makes pregnancy possible, and it owes its origin to viruses.[64]

Innovation through Threshold Effects

There are many situations in biology where a quantitative difference becomes qualitative, as a threshold is crossed and what would have activated one set of reactions now activates others.[65] This threshold concept is well-known for morphogens such as the bicoid transcription factor, which plays a critical role in the formation of the anterior–posterior axis in insects. At high concentrations, bicoid induces heads, at medium concentrations it induces thoraxes, and in low concentrations it leads to the formation of abdomens.[66]

Sometimes novelties arise when the regulatory network or cellular interactions of developing organisms experience such threshold effects. Here, the amount of some activating or inhibiting substance, or the number or volume of cells in a tissue, passes some critical threshold, tipping the regulatory network toward a new phenotypic outcome, and an entirely new organ can result. Polydactyly provides an example.

Humans, and many other vertebrates, including cats, dogs, and guinea pigs, are occasionally born with more than five fingers or toes on each hand or foot (having five is the "ground state"). The condition can be triggered by a point mutation in a *cis*-regulatory element that affects the sonic hedgehog (*Shh*) gene (the same gene whose overexpression caused the absence of eyes in cave fish; see chapter 8).[67] As with the dung beetle's ectopic eye, the extra digits are often fully functional, equipped with bones, muscles, nerves, and blood vessels. The mutation affects the expression of *Shh* in a manner that leads to a continued proliferation of cells in the embryonic limb bud: more cells means more raw material for making fingers and toes.[68] However, individuals with the point mutation can have two, four, six, or even eight extra digits, and genetics alone does not tell us how many. Again, we need to consider higher-level developmental features—here, the dynamical interactions among cells in tissues—to explain this.

As described in chapter 5, mathematician Alan Turing used reaction-diffusion models to show that two or more diffusing substances can interact to generate wavelike patterns that are able to explain a host of features in nature, including spots, stripes, and vertebrate limb development.[69] The diffusion of interacting morphogens in the developing limb bud generates waves of cell activity, leading to the production of stripes of precartilage, from which cartilage and bone will later form. Because a dynamical system with a fixed wavelength constrained to a bounded region must produce a discrete number of waves, the result is a discrete number of digits. There cannot be half of a finger or toe: either there will be enough cellular material for a whole new digit or none is produced, although extra digits can be small in size.[70] Such "threshold effects" are characteristic of the interactions between cellular and intercellular signals, which generate a host of patterns in animal tissues.[71]

Vertebrate digit patterning has been subject to extensive experimentation and theoretical analyses using Turing models, which provide strong support for this "self-organization" explanation.[72] In addition to increasing the amount of limb-bud tissue to be divided up into digits, a second way in which more fingers and toes can arise is by reducing the wavelength of the Turing pattern and thereby generating thinner digits. Sure enough, in embryonic mice, experimental manipulation of the dose of distal *Hox* genes, which modulates the wavelength of the Turing-type mechanism, was found to generate progressively more severe polydactyly.[73] Other experimental investigations implicate the genes *Bmp* and *Wnt* as playing activating and inhibiting roles in the patterning,[74] while theoretical models based on data from polydactylous cats suggested that the sensitivity of cells to switching state may be the principal determinant of how many digits are added to the ground state.[75] In the latter case, increasing the sensitivity of cells to change in state and advance cartilage formation led to a threshold being passed, and the model suggests that one or more extra digits would be generated.[76] Ultimately, the interactions between cells determine cell reactivity and hence the number of digits produced.[77] In simulations, manipulating the wavelength or number of cells produces a shift from the small number of digits found in most mammals, through the larger number of elements found in polydactylous limbs, to the very large number of skeletal elements in dolphin fins.[78] Other studies suggest that tweaking Turing reaction-diffusion systems can alter the number of cusps in mammalian teeth, the stripes of angelfish, and the coat patterns of mammals.[79] A Turing dynamic patterning system is also what is responsible for the unending variation among human fingerprints.[80] Thus, it appears that a self-organizing Turing-type mechanism is deeply conserved in tetrapod phylogeny.[81]

There are several important aspects of these findings. First, they go against the expectation, dating back to Darwin, that changes must be continuous and of small effect to be viable.[82] For instance, a gradualistic account might anticipate that any increase of limb bud cell number or reaction rate would result in the existing digits become longer or wider, or that a tiny rudiment of an incipient digit would appear upon which selection could act and eventually lead to an additional digit after long periods of selection. Instead, extra, full-size, and often functional digits appear instantaneously.

Second, genes do not directly specify the number of digits. While a mutation can trigger the production of extra digits, it is not directly responsible for their construction, which involves many genes and signaling pathways, as well as environmental factors; nor does the mutation dictate how many fingers or toes there will be, which varies among individuals with the same genotype.[83] Rather, the mutation throws a switch in the existing developmental architecture, making a difference in the final outcome; it is the cause of the difference, not the cause of fingers and toes.[84] The integrated ensemble of bone, muscle, tendon, nerves, and blood vessels that constitutes these organs is produced by a complex regulatory system, and control over the number of digits lies at the level of communication between cells, and interactions between internal and external regulatory elements. Any environmental perturbation that increases the number of cells in the limb bud would have the same effect as the mutation, while environmental factors that kill cells commonly lead to fewer fingers.[85]

Third, phenotypic variation in digit number is extremely biased in its distribution. For instance, it is more likely to have two extra digits per individual than one, since these changes mostly occur symmetrically on both the left and right sides of the body. In addition, two extra digits is more likely than four, and four more likely than six or eight. Developmental mechanisms can explain these patterns.[86] A smaller increase in cell reactivity is required to generate two extra digits than four, four than six, and so forth for larger numbers of fingers and toes.[87] In chapter 12 we will describe theoretical analyses showing that natural selection will strengthen the epistatic interactions between genes that produce useful symmetrical correspondences in limb properties, which is why odd numbers are hard to generate.[88]

Fourth, these biases, combined with developmental mechanisms, help to explain the taxonomic distribution of digit number, and patterns of gain and loss. In many vertebrates digits become smaller, but below a threshold size the affected digits do not typically decrease in size gradually; rather, there is a single-step deletion of entire digits. In repeated independent lineages of Australian skinks, for instance, evolutionary reductions of limb size seem to have occurred systematically through a stepwise disappearance of complete individual digits, which always followed a specific sequence that is the reverse of the pattern of their appearance in reptilian development.[89] Digit loss in crocodiles and birds can be understood in a similar way.[90] These findings help explain why certain limb patterns and digit numbers are more reachable than others, as well as the sequence and position of gains and losses, and sheds important light on the evolvability of these characters.[91] We return to these issues in chapter 12.

Understanding Macroevolutionary Patterns

That virtually all major (i.e., phylum-level) body plans were established in the Cambrian explosion, just over five hundred million years ago, has been one of the most intriguing evolutionary conundrums for decades.[92] An understanding of regulatory interactions in development is beginning to help scientists to explain this, together with some other enigmas concerning the emergence of phenotypic diversity over time.[93] Despite the immense diversity of animal forms that has arisen since the Cambrian, no radically different animal morphologies have appeared. In 2006, Davidson and Erwin suggested that the rapid evolutionary diversification of body plans during the Cambrian was caused by the evolution of particular GRNs.[94] They were able to identify component parts, or modules, of animal GRNs (called *kernels*), which play key roles in the development of body parts and as a consequence became highly conserved. The argument was received with skepticism by some commentators, in part because it was not clear why major classificatory divisions should necessarily occur early in the tree of life.[95] However, computational analyses of regulatory networks have subsequently suggested a possible explanation for this conundrum. Simple regulatory networks typically produce bigger differences among common phenotypes than do more complex networks,[96] confirming that the diversification rate should have been highest early in evolutionary time when regulatory networks were small and less elaborate. If this theory is correct, developmental bias will increase over evolutionary time, as networks get "locked in" to produce particular classes of phenotype, and lineages are predicted to become increasingly clumped as evolution progresses, with the greatest divergences appearing early and at higher-level (i.e., deeper) taxonomic grades.[97] These predictions are consistent with the early bursts of radiation seen across several metazoan taxa, including tetrapods and arthropods, with a similar pattern observed in some plants.[98]

The same reasoning can be applied more generally to explain why particular features of organisms are conserved and how developmental regulation channels phenotypic variation.[99] As explained above, the evolution of novelties, such as shells, limbs, eyes, feathers, wing patterns, or horns, is associated with rewiring existing developmental building blocks into new regulatory networks.[100] The theory in the previous paragraph predicts that, once novelties appear, their diversification should proceed rapidly at first, and slow down as their regulation becomes more developmentally entrenched.[101] Consistent with this, the shape of bird bills diverged rapidly during early avian radiation, while bill shapes within major bird lineages have subsequently occupied only limited regions of morphospace.[102] Mathematical analyses of bird bills and experimental manipulation of bill growth confirm that much of the observed diversity in shape can be explained by changes in a small number of regulatory interactions among key genes,[103] suggesting that morphospace may remain largely empty as a consequence of how bill development is regulated.[104]

Some novel traits have a wide ecological impact.[105] Many organisms actively modify their environments, including by both trophic and nontrophic activities, sometimes with dramatic impacts that trigger ecological and evolutionary changes in other species.[106] Some innovative forms of such niche-constructing activities can create opportunities for other species, leading to evolutionary change, taxonomic diversification, and/or changes in the structure of ecological networks.[107] Experiments have shown that niche construction can evolve rapidly, under a broad range of conditions,[108] leading to the creation of new niches. For instance, an experimental evolution investigation in bacteria by Magdalena San Roman and Andreas Wagner at the University of Zurich showed that huge biodiversity could emerge in a completely homogenous environment through niche construction.[109] They found that bacteria created new ecological niches when they excreted nutrient-rich waste products that could sustain other bacteria, and they identified thousands of such niches that had been created in this manner. Rather than lineages simply diversifying to exploit available opportunities, niches themselves were diversifying.[110] In addition, theory shows that niche-constructing effects that persist in the environment as an ecological inheritance can cause evolutionary time lags, and generate long-term, macroevolutionary effects; for instance, increasing the branching patterns of phylogenetic trees, with profound effects on the rate of genetic change.[111]

For decades biologists have debated whether macroevolutionary patterns can be understood solely through microevolutionary processes,[112] with different academic fields (e.g., genetics, paleontology, developmental biology) often reaching opposite conclusions. Punctuated equilibria in the fossil record are a case in point: evolutionary geneticists are frequently happy to interpret stasis as resulting from stabilizing selection, while paleontologists often explain the same patterns as products of developmental constraints. Computational analyses of simulated regulatory networks reveal that these should not be considered alternative explanations, and that both processes likely operate together.[113] Stabilizing selection will push evolving populations to regions of genotype space where changes in GRN topology do not affect the phenotype, which generates a bias for established phenotypes that reinforces selection and supports stasis. At the same time, disruptive selection shifts populations to regions of genotype space that allow rapid change,[114] generating a developmental bias that promotes phenotypic change and allows stasis to be punctuated with bursts of greater change than is likely to arise from the natural selection of unbiased variation.[115] The network models, which are shedding new light on the origins of evolutionary innovation,[116] make it difficult to believe that selection would fail to generate developmental bias, and that developmental bias would fail to channel selection. Both arguments are probably at least partly correct, but a complete explanation is likely to be richer than is encompassed by either.

Peeling the Onion

Perhaps the most important contribution that the field of evo-devo has made to the study of evolutionary novelty is to recognize novelty as a distinctive and significant problem in its own right, one that demands serious scientific attention.[117] This developmentalist tradition within evolutionary biology is exemplified by the work of Conrad Waddington, Stephen Jay Gould, and Mary Jane West-Eberhard. It is in marked contrast to the population and quantitative genetic tradition, which has tended to assume that continuous genetic variation, expressed in gradual genetic change, accounts for phenotypic evolution at both micro- and macroevolutionary scales.[118]

Once homologous features have been identified, a great deal of the mystery associated with evolutionary novelty is resolved. Before evo-devo, the traits of organisms tended to be regarded as either homologous or not. However, diving into the black box and investigating the mechanisms and organizational principles of development has revealed unanticipated complexity in the historical relations among species. Characters became like onions—layered, from genes to pathways to cell types to tissues to organs—with homology potentially manifest at any and every level.[119] Homology is a challenging concept, and can be pinned down only when how organisms build themselves during development is carefully unpacked.[120] What are clearly nonhomologous traits—for instance, beetle horns and insect legs—can involve the co-option and redeployment during evolution of the same structures and circuitry. This "borrowing" can create striking evolutionary novelty.

12

The Developmental Origins
of Evolvability

Nearly three billion years elapsed from the origin of life until complex multi-cellular organisms first appeared on Earth, approximately six hundred million years ago. Given this slow start, the pace of evolution ever since has been breathtaking. Today, on directly observable time scales, antibiotic and pesticide resistance are reminders that evolution can be both fast and relentless, yet the current wave of extinctions arising from human activities also shows us that organisms may fail to evolve fast enough to adapt.

These observations raise two major questions. First, do organisms or their characters differ in the potential for adaptation and diversification—that is, do they differ in their *evolvability*?[1] Second, does evolvability itself evolve, such that organisms can become better or worse at evolving? In general terms, evolvability refers to *the capacity, ability, or potential of systems to evolve.*[2] However, the capacity for adaptive evolution and diversification is far too complex to be captured by a single variable, and, perhaps as a result, evolvability has multiple meanings.[3] Concern about the prospects of a population facing climate change, where evolutionary rescue may be possible through an adaptive change in allele frequencies or the phenotypic mean, ties evolvability closely to the heritable variation that is present in that population. At the same time, as we saw in chapter 11's discussion of novel phenotypes, evolvability appears more to do with how readily developmental processes allow transformation of an existing phenotype into something new.

The properties that confer evolvability are not static, because they are themselves products of past evolution. There are many dimensions to evolvability: a population rescued from extinction by its response to selection will have lost genetic variation; the evolution of a rigid shell may have blocked off

189

some evolutionary paths for turtles, but opened up others; the origin of multicellularity created entirely new entities capable of undergoing evolution by natural selection; evolution has given humans capacities for social learning and cultural transmission that allow us to adapt through cumulative cultural change without individual differences in fitness. For "evolvability" to capture all these related ideas, acknowledging diverse mechanisms and exploring their impact on evolutionary change will be necessary.

In earlier chapters, we discussed how development gives directionality to adaptive evolution by structuring the phenotypic variation available for selection. In this chapter, we will explore further this relationship between developmental processes and the heritable phenotypic variation that enables a population to respond to selection. Evolvability concepts that foreground the capacity of populations to respond to selection have often been treated separately from evolvability concepts related to the generation of adaptive phenotypes in development, sometimes justified by the argument that they concern different time scales. However, population and developmental perspectives are inevitably linked, even on short time scales, since developmental processes determine how heritable variation translates into potentially adaptive phenotypes. Moreover, as we explain in the second part of this chapter, connecting developmental and population concepts of evolvability is critical to understanding how some phenomena discussed in previous chapters—including developmental bias, phenotypic plasticity, niche construction, and extragenetic inheritance—contribute to the capacity of organisms to evolve, adapt, and diversify.

The Introduction of Genetic Variation

The capacity of a trait to evolve is commonly thought to depend primarily on how much additive genetic variation there is for that trait, in which case evolvability reduces to the concept of heritability, or to related concepts such as the genetic coefficient of variation.[4] Given that molecular genetic studies typically find ample standing genetic variation underlying traits in natural populations,[5] it may seem reasonable to assume that the direction of adaptive evolution would be little affected by the processes that introduce variation.[6] This conclusion would be misleading, however, since not all phenotypic variants are possible, or even likely. The amount of standing genetic variation in each trait does not provide much information on how likely particular *trait combinations* are,[7] since, as laboratory selection demonstrates, it is possible to get responses to selection for most traits, but not for most combinations of traits—for example, the selection for eyespot size and color in studies of *Bicyclus* butterflies (see chapter 7).[8]

Given that adaptive change relies on heritable phenotypic variation in the direction of selection, it becomes important to quantify not just how much variation there is, but also how traits vary together.[9] As explained in chapter 2, evolutionary genetics summarizes the extent to which different traits vary together in a heritable manner using a variance-covariance matrix (i.e., $\mathbf{G}$).[10] Because the off-diagonal elements of $\mathbf{G}$ represent the heritable covariance between two traits, their magnitude will be relevant for their joint response to selection. One measure of the capacity for adaptive change, or evolvability, is thus the response of the population mean phenotype in the direction of the selection gradient. If traits are uncorrelated, the direction of change in one trait is fully determined by selection on that trait, whereas that trait's response will be expected to follow a different trajectory if it is correlated with another trait. High covariance has therefore been interpreted as indicating how one trait "constrains" the evolution of another trait under directional selection.[11]

Heritable trait variances and covariances can be estimated by recording the trait values of related individuals, and animal and plant breeders have made good use of the $\mathbf{G}$ matrix when predicting responses to selection. The success in predicting evolutionary responses in the wild has been more mixed,[12] partly because a statistical summary of the heritable phenotypic variation can be affected by short-term changes in the genetic composition of populations, due to inbreeding, migration, mutation, recombination, or selection.[13] Moreover, the consequences of genetic change for the phenotypic variation that is available to selection depend on the mechanistic details of development. For example, in chapter 7 we saw that the biology of neural crest cells ties together a suite of characters that otherwise might seem to be completely unrelated.[14] This is significant, because the same pattern of statistical covariation between traits could emerge from very different developmental mechanisms;[15] however, it does not follow that these mechanisms have the same capacity to generate phenotypic variation (i.e., they may differ in their variational properties). Yet, genetic evolvability depends not only on the standing variation in a population but also on how mutation translates into phenotypic variation (recall that, in quantitative genetics, this is referred to as the $\mathbf{M}$ matrix; see chapter 7). Plausibly, as selection uses up standing genetic variation, $\mathbf{G}$ will increasingly resemble $\mathbf{M}$, and hence $\mathbf{M}$ may provide a better estimate of evolvability over longer time spans.[16]

A remarkable study of mutation and selection in fruit flies illustrates how such developmental biases can direct evolutionary divergence over millions of years. David Houle and colleagues collected data on over fifty thousand fruit fly wings and found strong relationships between the variation produced by mutation, the standing genetic variation, and the rate of evolution in *Drosophila* wing patterns over the last forty million years. Although virtually all features of the *Drosophila* wing are variable in nature,[17] mutations disproportionately generated correlated effects on different parts of the wing, resulting in some shapes being more readily produced than others.[18] Moreover, the principal axes of variability were the very same dimensions along which fly wings had evolved. Fly wing evolution could be predicted with surprising accuracy from knowledge of patterns of phenotypic variability—the wing regions with the most mutational variation were the ones that evolved the most.[19] The authors describe their findings as "an important challenge for evolutionary theorists."[20] Evolution appears to have followed pathways reflecting the rate of introduction of phenotypic variation. Why should this be?

The Genetic Bases of Reciprocal Causation

The mutational effects that compose the **M** matrix, which feed genetic variation into a population, depend on patterns of pleiotropy and epistasis.[21] These are themselves largely properties of developmental systems and hence are evolvable characters. This means that trait covariance can change in response to selection, in ways that affect not just the rate at which new mutations alter a single trait but also the probability that traits vary together. At first sight, the existence of adaptive developmental bias may seem unlikely, as the selective advantage of responding in any particular way to random genetic change is small on short time scales.[22] Thus, even though it may be useful in the long run, natural selection for evolvability should be rather weak. However, developmental interactions can evolve as a result of consistent selection for particular combinations of traits or trait values. This, in turn, will influence how genetic variation translates into phenotypic covariation.

The evolution of phenotypic variability has been modelled using a diverse set of tools, ranging from population and quantitative genetics models to computational models of the evolution of regulatory networks. These models demonstrate that various selection regimes can produce patterns of pleiotropy and epistasis that align the phenotypic effects of mutations with the direction of the fitness landscape.[23] In other words, traits that are selected together because they function together will tend to become genetically and developmentally integrated, forming correlated clusters whose variability significantly impacts fitness.[24] This suggests that when new mutations arise, even though they occur at random, they may nonetheless bias the phenotype toward traits and trait combinations favored by past selection.[25] For instance, mutations might increase or decrease the length of both forelimbs, perhaps with correlated effects on both hind limbs, but are less likely to generate left and right limbs of different length. This may seem obvious, but there is of course an evolutionary explanation: past selection of longer or shorter limbs probably strengthened the epistatic interactions between genes that produce useful symmetrical correspondences in limb length, while reducing the strength of gene interactions that produce dysfunctional asymmetries.[26]

Models of the evolution of regulatory networks also suggest that selection can modify the degree of modularity in development. One consequence of this is that key "sensor" genes that mutate to produce adaptive phenotypic changes will evolve to become hubs of the GRN, whereas other genes will evolve to assume more peripheral roles.[27] Mutations generating changes in the level or timing of gene expression can potentially lead to dramatic changes in phenotype, while the majority of mutations have little effect on phenotypes. Several lines of theory show that the sensitivity of a trait to mutation can evolve in this way,[28] and experimental studies of evolution in yeast and bacteria confirm these findings, with at least one study reporting that the GRNs of laboratory populations evolve to become more efficient at generating adaptive mutations.[29]

The theory on the evolution of evolvability has profound implications. First, although mutations occur at random (i.e., *not* in response to need), there should be a tendency for phenotypic variation to be biased toward what was useful in the past.[30] Second, if the above theory is broadly correct, selection can modify regulatory networks in ways that allow extensive adaptive change to be achieved with a small number of regulatory mutations, rather than a long sequence of small changes.[31] These biases mean that organisms can generate new adaptive heritable phenotypic variation; not all the time, but nonetheless frequently enough to enable populations to evolve.[32]

Consider the famous adaptive radiation of Darwin's finches. Evolutionary developmental biologist Arkhat Abzhanov and his colleagues demonstrated a striking relationship between the amount and timing of expression of a single gene—*Bmp4*, which codes for a protein involved in bone and cartilage development—and the beak shape of fourteen species of Darwin's finches.[33] Species with deeper, broader beaks express *Bmp4* in their embryonic beak tissue at higher levels and at earlier stages. Changing *Bmp4* expression experimentally in a chicken model produced the same pattern of effects. A second study established that the level of expression of another gene (calmodulin) in the embryonic beak was strongly correlated with adult beak length.[34] By demonstrating a relatively simple developmental mechanism for integrated evolutionary transformations in beak depth, breadth, and length, Abzhanov and his colleagues helped to explain why this trait is so evolvable.[35] However, the renowned evolvability of bird beaks is perhaps no coincidence. The above network analyses imply that we may expect key developmental genes to be hubs in GRNs, and that the structure of these particular networks reflects the history of selection of different beak shapes among the ancestors of Darwin's finches. That beaks can evolve rapidly within this morphospace might seem impressive, but the same logic also suggests that in other instances developmental biases can make it difficult for evolution to find adaptive solutions that are very different from those that have been selected in the past. That forty million years of fly wing evolution should have proceeded along narrowly defined developmental lines is perhaps one consequence of this.

Network analyses further imply that evolvability commonly evolves through a restructuring of regulatory networks. It is not only the genetic sequence but rather a higher-level property—the *regulatory network* (a kind of *G-P map*; see chapter 6)[36]—that determines evolvability. The balance of positive and negative influences in the regulatory network has been found to influence whether mutations alter network dynamics, and hence what phenotype is generated.[37] Thus, a developmental property—the structure and dynamics of the underlying regulation of a trait—is a likely principal factor in a trait's evolvability.[38] Perhaps surprisingly, analyses using simulated GRNs also suggest that organisms may be able to produce viable phenotypes even when confronted with novel conditions. For instance, developmental interactions can facilitate adaptation to more extreme environments than hitherto encountered by a population because correlational selection in the environments to which the population has adapted makes it relatively easy for development to generate phenotypes that are fit in those extreme conditions.[39] Moreover, when challenged with multiple environments that have distinct features, network models evolve a modular structure, allowing them to adapt to new combinations of features, and to generalize flexibly from phenotypes selected in the past.[40] Provided the novel environments share some structural regularities with those previously encountered, the system can generate new but nonetheless adaptive phenotypes.[41]

In sum, theory suggests that evolvability can itself evolve, and organisms really can get better at adapting, but it may be easier to re-evolve past adaptations than to generate phenotypes suited to radically new environments. The selective history that biases both genetic and phenotypic variation along major axes of correlated phenotypic variation renders certain regions of phenotype space relatively easy to access, but large regions almost impossible to reach.[42] It may be easier to evolve more extreme phenotypes in line with existing clusters of correlated traits than to come up with phenotypic alterations that are uncorrelated with existing variation, with developmental biases generating differences in trait evolvability and channeling future selection. As we saw in chapter 7, this is exactly what has been found in mycalesine butterflies, where most of the three hundred species fall along a developmentally favored axis of correlated eyespot color patterns, but one genus, *Heteropsis*, has evolved the ability to control each eyespot's color independently.[43] Evolvability theory may also help to explain why there appears to be so much parallel evolution. An example is provided by the repeated adaptation to postglacial lakes of three-spined sticklebacks via pelvis loss, which arises through recurrent alterations in a key developmental control gene.[44] Theoretical findings lead us to speculate that ancestral selection in a population subject to repeated invasions into freshwater may have shaped stickleback GRNs so that, today, a dramatic change in phenotype can be achieved in a single mutational step.[45]

Traditionally, the biasing effects of mutation were conceived of as transient brakes on adaptive evolution.[46] In practice, because phenotypic variation is clustered, there is reduced dimensionality to selection,[47] but this is not well described as constraint (chapter 7). Selection is not powerful *in spite of* genetic correlation; rather, selection can be powerful *because* developmental bias evolves, leading to facilitated variation and enhanced evolvability in what were historically adaptively relevant dimensions.[48] Moreover, the theoretical and empirical findings discussed in this section highlight an additional important general point made repeatedly in this book—there is an important role for feedback in evolutionary explanation. Variation is the raw material of selection, but it is more than that: it also determines the direction and speed of response to selection. At the same time, natural selection modifies genetic and developmental interactions to generate major axes of phenotypic variation aligned with ancestral selection. Neither evolution nor evolvability can be attributed solely to selection, or solely to variational biases. Variation and selection are reciprocal causes of each other, and it is because of this interdependency that evolvability is evolving.

Some Additional Determinants of Evolvability

We take it as given that genetics plays a central role in conceptions of evolvability.[49] Here we go on to consider some relatively neglected components of evolvability that focus more broadly on the developmental properties of organisms. While their impact on the response to selection is not yet well understood, their investigation is likely to be critical to the integration of developmental and population conceptions of evolvability.

The Role of Plasticity

Environmental changes are also an important source of developmental bias and variation in evolvability.[50] Even if environmentally induced phenotypes are not heritable, plasticity has the potential to expedite adaptation and affect evolvability by facilitating the accumulation and subsequent expression of cryptic genetic variation, and by influencing when and where phenotypes arise and which traits are expressed together.[51] For instance, in previous chapters we discussed how marine three-spined sticklebacks reared on typical freshwater bottom-dwelling and mid-water foods develop a morphology that resembles the freshwater forms, with the former exhibiting a shortening and the latter a lengthening of the jaw.[52] We also saw how sticklebacks respond to temperature changes in a virtually identical manner, with longer jaws elicited by increases in temperature and shorter jaws by colder conditions.[53] This repeated reliance on the same developmental pathway may explain why there often appears to be a restricted number of phenotypic outcomes to both genetic mutation and environmental challenges.[54] Likewise, the repeated adaptive radiations of cichlid fishes in the African Rift Valley lakes—a striking coincidence when the explanation is restricted to convergent selection acting on unbiased variation[55]—can be readily understood as highly evolvable forms arising from consistent plastic responses to diet and other environmental features.[56]

The blind Mexican cave fish discussed in chapter 8 provide another example. Recall how when their sighted surface fish cousins were reared in the dark, some of their traits shifted in the direction of the blind cave fish phenotype (e.g., metabolic rate), while others shifted in the opposite direction, but with increased phenotypic variation overall (e.g., eye size).[57] Either way, plasticity shifted the distribution of selectable phenotypes in adaptively relevant dimensions,[58] and thereby incremented the capacity to respond to selection, as selection experiments showed.[59] Much attention has been given to how plasticity can allow a population to survive an environmental challenge, giving time for adaptations to evolve, such as when surface fish suddenly find themselves in the dark.[60] The fact that the evolutionary response to selection in the cave fish populations was aligned with the axes of maximum phenotypic variability in the plastic response of the surface fish suggests that plasticity does more than buy time—it contributes to the evolvability of fish trapped in caves. What is more, that the same patterns of correlated variation in developmentally codependent morphological, physiological, and behavioral traits in response to dark rearing is found in multiple experimental populations suggests a regularity to the plastic response.[61] A predictable pattern of trait variation, fitness differences, and heritability arises, which helps explain the virtually identical evolutionary response of over twenty independent cave fish populations.[62]

We can also see evidence of key regulatory genes operating as central hubs in GRNs that generate dramatic switches in phenotype in response to environmental challenges in ways that almost exactly mirror the effects of mutation. For instance, during blind Mexican cave fish development, the sonic hedgehog gene (Shh) is overexpressed, triggering a cascade of consequences that result in the eyes failing to form.[63] However, environmentally induced overexpression of Shh in surface fish also produces a blind cave fish phenotype.[64] Indeed, the same molecular mechanisms can underlie plastic and evolutionary responses.[65] Many of the cave fish traits that respond to selection have in common that they are connected by shared developmental mechanisms; most obviously, they respond plastically to the dark. This plastic response to the cave alters which cluster of variation contributes most to fitness, thereby biasing the selection that ensues.

Computational models that represent development as regulatory systems, and incorporate genes sensitive to environmental cues, suggest that this alignment of the plastic response, fitness, and selection is no coincidence.[66] These analyses reveal that selection acting on a phenotypically plastic population leads to the evolution of mutational variance that is exaggerated in the direction of the plastic response.[67] This implies that the evolution of phenotypic plasticity is itself an important determinant of patterns of mutation (i.e., the properties of the **M** matrix), patterns of genetic variance and covariance (i.e., the properties of the **G** matrix), and of the responsiveness of the population to selection (i.e., evolvability).[68] In fact, many empirical studies report that plasticity in an ancestor, or ancestor-proxy, generates intraspecific variation that matches interspecific variation across the clade.[69] A meta-analysis of twenty-one published studies involving nineteen species of insects, fish, and plants, found that plastic responses to novel environments tended to occur in phenotypes that exhibited substantial amounts of heritable variation.[70] Thus, plastic responses may represent highly evolvable dimensions of correlated phenotypic variation.[71]

Particularly germane to how plasticity relates to evolvability are those mechanisms of plasticity that use exploratory developmental processes to find adaptive solutions.[72] For instance, in chapter 11 we described how plastic adjustments to the skeleton, musculature, and behavior may have been critical for animals to negotiate the severe environmental challenge of the transition from water to land, as illustrated by experiments on bichir fish.[73] Learning provides another example.[74] We have seen how animals commonly learn to tune their behavior to environments, including novel environments, and through social learning these behaviors can then be propagated to other individuals, including nonrelatives, and across generations.[75] Extensive theory has demonstrated that learning accelerates adaptation in complex and changing environments, including by allowing individuals to adjust to rates of change too rapid to produce adaptation by the selection of genetic variation alone (although it may delay the response to selection in simpler environments).[76] Various forms of plasticity operate in a functionally equivalent manner to learning, by relying on a combination of exploratory and selective processes (e.g., adaptive immunity, the vascular system, the nervous system),[77] which raises the possibility that these findings concerning learning may be more broadly applicable to adaptive plasticity.

By learning how to discover and exploit new foods, or devising novel means to escape a threat, animals can introduce new adaptive behaviors into their repertoires. Species vary in how readily they generate such "behavioral innovations";[78] highly innovative birds have been found to be more likely to survive when introduced into novel habitats, implying that innovativeness is commonly adaptive.[79] Moreover, lineages of innovative birds are more species rich, which would make sense if innovation allowed animals to exploit new niches and diversify.[80]

Niche Construction as a Form of Developmental Bias

In chapters 8 and 9 we described experiments showing that dung beetle niche construction causes and strengthens the relations between a number of dung beetle traits and fitness.[81] Mothers construct and bury a brood ball containing a microbiome-laden package that the hatching larvae then process. Through their niche construction, beetles generate covariation among phenotypic traits, and directional epistasis in the genetic variation underpinning them.[82] This links variation in body size, trait size (e.g., eye, foretibia), life history (e.g., growth rate, time to adulthood), adult behaviors (e.g., dung ball manufacture, burying, egg laying), offspring behaviors (e.g., larval feeding, brood ball processing), sexually selected characters (e.g., horn size), and fitness (e.g., survival). Thus, niche construction creates clusters of associated phenotypes that vary together, and adaptive evolution is channeled along these axes of covariation. Consistent with this, beetle species have diverged in their reliance on niche construction, with corresponding differences in associated traits.[83] We also saw how beetle niche construction is a key determinant of selection by shaping population-specific reaction norms.[84] Here, niche construction strongly influenced the relationship between key secondary sexual traits (including horn size) and fitness.

By creating conditions that lead to traits being coexpressed, niche construction makes it possible to select those traits together, and hence for selection to modify the epistatic interactions that support them.[85] As such, niche construction can be viewed as a form of developmental bias (i.e., it generates correlated variation).[86] Yet it would be equally justifiable to view causation in the opposite direction: beetle niche construction creates the associations between functionally related traits and fitness, making selection coincide with these axes of covariation (i.e., modifying selection). The principal difference, compared with examples discussed previously, is that here the epistatic interactions arise via changes that organisms make to their external environments, rather than internally through developmental integration.[87] Such cases illustrate the chicken-and-egg nature of evolutionary causation: evolutionary episodes may be triggered by changes in variation or selection, which are inextricably linked and feed back to influence each other.

We have discussed many examples where the form of niche construction shapes the environment and individual experiences.[88] For instance, when birds and mammals build nests and burrows, they generate covariation among nest-building activities, antipredator behavior, and offspring growth patterns.[89] The epistatic effects of alleles underlying niche construction have been similarly demonstrated in plants, where genetic variation affecting germination time became statistically associated with flowering time by partly controlling the seasonal environment that plants experienced.[90] Habitat selection operates in a similar way. For example, pea aphid genotypes differ in habitat preference, leading to positive epistasis between alfalfa-preferring genotypes and specializations on alfalfa, and clover-preferring genotypes and clover specializations.[91]

Niche construction can even affect the evolvability of other species by providing exploitable environmental resources. This occurs both externally and by symbionts within a host, creating patterns of interspecies covariation that can form the bases of mutualistic relationships, such as the cow's rumen or the bobtail squid's light appendage.[92] Thus, when niche construction is involved, patterns of covariation between traits in entirely different species, and statistical associations between genes in different populations, can arise without a shared genetic basis but instead through association with a constructed environmental resource.[93]

As sources of variation and selection, niche-constructing traits can affect trait evolvability through either means.[94] Through their niche construction, organisms control key aspects of their environments, such as nests, buffering environmental variation and generating consistency in the resulting selection. For instance, a meta-analysis of selection gradients reported compelling evidence for reduced temporal and spatial variation and weaker selection in response to organism-constructed sources of selection such as nests and burrows compared with other sources, such as climate.[95] Plausibly, niche construction and habitat choice might be associated with increases in cryptic genetic variation in the traits buffered by organismal activities, which may be released when novel, stressful, or extreme environments are encountered.[96] As organisms modify environments in nonrandom ways consistent with their existing adaptations,[97] and as their artifacts and choices are major sources of selection,[98] niche-constructing traits will commonly be aligned with, and potentially help to identify, major multivariate axes of covariation and selection, and may exert a disproportionate influence on the direction of evolution.

At a macroevolutionary level, there is evidence that niche constructing can affect the propensity of lineages to diversify. Experimental, field, and theoretical studies reveal that organisms frequently open new niches that they and other species exploit, promoting diversification.[99] For instance, in chapter 11 we described an experimental study of evolution in bacteria, which found that extensive biodiversity could emerge in a completely homogenous environment, because bacteria create new ecological niches that sustain other bacterial populations.[100] Likewise, in tundra plant communities, forbs and grasses, which have a strong effect on ecosystem processes, are the least abundant growth forms but the strongest predictors of plant diversity.[101] Other studies suggest that diversity can arise through organisms' niche-constructing activities, including bioturbation, deposition of shell beds, and oxygenation of the ocean,[102] sometimes with long-term, macroevolutionary consequences, such as increased rate of branching of phylogenetic trees.[103]

Extragenetic Inheritance and Evolvability

Genes are not the only form of heritable variation subject to natural selection. In chapter 10 we made the case that extragenetic inheritance operates as a fast-response capability, allowing organisms to survive and adapt to novel, challenging, or changeable conditions.[104] There is extensive evidence, both experimental and from natural populations, for adaptation occurring through the selection of epigenetic, symbiotic, cultural, and other phenotypic variation.[105] Such phenomena are not analogous to biological evolution: they *are* biological evolution. For example, we described experimental studies in yeast, nematode worms, and *Arabidopsis* that clearly demonstrate that the natural selection of epigenetic variation can underlie adaptation (figure 14).[106] Likewise, studies of adaptation to a toxin in genetically uniform populations of yeast provided compelling evidence that multigenerational heritable silencing through epigenetic inheritance can be a powerful contributor to evolutionary adaptation.[107] Without an epigenetic switch, most of these populations went extinct, eventually resulting in genetic assimilation of mutations that enhanced the effectiveness of epigenetic silencing.[108] We also described how animals adjust to the consumption of novel diets through reliance on their microbiome to detoxify plant secondary compounds, or through the spread of cultural knowledge, as, for instance, with the lobtail feeding of humpback whales.[109] If evolvability is "the ability of a population to respond to selection,"[110] then adaptation that occurs through the selection of extragenetic variation surely needs to be considered, although this possibility has attracted limited attention.[111]

Genomes, parental effects, epigenetic modifications, the inherited microbiome, and cultural knowledge may all contribute to organismal adaptation.[112] Heritable extragenetic variation is now widely implicated in the ability of diverse species to invade new environments, cope with change and stress, evolve new phenotypes, and persist until "rescuing" genetic mutations appear and spread. Populations can't evolve unless they survive the initial challenge, and therefore this task of establishing a beachhead that selection can build on is highly germane to evolvability.

FIGURE 14. Epigenetic variation in the flowering plant *Arabidopsis thaliana* has been found to alter many ecologically important traits, including flowering time and root length. Selection experiments have established that inherited differences in morphology are linked to epigenetic variation. Selected populations showed a reduction in epigenetic diversity, including changes in methylation state, that were associated with the phenotypic changes, but no evidence for genetic changes. The authors conclude "epigenetic variation is subject to selection and can contribute to rapid adaptive responses" (Schmid et al. 2018b, 1).

This adaptive role for extragenetic inheritance has profound implications for how evolution is defined and understood. The evolution of different mechanisms of inheritance—themselves major innovations in evolvability—can lead to variation in how evolution by natural selection occurs.[113] We will save discussion of these issues for chapter 14, but note here that for many animals, humans included, culture is a vital and central aspect of evolutionary adaptability.[114] The same considerations apply to heritable epigenetic and symbiotic variation. Epigenetic regulation is thought to have originally evolved to control transposable elements, but, by allowing for cell lineages, it became critical in the origins of multicellularity, which afforded the opportunity for transgenerational epigenetic inheritance to evolve and to contribute to evolutionary adaptation.[115] More generally, there is now compelling evidence that major evolutionary transitions were critically dependent on extragenetic inheritance.[116] For instance, it is now widely accepted that the first eukaryote was formed through the incorporation of bacteria into prokaryotic cells to form mitochondria and chloroplasts,[117] and that multicellular organisms have evolved both through the uptake of whole microbes and through horizontal gene transfer of genes from microbes into both the microbiome and host genome.[118] There is also compelling evidence that developmental symbioses facilitated the evolution of herbivory in insects, dinosaurs, and mammals.[119] Most animals cannot digest plant cell walls, but once plant-wall-digesting and plant-toxin-neutralizing symbionts were established, herbivory (e.g., grazing) and adaptive structures for eating plants (e.g., molar teeth to grind up coarse material) could be favored,[120] with dramatic consequences for the structure of ecosystems.[121] Indeed, if meiosis were originally induced in the ancestors of animals by resident bacteria,[122] then a major basis of eukaryotic evolvability (i.e., sexual recombination) is predicated on symbioses.

The Evolution of Evolvability

The evolutionary process itself evolves. At a microevolutionary level, quantitative genetic theory reveals how evolvability evolves through the natural selection of epistatic interactions among functionally related traits.[123] Theory suggests this can align patterns of covariation with fitness, making the evolutionary process more efficient with time.[124] As the mycalesine butterfly example illustrates (see chapter 7), the structure of the genotype-phenotype relationship and pattern of genetic variation available to selection evolve even on microevolutionary time scales. There are many possible reasons for such changes in developmental regulation, including rare and dramatic ones, like gene or genome duplication. Nevertheless, organisms have been fine-tuning the relationships between their traits for almost four billion years, and as a result are now very good at evolving. Mutations do not occur when needed and are rarely directed,[125] but they nevertheless bias the phenotype toward dimensions that can be understood as outcomes of past evolution by natural selection, and as a result they generate useful variation more frequently than might be expected. However, that does not mean that all adaptive evolution can ultimately be traced back to natural selection alone, since the direction of selection is itself shaped by developmental biases.

The capacity to evolve has evolved at a macroevolutionary level too. Ultimately, organisms vary in their ability to respond to selection because of differences in the complexity of their development. Prokaryotes cannot adapt through the addition or deletion of coadapted multicellular packages, bacteria do not have the option of adjusting to stressors through cultural learning, and mammals can't exchange genes horizontally as frequently as bacteria do. Developmental processes both determine the evolvability of individual traits and explain differences in evolvability between traits and between organisms.

For many organisms, and certainly all multicellular organisms, there is more to evolvability than the availability of genetic variation, although these wider aspects of evolvability are currently less well understood. Extragenetic inheritance allows organisms to survive and adapt to novel or rapidly changing conditions through the selection of nongenetic variation. Moreover, past adaptive evolution has structured developmental processes into modules that can readily be strung together in new combinations through modest regulatory change. Yes, evolvability is "an organism's capacity to generate heritable phenotypic variation," but it is more than that: it also encompasses the ability of organisms to organize this variation to become useful (i.e., facilitated) variation.[126] Certainly, evolvability is "the capacity to respond to selection," but again, it is more than that and encompasses the ability to generate adaptation rapidly through the co-option of developmental modules into highly selectable packages. The distinction between conceptions of evolvability from quantitative genetics and evo-devo runs deeper than a mere difference in level of analysis, and concerns the complexity and diversity of the underlying mechanisms.[127] A major objective of this book is to illustrate how knowledge of developmental processes adds depth to evolutionary understanding. The evolutionary genetic view of evolvability is useful in quantifying, say, the evolvability of the beaks of Darwin's finches, or of vertebrate digits. However, in the absence of knowledge of developmental biology, it is difficult to explain—except in abstract statistical terms—*why* finches' beaks should be so evolvable, *why* beetle horns frequently are sexually selected characters, or *how* an entirely new organ can be generated with a single genetic change. Nor are these solely issues for a different field; they are relevant to evolutionary biology and its interest in evolvability.

Knowledge of development brings to evolutionary biology richer explanations and a capacity for prediction. Take polydactyly. Findings from evolutionary developmental biology not only help to explain why two extra digits are more common than four, or why five digits is robust, but also help to predict which digit numbers will be common in vertebrates, or which digit will be lost when losses occur.[128] The details of regulatory interactions in development are required to understand why some characters evolve readily and reliably, whereas others are rare or indeed never seen.[129] The variability of those mechanisms is itself a product of historical selection that directly or indirectly could have favored evolvability, thus emphasizing the reciprocal nature of evolutionary causation.[130]

Likewise, the intuition that only small variations would be viable, since variations of large effect would inevitably disrupt development,[131] is almost certainly wrong. The finches' beak, the stickleback's pelvis, the cichlid jaw, the bat's wing, the turtle's shell, and the polydactylous limb all attest to the experimentally verified fact that properties of developmental systems, ranging from modularity to compartmentation to exploratory mechanisms, can sometimes render mutations of large effect not only viable but capable of introducing rapid, large-scale change in continuous variation (shifts in beak shape), discontinuous change (extra fingers), and even major innovation (a whole new functional eye). Few researchers today would envisage that the evolution of an entirely new organ would require repeated bouts of selection to coordinate numerous small, independent changes in bone, muscle, nerve, blood vessel, and other tissues, all independently necessitating genetic change, which over eons gradually shape one another into a harmonized functional complex. That perspective made logical sense only in the absence of relevant understanding of development. We now know that developmental processes, and the laws of physics and chemistry, do much of the heavy lifting in translating mutational or environmental perturbations into functional phenotypes.[132] Multiple independently varying parts do not have to coevolve through natural selection because this is done by developmental systems, including through selective developmental processes, which allow small changes in regulatory input to generate harmonized refinement of interacting tissues.[133] If the capacity to produce large phenotypic differences already exists in the organism as self-inhibited alternate states that can be triggered by appropriate signals, as Gerhart and Kirschner propose, then large evolutionary strides can be made with modest genetic change.[134] Mutations of large effect do not necessarily lead to hopeful monsters; they, surprisingly frequently, generate well-conditioned beauties.[135]

13

Human Evolvability

Charles Darwin described the challenge of understanding human evolution as "the highest and most interesting problem for the naturalist."[1] Yet he was famously apprehensive about mentioning humans in the *Origin of Species*, and some of that reticence is still detectable among evolutionary biologists today.[2] Certainly, there exists a thriving field of human evolutionary studies, but it remains distinct from, say, evolutionary research in bacteria, yeast, plants, and insects. Indeed, human evolution is more often studied by physical anthropologists than zoologists. This is not because there is any controversy among reputable biologists concerning the fact of human evolution but rather because, even among scientists, our culture and mental capacities are often thought to make us "special."[3] At the same time, understanding and acceptance of evolution by the public is often disappointingly low. While the reasons for this are complex,[4] it is perhaps timely to consider the possibility that traditional accounts of human evolution, which rely on the natural selection of random genetic variation, and in which humans are often uncompromisingly portrayed as evolving through the same processes as bacteria and viruses, might to some people appear a little thin.

Conversely, acceptance that the evolutionary process itself evolves allows rich explanations for human evolution to be based on scientifically validated and widely observed natural processes. What is more, recognition that evolvability evolves situates human evolution within a wider explanatory framework without recourse to human exceptionalism. Humans have culture, but so do many other animals, whose phenotypic plasticity has also helped to direct their genetic evolution, and whose extragenetic inheritance has also been important.[5] We possess big brains and complex cognition, but recognizing that selection acts on developmentally biased variation allows this to be understood as both exceptional and consistent with mammalian expectations.

And that our technology is off the scale is not unexpected once the manner in which dynamical feedbacks between hands, brains, and social networks construct physical and learning environments is appreciated.

In this chapter we use human evolution to illustrate how a broader causal structure for evolutionary biology accommodates this most challenging of apparent exceptions. Our objective is not to provide a comprehensive treatment of the history of our species but rather to demonstrate that key aspects of the human story fall naturally into place, and can be better understood, in the context of the broader set of explanatory tools we have emphasized.[6] Human evolution is often characterized as comprising four key innovations: (1) bipedality and associated changes in morphology, (2) large brains allied with complex cognition and cooperation, (3) tool use, and (4) language. We use these topics to highlight some key points.

Bipedality and Gracile Morphology

Chimpanzees, bonobos, and other apes will stand and occasionally walk on two legs, suggesting that bipedality is an ancient capability. However, by five million years ago our ancestors were routinely walking on two legs (although their gait would have differed from ours, and they probably still spent significant time in trees), and by two million years ago they had long legs, an efficient stride, and probably the ability to run.[7] This demanded reshaping of the pelvis, lower limbs, and spine to bring the knees and feet directly under the body's center of gravity, with the head balanced above.[8] These changes took place over millions of years. However, there are lessons to be learned from the striking examples of developmental plasticity we discussed in the bichir fish that were trained to walk on land, or from the cases of two-legged goats and dogs.[9]

The key insight is that plastic adjustments to the entire skeleton, musculature, and behavior would have taken place in a coordinated fashion, with the bones and muscles of early bipedal apes changing shape during development, which enabled walking. While long periods of selection stabilized these changes and enhanced the efficiency of locomotion, plasticity almost certainly gave these adjustments a head start. This is not just speculation: the vertebrate exploratory mechanisms that allow muscles, tendons, nerves, and blood vessels to attach flexibly to cartilage and bone are such well-studied systems that it is hard to imagine how such plastic adjustments could not have taken place.[10] These empirical data are supplemented by theory, described in chapter 12, which suggests that, over time, functionally related traits that are selected together will coevolve through the selection of pleiotropic or epistatic interactions.[11] This body of theory allows certain testable predictions to be made. For instance, selection should have left the lower limbs and pelvis genetically and developmentally integrated, with developmentally biased variation aligned along a major axis of correlated trait variation. In addition, humans should exhibit weaker integration between arm and leg bones than other primates. These exact patterns have been confirmed by genetic and morphometric studies.[12]

Although humans are poor sprinters, they perform remarkably well at endurance running and possess a suite of specialized distance-running adaptations, from springing arches in their feet to a high proportion of slow-twitch muscle fibers in the legs to increased musculature supporting the spine.[13] We also sweat profusely, which cools the skin through evaporation and is critical since running can generate a huge increase in heating—most mammals cannot run for long before they overheat.[14] Running is thought to have evolved in the genus *Homo*, allowing hunters get close enough to prey to throw projectiles, run mammals to exhaustion ("persistence hunting"), or to facilitate scavenging, as has been documented among some modern foragers.[15] However, for this strategy to be productive, sweat-based thermoregulation—which involved the modification of hair-producing glands in our skin into sweat glands[16]—must have coevolved with, or after, cultural evolution had generated the know-how for making water carriers (e.g., from ostrich egg shells), locating water sources, and reading the tracks of animal prey.[17] Theory suggests that natural selection would have acted on variation in a suite of standing-walking-running traits rather than "tinkering" with each trait individually (see chapter 12).[18]

Childbirth is a long and painful affair for humans compared with other mammals, and is associated with unusually high rates of mortality in both mothers and infants.[19] This may be because the evolution of bipedality created an "obstetrical dilemma," in which selection for efficient bipedal locomotion, upright posture, or both, favored a small pelvis that, sometime later, conflicted with selection for larger brains or body sizes in human babies, although this remains contentious.[20] The shape of the human pelvis bone is thought to be a compromise among these various demands—a bigger birth canal would compromise standing, walking, and running, whereas a smaller one would lead to greater mortality in childbirth, generating an n-shaped fitness function.[21] This dilemma is partly resolved by culturally inherited childbirth practices, but also by evolved correlations between pelvis shape, stature, and head circumference.[22]

Women with large heads tend to give birth to large-headed babies, but they also tend to possess rounder birth canals that are better shaped to accommodate large-headed neonates. Short women, who are at increased risk from babies that are large headed relative to the size of mother's pelvis, also tend to have a rounder birth canal than tall women.[23] Again, phenotypic variation is biased in the predicted direction.[24]

Compared with the male pelvis, the human female pelvis is bigger and wider, with a larger outlet.[25] Interestingly, "strikingly similar sex differences" are observed in common chimpanzees, despite the fact that they have no obstetrical dilemma and despite clear differences between chimpanzees and humans in pelvis shape.[26] These and other comparative data suggest that the human pelvis has evolved along preexisting dimensions of correlated trait variation regulated by a conserved endocrine system.[27] In mammals, the pattern of pelvic sex differences is strongly influenced by the spatial distribution of estrogen, androgen, and other hormone receptors, as well as by hormone-induced bone remodeling, which means that pelvis shape will be partly determined by the effects of these hormones on multiple tissues, including pelvic soft tissue.[28] As a result, "when the size of the neonatal brain increased substantially in the human lineage during the Pleistocene, the genetic and developmental mechanisms to evolve a more spacious female pelvis were already in place."[29] A taxonomically widespread developmental bias helps to explain how the human pelvis can both conform to a mammalian template and be exceptional in its sexual dimorphism.

Several other changes in morphology characterize our species since the divergence from chimpanzees. Compared with early hominins, modern humans have thinner, more gracile bones, smaller teeth, and reduced sexual dimorphism for body size. We have also retained some juvenile physical and behavioral features into adulthood, a phenomenon known as *neoteny*. For instance, human heads are shaped like those of juvenile chimpanzees—we have flat faces with a forehead and no muzzle—and we experience a prolonged period of juvenile development. Intriguingly, all of these trends fall within the rubric of a single explanation, one that Darwin himself noted but couldn't satisfactorily explain: they are all characteristics of domesticated species.[30] As described in chapter 7, what connects the curious assortment of traits exhibited by domesticated species is joint developmental regulation through the neural crest cell gene regulatory networks.[31] This shared developmental origin, which results in statistical associations among these traits, is precisely the kind of outcome expected of the correlational selection discussed in chapter 12.[32]

We also described how wild foxes could adapt to urban settings with reduced muzzle length, diminished sexual dimorphism, and other domestication-syndrome-like changes in anatomy and behavior, and how similar patterns of covariation were reported across canid species.[33] There is nothing artificial about the association among domestication-syndrome traits—selection for tameness in domesticates merely exploits preexisting covariation in a cluster of traits observed in nature. That same developmental bias has probably been shaping the evolution of mammals, including humans, for millions of years. Selection for larger brains or a more cooperative disposition among our ancestors would almost certainly have brought about correlated changes in other traits such as face shape and bone thickness. If cultural activities contributed significantly to the cooperative endeavors underlying this selection in our ancestors for large brains or reduced aggression, then, just as in the blind cave fish and dung beetle examples discussed in chapters 8 and 9, plasticity would have shifted the cluster of trait covariation that dominates fitness, becoming a *cause* of natural selection in the sense outlined in chapter 9. While hominins have been subject to complex, diverse, and often conflicting patterns of selection, we suggest that the correspondence between the set of traits that characterize domestication syndrome and the evolutionary changes in our lineage is too much of a coincidence for this developmental bias not to have played an important role in human evolution.

Recent evolutionary changes in human dentition can also be understood, and even predicted, with knowledge of development. For instance, in chapter 2 we described an "inhibitory cascade" model of tooth development that drew on developmental mechanisms, identified through experimental investigations, to successfully predict the size and number of teeth in a sample of mammalian species.[34] Evolutionary changes, such as mammals losing teeth or teeth changing shape, are now reasonably well understood to be due to changes in the levels of various signaling molecules during development. As molars develop in a front-to-back sequence, from the inhibitory cascade model we expect that if jaws become smaller, the inhibitory effect on the last molar will become substantial, and the rearmost tooth may be lost. This is exactly what we see in humans, with many people lacking wisdom teeth, or having them removed because there is insufficient room for them in their mouths.[35] Reduction in consumption of hard foods may have contributed to smaller jaws, tooth crowding, and crooked teeth.[36]

Harvard evolutionary anthropologist Richard Wrangham has emphasized how humans have an unusual digestive system: our mouths, teeth, and jaw muscles are unusually small, our stomachs have only a third of the surface area, and our colons are far too short for a primate of our size.[37] This odd physiology is viable only because it has coevolved with culturally transmitted know-how related to food processing.[38] Chopping, mashing, pounding, soaking, leaching, marinating, but above all *cooking*, externalizes the digestive process, softening foods and reducing the work for our mouths, stomachs, and colons. Control of fire and cooking are now accepted to have played a central role in human evolution.[39] Cooked food provides more energy than raw food because it is easier to digest.[40] Cooking allowed our ancestors to spend more time hunting and gathering and less time grazing and chewing.[41] The use of fire, cooking, and food-processing are culturally inherited skills first recorded in our ancestors as far back as two to three million years ago, although this date remains a point of contention.[42] This culturally transmitted knowledge is what allowed smaller teeth, jaws, and stomachs to be viable. Sure enough, the appearance of cooking appears to coincide with the selection of genetic variants that are expressed in the brain and digestive tract and are involved in the determination of tooth size and the reduction in jaw muscle.[43]

Human Brains and Cognition

Perhaps the most distinctive feature of human evolution is the increase in brain size.[44] Human brains have more than triple the volume and number of neurons as chimpanzees' brains, largely because, unlike other apes, we experience a substantial period of infant brain growth.[45] Much of this growth is thought to come from the loss or repression of genes that in other animals limit neural proliferation.[46] Ultimately, more neurons lead to greater processing power.[47] However, that excess brain tissue comes at a price: humans consume 400–820 kcal of energy per day more than other great apes, with the brain a major consumer.[48] To support a larger brain an animal must either accrue more energy (e.g., by foraging for longer, foraging more efficiently, or accessing higher-quality food) or reduce the size and hence energy requirements of other expensive tissues, such as the gut.[49] To do either is difficult. Thanks to their culture, and suitable environmental conditions, humans have done both.[50]

Culture is widespread in animals but reaches its pinnacle in the great apes, which are renowned for their social learning, tool use, and extractive foraging.[51] There is extensive empirical evidence that these primates' ability to acquire energy-rich foods, such as nuts, termites, and honey, is critically dependent on culture.[52] For selection to favor a larger brain, the cognitive advantages it confers must result in an improved energy balance.[53] Through social learning and collaborative foraging, some primates are able to gather the calorie-dense, large-sized food resources necessary to grow and maintain a large brain.[54] Consistent with this, quantitative measures of the reliance on social learning, behavioral innovation, and tool use correlate with brain size measures in primates.[55] The same explanation applies to humans: University of New Mexico anthropologist Hillard Kaplan and colleagues have compiled compelling evidence that large brains, complex cognition, and longer lifespans coevolved because human cognitive abilities allowed us to exploit highly nutritious but difficult-to-access foods, such as animals that have to be hunted, or plant parts that are difficult to extract (e.g., nuts).[56] The energy obtained from these foods enabled brain growth, with longer lifespans favored because they allowed more time later in life for individuals to exploit the energetic bonanza, and more opportunities for transgenerational knowledge transfer.[57]

Not only are our brains exceptionally large for a primate of our size, our guts are unusually small.[58] In humans the energy savings are further enhanced by food preparation and cooking, which, like foraging, rely heavily on cultural knowledge. Culture is not the only important driver of brain evolution, but without recognizing its key role it is difficult to explain the large brains and complex cognition of the apes, and particularly hominins.[59] Here too, culture operated as a critical *cause* of natural selection, shifting the balance of fitness costs and benefits from favoring relatively larger guts and smaller brains to favoring relatively small guts and large brains.[60]

There are good reasons for thinking that brains might also have been subject to correlational selection generating developmentally biased covariation, perhaps also associated with face shape.[61] In mammals, the covariation among brain parts is extremely high, and much of the variation in size of mammalian brain components can be understood as arising through changes in overall brain size, guided by conserved features of neurogenesis.[62] In vertebrates, evolution has led to larger brains by prolonging brain development while keeping the sequence of developmental events broadly consistent across species.[63] In recent years, cognitive neuroscientists have distinguished between "easy" and "difficult" modes of brain evolution, with the former constituting selection on the duration of brain development, which generates concerted evolution of the whole brain, and the latter requiring independent enlargement of specific functional components.[64] As the labels imply, concerted brain evolution is thought to dominate.

Cognition appears to mirror this pattern of "concerted evolution," at least in primates.[65] Comparative analyses have established that traits thought to require complex cognition (including tool use, social learning, and behavioral innovation) have coevolved with one another and with absolute brain size in primates, and that those abilities that are correlated with brain size are more evolvable than those that aren't.[66] Neuroscience reinforces this picture.[67] For instance, stone-tool use appears to be associated with distributed brain circuitry that overlaps considerably with the circuitry involved in speech and the propensity to predict future events.[68] Perhaps the easiest way for a primate to evolve enhanced cognition is by having the whole brain grow for longer, while co-opting existing neural circuitry to new functions.[69] It used to be thought that each brain region had a highly specific and dedicated function.[70] However, it now appears that few parts of the human brain, and few cognitive domains, rely on their own independent and exclusive circuitry.[71] For instance, Broca's area—long associated with speech and language—is also critically involved in movement preparation, action sequencing, action recognition, and imitation.[72] The widespread co-option and reuse of neural circuitry in the evolution of the human brain is consistent with the arguments made in chapters 11 and 12 that innovation and evolvability depend in large part on the ability to borrow and reuse existing developmental pathways and networks in new ways.[73] The idea that human cognition is dominated by localized neural specializations that were species-specific adaptations—similar to the avian song nuclei or the enlarged hippocampi of food-storing birds[74]—is now being reconsidered.[75] The computational structures supporting human learning are diverse, interconnected, and distributed across the entire brain.[76]

Increases in brain size bring concomitant alterations in brain organization.[77] For instance, as mammalian brains evolved to become bigger, more subdivision emerged and larger regions became disproportionately connected and influential.[78] Biological anthropologist Terrence Deacon proposed a rule specifying that if brain regions become disproportionately large, then, as they evolve, they would tend to "invade" and become connected to regions that they did not innervate ancestrally.[79] This would increase the influence of the enlarged areas over other brain regions and make them more important to brain functioning.[80] One consequence of evolution creating larger brains by "stretching" brain development is that brain regions that mature relatively late become disproportionately large.[81] The two largest (and among the latest-developing) structures in the human brain are the neocortex and the cerebellum and, as predicted, they became increasingly embedded in complex neural networks during the course of human evolution, and exert considerable influence over them.[82] This evolutionary change includes increased connections between the neocortex and the motor neurons in the brain stem and spinal cord, a phenomenon known to be associated with increased manual dexterity and motor skill.[83] As a direct consequence, selection of a larger neocortex or cerebellum brought with it greater flexibility and dexterity, with these two brain regions exerting greater control over our arms, legs, face, and hands. To a large extent, the cognitive ability that allowed our ancestors to manufacture and use tools came automatically with the physical dexterity of precise hand movement, while the ability to learn a language probably went some way toward enhancing the flexible use of our mouth and tongue.[84]

The duration of human neurodevelopment is highly predictable from our knowledge of general mammalian development.[85] Humans have an unusually large neocortex, but it is exactly the size it "should be" given the size of our brains—that is, it is allometrically aligned with the brains of other mammals. Much research into comparative brain organization has sought to answer the question *What makes human cognition unique?*, and just about any feature of human brains not in accord with the mammalian expectation has been given as an answer.[86] Perhaps most important, however, are adaptations that arise through selection on whole-brain size, duration of neurogenesis, or timing of developmental events such as birth and weaning—but attempts to identify cognitive adaptations as *departures* from taxon-general expectations typically fail to take these into account![87]

Some (*precocial*) animals, such as sheep or antelopes, are ready to run shortly after birth, not because of any advance in the rate of brain maturation or acceleration in the neural control of their limbs, but by retaining the same conserved sequence of brain development and delaying birth until the animal is ready to run.[88] By contrast, puppies, kittens, and other *altricial* animals are born relatively immobile, often with closed eyes and unable to fend for themselves, but again the development of their brains follows the same conserved sequence of events as in other animals—it is just that they are born early.[89] Primates, in general, are born at a middle stage of neural maturation, but humans are the least mature—or most altricial—of primates. This is perhaps a consequence of the "obstetric dilemma" arising from bipedality, which requires humans to be born before their heads get too large, although other hypotheses remain credible.[90] However, our early birth places us in an unusual developmental niche, in which we are receptive to inputs from the external

environment for an unusually long period of time, which contributes importantly to our enhanced cognition.[91] Our "helplessness is coupled with an ability to soak up information about the environment at a remarkable rate and to
use it to manipulate the world."[92]

Thus the human brain has become specialized for an unusual degree of
plasticity.[93] Human brain anatomy and function have evolved to be highly
responsive to features of the environment, especially social interactions,[94]
which allows for experience- and usage-based neural reorganization, intercon-
nection, and pruning.[95] Numerous aspects of human brain development show
evidence of specialization leading to, or resulting from, increased plasticity,[96]
many of which are also evident in fossil hominins, and from analyses of ancient
DNA and genetics.[97] Moreover, the human striatum, a brain region that contributes to social behavior, exhibits a dramatically different neurochemical
profile from other primates, with elevated dopamine and reduced acetylcho-
line.[98] These changes in the hormonal profile of the human brain are thought
to have amplified sensitivity to social cues, contributing to the evolution of
conformity, empathy, cooperation, and language,[99] and perhaps susceptibility
to Parkinson's disease.

In sum, early birth and extended lifespan, which correspond to an extended childhood, combined with the enhanced plasticity of an unusually
large and powerful brain, result in a long period of development during
which humans are exceptionally well positioned to learn about their world,
particularly from an array of tolerant and invested caregivers.[100] Experiments
conducted over recent decades have revealed many similarities between the
cognitive abilities of humans and those of other animals, and yet there is
broad acceptance within the field of comparative cognition that in important
ways humans are cognitively unique.[101] Our superior brain power arises
through interactions and reinforcement among several cognitive domains in
which we excel—including memory, planning, tool use, problem solving,
social cognition, and communication.[102] For instance, human understanding
and use of tools come largely from copying others and through language,[103]
while our proclivity for language learning is critically dependent on our capacities for joint attention and learning sequences of actions.[104] Much that is
exceptional about human cognition results from trait interactions and feedbacks, with culturally scaffolded developmental experiences building upon
and reinforcing evolved biological differences.[105] Without cultural learning,
humans could never have evolved such a large brain, nor such a long period
of juvenile dependency.

Technology, Tool Use, and Control of the Environment

Tool use has long been regarded as pivotal in our species' evolution. By 2.5 million years ago, the members of our genus were skilled stone knappers. *Homo habilis*, and some Australopithecines too, are known to have been capable of producing sharp cutting flakes from a cobblestone core by striking it with a hammerstone (termed the *Oldowan*, or Mode 1, technology).[106] The surprising complexity of these tools, along with present-day tool-making experiments, suggests that Oldowan technology was learned and required considerable practice,[107] while its long prevalence, wide geographic spread, and regional variations all suggest that tool manufacture and use were socially transmitted and that cultural inheritance played a substantial role.[108]

Archaeologists believe that significant fitness benefits accrued from the ability to make and use stone tools and to transmit these skills rapidly, establishing a coevolutionary relationship between toolmaking and cognition.[109] In support of this hypothesis, changes to hominin morphology, including increased brain size, coincide with advances in stone toolmaking.[110] Experiments with contemporary humans find that the social transmission of Oldowan technology is enhanced by teaching and language,[111] and that the manufacture of bifacial (*Acheulian*, or Mode 2) hand axes—associated with *Homo erectus, Homo ergaster,* and later hominins 1.8–0.1 million years ago—excites the same brain regions as are involved in language.[112] These findings suggest that stone-tool manufacture generated selection for increasingly complex communication, teaching, and eventually language.[113] Of course, making stone tools is just one of several technical skills thought to have been culturally transmitted among our ancestors, and potentially instrumental in these feedbacks.[114]

The dynamic interplay between genetic and cultural processes is now regarded as central to human cognitive evolution.[115] Recent explanations stress a positive feedback loop in which accurate and efficient social learning enhanced the payoff for technical competences, including tool use, which benefited not just learners but also their relatives and cultural group members. In contemporary human societies, successful tool use comes largely from observational learning and linguistic communication.[116] High levels of manual skill and coordination are required for successful extractive foraging, food processing, and cooperative hunting. The ability to use tools flexibly is observed in just a handful of primate species, and has long been associated with intelligence.[117] Primate species that are flexible tool users tend to have broad omnivorous diets and exhibit both enhanced dexterity and a tendency for object manipulation.[118] Likewise, innovative tool use is strongly correlated with brain size in both birds and primates.[119] There is extensive evidence that ecological, technical, and cultural capabilities coevolved in primates, including humans.[120]

Historically, tool manufacture and hunting have been characterized as male activities, but this may be overly simplistic.[121] For Darwin, accounting for distinct gender roles was central to understanding human evolution, although commentators increasingly regard Darwin's writings to be biased by his Victorian values.[122] Despite the popular "man the hunter" stereotype, there is little evidence for gender roles in toolmaking or tool usage throughout most of the two-million-year history of our genus, and convincing archaeological evidence for gendered hunting appears only after ten to twenty thousand years ago.[123] Gender roles become more distinct with the advent of agriculture and settlements, including in forager societies, but a number of studies suggest that modern gender roles often do not reflect, or have become more pronounced than, past ones, a position supported by the discovery of early Holocene female big-game hunters and Viking women warriors.[124]

The human transition from hunting and gathering to food production—which is, of course, a cultural change—is now understood to have affected recent human evolution.[125] Incorporation of domesticated plants and animals into our diets triggered selection for genes that metabolize these foods. For human genes that have been subject to recent selection, the source of selection is often thought to be a switch to an agricultural diet.[126] Genetic variants expressed in the metabolism of dairy products, starch, proteins, lipids, phosphates, plant secondary compounds, and alcohol, and gene variants affecting jaw, facial muscle, and tooth characteristics, are all thought to have been favored as a result of this human cultural niche construction, possibly during times of disease or stress.[127] The available evidence strongly suggests that culturally mediated changes in diet preceded these genetic changes, and constitute examples of plasticity-led evolution.[128]

Agriculture also generated natural selection for genetic resistance to diseases by supporting population growth, thereby inadvertently exposing human societies to infectious diseases such as typhoid, tuberculosis, and cholera, and by creating conditions in which disease vectors could thrive.[129] For instance, the slash-and-burn agriculture of some West African populations created puddles of standing water when it rained, which became breeding sites for mosquitos, facilitating the spread of malaria. Subsequently, this triggered selection favoring the hemoglobin sickle cell (*HbS*) allele, which confers resistance to malaria in some (heterozygous) carriers.[130] That neighboring populations with different food-procurement practices do not show an increase in *HbS* supports the conclusion that cultural activities—clearing forests for crop cultivation—have driven this feature of human genetic evolution.[131] Ecological inheritance also plays a role, with the genetic evolution of generations of humans being affected by the activities of their forest-clearing ancestors many generations before.

Similar relationships that link cultural traditions, disease incidence, and resistance allele frequencies are reported for urban living, tuberculosis, and an allele of the *SLC11A1* gene.[132] The Irish potato blight famine and associated diet, iron deficiency anemia, and the frequency of the *HFE* gene variant *C282y*, which promotes iron absorption, are also linked.[133] In both cases, the genetic response to selection was contingent upon cultural and ecological inheritances.[134] Other studies show how the cultivation of crops in Bali and Polynesia, like animal domestication, not only imposed selection on genes underlying human digestion and disease resistance, as well as on the genes of domesticates, but also had long-lasting effects on the environment.[135] This influenced the rate and trajectory of agricultural development, shaped local ecosystems, and both facilitated and hindered technological innovation.[136]

Such examples showcase how human populations actively construct niches—including technological niches[137]—and how these constructed niches, in turn, affect the future transformation of these populations, their social systems, and their ecologies.[138] By promoting population growth, cultural niche construction increased the size and connectivity of social networks, and allowed for more effective exchange of knowledge and resources to generate a "collective brain."[139] While our focus on regulatory networks highlights the central role of the genome, we have stressed that many regulatory molecules and contextual factors contribute to phenotypic outcomes. This includes the environment, particularly when its features are regularly constructed by the organisms themselves, and hence provide consistent inputs to development. Human-constructed external environments often contain reliably present resources (e.g., cultivated crops), important hereditary and regulatory information (e.g., the maternal cue provided by famine), "lock-in" stabilizing features (e.g., irrigation methods), but also human-derived sources of selection (e.g., on *HbS*) and disease vectors (e.g., malaria-carrying mosquitos). Any of these can influence the capacity for, and trajectory of, evolution.

Archaeological and paleo-environmental records from eastern North America, Amazonia, the Near East, and China provide clear support for plasticity-led evolution, with key roles for cultural and ecological inheritance, and human niche construction, in shaping not just our own recent evolution but also the evolution of domesticates, commensals, and other species.[140] These data challenge the historical narrative that environments change and species respond passively to changed environments through selection. Our species has always transformed landscapes, altering ecosystems and creating new evolutionary trajectories through a combination of environmental control and disturbance, settlement, pollution, habitat destruction, predation, competition, introductions, and extinctions.[141]

A key technological advance was human control of fire, originating 1.5–0.4 million years ago.[142] The significance of this is generally thought to be the invention of cooking, but more recently it has led to the large-scale burning of fossil fuels such as coal, gasoline, and oil, to generate heat and light as well as smelting iron, powering combustion engines and turbines, fixing nitrogen, and manufacturing fertilizers.[143] These activities released a vast bank of energy, hitherto stored deep in the ground, into the biosphere. Although the energetic constraints on evolutionary change are rarely considered, it is a truism that populations cannot evolve without the energy to reproduce.[144] Our use of fire and exploitation of fossil fuels have increased food supplies, which has not only supported massive growth in human populations but affected the entire biosphere.[145] Exploitation of this energy supply has triggered profound transformations in human life history, diet, and subsistence, leading to far higher crop yields and huge numbers of cows, pigs, dogs, and chickens.[146] Human control of fire has had dramatic impacts on our own and other species' evolution, perhaps for several hundred thousand years.[147]

Our global impact is now so devastating that scientists claim we are in a new geological epoch, the Anthropocene, and speak of mass extinction. However, it is not our rapid genetic change that is threatening the biosphere but rather our rapid cultural evolution, combined with technological development and powerful niche construction. Other species struggle, but, if population numbers are any indication of the health of a species, humanity has been able to adapt to the self-imposed changes in conditions.[148] This is because humans possess a uniquely powerful culture that both elicits environmental change and enables rapid accommodation to it. In fact, the rate of evolutionary change experienced by our lineage appears to have accelerated recently.[149] This is no coincidence. Culture allows animals to exploit their environments with greater efficiency,[150] and as the hominin capacity for culture expanded, bolstered by complex forms of teaching, cooperation, and language, culture eventually began to dominate our evolutionary dynamics. However, by controlling fire, domesticating plants and animals, aggregating into dense populations, and changing environments in a myriad of other ways, our ancestors set themselves adaptive challenges to which they sometimes responded genetically. Such cases, and other reciprocal interactions between genetic and cultural processes, are known as gene-culture coevolution.[151] Mathematical modelling shows that gene-culture coevolution should be faster than conventional biological evolution, in part because cultural evolution occurs at faster rates than genetic evolution.[152] The contribution of gene-culture coevolution to human adaptation was probably initially modest but became more important over time, as our cultural capacity expanded and our control of the environment increased. The initial result was higher rates of morphological evolution in humans than other mammals: human genetic evolution is claimed to have accelerated more than a hundredfold in recent millennia.[153] However, in the modern era, cultural evolution dominates completely, as increasingly powerful culturally transmitted practices provide means to solve culture-derived problems. Cultural practices can have negative effects on humans, but these negative effects can be alleviated by further cultural activity before genetic evolution gets moving.[154] Our culture hasn't stopped genetic evolution but has overridden it. Today, human evolvability is dominated by cultural evolution, with genetic adaptation probably relatively less important in our species.[155]

The Evolution of Language

Charles Darwin recognized the huge evolutionary challenge posed by human language, acknowledging that it is "one of the chief distinctions between man and the lower animals."[156] In *The Descent of Man* he tried to bridge that gap by highlighting the similarities between animal and human communication.[157] At the time, prominent linguists took issue with this comparison, and whether human language evolved out of the communication systems of animals remains contentious to this day.[158] Experts in language evolution recognize a distinct gap between human communication and that of nonhuman primates, irrespective of whether language is deemed to have originated in vocalizations or gestures.[159]

Some primate vocalizations—known as *functionally referential calls*—are thought to symbolize objects in the world. Famously, vervet monkeys possess three such calls, which are thought to be labels for avian, mammalian, and snake predators.[160] Similar claims have been made for the calls of other mammals and birds.[161] However, whether monkey calls are genuinely referential remains contentious; primate vocalizations consist largely of single signals that are rarely put together to transmit complex messages, and there is little evidence that nonhuman primates can learn new vocalizations or imitate sounds.[162] Animals, including apes, dolphins, and parrots, have been taught to use symbols,[163] but they show little evidence for the comprehension of tense, syntax, or grammar.[164] Likewise, while apes possess a rich repertoire of gestures, some of which are learned, these gestures exhibit neither symbolism nor syntax.[165] Not only has science struggled to find compelling continuity between human and animal communication, but many linguists suggest the primary function of language is not communication at all, but organizing thought, and stress the human capacity to generate an infinity of phrases and their organization into a hierarchical syntactic structure.[166]

Fortunately, neuroscience is starting to provide clues to the origins of language. The brain circuitry of speech and language show extensive overlap with the circuitries associated with stone-tool use, imitation, learning visuomotor sequences, and planning.[167] These and other data suggest that the context for the evolution of language was set by the co-option and reuse of neural circuits involved in the social learning of hierarchically organized and recursive temporal sequences.[168] Many researchers have noted the parallels between the computational skills necessary for language and those underlying complex tool manufacture and use, as well as foraging and food processing, and have suggested that our ancestors' ability to process serial order and hierarchical structure in language, including long-distance dependencies in space and time, arose from these behaviors.[169] Through an evolutionary history of acquiring long sequences of tool-using and food-processing actions through social learning, as well as the capability to process hierarchical and recursive subroutines, combined with some referential communication, humans may have become cognitively predisposed to learn and process long strings of symbols. Language may have originated in the context of teaching foraging skills, food processing, and tool use to relatives, and may be linked to our ancestors working together in collaborative foraging and other forms of joint action.[170]

Interestingly, recently evolved brain functions, which include language, have been found to involve a greater number of widely scattered brain circuits than evolutionarily older functions.[171] This makes sense if, as mentioned previously, there has been extensive reuse of brain circuitry over evolutionary time, because recent functions would be more likely to encounter preexisting circuitry that could be incorporated usefully.[172] Humans may be unusually cognitively sophisticated partly because this extensive reuse of neural circuitry has generated greater neural integration than has occurred in other animals.[173]

Further clues come from the observation that learning to read—a behavior that first appeared too recently in human history to have evolved through the selection of genetic variation—results in a significant reorganization of connectivity and modularity in the brain.[174] This demonstrates the plastic responsiveness of the brain to new tasks, and its ability to organize novel inputs into functionally coherent *cognitive gadgets* that become well integrated with evolved structures.[175] Similarly, humans who have undergone several months of training in stone-tool manufacture exhibit marked changes in brain connectivity.[176] This flexibility in neural development can produce functional efficiencies in neural architecture during development,[177] and the extended duration of developmental plasticity leaves humans unusually sensitive to such life experiences. These examples illustrate how self-organizing developmental systems extend into the brain, and suggest that the neural underpinnings of language may have relied on the initially plastic co-option and reuse of existing circuitry.[178]

Darwin wrote that "the survival of certain favoured words in the struggle for existence is natural selection,"[179] and experimental research in the field of language evolution suggests that this intuition may have been correct. If linguistic structures are to persist, they must survive the process of being learned, spoken, and adopted by others. Words or phrases that are difficult to learn or speak will be at a disadvantage relative to alternatives, and experiments have shown that languages can adapt over time in this manner.[180] Many researchers have been impressed by the ease with which children acquire languages and have assumed that the human brain includes a dedicated language-learning capability.[181] However, children may be readily able to decipher the rules of syntax in part because languages have evolved rules that make them easy to learn.[182] Both experiments and mathematical models show that languages evolve culturally to become easier to acquire and more structured over time.[183] This shifts some of the explanatory burden away from the natural selection of genetic variation, and helps to make the challenge of explaining the origins of language more manageable. Once our ancestors evolved a socially transmitted system of symbolic communication, several other features of language arose concomitantly.[184]

Other researchers have suggested that language use imposed natural selection on human anatomy and cognition; for instance, favoring traits such as infant-directed speech, cognitive inference, and meta-cognitive awareness, and perhaps also furless faces making expressions easier to read and prominent whites of our eyes that render others' gazes easier to follow.[185] Language, along with imitation and other aspects of cultural learning, are strongly linked to the evolution of music, dance, and the arts.[186] There is also compelling evidence that language, along with social learning and teaching, played a central role in the evolution of various forms of cooperation.[187] The unprecedented scale of human cooperation—which is also manifest in warfare and conflict—builds on human reliance on cultural norms.[188] Young children acquire such norms readily and spontaneously enforce them.[189] Whether or not human cooperative tendencies reflect evolved altruistic social preferences, humans show an unusual proficiency at working together to achieve joint goals, helping others through teaching, and taking others' perspectives and intentions into account.[190] Languages, norms, and punishment of nonconformity stabilize group differences, and some scholars claim that this may have produced a form of group selection that strongly influenced human evolution,[191] and may be regarded as an evolutionary transition.[192] In a field that remains highly contentious, one point of broad consensus is that cultural inheritance and cultural evolution have played a key role in the origins of language, with major ramifications for human societies.

The Wealth of Human Evolutionary Mechanisms

The challenge of communicating evolutionary science to an often-skeptical public may occasionally have led to simplistic storytelling. The truth of the matter is that humans do not evolve like yeast, bacteria, or fruit flies, and the human story is far richer than any account based solely on the natural selection of genetic variation could capture. While we have not emphasized it here, there is no doubt that natural selection and genetic inheritance remain central to human evolution, as do drift, mutation, gene flow, and other standard evolutionary processes. However, other less-appreciated evolutionary processes have also left important marks.

In chapters 7, 8, and 12 we illustrated how developmental processes can integrate morphological, physiological, and behavioral traits, thereby creating strong patterns of covariation,[193] and how the resulting developmentally biased phenotypes can affect evolutionary outcomes by pre-specifying the options for natural selection.[194] This developmental bias has also played a critical role in human evolution. Developmental bias was manifest in the evolution of bipedality and in the resolution of the obstetrical dilemma created by our upright stance and large brains, by correlated changes in our ancestors' legs, arms, pelvises, and heads. Natural selection was not sorting between all combinations of the variants of each trait independently but was channeled toward pelvis designs tailored to the mother's, and hence offspring's, head shape and stature, integrated with symmetrical limbs, and allometrically proportioned arms and legs. We also suggested that the shift toward gracile bones, a neotenous face, smaller teeth, and reduced sexual dimorphism may have been collectively favored through a mild decrease in neural crest cell contributions to various parts of the body, as in several animal domesticates.[195] We described evidence for the concerted evolution of brain components, which in primates is paralleled by changes in cognition, for the co-option and reuse of neural circuitry in human brain evolution, and for a shift in the balance of hormones in the brain associated with a suite of human social behaviors. This integration of phenotypes into major clusters of correlated variation may have accelerated human evolutionary change and enhanced human evolvability.[196]

Some of the strongest evidence that plasticity is important to evolution is found in our own species. For example, theoretical models and ancient DNA analyses leave little doubt that the cultural practices of dairy farming and milk consumption came first, and only subsequently were alleles for adult lactase persistence favored by natural selection.[197] Few researchers now contest that our ancestors' domestication of plants and animals, and associated cultural changes in diet, imposed natural selection on our digestive system, leading to genetic change.[198] A combination of empirical evidence and theory suggests that social learning, teaching, and tool use played critical roles in the evolution of the brain, cognition, manual dexterity, and cooperation. The evolution of human life history, including the timing of birth, weaning, and adolescence and aspects of human cognition, is linked to the extended plasticity of our brains, while plastic adjustments of the skeleton were likely to have been important in the evolution of bipedality.

The actions and behaviors of humans strongly bias the selection that they, and other species, experience.[199] Researchers now suggest that niche construction played a central role in the evolution of cognition and language, the origins of domestication and agriculture, and other aspects of human evolution.[200] We have presented evidence that the development of agriculture and urbanization facilitated the spread of diseases, triggering selection for genetic resistance. We have also seen how our ancestors constructed a socially transmitted foraging, hunting, and food-processing niche in which a large brain, complex cognition, and technical competence were advantageous, and in which language skills became valuable. Humans' potent capability to regulate, construct, and destroy environments has also generated many current problems, ranging from deforestation and urbanization to climate change,[201] while driving evolutionary change in animal and plant domesticates, commensals, and urban invaders.[202]

While selection of epigenetic variation and inherited symbionts may also have been important, and an aspect of human evolvability,[203] culture is surely the form of extragenetic inheritance that has had the greatest impact on the evolution of our species. Without the capacity for cultural transmission our ancestors would almost certainly not have been able to forage—not to mention hunt, process food, and cook—with the efficiency necessary to fuel energy-demanding brains, or to survive with diminutive guts, or to tolerate reductions in our face, jaw, and tooth sizes. Without culture, there would have been no stone toolmaking traditions, no control of fire, and no animal or plant domestication. Our species has exhibited unusually high levels of evolvability, manifest as high rates of genetic change,[204] probably because cultural activities exposed our ancestors to new challenges to which they adapted genetically.[205] Culture allows human behavior and technology to change extremely fast, which extends human evolvability both by providing a means to respond directly to selection through cultural change and by indirectly affecting the evolvability of noncultural traits, as we saw for brain and gut size. For decades, traditional accounts of human evolution depicted hominins as adapting to external conditions, such as climate change, predation, or disease, in a manner little different from other organisms, and with human culture seen as a product of evolution but not a significant cause of genetic change.[206] However, the material reviewed in this chapter supports a perspective on human evolution according to which cultural activities are recognized as a potent force that could have elicited important genetic change in our species.[207] Exogenous factors remain important sources of natural selection, but we can now understand that many of those factors are modulated by human cultural activities.

Humans are often portrayed as unique, with that uniqueness sometimes seen as a challenge to evolutionary explanation. Once the full range of explanatory tools is brought to bear on the topic of human evolution, it becomes possible to view human "exceptionalism" as fitting general patterns that are characteristic of primate-, mammalian-, or vertebrate-wide developmental mechanisms. Conversely, a failure to recognize the breadth of evolutionary mechanisms, or the multiple interacting processes of human inheritance, can give rise to crude reductionism in which almost any human trait, and almost any property of human society, can be viewed as genetically determined.[208] Human evolution is an undeniable scientific fact, but it involves more diverse and complex processes than are represented in most treatments of Darwinian biology.[209]

14

The Structure of
Evolutionary Theory

The remarkable green sea slug leads its adult life as a nomadic leaf.[1] This strange animal feeds on yellow-green algae, extracting the chloroplasts and incorporating them into the cells that line its digestive tract, and then engages in photosynthesis. Thereafter, the sun provides it with virtually all the metabolic energy it needs to survive and develop.[2] The green sea slug is unusual because, unlike, say, coral, the chloroplasts are incorporated as naked organelles that somehow remain functional, so it is the animal's own cells, rather than its symbionts, that carry out photosynthesis.[3] The chloroplasts are not transmitted to the slug's offspring in the eggs or sperm—the larvae must seek out algae and feed for several days to acquire their own. The ingested chloroplasts turn the animal green, allowing it to live as a plant. Here, the organism's essential functionality—its way of life—is both self-constructed and environmentally acquired.[4] Are the chloroplasts part of the animal's environment or are they part of the organism? The answer is *both*.[5]

The green sea slug's mode of existence may be unusual, but the tight organism-environment relationship that it exemplifies is a universal feature of life. Nature and nurture are always inseparable components of the development of an organism. More to the point, the green sea slug's very idiosyncrasy is itself representative of evolving organisms. Once biologists dig a little deeper, and consider *how* phenotypic variation is generated, *how* fitness differences arise, and *how* inheritance occurs, the details matter. Just as the woodrats in chapter 1 survive by reliably extracting symbiotic bacteria from their environment, so the sea slug's fitness is contingent on its ability to root out and exploit chloroplasts. Such activities as these generate interdependencies between the processes generating fitness and phenotypic variation.[6] Similar interdependencies arise in the killer whales and fear-conditioned mice that we also met in chapter 1, where animal culture or epigenetic inheritance fundamentally modify both fitness and heredity. Likewise, our ancestors' domestication of plants and animals, and the associated cultural changes in diet, imposed selection on our digestive and immune systems, while their social learning and cultural transmission were critical to the evolution of brains, guts, jaws, and teeth (chapter 13).[7] Humans do not evolve like woodrats, mice, or sea slugs; the biological details matter. Just as the human story is richer than any account based solely on the natural selection of genetic variation, so for every species there is a richer story.

A New Vision

In the mid-twentieth century, the founders of the modern evolutionary synthesis characterized development as the unfolding of a genetic program, and for decades this metaphor minimized the relevance of development and virtually all proximate causes in evolutionary analyses.[8] Part of the beauty and simplicity of that version of Darwinism was that all organisms could be regarded as evolving in the same way through natural selection. In this final chapter we synthesize the research described in this book to show how contemporary evolutionary theory is starting to recognize additional roles for developmental processes in evolution, and how this enriches evolutionary explanations and spawns new research avenues.

Now a new vision of adaptive evolution, possessing its own elegant logic and richness, is starting to crystallize. Fundamental to this perspective is the recognition that development is not "programmed" but rather "constructive," with organisms continuously responding to, and altering, internal and external states to shape their own developmental trajectories.[9] And as organisms evolve they change their own future course of adaptation and diversification. Natural selection is not something that just happens to organisms: their activities and behaviors contribute to whether and how it happens.

Central to any understanding of evolution is a conception of how organisms work.[10] We have suggested that developmental processes *cause* adaptive change, by determining phenotypic responses to genetic and environmental perturbations, as well as which traits are expressed together and what impact they have on reproductive success or survival. We have also emphasized how natural selection will indirectly favor developmental interactions that promote functional integration, and that stabilize and refine traits. Selection may thereby *cause* some of the properties of development that provide organisms with their evolutionary potential.[11] From the origin of life, developmental processes have biased the phenotypic variation that is subject to natural selection—an imperative that follows from the fact that living organisms comprise molecules, cells, and tissues subject to the laws of physics and chemistry,[12] laws that impose structure on living organisms, and render some phenotypes more likely to arise, and be stable, than others. And from the beginning of life, developmental mechanisms have themselves evolved through natural selection, producing further developmental bias and new capacities for generating useful variation.[13] Through continual interactive cycles of such reciprocal causation, developmental processes have channeled and directed evolution by natural selection, and evolution by natural selection has sculpted those developmental processes.

Of course, many factors influence the dynamics of natural selection and other evolutionary processes, by affecting their strength or potency—for instance, population size. Such factors are generally regarded as modulators of, or constraints on, evolutionary processes, rather than evolutionary causes. Why should development be treated differently? The answer is that developmental processes do more than just modulate; they help to determine the *direction*, *rate*, and *equilibrium states* of adaptive evolution. They do so by determining which phenotypes become selected, by ensuring that adaptive features are passed on to future generations (or not), and by influencing which combinations of traits make the greatest contribution to fitness. As the turtle shell, blind cave fish, and dung beetles illustrate, developmental insights can explain *why that particular trait*, as opposed to plausible alternatives, evolved, shedding new light on *how novel features originate*. As for the domestication syndrome, developmental experiments can reveal *why that particular cluster of traits* has been subject to selection; shared developmental regulation explains why aspects of phenotypic variation should be correlated, and why apparently neutral or maladaptive traits might be subject to correlated selection. As the molar tooth example illustrates, knowledge of development allows a priori prediction of traits that are most likely be subject to selection, thereby helping to explain broad patterns of taxonomic diversity. And as the humpback whales' rapid adaptation to a new prey elucidates, broader views of inheritance help account for some challenging features of evolutionary adaptability. Consideration of population size, by contrast, sheds little light on any of these issues. We regard *What are the causes of fitness differences?*, *Which traits will be directly favored by selection?*, and *Which traits will be indirectly favored through correlated selection?* to be core questions for evolutionary biology. Like other evolutionary biologists, we want to know: *What explains the origin of novelties?*, *What are the drivers of adaptive change?*, and *What makes organisms so good at evolving?* A developmental perspective offers detailed, predictive, robust, and generalizable answers to such questions, complementing established statistical approaches.

The evolutionary processes commonly listed are those thought to change gene frequencies directly, such as natural selection, mutation, or random genetic drift.[14] Developmental processes, such as bias or plasticity, do not change gene frequencies directly, and would not qualify as causes of evolution from that perspective.[15] Equally, from a traditional standpoint, whether the natural selection of epigenetic or cultural variation would qualify as evolution is also questionable. As explained in chapter 4, and elaborated upon below, this traditional view is a consequence of a particular idealization of evolution by natural selection, not a fundamental property of how biological evolution works. Including developmental mechanisms in explanations for evolutionary change therefore requires conceptual change, rather than just the addition of more detail to well-established frameworks. Arguably, such change will eventually arise solely from the motivation to understand evolutionary phenomena such as rapid adaptation, parallel evolution, or transitions in individuality. Yet, there is also value in examining the structure of evolutionary theory more directly, to elicit more powerful explanations, identify new problems, and stimulate debate.

The Evolution of Evolutionary Biology

For many years the dominant view was that developmental processes are *proximate* but not *ultimate* causes, a dichotomy advocated prominently by Ernst Mayr.[16] Distinguishing between different types of causal explanation is valuable, as we have stressed in chapters 2 and 3. Nonetheless, philosophers and biologists have long recognized that Mayr's dichotomy is not the only way to describe causation.[17] Indeed, many researchers have expressed the concern that the proximate-ultimate distinction may have impeded recognition of certain important classes of evolutionary explanation.[18] Mary Jane West-Eberhard, for instance, writes: "The proximate-ultimate distinction has given rise to a new confusion, namely, a belief that proximate causes of phenotypic variation have nothing to do with ultimate, evolutionary explanation."[19] While some aspects of evolutionary dynamics—for instance, sexual selection or coevolutionary interactions—are understood to involve reciprocal causation through selective feedback between coevolving entities, developmental processes have not usually been recognized as evolutionary causes.[20]

These issues are not new, but in recent years omission of development from evolutionary explanations has become increasingly difficult to reconcile with what is known about biology. Few biologists would now contest the claim that developmental processes bias the phenotypic variation that is subject to selection. Historically, however, developmental factors were thought to play only a minor role in evolution: for instance, they were often invoked to explain the apparent absence of evolutionary change and adaptation, but rarely to account for how and why organisms are able to adapt and diversify.[21] Why evolutionary biology should have arrived at this position is a complex issue, but a major factor is its reliance on a small number of assumptions about biological evolution.[22] These assumptions had heuristic value, but now the field has moved on. Foregrounding these assumptions, and illustrating how they are now starting to be reevaluated, helps us to understand both why development was not part of the evolutionary mainstream and how the field is evolving. These assumptions included that:

1. *Developmental and evolutionary biology largely address distinct questions, and analysis of adaptive evolution does not require knowledge of development.* From the model of molar development in mice used successfully to predict patterns of diversity in tooth and phalange size across vertebrates (chapter 2), to experimental investigations of the development of butterfly wing spots that produced explanations for diversity across the clade (chapter 4), to developmental investigations of the formation of embryonic tissues that have helped to explain why the beaks of Darwin's finches are so evolvable (chapter 12), there is now extensive evidence that knowledge of development can provide valuable evolutionary insights, enrich evolutionary explanations, and enhance prediction.[23] Certainly, it is possible to study evolution without detailed understanding of development, but these examples illustrate the potential benefits of its close integration into the analysis of evolution.

2. *Genetic variation is pervasive, and hence selection will be powerful and bias transient.* Key theoretical assumptions in population genetics, and the existence of extensive standing genetic variation in natural populations, led to the view that the direction of adaptive evolution would be

little affected by the availability of phenotypic variation and would be largely determined by selection.[24] However, selection rarely acts on single traits in isolation, and selectable phenotypic variation can often be described by a few sets of traits with strong covariation within each set.[25] Theory suggests that evolution follows paths created by development, and stops at the fitness peaks of those paths rather than peaks in the adaptive landscape.[26] Whether one considers the structure of RNA, the blind Mexican cave fish phenotype, the color of butterfly eyespots, or the shape of *Drosophila* wings—developmental bias can exert a powerful influence on patterns of phenotypic diversity.

3. *Development "constrains" evolution.* A number of assumptions—that development is best understood as hindering evolution, that bias results primarily from physical or material limitations on form, and that constraint provides an alternative explanation to natural selection—have impeded understanding of the role of development in evolution.[27] However, developmental bias is not the opposite of developmental constraint.[28] Developmental processes help to explain why particular phenotypic outcomes occur, including adaptations, but also help to explain why parallel evolution can be seen in independent lineages (e.g., cichlid jaw shapes), why particular patterns of trait covariation arise (e.g., floppy ears and piebald markings in domesticated animals, or fragmented bones in blind cave fish), and why traits might evolve even when they have little effect on fitness. Selection is not powerful *in spite of* genetic correlation; it is (often) powerful *because of* developmental bias.[29]

4. *Development is genetically programmed.* That genomes, and other heritable resources, have evolved to enable development to adjust to environmental conditions is now widely recognized. Yet the concern remains that such "interactionism . . . [while] virtually universally adopted . . . has become conceptually vacuous."[30] The characterization of development as under the control of a genetic program, and of phenotypes as genetically specified reaction norms, can also be viewed as subtle forms of genetic determinism. In reality, organisms help to construct their own developmental trajectory through constant interactions with their internal and external environment, while reaction norms are products of developmental systems (chapters 5 and 6).[31] Even traits that might appear genetically determined are the product of developmental systems and will, at the very least, be

malleable to environmental factors outside specific ranges. Yes, there is structure in DNA sequences that can develop into information, but cellular and extracellular regulatory mechanisms, together with environmental cues and forces, provide additional sources of developmental information.[32] Tracing causality to genes is sometimes useful, but it is still just a convention; in reality, "control" of the phenotype is distributed across multiple levels of biological organization, and not confined to a single location or level.[33]

5. *Heredity reduces to genes.* As the woodrat, whale, and mouse examples of chapter 1 illustrate, and the review in chapter 10 confirms, restricting heredity to transmission genetics led to a seriously incomplete description of the inheritance of phenotypes and of differences among them. It is now clear that extragenetic inheritance plays a vital role in short-term evolutionary adaptation, and perhaps may have evolved to do so; however, to be effective in this role some level of autonomy from genes is necessary. Extragenetic inheritance is ubiquitous throughout nature, and there is good evidence that it is a source of selectable phenotypic variation. It can also facilitate population survival and adaptation, redirect trajectories of genetic change, and contribute to evolvability. As recent work demonstrates, the dynamics, including stable states, of evolution are better understood with broader characterizations of trait inheritance.[34]

6. *The causes of phenotypic variation, differential fitness, and heredity are largely autonomous.*[35] Lewontin identified three criteria necessary for evolution by natural selection—there must be *phenotypic variation, differential fitness,* and *heredity*[36]—however, a rarely discussed additional assumption is that the subprocesses responsible for generating these three requirements are mechanistically autonomous (chapters 1 and 4).[37] This assumption led to the inference that directionality in adaptive evolution could ultimately be attributed to fitness differences.[38] However, the subprocesses responsible for natural selection often interact in ways that fundamentally affect their mechanisms. For instance, we saw how whale culture not only generates phenotypic variation but also contributes to differential fitness (learned habits affect survival) and heredity (learned behavior is socially transmitted to descendants). These interactions entail that the three subprocesses are causally intertwined, while each contributes importantly to adaptation. As a result, the selection of genetic variation (or indeed of

extragenetic variation) is not the only cause of adaptation. We elaborate on this point below.

These six influential assumptions reinforced one another, collectively assigning explanatory privileges to genes and natural selection. Let's return to the question of how it can be accepted that developmental processes bias the phenotypic variation subject to selection and yet not be recognized as causes of evolution. Here, the twin assumptions that *selection but not mutation is powerful* and that *development "constrains" adaptive evolution* tend to play down the significance of developmental bias. At the same time, the assumptions that *developmental and evolutionary biology largely address distinct questions*, that *development is genetically determined*, and that *heredity reduces to genes* focus causality on genes and natural selection. No one disputes that the mechanisms of development have themselves evolved, so it is easy to recognize that developmental bias is a product of selection. However, the assumption that *the causes of phenotypic variation, differential fitness, and heredity are largely autonomous* makes it appear as if the adaptive directionality imposed by development could—and should—be replaced by selection in evolutionary explanations. This makes it easy to play down or overlook the fact that selection is no less shaped by developmental bias, on the grounds that the bias arose through earlier selection. The reasoning is flawed but achieves the satisfying simplicity of linearizing descriptions of causation, which makes explanations less complicated.

Why the reasoning is flawed can be understood by analogy with a population in mutation-selection balance. At this equilibrium, the rate at which a deleterious allele is eliminated by natural selection is balanced by the rate at which that allele is introduced into the population through mutation. The mutation rate may, of course, have been modified—likely reduced—by natural selection in the past, but this does not prevent mutation being recognized, with selection, as a cause of the balance in the present. Recognition of the effects of both processes is required for a satisfactory explanation for the balance.[39] The reasoning would apply even if there were no balance; to predict the dynamics of the system accurately at any point in time it would still be necessary to recognize the causal influence of both mutation and selection. To track the frequency of any changing entity, we need to know the rates of arrival and removal.[40]

The same logic applies to developmental bias and selection. For example, we saw when discussing RNA secondary structure in chapter 7 that developmental mechanisms can strongly influence the rate at which phenotypes arise through mutation, as well as the properties of those phenotypes. Here again, knowledge of the magnitude and direction of this bias in the rate of "arrival" of phenotypes, together with knowledge of the relative fitness of the resulting phenotypes, is necessary to predict accurately evolutionary change in RNA structure.[41] The same logic applies to other phenotypes, where the introduction of new variants may even impose adaptive directionality on evolution. Consider again the lobtail feeding of humpback whales discussed in chapter 1.[42] This adaptive trait is a behavioral innovation and, for its dynamics to be predicted, the rate at which this behavior increases in frequency through social learning is important, as are changes in its frequency arising through whale births and deaths. The same is also true for evolutionary rescue arising through the selection of epigenetic variation,[43] or adaptation arising through changes in symbionts.[44] Developmental processes, and biases in them, must be recognized as causes of the production of phenotypic variants—not just affecting rates but also affecting the direction of change and equilibria reached—if the evolutionary dynamics are to be accurately predicted, irrespective of whether or not ancestral selection has affected these processes in the past.[45]

However, developmental processes—including developmental bias, phenotypic plasticity, niche construction, and extragenetic inheritance—do more than just bias the form of variation and rate at which it is produced (Lewontin's first criterion for natural selection); they are also sources of fitness differences (Lewontin's second criterion) and trait inheritance (Lewontin's third criterion). Yet, by contributing to two or more subprocesses *simultaneously* rather than separately, and by modifying how each of the subprocesses operates, developmental processes violate the assumption that Lewontin's components are autonomous. We saw this in chapters 1 and 4, but it is instructive to return to it one last time, in the light of the experiments on dung beetles,[46] which exemplify how developmental mechanisms generate interdependences between the subprocesses (figure 15).

A traditional evolutionary explanation for the adaptive fit between dung beetles and their environments would invoke mutations that change, say, the beetles' brood-ball processing in a manner that enhances fitness, and are inherited by the next generation (figure 15, *top*). This allows brood-ball processing to be construed as a proximate mechanism and—since the generation of variation by mutation is assumed to be random and inheritance is assumed to be unbiased—allows fitness differences to explain the brood-ball processing

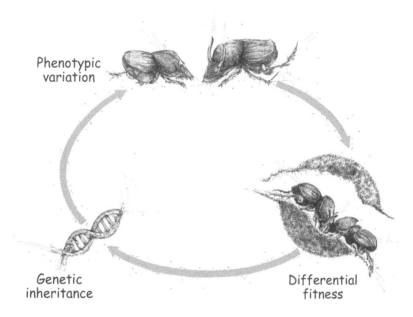

Phenotypic
variation

Genetic
inheritance

Differential
fitness

FIGURE 15. The intertwining of variation, fitness, and inheritance. *Top.* A
traditional evolutionary explanation for the adaptive fit between dung beetles
and their environments. Mutations change the beetles' brood-ball processing in
a manner that enhances fitness, allowing brood-ball processing to be construed
as a proximate mechanism and fitness differences to explain adaptation. *Bottom.*
However, the underlying causation is more complicated. The extent to which
some dung beetle traits, including offspring body size and developmental time,
contribute to fitness depends on phenotypic variation in the niche-constructing
behavior of mothers and larvae. Inheritance is dependent on phenotypic
variation and fitness differences, as the brood ball is a parental effect that is
ecologically inherited by the larvae. Furthermore, phenotypic variation depends
on inheritance, since beetles develop in the beetle-constructed environment of
the brood ball, which influences both the developing larvae's traits and relationships among them. Through these interactions developmental mechanisms bias
phenotypic variation, inheritance, and fitness. Attributing the complementarity
between beetles and their environments solely to fitness differences results in an
impoverished understanding of adaptation.

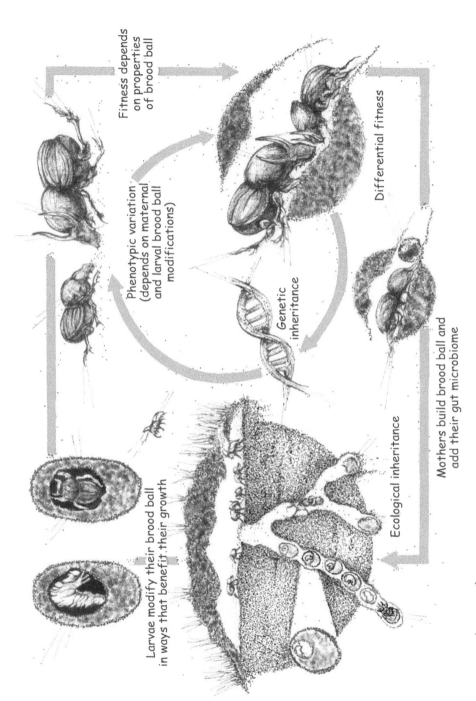

Fitness depends on properties of brood ball

Phenotypic variation (depends on maternal and larval brood ball modifications)

Differential fitness

Genetic inheritance

Larvae modify their brood ball in ways that benefit their growth

Ecological inheritance

Mothers build brood ball and add their gut microbiome

FIGURE 15 (*continued*)

adaptation. However, evo-devo experiments reveal that the underlying causation is more complicated (figure 15, *bottom*).

The extent to which some dung beetle traits, including offspring body size and developmental time, contribute to fitness depends on phenotypic variation in the niche-constructing behavior of both mothers and larvae.[47] Niche construction not only modifies trait fitness, it encompasses qualitative changes in the traits that contribute to fitness. Experiments quantify how the contribution to fitness of offspring traits depends critically on the properties of the brood ball, which is constructed by the mother and modified by the larvae to act as an external rumen (i.e., depends on niche construction). Also, inheritance is not independent of phenotypic variation and fitness differences, as the brood ball is a parental effect that is ecologically inherited by the larvae. Again, niche construction brings about qualitative changes in how inheritance occurs. For instance, experiments show that this inheritance depends critically on whether or not the mother incorporates into that brood ball a pedestal containing a sample of her microbiome. The inherited microbiota help to mitigate the effects of environmental stress, buffering selection on traits that are responsive to temperature and desiccation. Further, phenotypic variation is not independent of inheritance, since beetles develop in the beetle-constructed environment of the brood ball, which experiments show fundamentally influences both the developing larvae's traits and relationships among them. The interactions between the subprocesses of natural selection entail that beetle niche construction is no longer just a proximate mechanism. Rather, phenotypic variation is biased by development, as are inheritance and fitness differences. This means attributing the complementarity between beetles and their environments solely to fitness differences is questionable.

Here a digression to consider two different classes of explanation may be helpful. Richard Lewontin and Elliott Sober, among others, have emphasized the distinction between *variational* and *transformational explanations* for change in a population.[48] For instance, imagine a class of school children who perform better in their subject at the end of the school year compared with the beginning. That could be because good teaching has, on average, improved the pupils' knowledge and understanding so that the mean score has increased—a transformational explanation. Alternatively, the increase might have occurred because lower-scoring pupils were disproportionately likely to drop out during the school year, or were sent to another class for remedial schooling—a variational explanation. In the latter case, no individual student needs to have improved for the average score in the class to increase. Darwin's theory of evolution by natural selection provides a variational explanation for change in species over time, while, in marked contrast, Lamarck's earlier account of evolution offered a now-discredited transformational explanation.[49] However, as the classroom example illustrates, both types of explanation may be reasonable.

The distinction between variational and transformational explanations is relevant here because close inspection of the interactions between the subprocesses of natural selection reveals a significant but poorly recognized role for transformational explanations in adaptive evolution, alongside the established variational explanation provided by fitness differences. The "fittedness" or "match" of dung beetles to their immediate local environment arises partly because those beetles experiencing a poor fit have died or failed to reproduce, but also partly because individual beetles inherit a maternally modified brood ball that is well suited to other aspects of the larval phenotype, and partly because how the larvae develop inside that brood ball is sensitive to the brood ball's properties. Thus, three processes operate here to bring about an organism-environment match: the standard *variational explanation* of the selective survival of fit individuals, the *transformation* of the developmental environment experienced by the larvae arising through both maternal and larval activities, and the *transformation* of the larvae through the development of the focal phenotype in a specialized organism-constructed medium.[50] Moreover, these three processes interact, both in the present and in the past, and cannot be traced to just one original cause.

This interdependence of phenotypic variation, fitness differences, and heredity is now starting to be formally recognized in mathematical evolutionary models. For instance, evolutionary theoreticians Laurel Fogarty and Michael Wade developed a quantitative-genetic model of the evolutionary consequences of niche construction.[51] They showed that the response of a phenotype to selection and its rate of evolution are strongly affected by the manner in which organisms alter selective environments, especially if they modify environments for descendants, creating an "ecological inheritance." The analysis leads them to recognize three separate ways by which niche construction can influence evolution: "First, NC [niche construction] can plastically change the value of an expressed trait" (i.e., affect phenotypic variation). "Second, NC can alter the fitness landscape" (i.e., modify fitness differences). "And finally, NC can change breeding values, thereby changing the nature of trait heritability" (i.e., alter heredity).[52] Fogarty and Wade go on to point out deficiencies in how heritability is generally understood and warn of "potentially serious consequences" of failing to extend the concept in ways that incorporate niche construction and ecological inheritance.[53] Similar points have been made concerning the contributions of extragenetic inheritance to evolution,[54] while other models show that developmental processes, including plasticity and niche construction, typically have long-term rather than only transient effects on evolutionary dynamics.[55] Cases like the dung beetles (or killer whales, desert woodrats, or other phenomena discussed in this book), supported by theory,[56] illustrate that how organisms develop and what organisms do partly determine the strength and direction of natural selection, the resulting dynamics, and the equilibria reached.[57] It is in this sense that we characterize developmental processes as causes of natural selection. Developmental mechanisms are not just a product of evolution; they are also part of the evolutionary process.

This assessment of the evolutionary role of development entails that natural selection on living organisms has itself evolved.[58] The properties of natural selection have evolved through changes in each of the three component subprocesses; that is, through changes in how phenotypic variation is generated, changes in how fitness differences are constructed, and changes in the channels of inheritance. As a result, organisms vary in their ability to generate selectable phenotypic variation, *and* in their capacity to modify fitness differences, *and* in their means to contribute to inheritance (chapter 12). However, different forms of interdependence among these three subprocesses have *also* evolved, and these interactions also vary across taxa and at different times in the history of life on Earth. Here again, the evolution of new developmental mechanisms has afforded new opportunities for each subprocess of natural selection to affect the inner workings of the others. The power of the evolutionary process to modify itself is apparent in the origin of new entities capable of undergoing evolution by natural selection, such as the transition from unicellular to multicellular organisms or from multicellular to colonial organisms like ants and termites.[59]

We have suggested that these considerations demand changes in how the process of evolution is understood. Whether a revised structure of evolutionary biology will emerge and what it will look like remain to be seen. However, we anticipate that biologists will eventually embrace a wider interpretation of evolutionary causation, in which the organism once again takes a more central role, both as object and subject of evolutionary change.[60] This will require recognition that (1) evolutionary causation is inherently reciprocal, with development processes and natural selection codetermining adaptation and diversification; (2) developmental processes impose directionality on natural selection, through biases in the processes responsible for phenotypic variation, fitness differences, and heredity, and through their interaction; and (3) adaptation and biological diversity can arise through the natural selection of heritable extragenetic variation.[61] Although some other nontrivial conceptual challenges might have to be assessed, we nonetheless submit that the revisions outlined above will be useful in the longer term.[62]

A Research Agenda for the Twenty-First Century

Most evolutionary geneticists would agree that the major problems of the field have been solved. We understand both the nature of the mutational processes that generate novel genetic variants and the population processes which cause them to change in frequency over time—most importantly, natural selection and random genetic drift, respectively.... [W]e will never again come up with concepts as fundamental as those formulated by the "founding fathers" of population genetics.

—BRIAN CHARLESWORTH, (1996, 220) "THE GOOD FAIRY
GODMOTHER OF EVOLUTIONARY GENETICS"

From a developmental perspective, matters look quite different. If "the field" is taken to be evolutionary biology, there remains much important work to do, as can be illustrated again by considering Lewontin's requirements for evolution by natural selection, and the ways in which the evolutionary process can itself evolve. Here we focus on the contribution of developmental processes to adaptive evolution via natural selection, by highlighting some example questions. Some aspects of this research agenda are already being vigorously investigated, and others remain relatively unexplored, but all merit further attention. Moreover, the mechanisms of development also play potentially important roles by affecting other processes, including random genetic drift, mutation, and gene flow.[63]

1. **Phenotypic variation.** *What are the developmental rules and regularities by which phenotypic variation is generated? What distributions of phenotype variation will arise? What developmental biases are operating?*

    If genetic mutation and recombination are regarded as the sole sources of heritable phenotypic variation, and if these are thought to generate phenotypic variation in a ubiquitous and broadly unbiased manner, there might indeed be few fundamental problems in evolution left to resolve. Conversely, if it is accepted that genetic and phenotypic variation are biased, then specifying the nature of those biases determining under what circumstances they can be understood through the study of developmental mechanisms, and to what extent such analyses can explain the patterns of phenotypic variation observed in nature become important challenges for the field. Experiments exploring the relationships among the variation produced by mutation, the standing genetic variation, and the rate of evolution become central; for instance, those on the *Drosophila* wing (chapter 12).[64] However, so do experiments on the mechanisms of development; for instance, the study of mammalian teeth (see chapter 2).[65] These explain the existence of those biases, support predictions about patterns of taxonomic diversity and of diversity in other traits with similar developmental mechanisms, and relate genes and their interactions to development. Once bias is recognized, formal theory exploring how developmental processes connect genetic to phenotypic evolution becomes important.[66] The roles of developmental plasticity, niche construction, and extragenetic inheritance in generating biased phenotypic variation become pivotal, as do how developmental bias generates evolutionary novelties, how it impacts robustness and evolvability, how it is shaped by modularity, and whether and how it can account for macroevolutionary trends and patterns.

2. **Fitness differences.** *What are the developmental causes of fitness differences? Can rules and regularities of development be identified that allow a priori specification of the phenotypes subject to selection, and the relationship between traits and fitness? Can knowledge of developmental mechanisms be used to predict correlational selection and responses to selection?*

If the source of natural selection is assumed to be exogenous to the organism, and if the manner in which developmental processes transduce environmental inputs is ignored, finding the causes of fitness differences might appear to be impossible. Such assumptions may allow the traits that contribute to fitness to be retrospectively identified, but do not explain why those traits and not others were favored by selection. Conversely, as experimental investigations of dung beetles and blind Mexican cave fish show (chapters 8 and 9), organisms themselves construct important components of their environments, and develop plastically in response to encountered conditions, in ways that can be studied experimentally. These studies reveal the causal connections between traits and fitness that are outcomes of what organisms do and how they develop. Shared developmental regulation, including that arising indirectly through niche construction, can create covariation between phenotypic traits within and across generations.[67] Furthermore, through niche construction and developmental plasticity, organisms can help to determine which clusters of trait variants make the greatest contributions to fitness,[68] since the environment both "instructs" and "selects" the organism.[69] The research described in this book demonstrates that developmental causes of fitness differences are, at least sometimes, open to experimental and theoretical investigation.

3. **Inheritance.** *How do different inheritance channels interact? How should we model extragenetic inheritance, and what difference will it make to evolutionary dynamics? How does extragenetic inheritance contribute to trait robustness, short-term adaptation, evolutionary rescue, and evolvability? How can we explain taxonomic variation in inheritance mechanisms?*

If inheritance includes only transmission genetics, then designing evolutionary experiments and constructing evolutionary models can be relatively straightforward. However, additional forms of inheritance are common throughout nature, may have strong impacts on fitness, can be subject to natural selection, and can involve transmission that is independent of genetic variation. The existence of extragenetic inheritance raises some major questions for evolutionary biologists, most of which are underexplored. Experiments on epigenetic inheritance in the plants *Arabidopsis* and *Polygonum*, yeast, and nematode worms; on the inherited microbiome in desert woodrats and aphids; and on animal culture (see chapter 10) show that extragenetic inheritance can be tractable to empirical investigation. Likewise, well-established traditions for the mathematical modelling of cultural evolution and gene-culture coevolution provide starting points for exploring how inheritance systems coevolve.[70]

4. **Evolution of the evolutionary process.** *How do the three subprocesses of natural selection interact? When do they evolve independently and when do codependencies appear? What difference do interactions among them make for evolutionary experimentation and theory? How do new evolving entities arise, and what roles do developmental processes play in major evolutionary transitions?*

As we have seen, if the processes responsible for generating phenotypic variation, fitness differences, and inheritance are assumed to be mechanistically independent, then evolutionary causation is commonly traced back to differences in fitness.[71] Matters become more complicated when these processes interact, with one or more processes modifying the mechanism of another. Under such circumstances, transformational explanations for adaptation become plausible. These interdependencies are important because they influence evolvability; indeed, without such interactions the potential for evolvability to evolve would be severely limited. More research is required to establish which forms of inheritance (e.g., cultural but not genetic?) and which circumstances (e.g., organism-constructed but not autonomous environments?) generate these codependencies. Several studies that we have discussed show that interdependences between subprocesses are amenable to experimentation, and theoretical work is beginning to explore these issues.[72]

Developing Evolutionary Theory

For at least a century, developmental perspectives on evolution have existed alongside a dominant mainstream that emphasized genes and external sources of natural selection. Developmentalist thinking never quite made it into the mainstream, partly because the developmentalists' arguments were in some respects ahead of their time,[73] but also because understanding of development and the relevant experimental and theoretical methodologies were not at the time sufficiently advanced to render the developmentalists' arguments completely convincing. In recent years that has started to change. The organismal perspective is once again emerging, having gained traction under the labels of "evolutionary developmental biology," "ecological developmental biology," and the "extended evolutionary synthesis." The arguments laid out in this book can be seen as a summary and continuation of this tradition. Just like the evolutionary process itself, understanding of the causes of evolution is evolving.

Ultimately, the case for a developmental perspective is a pragmatic one. Evolutionary research benefits from incorporating developmental mechanisms and findings. We have seen that developmental insights constitute powerful sources of prediction and explanation for evolutionary biology. The studies we have described demonstrate that while researchers may not *need* to know which biases are operating in the generation of phenotypic variation, or what causes fitness differences in order to generate evolutionary explanations, they can often do a better job at both prediction and explanation if they do. Where researchers go beyond identifying the correlates of fitness differences to specify causes of natural selection, and where they propose and test clear hypotheses concerning how developmental mechanisms lead to the evolution of particular phenotypes, they can generate more powerful explanations of how evolution works, as well as for patterns of adaptation and diversification.[74]

This emerging view of evolutionary causation requires major inputs from other disciplines, not just from developmental biology but also from paleontology, molecular biology, epigenetics, ecology, physiology, behavioral biology, and the human sciences, to reveal and explain the processes and dynamics of evolution.[75] Unlike Charlesworth, we do not believe that genetics has little more of fundamental significance to contribute to understanding biological evolution;[76] rather, genetics will retain a vital and central role in the conceptual development of a wider framework for evolutionary theory. However, a concerted interdisciplinary effort will be needed for a true understanding of how evolution evolves.

Our suggestion that evolutionary biology is now benefitting from greater recognition of the roles that developmental processes play in evolution partly reflects extensive current interest within the field given to topics such as plasticity-led evolution, developmental constraint, and epigenetic inheritance. Does this imply that the contribution of development to evolution is already well recognized? Yes and no. There has been a swell of interest in these topics, particularly in the past twenty years, but not always in a manner that highlights development. For example, although the role of developmental plasticity in evolution has often been portrayed as a challenge to traditional evolutionary explanations,[77] if evolutionary biology has accommodated this hypothesis it is largely *without* conceptual change.[78] This has been possible by treating plasticity as properties of genotypes (graphically represented as reaction norms), which allows plasticity-led evolution to be reformulated and modelled mathematically as selection on standing genetic variation.[79] This heuristic fits comfortably with the genetic idealization of evolution by natural selection, and renders plasticity-led evolution a modest "add on" to existing theory.[80]

While there may be heuristic value in representing plasticity in this way, we suspect a hidden price is paid for the convenience of this abstraction. Reducing plasticity-led evolution to selection on environmentally sensitive genotypes does capture some aspects of the mechanics of the process, but it diminishes the role of development by stripping it of many of its most interesting ramifications.[81]

Were development programmed, phenotypes could not lead and genotypes could not follow, since the latter would be prerequisites for the former. However, West-Eberhard, who brought the hypothesis of plasticity-led evolution to prominence, places considerable emphasis in her writings on exploratory mechanisms, stressing how they can generate functional phenotypes without central (i.e., genetic) coordination.[82] To West-Eberhard, developmental plasticity should not just be made to fit variational explanations of adaptive change, but rather it motivates the integration of variational and transformational explanations.[83] In chapter 5 we discussed how phenotypes result from complex interactions among genotype, epigenotype, maternal environment, current environment, and sometimes learned behavior. Treating phenotypes as a property of the genome not only privileges genes over other informational sources but ignores how developmental systems introduce phenotypic novelty.[84] Conversely, once plasticity is recognized as a property of a developmental system, nontrivial ways in which it can guide adaptive evolution can be investigated.[85] When phenotypes are seen to derive from constructive, rather than programmed, development, how plasticity can generate bias, structure genetic evolvability, and facilitate adaptation and diversification can be understood.

The same point might be made with respect to Richard Lewontin's assertion that "organisms do not adapt to their environments; they construct them out of the bits and pieces of the external world."[86] There is a radical quality to Lewontin's essays on niche construction and adaptation, which argue that the organism itself, including how it develops and behaves, is an active determinant—a *cause*—of natural selection. Again, once niche construction is understood to result from the constructive processes of development, it is easier to envisage how it can affect the direction of adaptive evolution. Just as the spirit of West-Eberhard's writings might sometimes be missing in contemporary analyses of plasticity's role in evolution, so Lewontin's admonitions are often neglected in some contemporary discussions of, for example, "extended phenotypes" or "eco-evolutionary dynamics." The same argument might be made with regard to Conrad Waddington's conceptions of epigenetics, genetic assimilation, and the exploitive system.[87] These pioneers shared a vision of how to understand evolution while fully recognizing the role of development, and that vision remains a powerful inspiration to us.

Like Waddington, Lewontin, and West-Eberhard, we suggest that research in evolutionary biology should embrace the inherent reciprocities involved in evolutionary causation, and explore how developmental processes impose directionality on selection and other evolutionary processes. If evolutionary biology is to discover the mechanisms by which new phenotypes come into existence, succeed, and diversify—mechanisms in which developmental processes play key roles—then the field is only just getting started. As recent research into extragenetic inheritance demonstrates,[88] once the diversity of inheritance mechanisms is recognized, the study of adaptation at multiple temporal and spatial scales and the exploration of new forms of evolvability offer fresh opportunities for major discovery. That is because evolvability is inextricably tied to development. Even within the confines of the genetic representation of evolution by natural selection, there is plenty of valuable work to be done, yet there is no doubt in our minds that going beyond this heuristic offers the more exciting possibilities. Once it is accepted that the evolutionary process evolves, it follows evidently that developmental mechanisms must matter to evolution; and, conversely, it is because development matters that evolution is evolving.

NOTES

Notes to Foreword

1. Mayr 1961.
2. Bonduriansky and Day 2018, 6.
3. Bonduriansky and Day 2018, 7.
4. Soto and Sonnenschein 2018, 499.

Chapter 1

1. The technical names for desert woodrats and creosote bushes are *Neotoma lepida* and *Larrea tridentata*, respectively.

2. Creosote resin contains nordihydroguaiaretic acid, a phenolic compound that causes kidney cysts and liver damage in laboratory rodents (Kohl et al. 2014).

3. Kohl et al. 2014.

4. Kohl et al. 2014.

5. Kohl et al. 2014. Conversely, antibiotic-treated Mojave woodrats given a control diet not containing creosote showed no such decline.

6. These woodrats were from the Great Basin Desert. This finding is in contrast to Great Basin Desert woodrats that were not inoculated with the microbiota of Mojave woodrats (Kohl et al. 2014). Subsequent experiments by the same team have established that laboratory rats inoculated with tannin-degrading bacteria are able to consume a tannin-rich diet (Kohl et al. 2016), and rats inoculated with oxalate-degrading bacteria are able to consume an oxalate-rich diet (Miller et al. 2016). These findings suggest that many animals may be able to acquire the capacity to consume otherwise toxic foods through external access to symbionts; for instance, by coprophagy.

7. Kohl et al. 2014.

8. Creosote bushes invaded the southern portion of the range of the desert woodrat around 17,000 years ago, putting an upper bound on the inception of this dietary niche (Kohl et al. 2014).

9. J. Allen et al. 2013. Humpback whales are *Megaptera novaeangliae* and the sand lance is *Ammodytes hexapterus*.

10. J. Allen et al. 2013.

11. J. Allen et al. 2013.

12. Dias and Ressler 2014.

13. For instance, Ernst Mayr (1982, 828) writes: "The proteins of the body cannot induce any changes in the DNA. An inheritance of acquired characters is thus a chemical impossibility."

14. Obviously, we do not condone Weismann's experiment, which would be regarded as unethical by contemporary standards.

15. Francis 2011. The term "epigenetics" is often deployed to refer to alterations that don't involve changes in the DNA sequence but nonetheless influence gene activity.

16. A methyl group attachment has the chemical formula—CH_3.

17. Dias and Ressler 2014. In vitro fertilization and cross-parenting showed that this transgenerational inheritance was mediated through the gametes and not by cues from the adults.

18. Jablonka and Lamb 1995, 2014; Klosin and Lehner 2016; Jablonka 2017; Bonduriansky and Day 2018; Danchin et al. 2018; Adrian-Kalchhauser et al. 2020; Anastasiadi et al. 2021.

19. Cossetti et al. 2014; A. Sharma 2014; Klosin and Lehner 2016. *Noncoding RNAs*, both small and long, which regulate gene expression posttranscriptionally, often by binding to and thereby silencing RNA molecules, are known to be epigenetically inherited (Jablonka and Lamb 1995, 2014; Klosin and Lehner 2016; Danchin et al. 2018).

20. Moore et al. 2019. The inheritance of noncoding RNAs has also been shown to result in enhanced cognition after environmental enrichment in mice (Benito et al. 2018).

21. Kohl et al. 2014. While woodrats from regions where the species doesn't consume creosote were able to consume this deadly food when inoculated with the microbiota of Mojave woodrats, they did not achieve the same level of consumption or growth. This implies that Mojave woodrats have evolved detoxification adaptations that bolster their specialization on this diet (Kohl et al. 2014).

22. S. Gilbert et al. 2012; Roughgarden et al. 2018; S. Gilbert 2019; N. Moran et al. 2019.

23. Kamra 2005; S. Gilbert et al. 2012; Roughgarden et al. 2018; S. Gilbert 2019.

24. S. Gilbert et al. 2012; Funkhouser and Bordenstein 2013; Roughgarden et al. 2018.

25. For illustration, in humans where microbiome inheritance occurs largely through "intimate neighbor" transfer, more than 72% of marker alleles present in mothers' microbiomes were also found in their offspring (Nayfach et al. 2016).

26. Funkhouser and Bordenstein 2013.

27. Hoppitt and Laland 2013; Whiten et al. 2017; Oudman et al. 2020.

28. Social learning is also known to be responsible for the development of foraging strategies by dolphins and killer whales, as well as song types in male humpback whales and migratory routes by southern right whales (Rendell and Whitehead 2001; Whitehead and Rendell 2015; E. Carroll et al. 2015). Socially learned migrations have also been demonstrated in moose and bighorn sheep (Jesmer et al. 2018) and in several migratory species of large birds such as geese, storks, and cranes (Sutherland 1998; Oudman et al. 2020). There is extensive evidence that foraging behavior, song type, nest building, courtship, and predator recognition are influenced by social learning in many bird species (Slagsvold and Wiebe 2007; Aplin 2019). For fish, there is evidence that social learning plays a crucial role in movement decisions concerning foraging and migration (C. Brown and Laland 2001), as well as in mating preferences (B. Jones and DuVal 2019). The social learning of mating preferences and of foraging and colony location choices are also observed in invertebrates (Leadbeater and Dawson 2017; B. Jones and DuVal 2019).

29. Termite fishing: Goodall 1986; Whiten et al. 1999, 2001. Milk bottles: J. Fisher and Hinde 1949; Lefebvre 1995. Songs: Marler and Tamura 1964; Rendell and Whitehead 2001; Whitehead and Rendell 2015.

30. Éadin O'Mahony and Luke Rendell, personal communication.

31. Laland 2017a.

32. Whitehead et al. 2019.

33. Ecotypes: Foote et al. 2016; Hoelzel and Moura 2016. Reproductive isolation: Riesch et al. 2012; Moura et al. 2014, 2015. The killer whale is *Orcinus orca*.

34. Plasticity-led evolution: West-Eberhard 2003.

35. Jablonka and Lamb 1995, 2014; Jablonka and Raz 2009; Francis 2011; Heard and Martienssen 2014; Klosin and Lehner 2016; O'Dea et al. 2016; Adrian-Kalchhauser et al. 2021;

Anastasiadi et al. 2021; Stajic and Jansen 2021. These terms are explained in chapter 10. For the moment it is necessary only for the reader to be aware that there are many distinct mechanisms of epigenetic inheritance.

36. Jablonka and Raz 2009; Anastasiadi et al. 2021.

37. Stajic et al. 2019.

38. Jablonka and Lamb 2005, 2014; Bonduriansky and Day 2009, 2018; Badyaev and Uller 2009; Danchin et al. 2011, 2018; Uller 2013, 2019; Uller and Helanterä 2017.

39. Bonduriansky and Day 2018. Ernst Mayr described soft inheritance as a "misconception," the refutation of which led to major advances in evolutionary thinking (Mayr 1982, 570).

40. Mayr 1980, 15.

41. Soucy et al. 2015.

42. Organisms as communities: S. Gilbert et al. 2012. *Genotype* refers to the genetic constitution of an individual, while the *phenotype* is the set of observable characteristics of an individual—its traits. This distinction was first made by Wilhelm Johannsen (1911), and has clear heuristic value when inherited genes are the sole or principal source of heredity and phenotypes are viewed as resulting from interactions between inherited genes and the offspring's developmental environment. However, matters become more complicated in the light of extragenetic inheritance, since the accrual of hereditary information continues throughout the life cycle. For the "genotype" to comprise an organism's full hereditary information, researchers would need to recognize that the genotype concept extends well beyond information encoded in the genome sequence (for instance, encompassing epigenetic and learned information). Conversely, were the "phenotype" to remain an organism's actual observed properties—such as morphology, development, or behavior, some elements of which are inherited by nongenetic means—then some traits (e.g., skills an individual learned from its parents) could simultaneously qualify as both genotype and phenotype. Hence, in the light of extragenetic inheritance, the developing organism cannot so easily and cleanly be separated into genotype and phenotype components. Rather, the developmental system responds flexibly to a wide variety of internal and external inputs, many of which are inherited at different points in time. See Pontarotti et al. (2022) for further discussion.

43. Oyama 2000.

44. Amundson 2005. For instance, Ernst Mayr (1961) distinguished between proximate and ultimate causation, and claimed (1980, 9–10) that "the clarification of the biochemical mechanism by which the genetic program is translated into the phenotype tells us absolutely nothing about the steps by which natural selection has built up the particular genetic program."

45. For instance, leading British evolutionary biologist John Maynard Smith (1982, 6) writes: "One consequence of Weismann's concept of the separation of the germline and soma was to make it possible to understand genetics, and hence evolution, without understanding development."

46. Amundson 2005. See S. Gilbert (2003a) for a paper-length historical overview and S. Gilbert and Epel (2015, appendix C) for a brief precis. The field of evolutionary developmental biology is the most obvious and successful disciplinary integration of evolutionary and developmental biology (Hall 1999; S. Carroll 2005; Stern 2010).

47. Among the most influential of these attempts was the "organicist" movement, in particular the writings of Conrad Waddington, which we discuss in later chapters.

48. See, for instance, the exchange in the pages of *Nature* (Laland et al. 2014b; Wray et al. 2014) for a brief summary of the issues, and S. Gilbert and Epel (2015), Laland et al. (2015), G. Müller (2021), Lange (2023) and Dickins and Dickins (2023) for more detailed treatments. How radical the developmental perspective on evolution that we advocate is regarded as is a moot point. While it is frequently asserted that these new ideas are merely extensions of classical

evolutionary theory (e.g., Wray et al. 2014; Dickins and Dickins 2023), in 2016 the ideas in this book were regarded as so "heretical" by several evolutionary biologists that they signed a petition to the Royal Society trying to stop a discussion meeting on the topic from taking place, likening the practitioners to "advocates of homeopathy"!

49. Lewens 2019.

50. Lewontin 1970.

51. Badyaev 2011, D. Walsh 2015, Uller and Helanterä 2019.

52. See also Badyaev 2011; Walsh 2015; Uller and Helanterä 2019; Uller and Laland 2019; R. Watson and Thies 2019.

53. Dobzhansky 1973, 125.

54. Bonduriansky and Day 2018, 6, 7.

55. Here we paraphrase Marc Kirschner (2015, 203), who writes: "nothing in evolution makes sense except in the light of cell, molecular, and developmental biology," and Jukka Jernvall (2006, quoted in S. Gilbert and Epel 2015, 379) who claimed: "nothing in variation makes sense except in the light of development."

Chapter 2

1. S. Gould 1989, 60.

2. Gould 1989, 60; Pineda-Munoz et al. 2017; S. A. Martin et al. 2016.

3. Kavanagh et al. 2007.

4. Kavanagh et al. 2007; Roseman 2020; Hayden et al. 2020; Christensen et al. 2023.

5. Kavanagh et al. 2007.

6. Kavanagh et al. 2007. In vitro experiments have also contributed importantly to this research (e.g., Kavanagh et al. 2007).

7. Kavanagh et al. 2007.

8. Kavanagh et al. 2007.

9. Salazar-Ciudad and Jernvall 2010; Harjunmaa et al. 2014. The computational analysis is in Salazar-Ciudad and Jernvall. See also Kavanagh et al. 2007; Evans et al. 2016. As tooth size constrains the possible number of cusps, the complexity of molars also follows a linear relationship from back to front (Couzens et al. 2021).

10. Harjunmaa et al. 2014; Morita et al. 2020; Hayden et al. 2020.

11. Researchers have found that organisms (whether living or extinct) actually occupy a surprising small portion of morphospace, which reflects both biases in the production of variation and selection against dysfunctional forms (e.g., Raup 1966).

12. Note that how the clustering of points is described depends on the number of dimensions considered. Strictly, the points can be said to fall along a line only when depicted in two-dimensional space. In practice, morphospace is multidimensional.

13. The way in which developmental processes create pathways through morphospace has been formally demonstrated with mathematical evolutionary models (S. Rice 2004, 2008a,b; González-Forero 2023, 2024). The analysis shows that rather than converge on fitness peaks in the adaptive landscape, as traditionally assumed, populations converge on path peaks.

14. Kavanagh et al. 2007. See also Roseman and Delezene (2019) and Bermúdez de Castro et al. (2021) for further exceptions, which primarily relate to within-species variation. Currently, the data suggest the inhibitory cascade model is most useful for predicting macroevolutionary patterns, for which it has strong support. When applied to within-species variation, the model appears to work well for species that in macroevolutionary (between-species) analyses lie on the predicted line, but less well for species that in macroevolutionary analyses lie away from the line. This observation is consistent with the explanation that the mechanisms of development have evolved in such species, modifying the variational properties of teeth.

15. Kangas et al. 2004.

16. More precisely, the bias imposed on the distribution of phenotypic variation, arising from the structure, character, composition, or dynamics of the developmental system, relative to the assumption of isotropic variation, is known as developmental bias (Uller et al. 2018, 949). See also Maynard Smith et al. 1985; Arthur 2004; G. Müller 2007; Brakefield 2011.

17. Navarro and Murat Maga 2018.

18. Harjunmaa et al. 2014; Morita et al. 2020; Hayden et al. 2020.

19. Variation across vertebrates: Kavanagh 2020. Predictable proportions: Kavanagh et al. 2013.

20. More specifically, 419–359 million years ago (Hall 2007; Shubin et al. 1997).

21. Kavanagh et al. 2013.

22. N. Young et al. 2015.

23. Kavanagh et al. 2013.

24. The "fingertips" being the distal-most phalanges in the series Kavanagh analyzed. Note, some researchers refer to the ungual (claw) phalanx as the distal-most. The elongated phalanx in the raptors is the distal phalanx before the claw (or the penultimate phalanx if the claw is counted). Kavanagh removed the ungual from the phalanx series analyzed in this paper on the grounds that it evolved as a separate module, a module added later in evolution from the basal "phalangeal cascade" seen in the earliest tetrapods.

Birds that are diggers, perchers, and raptors have diverged from the general pattern, but in an easily comprehensible departure from the basic model, in which a distal signal extends growth and delays segmentation (Kavanagh et al. 2013). Morphological variation remains highly restricted, with size proportions evolving as a module along predictable lines of morphospace. Further work has found that other segmented systems, such as limbs, show similar patterns of correlated developmental variation and biased macroevolution of size proportions. Note, by a *module* we mean that component parts of a developmental system possess their own intrinsic dynamics. Modularity is discussed in detail in chapters 5 and 6.

25. Even the deviations from normal development are patterned. For instance, all vertebrates develop eyes, and there are many developmental anomalies involving eyes. But the eyes, no matter how they form, are always in the head, and they are never anywhere else. This is because they arise from interactions between foregut and brain cells. Similarly, no vertebrate, even malformed, has its ulna and radius proximal to its humerus. There is, as Pere Alberch (1989) said, "a logic of monsters" (see also Diogo 2017; Diogo et al. 2019).

26. In the early twentieth century, researchers such as Thomas Morgan recognized that there were rules of heredity but struggled to find rules of development, and made the sensible decision to focus on the former (S. Gilbert 1978). Now, early in the twenty-first century, evo-devo is starting to reveal rules governing the generation of phenotypic variation and to demonstrate how they affect evolvability (N. Young et al. 2015). The inhibitory cascade model may even reflect a "universal design principle" (N. Young et al. 2015) of vertebrate development, capable of predicting both short-term responses to selection in population-level variation and longer-term evolvability and patterns of microevolutionary diversity. For discussions of whether or not one can have a theory of development, see essays in Minelli and Pradeu (2014).

27. In summary, it is because science relies on representations of reality, which highlight certain causal influences but neglect others.

28. For instance, **G** can be used to predict the short-term response to selection. If the genetic covariance between two traits is nonzero, selection on one is understood to constrain the response to selection of the other. Evolution occurs in the direction of the eigenvector $\mathbf{g}_{max}$ associated with the largest eigenvalue of **G** (e.g., Guillaume and Whitlock 2007).

29. González-Forero 2023, 2024.

30. Kavanagh et al. 2013. For reciprocal causation: Laland et al. 2011, 2012, 2013.

31. Evolution and development do not operate on separate time scales, nor is the latter irrelevant to the former. As Mary Jane West-Eberhard (2003, 11) states: "the proximate-ultimate distinction has given rise to a new confusion, namely, a belief that proximate causes of phenotypic variation have nothing to do with ultimate, evolutionary explanation."

32. Here the different perspectives of the two fields are relevant. Evolutionary geneticists commonly view constraints as there not being enough genetic variation in the population to support a response to selection. In contrast, developmental biologists often view constraints as there not being a mechanism to produce particular phenotypes. When evolutionary geneticists invoke "mechanisms" they usually mean natural selection, mutation, and drift, and are discussing the causes of changes in populations. When evolutionary developmental biologists invoke "mechanisms" they are usually discussing cellular and molecular processes. As evolutionary developmental biologist Rudy Raff said of the population-based evolutionary biologists: "They're interested in species; we're interested in bodies" (quoted in Amundson 2005, 167, 253; see also Riedl 1978; Raff 1996).

33. An example of such evolutionary geneticists: Marroig and Cheverud 2001. Quote from McGlothlin et al. 2018. In the last decade in particular, there has been extensive use of methods designed to shed light on the genetic architecture of traits, such as quantitative trait loci (QTL) mapping (e.g., Feller et al. 2020).

34. Schluter 1996. The **G** matrix is a matrix whose entries specify the additive genetic variances and covariances for a series of measured traits. The name $\mathbf{g}_{max}$ is given to the first eigenvector of the variance-covariance matrix, and as this represents the direction of trait space exhibiting the most genetic variation and covariation, populations are predicted to evolve readily in this direction (Schluter 1996).

35. Schluter 1996, 1766.

36. Hansen and Houle 2008; Bolstad et al. 2014; McGlothlin et al. 2018.

37. Quote from Bolstad et al. 2014.

38. Brakefield 2006.

39. Hallgrímsson et al. 2009; Armbruster et al. 2014.

40. S. Gilbert 2018.

41. Baedke et al. 2020; Uller et al. 2020.

42. Ylikoski and Kuorikoski's (2010) analysis is not the only philosophical analysis of explanatory power, but it is broadly representative of this literature and is well regarded in philosophy of science. Most analyses (e.g., Kitcher 1989; Morton 2002; Woodward 2006) focus on one or more of the "explanatory virtues" listed here. See also Nickles (1989) for background and Baedke et al. (2020) for a more recent treatment.

43. Kavanagh et al. 2013.

44. Schluter 1996, 1766. As pointed out by several authors (e.g., Hansen and Pélabon 2021), the fact that the **G** matrix has been widely used to predict evolvability for numerous quantitative traits in diverse organisms is a virtue of this approach. Our point here is that in any such instance, the identified $\mathbf{g}_{max}$ cannot reliably be generalized as applying to other traits, or other populations. In this respect, the findings of such predictive analyses using **G** from quantitative genetics do not show "nonsensitivity" (Ylikoski and Kuorikoski 2010), even if the methods are useful, robust, and widely applicable.

45. Hansen and Pélabon 2021; Pélabon et al. 2023. In chapters 7 and 12 we discuss cases where the **G** matrix may provide reliable longer-term prediction.

46. Møller and Jennions (2002) asked "how much variance can be explained by ecologists and evolutionary biologists?" Their article began (492): "The average amount of variance explained by the main factor of interest in ecological and evolutionary studies is an important quantity because it allows evaluation of the general strength of research findings." The subsequent analysis, which used data from 43 published meta-analyses in ecology and evolution,

revealed a mean amount of variance (r^2) explained of just 2.51%–5.42%. While other subdisciplines of biology also performed poorly, Møller and Jennions acknowledged that "the amount of variance explained decreased from physiology over ecology to evolution" (498). They concluded by encouraging researchers to consider, alongside improved designs and increased sample sizes, "the merits of different avenues of research" (498), which must include developmental approaches.

Chapter 3

1. Ammonites: Monks and Palmer 2002. Spiders: Zhang and Maddison 2013. Elephants: Stanhope et al. 1998. Y chromosome: Underhill et al. 2001.

2. In one sense, prediction has always been a feature of evolutionary biology. In *The Descent of Man* ([1871] 1981), for instance, Darwin used the similarity of humans to other apes to infer, correctly, "it is somewhat more probable that our early progenitors lived on the African continent than elsewhere" (199). Likewise, today's evolutionary biologists commonly test evolutionary models by predicting, for instance, that particular evolutionarily stable strategies will be observed in nature (Maynard Smith 1982). But this is a different use of the term "prediction" from, say, the use of mathematical models to predict the weather or stock market, and relates more to anticipating that particular historical patterns will be confirmed in the future, or putting to the test models of evolutionary processes.

3. Ramakers et al. 2019.

4. Neeman et al. 2015.

5. e.g., S. Gould 1989.

6. Elena and Lenski 2003; Lassig et al. 2017; Heckley et al. 2022; D. Kemp et al. 2023.

7. Potochnik 2017.

8. Potochnik 2017.

9. Potochnik 2017; De Regt 2017.

10. e.g., Lande and Arnold 1983.

11. Orenstein 2012.

12. Schoch and Sues 2015.

13. A. Burke 1991; Nagashima et al. 2009; Rieppel 2009; R. Rice et al. 2016.

14. More generally, evo-devo researchers have established that developmental mechanisms bias the production of phenotypic variation available to selection, with some phenotypes more likely than others (Beldade et al. 2002; Arthur 2004; Kavanagh et al. 2007). This developmental bias arises from many sources, including the differential reuse of developmental modules, which enables novel phenotypes to arise by rearrangements of ancestral elements, as in the case of the turtle shell. From this evo-devo perspective, the origin of variation cannot be dismissed or taken for granted, and developmental bias is part of the explanation for why the turtle got its shell.

15. Rieppel 2001.

16. Li et al. 2008.

17. Yong 2016.

18. Lyson et al. 2016.

19. Chiari et al. 2012; N. Crawford et al. 2012; Lyson et al. 2013; Field et al. 2014; Bever et al. 2016.

20. Zardoya and Meyer 1998; Chiari et al. 2012; N. Crawford et al. 2012; Lyson et al. 2013; Field et al. 2014; Bever et al. 2016; Lyson and Bever 2020.

21. See Wylie (2019) for further examples of how the assumptions of different fields shape research agendas.

22. S. Gould and Lewontin 1979.

23. This does not mean that scientific research does not, or cannot, produce good explanations about the world. Rather, what the turtle shell example shows is that, in spite of these deficiencies, with time, the scientific process can self-correct as new data and theories become available. However, given that such self-correction cannot be guaranteed, good practice involves maintaining a self-critical stance toward those assumptions and idealizations deployed.

24. This makes it all the more important that scientists are willing and able to change their minds about their theories when evidence goes against them. As Thomas Huxley (1870) noted, this gives science a tragic dimension, for it often proceeds by "the slaying of a beautiful theory by an ugly fact."

25. Provine 1971; S. Gould 2002.

26. Amundson 2005; Rheinberger and Müller-Wille 2017; Bonduriansky and Day 2018.

27. Likewise, the alternative forms of a gene at a particular locus were called *alleles* from the Greek word meaning "different."

28. T. H. Morgan 1910, 1926, quotes at 449 and 27. For a more detailed account see S. Gilbert (1998).

29. Johannsen 1911, 159.

30. Provine 1971; Sapp 1983. These developments allowed genes to be viewed as causes of traits at a distance, without having to trace a continuous causal path from zygote (i.e., fertilized egg) to adulthood, or identifying the multitude of other possible causal factors that might contribute to the trait. In the early twentieth century, researchers studying heredity tended to assume that a single universal mechanism would account for all forms of inheritance. Morgan (1919) entitled his book on Mendelian genetics *The Physical Basis of Heredity*. This too was significant, because it meant that any processes that were inconsistent with Mendelian genetics could be disregarded as evolutionary mechanisms. Morgan dismissed soft inheritance as being of little evolutionary significance on these grounds (Amundson 2005; Rheinberger and Müller-Wille 2017).

31. Provine 1971; S. Gould 2002; Bonduriansky and Day 2018.

32. Evolutionary biology was dominated by a conflict between the Mendelian and Darwinian biometrician schools. Early Mendelians were mutationists, who questioned whether natural selection acting on continuous variation could affect evolutionary change, while the Darwinians rejected Mendelism because of its saltationist leanings. R. A. Fisher's (1918) foundational paper resolved the conflict by showing mathematically that mutations of small effect could combine to form a continuous phenotype for which correlations between relatives could be computed. Population genetics could be both Mendelian and Darwinian (Provine 1971; Amundson 2005). Another key idealization was the Hardy-Weinberg law, independently developed by its name bearers in 1908. During the nineteenth century, heredity and Darwinian adaptation had been conceived of as in opposition, the key question being whether selection could overcome the conservative force of (blending) inheritance (Amundson 2005). The law established that, in the absence of selection or mutation, Mendelian inheritance does not change the frequency of alleles. The Hardy-Weinberg equilibrium presupposes that inheritance is exclusively genetic and Mendelian, an assumption that may not always hold. Nonetheless, the idealization allowed Fisher (1918) to demonstrate that even a marginal selective advantage for an allele would suffice to allow it to spread through a population.

33. Provine 1971.

34. X-rays: Muller 1929. Selection sufficed: S. Gould 2002.

35. S. Gould 2002, 504.

36. S. Gould 2002.

37. R. Fisher 1930.

38. S. Gould 2002; Stoltzfus 2021.

39. Provine 1971; S. Gould 2002.

40. S. Gould 1980a, 1982, 1983b,c; Beatty 1988; Smocovitis 1996, 2000; Amundson 2005. The hardening is also marked by increasing adamancy that macroevolutionary patterns were explained by microevolutionary processes (S. Gould 2002).

41. S. Gould 2002.

42. S. Gould 2002. If experimental findings could not explain the trend, that is not to suggest that they did not contribute to it. For instance, Dobzhansky's influential experiments with genetic inversions showed selection in lab populations (S. Gould 2002).

Eugenics is thought to have been popular because it allowed social elites to regard themselves as inherently superior, and provided a scientific justification for racist and antiredistributionist attitudes (Sapp 2003; E. Peterson 2016). The uncomfortable truth that some leading population geneticists, including Ronald Fisher, Hermann Muller, and Julian Huxley, were enthusiastic advocates of eugenics has been treated with an embarrassed "conspiracy of silence" (S. Gould 2002, 513; E. Peterson 2016). The popularity of eugenics waned gradually after the Second World War but nonetheless persisted into the second half of the twentieth century (G. Allen 1970; Louçã 2009).

For the reformulation of human genetics: Paul 1988; Beatty 1991, 1994a; S. Gilbert 2011b.

Smocovitis (1996) argues that selectionist attitudes, but not the view that important events arise through chance, fit with a post-Second-World-War frame of mind seeking to improve the world. See also Beatty (1991, 1994a) and S. Gould (2002, 542–43). Smocovitis writes (1996, 503): "If selection had enough agency . . . then all the more rapid and possible the 'improvement' of humans. . . . More strongly selectionist models would be favored by biologists who modelled themselves after physicists at the same time they pointed the way to the 'improvement' of humanity and painted a progressive and optimistic picture of the world."

43. When Dobzhansky (1951, 11) wrote that "the study of mechanisms of evolution falls within the province of population genetics," he was reflecting the dominant view that macroevolution could be explained by the accumulation of allelic changes within a species.

44. J. Watson and Crick 1953. Francis Crick's "sequence hypothesis" (that genetic information was encoded as nucleotide bases) and his central dogma reinforced the view that evolutionary change could be adequately described as changes in gene frequency. Programs or blueprints: Rheinberger and Müller-Wille 2017.

45. Trofim Lysenko (1898–1976) was a biologist who rejected Mendelism and instead embraced a theory of environmentally acquired inheritance. Backed by Stalin, he used his political influence to undermine Mendelian genetics within the Soviet Union. Many Soviet scientists who refused to renounce genetics lost their jobs and were imprisoned or killed. On the backlash in the West: E. Peterson 2016; Bonduriansky and Day 2018.

46. Mayr 1982, 47. Historians have concluded that, partly to promote what was distinctive and important about the field, Darwin was, to quote Smocovitis (2000, 279), "reinvented as the founding father of the Synthesis" in the celebrations and publications that marked the centennial anniversary of the Origin (see also Smocovitis 1996; Sapp 2003; Amundson 2005).

47. Mayr 1980, 9–10. Mayr's (1961) distinction between proximate and ultimate causation, and in particular his use of the label "ultimate" cause for natural selection, has similarly been viewed as an attempt to foreground his field in the face of the relentless march of molecular biology (Beatty 1994b; Haig 2013; Laland et al. 2013).

48. West-Eberhard 2003; Amundson 2005; Laland et al. 2011, 2012, 2013; Moczek 2020.

49. Evolutionary idealizations often involve assumptions about the time scales of different processes. Frequently it is helpful to assume that developmental or ecological processes occur on shorter time scales than evolutionary processes, or that one such process has little impact on another. Population genetics models inheritance but typically leaves out ecological detail, for instance, while evolutionary game theory ignores genetic detail but often incorporates ecological or social variables (Otto and Day 2007). The success of these methods is built upon the

exclusion of causal factors deemed noncentral: without idealizations (e.g., selectively neutral divergence in genomes, no horizontal gene transfer) many common evolutionary methodologies (e.g., the construction of phylogenetic trees) would not be possible (Felsenstein 2004). There is no question that such approaches have been successful. Consider, for instance, game theory's ability to explain how social interactions give rise to frequency-dependent selection even without knowledge of the proximate causes of that social behavior (Maynard Smith 1982; Gintis 2009), or the capacity of population genetic models to make predictions that sort between alternative explanations for data—for instance, providing support for a parasitic rather than adaptive explanation for the existence of transposable elements in animal populations (B. Charlesworth and Langley 1989).

50. Endler 1986; Hoekstra et al. 2001; Kingsolver et al. 2001, 2012; Siepielski et al. 2009, 2013.

51. Personal communication to KNL.

52. Examples of the exceptions: Dobzhansky was interested in ecology, and later MacArthur and Levins's work in the 1960s was very ecological. Some of Lewontin's early (1950s) work involved density dependence and niche construction (e.g., in fruit flies), but this was done in the laboratory. The tradition was slightly different in the United Kingdom, where evolutionary ecology was emerging. R. A. Fisher was one population geneticist to take note. Fisher was influenced by E. B. Ford, and analyzed some of his ecological data (R. Fisher 1930). See Slobodkin (1961), Provine (1971), S. Gould (2002), and Huneman (2019) for historical detail of American population genetics at that time.

53. Again: Slobodkin 1961; Provine 1971; S. Gould 2002; Huneman 2019.

54. S. Gould 2002. At the time, population geneticists primarily studied selection in the laboratory or through mathematical modeling (Provine 1971; S. Gould 2002).

55. Mayr 1942, 32.

56. Provine 1971.

57. Lande and Arnold 1983.

58. On measuring selection in the wild: Svensson 2023. On eco-evolutionary dynamics: Odling-Smee et al. 2003; Pelletier et al. 2009; Post and Palkovacs 2009; Loreau 2010.

59. Hutchinson 1957; Post and Palkovacs 2009.

60. Again, there were exceptions; for instance, research into density-dependent selection (e.g., MacArthur 1962).

61. Roughgarden 1995; Post and Palkovacs 2009; Loreau 2010; Schoener 2011.

62. Maynard Smith 1982, 6.

63. Bonduriansky and Day 2018, x.

64. Bonduriansky and Day 2018.

65. Jennings 1937.

66. Dias and Ressler 2014.

67. Vasquez Kuntz et al. 2022.

68. Jablonka and Lamb 1995, 2005, 2014; S. Gilbert and Epel 2015, appendices A and C; Bonduriansky and Day 2018.; MacCord 2024.

69. L. Buss 1987.

70. Bonduriansky and Day 2018. Mayr (1982) describes the inheritance of acquired characteristics as "a chemical impossibility." Bonduriansky and Day (2018) characterize the field's apparent refutation of soft inheritance as "among the most influential circular arguments in the history of science" (25).

71. Rheinberger and Müller-Wille 2017.

72. Rheinberger and Müller-Wille 2017, 68. Many noncoding yet functional DNA elements have been identified (ENCODE Project Consortium 2012), while alternative splicing, RNA editing, and posttranslational modification of proteins all undermine the suggestion of a simple

correspondence between gene and protein product. Even coding genes themselves are now thought to have arisen through the mix-and-match combination of preexisting "modules," which often correspond to intron-exon boundaries and frequently specify distinct functional domains within proteins (Rheinberger and Müller-Wille 2017).

73. ENCODE project consortium 2012; Griffiths and Stotz 2013; Rheinberger and Müller-Wille 2017.

74. The idea of a "genetic program" is that the phenotype of an organism is a direct readout of the genome and that therefore development is nothing more than a chemical computer algorithm that reads the genetic code. Originally, Ernst Mayr (1961) and Francois Jacob (1974) used the idea of a genetic program to circumvent the notion of there being a teleological goal for development and that a vital force was striving for this goal (see Peluffo 2015; Soto and Sonnenschein 2018). Combining evolution and development into the notion of a program, Mayr (1961, 1503–1504), wrote:

> The completely individualistic and yet also species-specific DNA code of every zygote (fertilized egg cell), which controls the development of the central and peripheral nervous systems, of the sense organs, of the hormones, of the physiology and morphology, is the program for the behavior computer of this individual. Natural selection does its best to favor the production of codes guaranteeing behavior that increases fitness. . . . The purposive action of an individual, insofar as it is based on the properties of its genetic code, therefore is no more or less purposive than the actions of a computer that has been programmed to respond appropriately to various inputs.

This view was enhanced and popularized by Richard Dawkins in *The Selfish Gene* (1976). The closely related concept of *genetic determinism* contends that all important human phenotypic traits, especially our behaviors, are directly controlled by nuclear genes.

75. Wray et al. 2014.

76. S. Gilbert and Sarkar 2000; E. Peterson 2016.

77. S. Gilbert and Sarkar 2000; Nicholson and Gawne 2015; Peterson 2016; Baedke and Fábrigas-Tejeda 2023.

78. Peterson 2016. See also Nicholson 2014; Baedke 2019; Baedke and Fábregas-Tejeda 2023. Organicism has been traced back to turn-of-the-twentieth-century pioneers such as Charles Otis Whitman, Conwy Lloyd Morgan, and John Scott Haldane (Peterson 2016).

79. Waddington was also a pioneer of the perspective now known as *niche construction theory* (Odling-Smee et al. 2003). He argued that biological evolution involves four major factors, one of which was what he called "the exploitive system," the process through which animals choose and modify their own niche (Waddington 1959).

Other members of the Theoretical Biology Club included Joseph Woodger, Joseph Needham, Dorothy Needham, Dorothy Wrinch, and John Desmond Bernal, while other intellectuals, including Karl Popper and Peter Medawar, regularly participated in the discussions (Peterson 2016). Other influential organicists include Paul Weiss, Jakob von Uexküll, and John Scott Haldane (Peterson 2016).

80. Waddington (1942) 2012; S. Gilbert 2011a,b.

81. Peterson 2016; Baedke 2021; Baedke and Fábrigas-Tejeda 2023. For instance, Ernst Mayr repeatedly discredited the work of the organicists, unfairly categorizing it as "Lamarckian" and even "Lysenkoist" (Peterson 2016; Baedke and Fábrigas-Tejeda 2023).

82. Sapp 1987; Peterson 2016; Bonduriansky and Day 2018. See earlier footnote on Lysenko. The dubious association of extragenetic inheritance with Lysenkoism has continued into the twenty-first century. Indeed, three of the authors of this book have been fallaciously labelled Lysenkoist sympathizers for no more than emphasizing that cultural processes may play important evolutionary roles.

83. Peterson 2016. The fields of evolutionary biology, genetics, and statistics have a disturbingly long-standing relationship with eugenics that lasted well into the second half of the twentieth century (G. Allen 1970; Louçã 2009). Francis Galton invented the term "eugenics," and founded the eugenics movement in the 1870s and 1880s (Allen 1970). Karl Pearson founded a scientific journal for the study of eugenics in 1925 (*Annals of Eugenics*) and was its first editor (the journal was renamed *Annals of Human Genetics* in 1954). Ronald Fisher was the founding chairman of the University of Cambridge Eugenics Society in 1911 and his appointment in 1933 at University College London was as professor of eugenics (*Inquiry into the History of Eugenics*, 2020 https://www.ucl.ac.uk/provost/sites/provost/files/ucl_history_of_eugenics_inquiry_report.pdf). The second edition of Fisher's classic book *The Genetical Theory of Natural Selection*, published in 1958, devoted five chapters to eugenics. Julian Huxley was president of the British Eugenics Society from 1959 to 1962. Hermann Muller was a leader of the American eugenics movement, retaining an influence up until his death in 1967 (Allen 1970).

84. Peterson 2016. In addition to organicism, there are several other historical subfields or intellectual movements that shaped or anticipated the ideas in this book. The process philosophy of mathematician and philosopher Alfred North Whitehead has some overlap with organicist ideas, and influenced the movement. Cybernetics, which is thought to have influenced Waddington and some other organicists, focused on regulatory feedback processes, including in biological systems, with notable work from Ross Ashby, Norbert Wiener, and John von Neumann. The structuralism of D'Arcy Wentworth Thompson, as seen in his classic book *On Growth and Form* (1917), was also a major influence on Waddington (Fabris 2021), as well as on Stephen Jay Gould. Richard Levins and Richard Lewontin (1985) are explicit in acknowledging a debt to dialectical materialism. In his *Factors of Evolution* (1949) book, Ivan Schmalhausen attempted to integrate embryology within evolutionary biology. The *Towards a Theoretical Biology* workshops and book series, organized and edited by Waddington (Waddington 1968, 1969a,b, 1970, 1972), anticipated, initiated, or significantly developed many of this book's themes, including reciprocal causation, evolvability, and the roles of developmental bias, plasticity, and niche construction in evolution.

85. Peterson 2016. Other researchers who adopted broadly sympathetic positions to the organicists include the American psychologist James Mark Baldwin (1861–1934), the Scottish biologist Edward Stuart Russell (1887–1954), the Chinese developmental psychologist Zing-Yang Kuo (1898–1970), the Russian zoologist Ivan Schmalhausen (1884–1963), and American comparative psychologist Daniel Lehrman (1919–1972). The works of Niles Eldredge (1985); Stephen Jay Gould (1980b, 1983a, 2000); Mae-Wan Ho and Peter Saunders (1984); Steven Rose, Richard Lewontin, and Leon Kamin (1984); Richard Levins and Richard Lewontin (1985); and Brian Goodwin (2001) can all be viewed as developmentalist perspectives on evolution.

86. S. Gilbert et al. 1996; Brakefield 2006, 2011; G. Müller 2007.

87. Kavanagh et al. 2007, 2013; Brakefield 2010; N. Young et al. 2015; Uller et al. 2018.

Chapter 4

1. Bonner 1982; S. Gould 2002; Amundson 2005; Lewens 2019.

2. Lewontin 1970.

3. Feldman and Cavalli-Sforza 1981; Boyd and Richerson 1985; Mesoudi 2011; Laland 2017a. For instance, bicycles vary in their designs (phenotypic variation), with some variants manufactured more and selling better than others (differential fitness), and with the production lines of bicycle manufacturers maintaining some continuity in design over time (heredity). As a consequence, the properties of the bicycle have evolved over time, including through motorization to produce motorcycles (Petroski 1992).

4. Badyaev 2011; D. Walsh 2015; Uller and Helanterä 2019. Philosophers describe these three processes as "quasi-independent" (D. Walsh 2015).

5. Hoekstra et al. 2006.

6. Mayr 1961.

7. Even in this example, however, chosen for its fit with traditional lines of thinking, there remains the possibility that mice actively choose habitats that match their coat color.

8. Inherited techniques: Foote et al. 2016; Hoelzel and Moura 2016. Sympatric groups: Riesch et al. 2012; Moura et al. 2014, 2015.

9. Foote et al. 2016.

10. Moura et al. 2015.

11. Foote et al. 2016.

12. Gene-culture coevolution: Whitehead et al. 2019. Central role in human evolution: Laland et al. 2010, Creanza and Feldman 2016.

13. Maladaptive culture in animals: Galef 1995; Laland 2017a. In humans: Feldman and Cavalli-Sforza 1981; Boyd and Richerson 1985; Richerson and Boyd 2005; Mesoudi 2011.

14. Feldman and Cavalli-Sforza 1981; Boyd and Richerson 1985; Richerson and Boyd 2005; Mesoudi 2011.

15. Avital and Jablonka 2000; Hoppitt and Laland 2013; Whitehead and Rendell 2015; Laland 2017a; Danchin et al. 2018; R. Kendal et al. 2018; Aplin 2019; Whiten 2019, 2021.

16. Darwin 1881.

17. K. Lee 1985, Turner 2000; Odling-Smee et al. 2003; Dempsey et al. 2013.

18. K. Lee 1985; Turner 2000; Odling-Smee et al. 2003; Dempsey et al. 2013.

19. Turner 2000; Odling-Smee et al. 2003; Dempsey et al. 2013.

20. Scott-Phillips et al. 2014.

21. Turner 2000; Odling-Smee et al. 2003; Laland et al. 2019.

22. Turner 2000; Odling-Smee et al. 2003; Laland et al. 2019.

23. Physiological analyses: Turner 2000. See also Odling-Smee et al. 2003.

24. Fogarty and Wade 2022; see also González-Forero 2023, 2024. More generally, recognizing niche construction as an evolutionary process is an attempt to describe evolutionary causation in a manner that does not dismiss or trivialize this causal interdependence (e.g., Odling-Smee et al. 2003; Laland et al. 2019).

25. Laland et al. 2019; Uller and Helanterä 2019; Odling-Smee 2024.

26. Uller et al. 2018.

27. These processes all create and modify evolutionary trajectories; for a formal demonstration, see González-Forero (2023, 2024).

28. For the debates: Laland et al. 2014b; Wray et al. 2014; Laland et al. 2015; D. Charlesworth et al. 2017; Futuyma 2017.

29. Such strategies may be examples of what philosopher Imre Lakatos (1978) called "auxiliary hypotheses," adopted to protect the "hard core" of a major hypothesis.

30. Pocheville 2019.

31. E.g., Scott-Phillips et al. 2014.

32. See Uller and Laland (2019) for a more detailed treatment.

33. This is one reason why examples of how plasticity and extragenetic inheritance contribute to adaptation continue to excite controversy within the field, and why they are considered some of the main challenges to the "standard" narrative for adaptive evolution (e.g., West-Eberhard 2003; Jablonka and Lamb 2014; Laland et al. 2015; Moczek 2015). When the component processes that underlie evolution by natural selection are causally intertwined, one cannot isolate differential fitness as the sole contribution to the adaptive fit between organism and environment. As a consequence, proximate causes like behavioral plasticity and social learning become important in explanations for the evolution of adaptive fit (Uller et al. 2020).

34. Shea et al. 2011; Dall et al. 2015; Uller et al. 2015; McNamara et al. 2016; Danchin et al. 2019; Proulx et al. 2019; Stajic et al. 2019; Stajic and Jansen 2021; Anastasiadi et al. 2021; Graeve et al. 2021a; Rampelli et al. 2021.

35. E.g., Dickins and Rahman 2012; D. Charlesworth et al. 2017.

36. E.g., Dickins and Rahman 2012; D. Charlesworth et al. 2017.

37. E.g., Dickins and Rahman 2012; D. Charlesworth et al. 2017.

38. Ghalambor et al. 2007; Lande 2009.

39. S. Gould 1980b, 2002.

40. S. Gilbert, interviewed for the Università di Padova, YouTube, September 2, 2019, https://www.youtube.com/watch?v=i41lkb1rL2o, quotes at approx. 1:40.

41. Potochnik 2017.

42. The opening sentences of the preface of R. A. Fisher's (1930, vii) book read: "Natural selection is not evolution. Yet, ever since the two words have been in common use, the theory of natural selection has been employed as a convenient abbreviation for the theory of evolution by means of natural selection, put forward by Darwin and Wallace."

43. Mayr and Provine 1980.

44. Laland et al. 2015.

45. Mayr 1961; Futuyma 2017, D. Charlesworth et al. 2017.

46. E.g., D. Charlesworth et al. 2017.

47. Kounios et al. 2016; R. Watson 2020. Arguably, explicit incorporation of development into conceptions of evolution by natural selection retains unnecessary biological detail if the aim is to generate understanding about the role of natural selection in maintaining alternative phenotypes. In this case, population genetics or game theory—with their highly idealized assumptions about variation and inheritance—are effective frameworks. In contrast, a developmental representation of evolution by natural selection might help to reveal the joint effect of natural selection and developmental bias on evolution (Uller et al. 2018). Evolution is fundamentally a recursive process, and evolution by natural selection is itself an evolving property (R. Watson et al. 2016), making it potentially misleading to assume that the phenotypic effects of mutation do not impose directionality on evolution (even when there has been no past selection for evolvability; R. Watson and Szathmary 2016; Uller et al. 2020). A similar case for the usefulness of a developmental view of evolution by natural selection can be made for understanding evolutionary transitions in individuality (R. Watson et al. 2016).

48. For example, from the 1980s, with the discovery of Hox genes, developmental genetics became hugely successful in part because it became possible to ignore downstream developmental processes and focus on genes and phenotypes. This allowed developmental geneticists to make sense of macroevolutionary patterns in terms of a common "tool kit" of master regulatory genes that govern the formation and patterning of bodies (S. Gilbert et al. 1996; S. Gilbert 2003a; S. Carroll 2005).

49. S. Gilbert 2001, 2003a; West-Eberhard 2003; S. Gilbert et al. 2015.

50. Schlichting and Pigliucci 1998; S. Gilbert 2001, 2003a; West-Eberhard 2003, 2005; G. Müller 2007; S. Gilbert et al. 2015.

Chapter 5

1. While we will not dwell on it, many of the issues we discuss in this chapter apply also to single-celled organisms, even if these are sometimes considered not to have development.

2. Barresi and Gilbert 2020, G-9.

3. Hamburger 1980; S. Gilbert 2003a; Amundson 2005. The title of this chapter—"Opening the Black Box"—has the intended implication that, unlike the historically dominant approach, in this book we will be considering the mechanisms of development and drawing on them to

shed light on evolutionary processes. We acknowledge that there is a little looseness in our use of metaphor here. Technically, a black box cannot be opened, as it is an overt strategy of ignoring the contents of the box. We hope the reader will forgive this deliberate slippage.

4. A comprehensive treatment of development would be beyond the scope of this book, but we hope that this chapter will stimulate readers to appreciate how an understanding of developmental biology can be useful, and encourage them to seek that grounding.

5. Our use of the term "epigenetics" in this chapter refers to developmental interactions above the level of the gene (Holliday and Pugh 1975; Riggs 1975). Included in this usage is the recognition that the phenotype is often dependent on higher-level interactions between genes and other elements in regulatory networks.

6. The fact that there is always an organized phenotype already in place at all stages of development West-Eberhard (2003, 29) calls the "continuity of the phenotype."

7. West-Eberhard 2003.

8. West-Eberhard 2003.

9. Readers should nonetheless be wary that this standpoint brings inherent dangers. It has misled even sophisticated readers into envisaging that the only important entity transmitted across generations is DNA, that the zygote's DNA contains all the instructions necessary for development with the environment playing little role, and that there is negligible structure to the egg and its surroundings. See, for instance, Dawkins's (1982, 259) characterization of the bridge between generations as entirely genetic, with each individual a "return to the drawing board."

10. See Barresi and Gilbert (2020) for further detail, as well as for comparable descriptions of plant life cycles.

11. In practice, the claim that all the cells in the human body contain the same genes is not quite correct, for three reasons. First, not all the cells in the human body are human—our bodies contain many symbionts. Second, some human cells, such as red blood cells, lack a nucleus and have no DNA. Third, cells acquire mutations during development.

12. Arendt et al. 2016; Barresi and Gilbert 2020.

13. Barresi and Gilbert 2020, 60.

14. Arendt et al. 2016.

15. Barresi and Gilbert 2020.

16. L. Wolf et al. 2009.

17. The three-spined stickleback is *Gasterosteus aculeatus*.

18. Shapiro et al. 2004; Y. Chan et al. 2010.

19. Peter and Davidson 2015.

20. Peter and Davidson 2015; Alon 2020.

21. Hovland et al. 2020.

22. For illustration, hedgehog—so named because the mutation originally identified in *Drosophila* caused larvae to be covered in pointy spikes—is a prominent paracrine family comprising a variety of factors (playfully named sonic hedgehog, desert hedgehog, and Indian hedgehog) that function as morphogens in a wide variety of vertebrate and invertebrate tissues. The hedgehog pathway is known to play a central role in vertebrate limb patterning, neural differentiation, and the formation of the face and eyes. The same pathway is also implicated in numerous medical conditions. Cyclops, of Greek legend, is the name of a condition that arises in mammals when the carrier expresses hedgehog mutants.

23. For instance, in a classical model known as the "French flag," high concentrations of a morphogen produce one cell type, medium concentrations a second type, and low concentrations a third, to generate tissue that resembles the colored bands of the flag (L. Wolpert 1969).

24. Recent cell lineage tracing studies suggest that a small proportion of neural crest cells are unipotent (Barresi and Gilbert 2020). Wnt is a portmanteau name derived from the wingless and *Int-1* mutations. Bmp stands for bone morphogenetic protein.

25. Technically known as placodal ectoderm.

26. Whether researchers wish to regard the entire complex as a single GRN or the induction, specification, and migration modules as constituting separate GRNs "wired together" into a larger complex is largely a matter of taste.

27. In reality, actual GRNs comprise not just a network of transcription factors but also their downstream targets and various other active elements.

28. When cells move together they often extend "arms" that temporarily connect the moving cells, allowing them to "follow the leader" (Barresi and Gilbert 2020).

29. Hovland et al. 2020.

30. Verd et al. (2019) point out that, in spite of its usefulness, structural modularity has a number of serious limitations. Some theoretical studies suggest that modularity is not necessary for evolvability (see, e.g., Crombach and Hogeweg 2008). Furthermore, it is notoriously difficult to identify structural modules and delimit their boundaries with any precision. One reason for this may be that the definition of (sub)system boundaries is fundamentally context- and problem-dependent (see, e.g., Chu et al. 2003; Chu 2011). More to the point, even the simplest subcircuits tend to exhibit a rich dynamic repertoire comprising a range of different behaviors depending on context (boundary conditions), quantitative strength of parameter values (determining genetic interactions as well as production and decay rates), and the specific form of the regulation-expression functions used to integrate multiple regulatory inputs (Mangan and Alon 2003; Wall et al. 2005; Ingram et al. 2006; Siegal et al. 2007; Payne and Wagner 2015; Ahnert and Fink 2016; Page and Perez-Carrasco 2018; Perez-Carrasco et al. 2018). Because of this, it is usually not possible to single out subsections of the network exhibiting specific behaviors and functions that are robustly independent of their network context. In cases like these, looking for structural modules is not the most fruitful approach to partition a complex regulatory network.

31. Gerhart and Kirschner 1997; West-Eberhard 2003; M. Kirschner and Gerhart 2005.

32. Davidson and Erwin 2006.

33. Wahlbuhl et al. 2012.

34. Hovland et al. 2020.

35. Salazar-Ciudad 2006; Borenstein and Krakauer 2008; Jaeger and Crombach 2012; Jaegar and Monk 2014; R. Watson and Szathmary 2016.

36. Developing organisms are complex systems, and computational models have established that most complex systems have more than one stable state (or *attractor*) at any time, a challenge to many simple population genetic models. At the same time, computational analyses of GRNs reveal that, despite their complexity, the number of phenotypic outcomes is often surprisingly small, with many different genotypes mapping to a small number of phenotypes (Borenstein and Krakauer 2008; Munteanu and Solé 2008; Jaeger and Crombach 2012). This is partly a direct outcome of the positive feedback loops manifest in developing systems, and helps us to understand why only a small subset of possible phenotypes are observed in nature (Raup 1966; Maynard-Smith et al. 1985; Brakefield 2006). We return to this issue in chapter 12, where we discuss theory showing that the phenotypic equivalence of vast domains of the space of possible regulatory networks enhances the evolvability of the system (Ciliberti et al. 2007; A. Wagner 2011b). Such analyses also illustrate how the phenotypic variation available for natural selection will typically be biased, sometimes in a functional manner, even when mutations are randomly distributed (R. Watson and Szathmary 2016; Uller et al. 2018). Computational models of GRNs also help us to understand how developmental systems sometimes show robustness, generating the same phenotype despite mutation or environmental change. Changes to the regulatory elements in the GRN will generate phenotypic change only when they push the system into a new basin of attraction, and the positive feedback built into many GRN modules has almost certainly evolved precisely because it confers robustness to perturbation. However, if the altered

gene-product concentrations of the GRN suffice to push the system toward a new attractor, dramatic change may ensue. This explains why developmental systems often exhibit threshold effects, where small perturbations have little or no effect, but sudden change occurs once perturbations reach a particular magnitude (Jaeger and Crombach 2012; Jaegar and Monk 2014). Thus robustness is not incompatible with evolutionary change (Ciliberti et al. 2007; A. Wagner 2011b).

37. Liu et al. 2019.

38. We return to the issue of what we mean by information and where it resides in chapter 6.

39. Newman 2010.

40. Waddington (1942) 2012. The DNA in cells is often clothed in a variety of chemically attached molecules that alter gene expression. Today, most commonly the term "epigenetics" refers to such attachments and alterations that don't involve changes in the DNA sequence but nonetheless influence gene activity.

41. Waddington (1942) 2012, 1957. Waddington's original use of the term "epigenetics" was as a portmanteau word combining "epigenesis" and "genetics." His usage referred to what we nowadays call "developmental genetics," but included in this concept was the recognition that outside agents were crucial to development, that the genome was both active and reactive, and that the phenotype was often dependent on higher-level interactions between genes and other regulatory elements in "networks."

42. Libbrecht and Rasmussen 2004.

43. Townes and Holtfreter 1955.

44. Steinberg and Takeichi 1994.

45. Forgacs and Newman 2005.

46. Newman 2010.

47. Chen et al. 2022; Pang et al. 2023.

48. Tallinen et al. 2016.

49. Tallinen et al. 2016.

50. Another good illustration of how physical resources can affect development is provided by the yolk content of zygotes (Barresi and Gilbert 2020). Yolk is a nutrient-rich resource that functions to supply the developing embryo with food. Patterns of cleavage vary widely across animal groups, with the sheer volume of yolk a major determinant of this variation. Yolk impedes cleavage, and hence when eggs undergo cleavage, the cleavage furrow starts in the least yolky region and gradually spreads across the egg. If the yolk is sparse and evenly distributed, as for instance in echinoderms, the cleavage typically extends through the entire egg, and divisions are symmetrical. Similar, if not identical, cleavage patterns occur in mammals that derive their nutrition through the placenta. However, in animal eggs where one region is relatively yolk-free, cellular divisions occur there at a faster rate than at the yolky part. Extreme examples are fish, reptiles, and birds, where just one small yolk-free region of the egg undergoes cleavage, creating a disc of cells adjacent to the yolk. That is because here most of the cell volume is made up of yolk, necessary to supply the animal with all the nutrients it requires throughout embryonic development. Many insects undergo a superficial form of cleavage in which the cells divide only on the external rim, with yolk in the center.

51. Newman 2010.

52. Tung and Levin 2020.

53. L. Wolpert 2011.

54. Tung and Levin 2020.

55. External signals are also necessary for daughter cells to retain the same state as the mother cell, rather than differentiate.

56. Typically, zygotic gene activity is not required until the blastula stage. When they have done their job, the maternally provided mRNAs are degraded. The destruction of maternal

mRNAs is itself initially accomplished by maternally encoded products, although this is accelerated by zygotic transcription. Such a maternal-to-zygotic transition is a feature of numerous vertebrates, invertebrates, and plants (Baroux and Grossniklaus 2015; Tadros and Lipshitz 2009).

57. Nishida and Sawada 2001.

58. Barresi and Gilbert 2020.

59. For instance, one prominent set of paracrine factors is known as the fibroblast growth factor (Fgf) family. Fgf8, a member of this family, is the paracrine factor centrally involved in lens induction during the formation of the vertebrate eye. Fgf8 triggers a specific signal transduction pathway, called Rtk. Here we sketch some details of this pathway to get across the sheer complexity of signal transduction. Fgf binds to an Rtk receptor on the cell membrane, leading to chemical changes in Rtk within the cell, which activate the Gef protein, which activates another protein (Raf), which brings about changes in an enzyme (Mek), which enters the cell nucleus in a new form (Erk), which activates transcription factors in the nucleus and finally elicits transcription of the cell's genes. Many elements in this chain also activate other pathways, hence causality is both diffuse and nonlinear. In spite of its complexity, the pathway produces reliable outcomes that are deployed repeatedly in the organs of diverse organisms, from nematodes to fruit flies to humans.

60. Dalle Nogare and Chitnis 2017. More recent work has identified a key role for a third cell type, the iridophore (Patterson and Parichy, 2019). See also Kondo et al. 2021.

61. Dalle Nogare and Chitnis 2017.

62. Dalle Nogare and Chitnis 2017.

63. Lange and Müller 2017; Lange 2020.

64. Martino et al. 2018; Ayad et al. 2019.

65. Shyer et al. 2017.

66. Herring 1993.

67. Eyal et al. 2015.

68. Herring 1993; Barresi and Gilbert 2020.

69. Aragona et al. 2020.

70. Hu and Albertson 2017.

71. Hu and Albertson 2017.

72. Hu and Albertson 2017. For other examples of how embryonic movement affects development, see G. Müller (2003a).

73. Hu and Albertson 2017. These authors report that the bone most affected is the retroarticular process, a critical bone for the action of jaw opening.

74. Hu and Albertson 2017.

75. Parsons et al. 2014; Hu and Albertson 2017; Zogbaum et al. 2021. Likewise, in several fishes the number of vertebrae can change in response to temperature (e.g., Corral et al. 2019), and this plasticity appears to have been lost in some other vertebrate groups.

76. Differences in feeding behavior: e.g. Hegrenes 2001; Parsons et al. 2011. Experimental manipulation: Parsons et al. 2016.

77. Similar effects have been demonstrated for vertebrate limb bone growth, where again embryo movement stimulates bone cell proliferation and alters limb proportions (A. Pollard et al. 2017). Limb magnitude depends partly on the size of the original cartilage condensation but also partly on changes in growth during ossification due to proliferation of the cartilage cells, which is linked to patterns of muscle contraction. Experiments show that prenatal developmental plasticity allows for variation in limb proportions to emerge via environmentally triggered alterations in movement, and that these effects are of sufficient magnitude to be a major source of interspecific geographical variation in body size (A. Pollard et al. 2017),

potentially contributing to global trends such as Allen's rule—the trend for smaller surface-area-to-body-size ratios in cooler climates, often achieved through reductions in limb length.

78. Other examples include voluntary movement to develop the brain's motor system and play in the development of behavior (West-Eberhard 2003). The mammalian visual system, for example, begins its development in utero, where light is limited and thus of little use to guide development. However, rather than the eyes forming autonomously, their development requires spontaneous activity that mimics the action potentials triggered by light after birth. These action potentials are endogenously generated by the ganglion cells, which fire spontaneously and synchronously with one another, generating "waves" of activity that travel across the retina (West-Eberhard 2003).

79. West-Eberhard 2003.

80. Standen et al. 2014.

81. Barresi and Gilbert 2020.

82. L. Wolpert 2011.

83. Brakefield and Reitsma 1991.

84. Rembold 1987; Rachinsky and Hartfelder 1990; S. Gilbert and Epel 2015. The differentiation of turtle gonads into either ovaries or testes is determined by the temperature experienced by the embryo as its gonads are forming. This dependency on the environment has important consequences for the effects of global warming (L. Hawkes et al. 2009; Tomillo et al. 2012).

85. Weber et al. 2020. Environmental influences on gene expression can also be mediated by gene methylation, as is the case for the effects of royal jelly diet on honeybee development into workers or queens (Lyko et al. 2010).

86. Moczek and Emlen 1999; Moczek 2005; Casasa et al. 2020.

87. Casasa et al. 2020.

88. Other examples include the massive changes to roots and leaves in terrestrial and submerged plants adapted to a semiaquatic lifestyle, and the distinctive morphology of water fleas reared in the presence or absence of predators (West-Eberhard 2003; S. Gilbert and Epel 2015). One of the most impressive examples of using the external environment as a developmental agent occurs during mammalian birth (Del Negro et al. 2018; Tan and Lewandowski 2020; Shi et al. 2021; S. Fu et al. 2023). Birth is not merely the "delivery" of a preformed baby. Rather, the acquisition of oxygen has to be shunted from the placenta to the lungs. The first breath of air changes the pressure balances throughout the body. This causes gene expression changes in the endothelial cells of the lungs, causing them to pump amniotic fluid out of the lungs and into the lymphatic circulatory system, thus clearing the lungs for breathing. Second, in the brain stem, the pressure changes initiate one of our most important unconscious behaviors, breathing. As Shi and colleagues (2021, 426) proclaim, "among numerous challenges encountered at the beginning of extrauterine life, the most celebrated is the first breath that initiates a life-sustaining motor activity." And third, the heart itself changes. The vessels that allowed circulation to the placenta are closed, such that the heart becomes divided into the chambers pumping blood to the lungs and the chambers pumping blood to the rest of the body.

89. Chiu and Gilbert 2015; S. Gilbert 2020b.

90. Bäckhed et al. 2005; Sender et al. 2016.

91. Kimura et al. 2020; Vuong et al. 2020. See S. Gilbert and Epel (2009) and S. Gilbert et al. (2012) for an overview.

92. Vertebrates: S. Gilbert et al. 2012. Invertebrates: Barresi and Gilbert 2020.

93. McFall-Ngai and Ruby 1991; Montgomery and McFall-Ngai 1994.

94. Essock-Burns et al. 2021.

95. Symbiotic bacteria may have played key roles in evolution by changing organismal development, potentially facilitating major transitions. For instance, some species of unicellular choanoflagellates can become multicellular when exposed to certain bacteria. Moreover, other

bacteria induce these cells to undergo meiosis and have sex. Remarkably, animals have evolved to expect bacterial contributions to their development. This has led to the hypothesis that animals evolve as "teams" or consortia of several species (Chiu and Gilbert 2015; S. Gilbert 2020b).

96. This is sometimes known as a "generate-test-regenerate heuristic" (D. Campbell 1960).

97. Gerhart and Kirschner 1997.

98. Edelman 1987; M. Kirschner and Gerhart 2005.

99. Gerhart and Kirschner 1997.

100. Hubel and Weisel 2005.

101. The mechanism is technically known as "instrumental" or "operant" conditioning (Staddon 2016)

102. Pulliam and Dunford 1980; Staddon 2016.

103. Gerhart and Kirschner 1997; West-Eberhard 2003; M. Kirschner and Gerhart 2005.

104. Klenerman 2017.

105. Kardon 1998; Rodriguez-Guzman et al. 2007.

106. Gerhart and Kirschner 1997; West-Eberhard 2003.

107. Gerhart and Kirschner 1997; West-Eberhard 2003; Hall 2015; Gordon 2023.

108. Plausibly, many aspects of gene regulation may involve stabilization of initially stochastic patterns of gene expression.

109. M. Kirschner and Gerhart 2005, 155.

110. Soen et al. 2015.

111. Schlicting and Pigliucci 1998; Moczek et al. 2011; Uller et al. 2020.

112. Gerhart and Kirschner 1997.

113. Gerhart and Kirschner 1997.

114. M. Kirschner and Gerhart 2005.

115. Liu et al. 2019.

116. Studies of regeneration—such as in salamanders that regenerate limbs, or in planarian flatworms that regenerate amputated parts of their body (Maden 2009, 2018)—raise the possibility that some developing organisms may nonetheless have a target morphology (Tung and Levin 2020). It seems that bioelectric signaling provides a template that is exploited to define developmental axes and direct the regrowth of limbs (Tung and Levin 2020). This implies that an important component of the control of development is biophysical: instructive patterning information is encoded by the bioelectric states of tissues. These bioelectric states are manifest at a higher-order (i.e., "epigenetic") level of organization: not only are they not a direct product of the genome, they are themselves an immediate cause of downstream gene expression and morphogenesis (Levin 2020).

117. Helms et al. 2005.

Chapter 6

1. Callebaut and Rasskin-Gutman 2005. For instance, concerted Wnt and Bmp action is required in all cases, and the same GRN is activated (Steinhart and Angers 2018). Moreover, Wnt and Bmp are also involved in many more developmental events that do not involve cells of the neural crest. For instance, Wnt is involved in bone remodeling and the formation of the anterior–posterior (head–tail) axis of the embryo, while Bmp is critical for heart formation and specifies the dorsoventral (front–back) axis of the organism (Barresi and Gilbert 2020).

2. Another classic example is the vertebrate limb. Early in development the embryo will produce a limb "bud," comprising a collection of multipotent mesodermal cells covered by an ectodermal layer. Strikingly, a limb bud can be transplanted to other parts of the body—or even between species—and still develop (Barresi and Gilbert 2020). This demonstrates that the initiation of a limb is quite independent of the development of the limb. Likewise, an

experiment that grafted a section of chick embryo spinal cord that would normally innervate wing muscles into the region that serves the legs found that the chicks activated both legs together, as though they were trying to flap their wings (L. Wolpert 2011).

3. Tomoyasu et al. 2009.

4. Zattara et al. 2017.

5. Barresi and Gilbert 2020.

6. Songs: Bolhuis et al. 2010. Courtship: Bolhuis et al. 2022.

7. For instance, a version of "massive modularity" of mind, which regards cognition as modular right through from perception to action, with modules rarely interacting, became popular among evolutionary psychologists but is unlikely to reflect how the brain works (M. Anderson 2010). More recent evolutionary psychology accounts, while retaining an emphasis on domain specificity, acknowledge this interaction between domains (H. Barrett 2015).

8. Schlosser and Wagner 2004.

9. Badyaev 2009; Koyama et al. 2013.

10. Gerhart and Kirschner 1997; M. Kirschner and Gerhart 2005.

11. M. Kirschner and Gerhart 1998.

12. Note, while our use of "epigenetic" here is broadly consistent with Waddington's original (1940, 1959) usage, for Waddington the "epi" did not mean "above" (as in "epidermis") but rather was short for epigenesis. Waddington's "epigenetics" is what we would now call developmental genetics, where the environment could also activate genes.

13. Waddington 1939; S. Gilbert 2003b; E. Keller 2010.

14. Wray et al. 2014.

15. Barresi and Gilbert 2020, 20.

16. Moss 2003; Bizzarri et al. 2019; Hallgrímsson et al. 2023.

17. Green and Annas 2008; D. Wheeler et al. 2008.

18. For us, "regulatory network" is an "epigenetic" (*sensu* Waddington 1940, 1959) rather than genetic phenomenon, so we avoid the terms "genetic architecture" and "G-P map," but acknowledge parallels with related arguments made by Hansen (2006, 2011), G. Müller (2007), Pavličev et al. (2023), and Hallgrímsson et al. (2023). For more detailed treatments of levels of biological organization consistent with our viewpoint, see Wimsatt (2021) and Griesemer (2021).

19. Individual structures such as bones can grow at different rates, allowing the shape and proportions of organisms to be changed substantially during evolution by heritable changes in the duration of growth. In the horse, for example, the central toe has evolved to grow faster than the other toes, resulting in the hoof, while the skeleton-forming cells of toy poodles have undergone fewer cell divisions than those of Great Danes. Form arises through cell death too—the orifices of our mouths, anuses, and reproductive organs all form through the programmed death of cells at key points in development, while, in humans, the death of cells between the developing digits' cartilage is essential for separating our hands and feet into individual fingers and toes (L. Wolpert 2011).

20. Arthur 2004.

21. Barresi and Gilbert 2020.

22. Barresi and Gilbert 2020.

23. Abzhanov et al. 2004. Here, BMP4, the same protein that was used to help specify neural crest cells, is being used in a different way at a different location.

24. Lewontin 1983; Oyama et al. 2001; Gerhart and Kirschner 2007; Hall and Hallgrímsson 2007; Laland et al. 2015.

25. Within evolutionary biology, development has been traditionally viewed as under the direction of a genetic program (e.g., "all of the directions, controls and constraints of the developmental machinery are laid down in the blueprint of the DNA genotype as instructions or potentialities" [Mayr 1984, 126]). While the terminology of contemporary biologists is often more nuanced, genetic "blueprint," "program," or "instructions" remain widely used metaphors

in evolutionary biology texts (Moczek 2012). Nor are developmental biologists immune from the use of this language (L. Wolpert 2011). However, "the individual's genotype can never be said to control development . . . [which] depends at every step on the preexistent structure of the phenotype" (West-Eberhard 2003, 29; see also Oyama 1985).

26. This claim holds even for the zygote, which is typically embedded within a wider cellular environment (Amundson 2005).

27. The development of an organism has been likened to a performance. The genome is the score. But the score is to be interpreted each time it is played. Improvisation is evolution, an experiment in changing development within certain parameters but not others (S. Gilbert and Bard 2014).

28. Oyama 1985; S. Gilbert and Sarkar 2000; Gilbert 2003a; E. Keller 2010, 2014.

29. Gerhart and Kirschner 1997, 2007.

30. The information required to assemble an organism does not reside solely in the genome. Many evolutionary biologists, from George Williams (1966) to Richard Dawkins (1976) have portrayed genes as units of information, which, by specifying the sequence of nucleotides in a stretch of DNA and hence the structure of proteins, control development and build bodies. However, "control" cannot be attributed to the DNA without also allowing other components to share it.

31. S. Gilbert and Bard 2014.

32. Griffiths and Stotz 2013.

33. Crick 1958. See Griffiths and Stotz (2013) for further detail and discussion of information concepts.

34. Griffiths and Stotz 2013.

35. S. Gilbert 1991; Nijhout 1999.

36. Stotz et al. 2006. As John Stamatoyannopoulos (2012, 1603), one of the leaders of the ENCODE project, recently concluded: "Although the gene has been conventionally viewed as the fundamental unit of genomic organization, on the basis of ENCODE data it is now compellingly argued that this unit is not the gene but rather the transcript. . . . On this view, genes represent a higher-order framework . . . creating a polyfunctional entity that assumes different forms under different cell states."

37. Susan Oyama (1985), a founder of developmental systems theory (Griffiths and Gray 1994; Oyama et al. 2001) has famously called this "the ontogeny of information."

38. Oyama 1985.

39. Griffiths and Stotz 2013.

40. Oyama 1985.

41. Oyama 1985.

42. Oyama 1985, 2000; Oyama et al. 2001.

43. R. Fisher 1930; Dawkins 1976.

44. Pavličev et al. 2011; Uller et al. 2018.

Chapter 7

1. Darwin 1868.

2. As we discuss in chapter 11, we now know that inheritance extends far beyond gene transmission, and in the light of this data Darwin's pangenesis theory does not look quite so far off the mark.

3. See Sánchez-Villagra et al. 2016.

4. Darwin speculated that the suite of traits associated with domestication might have arisen through the "gentler conditions of living" such as improved diets, or through the hybridization of different breeds. Later, in the second edition of this volume, he wrote: "Correlation is an important

subject; for with species, and in a lesser degree with domestic races, we continually find that certain parts have been greatly modified to serve some useful purpose; but we almost invariably find that other parts have likewise been more or less modified, without our being able to discover any advantage in the change" (1876, 346–347). What is particularly intriguing about mammalian domestication is that a selective advantage to the idiosyncratic constellation of traits is hard to rationalize (but see Gleeson and Wilson 2023), as Darwin himself recognized. If floppy ears are inherently advantageous, why are they absent in virtually all nondomesticated mammals? There is evidently selection for one or more traits during domestication, but the clear implication is that selection cannot be the whole story (Wilkins et al. 2014; Wilkins 2020). In a way, Darwin's conundrum arose because while natural selection could explain the "survival of the fittest," it failed to consider the "arrival of variation."

5. The foxes were of the species *Vulpes vulpes*, the red fox.

6. Trut et al. 2009. This extraordinary experiment is currently supervised by Belyaev's collaborator Lyudmila Trut.

7. Conditions where a change in one early developmental event (an insufficiency of neural crest cells or a mutation of the *Sox9* gene) causes many phenotypic effects (because neural crest cells migrate to many places, or the *Sox9* gene is used in constructing several organs) are called *pleiotropy*. In situations where many traits are reliably affected, the resulting alterations are called a "syndrome." Down syndrome, caused by an extra copy of chromosome 21 in humans, affects a person's heart, facial muscles, and brain.

8. Lord et al. (2019, 25) claim that "both the conclusions of the Farm-Fox Experiment and the ubiquity of the domestication syndrome have been overstated." They are concerned that the Russian farm-fox experiment has been mischaracterized as a case of de novo domestication, whereas these foxes were acquired from a population of Canadian farmed foxes that had been selectively bred for 50+ generations for greater docility (i.e., tameness). However, Lord and colleagues' conclusion that the general idea of a domestication syndrome is only weakly supported has been heavily criticized (Trut et al. 2020; D. Wright et al. 2020; Zeder 2020; Wilkins et al. 2021). In a subsequent reanalysis, Zeder (2020, 648) presents compelling evidence that "the large number of traits found in the Canadian foxes subjected to generations of breeding for greater docility, and their subsequent increase in the Russian Farm-Fox Experiment . . . support, rather than weaken, the case for domestication syndrome expression in both the Canadian farmed foxes and the Russian Farm-Fox Experiment." Likewise, D. Wright et al. (2020, 1060) show that for core domestication syndrome traits "abundant evidence exists that these are shared between almost all domesticated animals." Wilkins et al. (2021) show that there is extensive evidence for both a domestication syndrome in mammals and for the neural crest cell explanation that we favor. See also Zeder 2015.

9. Wilkins et al. 2014, 2021; Wilkins 2020.

10. Wilkins et al. 2014, 2021; Wilkins 2020.

11. By "neural crest cell genes" we mean any genes required for neural crest cell formation or development (not any gene expressed in a neural-crest-derived cell). These genes are likely involved in multiple functions, and hence should not be regarded as exclusively associated with neural crest functionality (Wilkins et al. 2021).

12. Hovland et al. 2020; Wilkins et al. 2021.

13. Wilkins 2020. See also Pendleton et al. 2018; X. Wang et al. 2018.

14. Sánchez-Villagra et al. 2016; Wilkins 2020.

15. Mutations in neural crest cell (NCC) genes could operate by decreasing NCC numbers, delaying NCC migration, or through altering rates of proliferation or differentiation of cells arriving at target sites (Wilkins et al. 2021).

16. At the time of writing, this suggestion is perhaps best described as a highly plausible hypothesis, and subject to extensive ongoing investigations (Wilkins et al. 2021). Moreover,

other selective hypotheses have been proposed for the domestication syndrome seen in animals (Gleeson and Wilson 2023). However, recently Rubio and Summers (2022) have tested and confirmed two major predictions of this neural crest model for domestication syndrome; namely, (1) that genes responsible for neural crest cell migration would show strong signals of positive selection in domesticated lineages (relative to closely related wild lineages) and (2) that this pattern would be specific for these genes and not others.

17. Wilkins 2020.

18. Consider, for instance, how the "AND logic" of GRN feed-forward circuits impacts development: because Tfap2a, Zic1, and Pax7 are all required for Snail 1 and Snail 2 gene expression, mutations in any of them can result in downstream reductions in precursor cells.

19. Parsons et al. 2020. Also consistent with the hypothesis that selection for tameness exploits a preexisting cluster of correlated trait variations is the observation that when domesticated mink become feral they evolve larger brains (Pohle et al. 2023).

20. Parsons et al. (2020) speculated that shorter snouts might confer a higher bite force, which could be advantageous in urban settings where foxes feed on stationary (as opposed to moving) foods. If they are correct and selection has indeed favored shorter muzzles, then increased tameness (or, perhaps, "boldness") could be an incidental evolutionary outcome of indirect selection arising through the developmental bias. Alternatively, no adaptive explanation may be required for the urban foxes' shorter snouts and smaller brains, which could be side effects of selection for boldness (or, rather, for low reactivity) arising naturally in foxes scavenging in an urban setting. If the neural crest cell explanation for domestication syndrome is correct, there is probably nothing special about selection for tameness in the evolution of domestication, and selection for shorter snouts, or, indeed, many of the other domestication traits, could plausibly have similar effects in generating a syndrome of apparently unrelated characters.

21. Parsons et al. 2020.

22. Powder and Albertson 2016.

23. S. Gould 2002.

24. Spandrels: S. Gould and Lewontin 1979. Punctuated equilibria: Alberch 1980; S. Gould 1980b. See also Bonner 1982. For a historical perspective on the development of paleobiology, see David Sepkoski's (2012) *Re-reading the Fossil Record*.

25. Maynard-Smith et al. 1985; Futuyma 2015.

26. E.g. S. Gould 1980b, 2002; B. Charlesworth et al. 1982; Brigandt 2015, 2020. Several classes of constraints on evolution are recognized (Hansen 2015), including *physical constraints* (the imperative that biological systems must obey the laws of physics and conform to the properties of materials), *phylogenetic* or *historical constraints* (where the history of a lineage restricts subsequent evolution), *functional constraints* (caused by selection on other traits and trade-offs between traits), and *developmental constraints* ("a bias on the production of various phenotypes caused by the structure, character, composition, or dynamics of the developmental system" [Maynard-Smith et al. 1985, 265]). See also Antonovics and van Tienderen 1991.

27. B. Charlesworth et al. 1982; D. Charlesworth et al. 2017.

28. E.g., Galis 1999.

29. Galis et al. 2006.

30. E.g. B. Charlesworth et al. 1982.

31. Gould's 1980 claims (S. Gould 1980b) were rebutted by B. Charlesworth et al. (1982, 477–478).

32. Badyaev 2011.

33. Amundsen 1994; Brigandt 2020.

34. Maynard-Smith et al. 1985; Arthur 2004; G. Müller 2007; Brakefield 2011; Uller et al. 2018.

35. The vertebrate neck bone example can also be used to make this point. Two groups of animals do break the rule that mammals have seven cervical vertebrae (Arthur 2011): manatees, which have six vertebrae, and sloths, which can have up to nine. These exceptions to an otherwise rigid canon observed in over 5,000 mammalian species illustrate that it is not impossible for mammals to evolve a different number of neck bones, only difficult. A probabilistic developmental bias operates that leaves exceptions to seven a rare possibility, rather than an absolute impossibility.

36. Brakefield 2006; Arthur 2011; Uller et al. 2018.

37. Falconer and Mackay 1996; M. Lynch and Walsh 1998. Strictly, the entries of **G** represent the additive genetic variances and covariances of the traits, and **z** is described as a vector of differences in trait means. This equation is sometimes known as the Lande equation. Quantitative genetics also makes widespread use of the breeder's equation, $\Delta \mathbf{z} = h^2 S$, where h^2 is the heritability and S is the selection differential.

38. Arnold 1992.

39. E.g., Schluter 1996.

40. E.g., Houle 1991; Gromko 1995.

41. West-Eberhard 2021, xiv.

42. Paulsen 1996; G. Wagner and Altenberg 1996.

43. How new variation arises: Lande 1980; Cheverud 1984. How it affects other traits: e.g., A. Jones et al. 2007; Chebib and Guillaume 2017.

44. For essentially the same reasons: many epistatic or pleiotropic interactions could generate particular associations in the **M** matrix.

45. Uller et al. 2018.

46. Hansen and Houle 2008; W. Hill et al. 2008.

47. Genotype-phenotype relationship: G. Müller 2007; Hallgrímsson et al. 2023. Models: Van Gestel and Weissing 2016; Milocco and Salazar-Ciudad 2020; González-Forero 2023, 2024.

48. Some quantitative geneticists dismiss the suggestion that their method fails to provide satisfactory answers to how development affects evolution (see, for instance, the foreword in B. Walsh and Lynch [2018]). This probably reflects, at least in part, the fact that, in contrast to population genetic models, quantitative genetics is specifically concerned with phenotypes. But quantitative genetics was designed on the basis of idealizations that are most useful to address the evolutionary consequences of fitness differences, with applications to agriculture and breeding, not the evolutionary consequences of development.

49. A. Jones et al. 2007, 2014; Hansen 2006; Hansen and Houle 2008; Hansen and Pélabon 2021.

50. G. Wagner and Altenberg 1996.

51. For reasons described in the preceding chapter, we eschew this term and will refer to either the *genotype-phenotype relationship* or *regulatory network* instead. In brief, our concern is that the phrase "genotype-phenotype map" subtly implies a direct, one-to-one, linear-causal, unidirectional and/or static relationship between genotype and phenotype, whereas typically the relationship is none of these.

52. Of course, there is also a vast number of possible ecological states that cause fitness differences between individuals, but this has not prevented biologists from measuring fitness or trying to understand its contribution to evolution.

53. Manrubia et al. 2021.

54. Fontana 2002; Manrubia et al. 2021.

55. Fontana 2002; Manrubia et al. 2021.

56. E.g., Houle et al. 2017.

57. Braendle et al. 2010.

58. Camara and Pigliucci 1999; Camara et al. 2000; Braendle et al. 2010; Houle et al. 2017. See also Schaerli et al. 2018.

59. Fontana 2002; Manrubia et al. 2021.

60. K. Dingle et al. 2015, 2020; Houle et al. 2017.

61. K. Dingle et al. 2015, 2020.

62. K. Dingle et al. 2015, 2020.

63. Houle et al. 2017. David Houle and colleagues collected data on over 50,000 fruit fly (drosophilid) wings and demonstrated unexpectedly strong and positive relationships between the variation produced by mutation, the standing genetic variation, and the rate of evolution in *Drosophila* wing patterns over the last 40 million years. P. Rohner and Berger (2023) show that the relationship between developmental bias and macroevolutionary divergence that Houle et al. (2017) revealed in *Drosophila* is also present in the sepsid fly (*Sepsis punctum*). The authors also show that developmental bias relates to not only genetic and macroevolutionary covariation but also plastic responses (e.g., to temperature).

64. Crombach and Hogeweg 2008; Parter et al. 2008; Draghi and Wagner 2009; Draghi and Whitlock 2012; R. Watson et al. 2014; Kouvaris et al. 2017; Rago et al. 2019.

65. Athey et al. 2017.

66. Stoltzfus and Norris 2016; Stoltzfus 2019, 2021. DNA nucleotides include two purines (A and G) and two pyrimidines (T and C). A transition is a nucleotide change within a chemical class (e.g., T to C) while a transversion is a change from one chemical class to the other (e.g., T to A).

67. Stoltzfus 2019. Population geneticists have reasoned that a deleterious allele is unlikely to reach fixation through mutation pressure alone, and will typically be very rare, since mutation rates are believed to be small relative to selection (Futuyma 2013).

68. While an adenine (A) nucleotide base is more likely to mutate into a guanine (G) than into a cytosine (C), it is not immediately obvious why this mutational bias should affect the distribution of adaptive phenotypic variation. As we saw in chapter 5, how genes actually participate in development strongly influences how any genetic change will affect phenotypes.

69. González-Forero 2023, 2024.

70. Raup 1966. See also McGhee 2007; Brakefield 2008.

71. K. Dingle et al. 2020.

72. K. Dingle et al. 2015, 2020.

73. K. Dingle et al. (2020, 6) write: "Natural selection mainly works by further refining parts of the sequence for function, rather than significantly altering the structures."

74. K. Dingle et al. 2020, 1.

75. G. Roth and Wake 1985; Salazar-Ciudad and Cano-Fernandez 2023.

76. Salazar-Ciudad 2021.

77. Amundson 2005, 239.

78. Uller et al. 2018. See, for illustration, Gould's use of heterochrony to account for a suite of changes in shell size, shape, and coiling in marine bivalves (S. Gould 2002, 1040–1045).

79. Subtribe Mycalesina (Nymphalidae: Satyrinae: Satyrini).

80. Brakefield and Roskam 2006; Brattström et al. 2020. For size of eyespots responding to selection: Beldade et al. 2002.

81. Brakefield 2010; Brattström et al. 2020. Not responding to antagonistic selection: C. Allen et al. 2008.

82. Braendle et al. 2010; Houle et al. 2017. See also Kiontke et al. 2007; Schaerli et al. 2018.

83. Felice et al. 2018; Jablonski 2020.

84. Navalón et al. 2020.

85. N. Barton 1990; Kirkpatrick 2009; B. Walsh and Blows 2009. Historically, quantitative genetics models often assumed that genetic variation was likely to be present in all selectable

dimensions (e.g., R. Fisher 1930; Lande 1979), which might imply that the biasing effects of developmental systems could only be transient. More recently, quantitative geneticists have emphasized that the claim that there exists substantial genetic variation in most traits may be misleading (Barton 1990; Kirkpatrick 2009; B. Walsh and Blows 2009).

86. Walsh and Blows 2009, 41. A recent extension of quantitative genetic theory that models genetic and phenotypic evolution simultaneously found that developmental bias exerts a long-term influence on evolutionary dynamics (González-Forero, 2023, 2024). That is because, as described in chapter 2, developmental processes create admissible pathways through morphospace, (or, more strictly, through a genotype-phenotype joint space) and populations converge on the fitness peaks of these paths rather than fitness peaks in the adaptive landscape (see chapter 9). See also Friston et al. 2023.

87. In chapter 12 we will discuss the evolution of developmental bias, and how this can facilitate adaptive evolution and promote evolvability.

88. Kavanagh et al. 2007; Brakefield 2010.

89. Brattström et al. 2020.

90. Brattström et al. 2020.

91. Brattström et al. 2020. This study illustrates that the notion of "absolute constraint" is overly simplistic, and few evolutionary biologists now think that way. While the majority of species fit the trends predicted from knowledge of developmental mechanisms, exceptional species are found with eyespots that do not match expectations.

92. A. Wagner 2011a,b, 2013.

93. Strictly, their "thumbs" are their preaxial digits and their "little fingers" are their postaxial digits. Last digits formed: Alberch and Gale 1985.

Decades of research have revealed how the development of the vertebrate limb skeleton is regulated (Hall 2015), which makes it possible to explain and predict correlated changes in digit length and the ordered loss or gain of digits over evolutionary time (e.g., Alberch and Gale 1985; Kavanagh et al. 2013). A compelling example was provided in 1985 by Pere Alberch and Emily Gale, who used chemicals to inhibit digit development in the limb buds of frog and salamander embryos. The treatment consistently caused specific digits to be missing, but this varied in the two groups: the frogs always lost their preaxial digits while the salamanders lost their postaxial digits. Intriguingly, the pattern strongly reflects evolutionary trends, with frog species far more frequently losing (and gaining) their "thumbs" than "little fingers," and salamanders repeatedly experiencing the reverse pattern, over evolutionary time spans. The difference can be understood by studying developmental mechanisms: for both frogs and salamanders it is the last digits to form that are most often lost. Here again, naturally occurring variation in developmental systems is shaped by species-typical developmental mechanisms, and knowledge of those mechanisms helps to explain differences in trait evolvability, patterns of parallel evolution, and the relative reversibility of changes in morphology.

94. Hoekstra et al. 2006.

95. This explains how a few transcriptional regulators can be candidate genes for many different phenotypes, in highly diverse organisms. A surprisingly small toolkit of regulatory factors—not just genes, but other elements, including microRNAs (Kittelman and McGregor 2019)—orchestrates the development of exceedingly diverse structures (Barresi and Gilbert 2020).

96. A. Wagner 2011a,b.

97. West-Eberhard 2003.

98. A. Wagner 2011a,b.

99. Uller et al. 2018.

100. Uller et al. 2018; Brun-Usan et al. 2022; Salazar-Ciudad and Cano-Fernandez 2023.

101. Teeth: Kavanagh et al. 2007. Phalanges: Kavanagh et al. 2013. The same potential exists for the neural crest cell account of domestication syndrome, which is starting now to lead to a

deeper understanding of evolutionary patterns across nondomesticated animals—not just *Vulpes vulpes* and its close relatives, but potentially canids in general, and perhaps other vertebrate families.

102. As with any field of science, quantitative genetics is practiced by diverse researchers with a broad spectrum of views, some of whom may take issue with this expectation. We recognize and appreciate the important contributions of developmentally minded quantitative geneticists to scientific understanding of the relationship between development and evolution (e.g. Cheverud 1982, 1984, 1996; Pavličev et al. 2011; A. Wagner 2011b, 2014; Pavličev and Wagner 2012; Hansen et al. 2023). González-Forero (2023, 2024) shows how developmental biases exert a long-term influence on evolutionary dynamics.

103. New metaphors and descriptions may be needed to replace the older ones. For instance, as we mention in chapter 4, rather than seeing development as a constraining force and natural selection as having creative agency, S. Gilbert suggests: "Development displays the creativity of the artist. Natural selection displays the creativity of the curator" (interviewed for the Università di Padova, YouTube, September 2, 2019, https://www.youtube.com/watch?v=i41lkb1rL2o, quotes at approx. 1:40).

104. Laland et al. 2011. See also the coda in S. Gilbert and Epel (2015).

Chapter 8

1. West-Eberhard 2003; Pfennig 2021. Phenotypic plasticity is also known as *developmental plasticity*.

2. McGaugh et al. 2019.

3. McGaugh et al. 2019; Kowalko et al. 2020.

4. McGaugh et al. 2019; Kowalko et al. 2020.

5. Kowalko et al. 2020.

6. Gross 2012; Ornelas-García and Pedraza-Lara 2016.

7. Kowalko et al. 2020.

8. Jeffery 2008; N. Rohner et al. 2013; McGaugh et al. 2019; Kowalko et al. 2020.

9. Cave animals have fascinated evolutionary biologists for well over a century (Culver and Pipan, 2009). Darwin ([1859] 1968, 179) described them, somewhat poetically, as "wrecks of ancient life," and presented loss of eyes in cave animals as a classic example of the evolutionary effects of disuse.

10. Jeffery 2008.

11. N. Rohner et al. 2013; Levis and Pfennig 2016; Bilandžija et al. 2020; Kowalko et al. 2020.

12. Bilandžija et al. 2020.

13. Plausibly, surface-dwelling fish may have evolved adaptive plastic responses to cave environments if they were encountered regularly.

14. Bilandžija et al. 2020.

15. Yoffe et al. 2020.

16. Bilandžija et al. 2020.

17. To test whether plastic changes would be maintained in the next generation of cave dwellers, dark-reared surface fish were bred and their offspring reared in darkness too (Bilandžija et al. 2020). The larvae showed a further shift toward lower metabolic rate and higher starvation resistance. The cave-fish-like plastic responses of the first generation were seemingly reinforced and refined when a second generation was exposed to darkness (N. Rohner et al. 2013; Bilandžija et al. 2020).

18. A third possibility is that both the genetic response to selection and plastic response to the cave environment are shaped by another process or factor.

19. Juan et al. 2010.

20. N. Rohner et al. (2013) treated surface *A. mexicanus* with Radicicol, which impairs Hsp90 chaperone activities during development. They measured the principal abiotic differences between river and cave water, including temperature, pH, oxygen content, and conductivity, and found that the biggest difference was the lower conductivity of cave water. Fish embryos that develop under low-conductivity conditions upregulate *Hsp90*, showing that they are in a state of physiological stress, and activate the same heat shock response genes as are activated by treatment with Radicicol, including *Bag3*, *Hsp27* (*Hspb1*), and *Hsp90*.

21. N. Rohner et al. 2013. There was no such effect on zebra fish, indicating that the cave fish's ancestors, but not necessarily other fishes, would have responded plastically to the cave environment.

22. Waddington 1953; West-Eberhard 2003; Pigliucci et al. 2006. Here, contemporary surface *A. mexicanus* are assumed to resemble the cave fish ancestors.

23. N. Rohner et al. 2013.

24. N. Rohner et al. 2013.

25. To quote William Jeffery (personal communication): "Larval fish are almost all eyes!" Borowsky (2023) found that cave fish and surface fish regularly hybridize, but that strong selection eliminates surface alleles affecting pigmentation and eye size from cave populations.

26. Jeffery 2008; Yoffe et al. 2020; Bilandzija et al. 2020.

27. Jeffery 2008; Sifuentes-Romero et al. 2020.

28. Barresi and Gilbert 2020.

29. Yamamoto et al. 2009.

30. Yamamoto et al. 2004.

31. Jeffery 2008.

32. Jeffery 2008; Sifuentes-Romero et al. 2020.

33. Jeffery 2008.

34. Jeffery 2008.

35. Sifuentes-Romero et al. 2020. At the molecular level, an expansion in the expression domain size of *Shh* and reductions in the expression domain size of *Pax6* and *Rx3* are each reported in multiple cave fish populations.

36. For instance, the *Shh* domain size expansion responsible for eye regression is also known to play a role in the enhancements of taste buds and jaws, and evolution of the brain, which are thought to be more plausible targets of selection (Sifuentes-Romero et al. 2020).

37. Powers et al. 2018, 2020a,b; Sifuentes-Romero et al. 2020.

38. Powers et al. 2020a, 458.

39. Powers et al. 2020a.

40. Yoffe et al. 2020.

41. Jeffery 2008; Yoffe et al. 2020.

42. Jeffery 2008.

43. Franz-Odendaal and Hall 2006a,b; Yamamoto et al. 2009; Bilandžija et al. 2018; Ma et al. 2021; Powers et al. 2020a,b; Sifuentes-Romero et al. 2020.

44. Jeffery 2008.

45. J. Baldwin 1896.

46. Bilandžija et al. 2020.

47. Recent origin: Niemiller et al. 2008; Klaus et al. 2013; Behrman-Godel et al. 2017; Bilandzija et al. 2020. General mechanism: Bilandzija et al. 2020.

48. West-Eberhard 2003.

49. West-Eberhard 2003.

50. Our claim here goes beyond the observation that genetic covariances among traits exist and shape the evolutionary response, as for instance anticipated by much quantitative genetic

theory. Rather, our point is that developmental mechanisms *create* those covariances, and do so differently in different environments, and that their investigation will produce better understanding and prediction of which particular traits will respond to selection, and why. We return to these issues in chapter 9.

51. P. Bateson and Gluckman 2011; P. Bateson 2017.

52. J. Baldwin (1896), but around the same time, Conwy Lloyd Morgan (1896) and Henry Osborn (1896) and, decades earlier, Douglas Spalding (1873) put forward similar ideas.

53. West-Eberhard 1989, 2003 (quote at 20). Simpson (1953) is an example of a skeptic.

54. West-Eberhard 2003, 28. Newman and Müller (2000) make a similar point.

55. Reduced plasticity: West-Eberhard 2003; Levis and Pfennig 2016, 2020. No change: W.-C. Ho and Zhang 2018; Radersma et al. 2020. A polyphenism is a form of phenotypic plasticity in which multiple discrete types of phenotypes (e.g., broad or narrow leaves) arise in response to different environmental inputs (e.g., dim or bright light).

56. For reviews: West-Eberhard 2003; Nijhout and Reed 2014; Schlichting and Wund 2014; Ehrenreich and Pfennig 2016; Levis and Pfennig 2016, 2020; Schneider and Meyer 2017; Nijhout et al. 2021. For a theoretical analysis showing the long-term effects of developmental plasticity: González-Forero 2023, 2024.

57. West-Eberhard 2003; Levis and Pfennig 2016; Uller et al., forthcoming.

58. Bilandzija et al. 2020.

59. As multiple individuals are likely to respond to any environmental challenge, plasticity-led evolution avoids the problem of the "hopeful monster": the aberrant individual who struggles to find a mate.

60. Nijhout 2003; S. Gilbert and Epel 2015; Pfennig 2021.

61. R. Anderson 1995; Via et al. 1995; Ancel 2000; Frank 2011.

62. N. Rohner et al. 2013; Bilandžija et al. 2020.

63. N. Rohner et al. 2013; Bilandžija et al. 2020.

64. Levis and Pfennig 2016.

65. West-Eberhard 2003.

66. C.f. A. Wagner 2014.

67. That plasticity can determine the direction of evolution has been demonstrated mathematically (González-Forero 2023, 2024).

68. Levis and Pfennig 2016.

69. Levis and Pfennig 2016.

70. Ledón-Rettig et al. 2008, 2009, 2010.

71. The first group of tadpoles commonly feed on detritus. In the experiments of Ledón-Rettig et al. (2008), this was mimicked using ground fish food. The second group feed on fairy shrimp in nature, but in these experiments brine shrimp were used.

72. Levis et al. 2020; Pfennig 1990.

73. Shorter gut and gene expression: Levis et al. 2018. Gut length: Ledón-Rettig et al. 2008.

74. Ledón-Rettig et al. 2008. *Spea* populations typically produce roughly equal numbers of omnivores and carnivores, but where two species of this genus (*multiplicata* and *bombifrons*) co-occur they tend to specialize, with *Sp. multiplicata* mainly developing into omnivores and *Sp. bombifrons* primarily becoming carnivores (Pfennig and Murphy 2000). Selection for character displacement to reduce competition between species is common, but what is relevant here is that the threshold for switching between morphs depending on environmental cues has been altered. Natural selection appears to be shifting populations along a major axis of phenotypic variability, ranging from the pure omnivore morphology, through a polyphenic mix, to pure carnivores.

75. S. Simpson et al. 2011; S. Gilbert and Epel 2015.

76. Y. Suzuki and Nijhout 2006.

77. Y. Suzuki and Nijhout 2006.

78. Rountree and Nijhout 1995.

79. Badyaev 2009.

80. Badyaev 2009. Other mechanisms include DNA methylation, transcription factor activation, and symbiont activities (S. Gilbert and Epel 2015). In horned beetles, for example, selection on the nutrition-dependent morphology of head horns was mediated by amplification or reduction of ancestral nutrition-responsive gene expression, combined with recruitment or loss of formerly nutritionally less-responsive genes into nutrition-dependent regulatory pathways (Casasa et al. 2020). See also Love and Wagner 2022.

81. Pfennig et al. 2010.

82. Cavalli-Sforza and Feldman 1981; Boyd and Richerson 1985; Laland 1994a,b; Lachlan and Slater 1999; Beltman et al. 2003; Masel 2004; Borenstein et al. 2006; Mills and Watson 2006; Lande 2009; González-Forero 2023, 2024.

83. Badyaev and Forsmann 2000; Wund et al. 2008; Ledón-Rettig et al. 2008; Muschick et al. 2011; see Pfennig et al. 2010, and references therein. The actual ancestors are rarely still present, so such studies will typically use another species or population that is thought to resemble the ancestor as an ancestor proxy.

84. The marine three-spined stickleback is *Gasterosteus aculeatus*.

85. Wund et al. 2008.

86. Pfennig et al. 2010.

87. See Pfennig et al. 2010, table 1, and references therein.

88. Pfennig and McGee 2010.

89. West-Eberhard 2003.

90. West-Eberhard 2003.

91. Muschick et al. 2011; Navon et al. 2020; Schneider and Meyer 2017. For the cichlid fishes of the African Rift Valley lakes: Meyer et al. 1990; Muschick et al. 2012; Brawand et al. 2014; Saltzburger 2018.

92. Darwin's finches appear to have diversified in a similar way, consistent with the flexible stem hypothesis (Tebbich et al. 2010).

93. Radersma et al. 2020. Radersma and colleagues analyzed data from 34 studies in which plants of two (or more) populations of the same species (let's call them A and B) were both grown in their own environment (which we will call *a* and *b*) and transplanted to the other population's environment. This created four experimental populations for each study (plant species A grown in its normal environment *a*, A grown in the novel environment *b*, and the reciprocal pair of plant B grown in *a*, and B grown in *b*). The plant populations were all locally adapted, which means that fitness was highest when plants were grown in their native location. Naturally, the pairs also differed in some aspects of their phenotype; for instance, leaf shape or flowering time. Multiple traits of each plant were measured, which is important because almost all traits can change, making it appear that phenotypes vary freely even when particular trait combinations are rare (Uller et al. 2018). Multivariate analyses provide far stronger tests of whether plastic responses coincide with adaptive evolution. The approach allowed the phenotypic differences between pairs of populations grown in their native environment (the difference between A in *a* and B in *b*) to be described by a vector, representing the phenotypic differences between the two populations. Likewise, the shift in the plants' traits due to plasticity could be captured with another vector, representing the phenotypic differences between plant populations of the same type reared in the two environments (for instance, comparing A in *a* with A in *b*). The authors reasoned that if plasticity were directing the course of adaptive evolution then the two vectors should be aligned. For example, imagine that population A is the ancestor of population B. If plasticity-led evolution had been responsible for the evolutionary change from A to B, then the locally adapted plants in the derived population (B in *b*) should resemble

the phenotypes of plant A when these were also grown in location b. The authors found that the vectors were indeed well aligned; that is, it was possible to predict the phenotypes of locally adapted populations (e.g., B in b) on the basis of plastic responses.

94. Radersma et al. 2020. The study found that the average angle between the phenotypic trait vector and the plasticity vector was 25° and, on average, 74% (95% CI: 55–93) of the length of the phenotypic difference vector could be accounted for by phenotypic plasticity.

95. Radersma et al. (2020) report that the two plasticity vectors were close to antiparallel (mean angle of 152°, 95% CI: 145–158) and of similar length.

96. Radersma et al. (2020) also compared the alignment between the plasticity vector and a third vector representing the adaptive fine-tuning of phenotypes to the novel environment after the plastic response (i.e., the vector comparing A in b to B in b). If the vectors are aligned, then adaptation through natural selection would move the population in the direction of the plastic response itself, perhaps exaggerating (alternatively reversing) the phenotypes that were environmentally induced, a response known as *genetic assimilation* (West-Eberhard 2003). For example, plants that are locally adapted to dry environments often grow longer roots and have a thicker cuticle than plants from wetter environments, and it is possible that these responses are exaggerated versions of ancestrally plastic responses. However, in the plant study, there was no evidence that genetic divergence moved phenotypes along such narrow pathways. Thus, for the traits commonly studied by plant evolutionary ecologists at least, plasticity makes certain phenotypes available to selection, but any subsequent selection on the regulation, form, or side effects of novel traits is in a different direction from the plastic response. This kind of evolutionary response is known as *genetic accommodation* (West-Eberhard 2003). Another meta-analysis, by Noble et al. (2019), in contrast, scrutinized the data from 21 studies of 19 species (including insects, fish, and plants) and demonstrated that plastic responses to novel environments did tend to occur along phenotypic dimensions that harbor substantial amounts of heritable variation. Noble and colleagues identified 21 studies where G and P (the phenotypic covariance matrix) could be compared (i.e., 32 environmental comparisons of G and P with $n = 64$ effects total). They used 32 studies ($n = 98$ effects) across 30 species to derive estimates of the quantitative genetic evolvability in the direction of plasticity. Collectively, these data suggest that plastic responses may, at least sometimes, represent particularly evolvable dimensions.

97. Ghalambor et al. 2015; W.-C. Ho and Zhang 2018.

98. Feiner et al. 2020. On each island, these lizards have diversified into a number of distinctive *ecomorphs* (i.e., forms exploiting different habitats, some living on tree trunks, others on bushes, and others in the crowns of trees). It has long been suggested that the lizards' plastic responses to their microhabitat may have facilitated the repeated convergent evolution of their distinctive morphologies across different islands (Losos et al. 2000; West-Eberhard 2003). Given that bone growth responds to mechanical stress, and the ways lizards move can affect limb growth (Losos et al. 2000), *Anolis* species might have evolved similar morphologies on different islands because the mechanical stress arising in each habitat made some functional morphologies more accessible to natural selection than others (Feiner et al. 2020). Plausible though this might appear, an analysis of the morphology of the locomotor skeleton of 95 *Anolis* species found that the most plastic aspects of their morphology did not align with the evolutionary divergence between ecomorphs. The most parsimonious explanation is that the morphological similarity of species on different islands within the *Anolis* radiation was not plasticity-led (Feiner et al. 2020).

99. Feiner et al. 2020. These criteria may also be more likely to be met when the environmental variables that change are those that developmental systems use as cues in adaptive plasticity (Feiner et al. 2020). When organisms are reliant on an evolved adaptive response to particular cues, most individuals in the population will respond in similar (yet variable) ways

when they encounter an extreme environment. These "extrapolated" phenotypes may nonetheless be reasonably fit, especially if the cue and selective environment are highly correlated (Kouvaris et al. 2017). Consequently, locally adapted phenotypes should resemble environmentally induced phenotypes, as was seen in the above meta-analysis of reciprocal transplant experiments in plants (Radersma et al. 2020). The phenotype dimensions that are adaptively plastic may in fact have particularly high evolvability, since theoretical and empirical studies suggest that those dimensions harbor particularly high levels of additive genetic variation (Draghi and Whitlock 2012; Noble et al. 2019; Brun-Usan et al. 2021).

100. West-Eberhard 2003.

101. Moczek et al. 2011.

102. Standen et al. 2014.

103. Lewontin 1982; Odling-Smee 1988; Odling-Smee et al. 2003. In fact, organisms can bias the action of natural selection in at least two ways (Laland et al. 2015). Developmental mechanisms can bias the phenotypic variation that is exposed to natural selection by producing some forms more readily than others. However, organisms can also bias the action of natural selection by behaving in ways that systematically modify their external environments. This niche construction (Lewontin 1982; Odling-Smee 1988, 2024; Odling-Smee et al. 2003), contributes to both forms of bias.

104. For general references, see Lewontin (1982), Odling-Smee (1988), Odling-Smee et al. (2003), and Sultan (2015). Animals can also modify social environments that affect their evolution (*social niche construction*), with examples including social resource networks in monkeys, mixed species flocks in birds, and social demographics in fruit flies (e.g., Flack et al. 2006; Harrison and Whitehouse 2011; Saltz and Foley 2011; Saltz and Nuzhdin 2014).

105. Odling-Smee et al. 2003; West-Eberhard 2003; Wade and Sultan 2024.

106. Hu et al. 2020.

107. Schwab et al. 2016.

108. Schwab et al. 2016; Hu et al. 2020.

109. Schwab et al. 2017. Interestingly, the same study demonstrates that suppression of larval niche construction leads to a dramatic reduction in sexual dimorphism in some morphological traits.

110. Hu et al. 2020; P. Rohner et al. 2022, 2024; P. Rohner and Moczek 2023.

111. Hu et al. 2020; P. Rohner et al. 2022.

112. Some, but not all, consequences of niche construction can be understood as indirect genetic effects (IGEs) of niche-constructing genotypes. IGEs occur when the genotype of an individual affects the phenotypic trait value of another conspecific individual (Bijma 2014). In dung beetles, the impact of maternal niche construction on larval growth can be understood as an IGE, but the impact of larval niche construction on larval development does not qualify. Yet, both forms of niche construction generate a developmental bias, and both bias the action of natural selection (Odling-Smee et al. 2003; Laland et al. 2015; P. Rohner et al. 2022).

113. Hu et al. 2020.

114. Hu et al. 2020; P. Rohner et al. 2021, 2022, 2024.

115. Hongoh et al. 2005.

116. S. Gilbert et al. 2012; Chiu and Gilbert 2015; S. Gilbert 2020a,b.

117. Vermeij and Lindberg 2000; S. Gilbert 2019, 2020a.

118. Flint et al. 2008; La Reau and Suen 2018; S. Gilbert 2019, 2020a.

119. Sander et al. 1959; R. Baldwin et al. 2018; S. Gilbert 2019; Chiu and Gilbert 2020.

120. Gilbert 2020a, 154.

121. Russell et al. 2009.

122. Bellwood 2003; Brocklehurst 2017; S. Gilbert 2020a.

123. S. Gilbert 2020a.

124. Sabeti et al. 2006, 2007; Voight et al. 2006; E. Wang et al. 2006; Nielsen et al. 2007; G. Perry et al. 2007; Laland et al. 2010; but see Evershed et al. (2022) for complications to the dairy farming and adult lactose absorption story.

125. Bosse et al. 2017.

126. Forss et al. 2016.

127. Laland et al. 2019. Learned behavior is often the result of an exploratory search conducted over multiple trials, with selective retention of rewarded outcomes, a process that resembles natural selection (Skinner 1938). When animals learn socially, this search encompasses the learning trials of multiple individuals, allowing for rapid and effective achievement of successful solutions (Rendell et al. 2010).

128. Birds: N. Davies and Welbergen 2009. Orangutans: Russon 2003. The orangutan is *Pongo pygmaeus*.

129. Behavioral innovations: Lefebvre et al. 1997; Reader and Laland 2003; Reader et al. 2016; Snell-Rood et al. 2018. Social learning: Laland et al. 2019.

130. Laland 2004; Rendell et al. 2011.

131. Blackbirds (*Agelaius phoeniceus*): Mason and Reidinger 1982. Sticklebacks (*Pungitius pungitius*): Coolen et al. 2003. Bats: Wilkinson 1992.

132. Parter et al. 2008; Draghi and Whitlock 2012; Kouvaris et al. 2017; Rago et al. 2019. Learning has an advantageous effect on adaptation in changing or spatially complex environments, allowing individuals to adjust to environmental changes that cannot be tracked by selection of genes (Cavalli-Sforza and Feldman 1981; Boyd and Richerson 1985; Borenstein et al. 2006; Mills and Watson 2006; Frank 2011). Similar conclusions are reached by analyses that simulate gene regulatory networks (Van Gestel and Weissing 2016; Kounios et al. 2016).

133. Cavalli-Sforza and Feldman 1981; Boyd and Richerson 1985; Borenstein et al. 2006; Mills and Watson 2006; Frank 2011.

134. West-Eberhard 2003; Gerhart and Kirschner 2007; Snell-Rood 2012.

135. Coolen et al. 2003.

136. M. Webster and Laland 2011.

137. M. Webster and Laland 2011; M. Webster et al. 2019.

138. Mineka and Cook 1988; Olsson and Phelps 2007.

139. Duckworth et al. 2018.

140. Cote et al. 2010; Duckworth et al. 2018.

141. Duckworth et al. 2018; Duckworth 2019.

142. Duckworth et al. 2018; Duckworth 2019.

143. Duckworth et al. 2018.

144. Nijhout 1999; Badyaev 2011, 2014; Duckworth et al. 2018.

145. Levis and Pfennig 2016, 2020; Uller et al. 2020.

146. Evolution of plasticity: Ancel and Fontana 2000; Hansen 2006; Draghi and Whitlock 2012. Evolvability is defined in different ways (Pigliucci 2008), ranging from broad (e.g., an organism's capacity for adaptive evolution) to narrow (e.g., the ability to generate heritable genetic variation). We discuss these issues in more detail in chapter 12.

Chapter 9

1. Ward 2016.

2. Deserts are typically defined by their aridity (Ward 2016). A place that receives less than 25 centimeters (10 inches) of rain per year is considered a desert.

3. Oskin 2013.

4. *Camelus* spp. Scientists now recognize three species of camel—the dromedary *C. dromedarius*, the Bactrian camel *C. bactrianus*, and the wild Bactrian camel *C. ferus*.

5. The oxidation of hump fat also yields additional water (Ward 2016).

6. Ward 2016.

7. Other camel adaptations to desert existence include their guts, which are set up as temporary water stores, and their thick, pale-colored coats, which reflect light and provide insulation from the heat of the midday sun, while keeping the camels warm at night.

8. Indeed, key genes in camelid adaptation to the desert have been identified and found to carry signatures of selection (Wu et al. 2014).

9. Sober 1984; Godfrey-Smith 1996.

10. Lev-Yadun et al. 2009; Sultan 2015. The desert rhubarb is *Rheum palaestinum*.

11. Lev-Yadun et al. 2009; Sultan 2015.

12. Lev-Yadun et al. 2009; Sultan 2015.

13. Deshmukh and Pathak 1989. These termites are of the subfamily Macrotermitinae.

14. Turner et al. 2006.

15. Turner 2000, 2005.

16. Hansell 1984, 2000; S. J. Martin et al. 2018.

17. The fungus is *Termitomyces*.

18. The combs are also hygroscopic, adsorbing water vapor at local humidities above roughly 80% and evaporating liquid water when local humidity falls below 80%. This has the overall effect of "clamping" nest humidity (at about 80% relative humidity) against short-term fluctuations of water flux between the broader soil environment and the nest.

19. W. West 1970; Abushama 1974; Sieber and Kokwaro 1982; Lys and Leuthold 1994.

20. Turner et al. 2006.

21. Turner 2000; Turner et al. 2006; Laland et al. 2014a.

22. Hansell 1984, 2000; Ward 2016.

23. Seely and Hamilton 1976.

24. Ward 2016. For instance, Shachak and Steinberger (1980) found that the desert snail *Sphincterochila zonata* restricted its activity to just 8–27 winter days annually, aestivating the rest of the year.

25. Ward 2016.

26. Williams et al. 2001, Ward 2016.

27. Oryx leucoryx.

28. J. Williams 2001; Ward 2016.

29. "Desert," encyclopedic entry, *National Geographic*, accessed August 2023, https://www .nationalgeographic.org/encyclopedia/desert/.

30. A. Bennett et al. 1984. The Cape ground squirrel is *Xerus inauris*.

31. E. Crawford and Schmidt-Nielsen 1967.

32. Odling-Smee et al. 2003; Sultan 2015.

33. Evolutionary processes have transformed our body hair into sweat glands that can cool us through evaporation (N. Jablonski and Chaplin 2000; Ruxton and Wilkinson 2011; Lu et al. 2016).

34. We distinguish between *adaptations* (characters favored by natural selection for their effectiveness in a particular role; e.g., G. Williams 1966) and *adaptive behavior* (functional behavior that increments reproductive success). Adaptations may or may not be adaptive and adaptive behavior may or may not be an adaptation.

35. Echavarri 2020.

36. Of course, many, but not all, instances of niche construction and phenotypic plasticity will also be adaptations, but our focus here is on the role of niche construction and phenotypic plasticity in modifying natural selection. The acquisition of symbionts constitutes another route to adaptation. We consider this in detail in chapter 10.

37. Odling-Smee et al. 2003; Sultan 2015.

38. West-Eberhard 2003.

39. By our definition of "adaptation" (see earlier note), not all adaptive traits are adaptations; for instance, some may be *exaptations* (characters that enhance fitness but were not the result of natural selection for their current role; S. Gould and Vrba 1982).

40. Lewontin 1970. In practice, these conditions will not always lead to evolutionary change, as selection is frequently stabilizing (Godfrey-Smith, 2009).

41. Darwin (1859) 1968.

42. Otsuka 2016; D. Walsh 2019.

43. D. Walsh 2019.

44. D. Walsh 2019.

45. D. Walsh 2019.

46. Lewontin 2000, S. Sober 1984, 59.

47. After all, in the *Origin*, Darwin wrote: "All those exquisite adaptations of one part of the organization to another part, and to the conditions of life, and of one distinct organic being to another being . . . follow inevitably from the struggle for life" ([1859] 1968, 114–115). Struggling for life is what individual organisms do (D. Walsh 2019). For all his prescience as a population thinker, Darwin also recognized that it was the (albeit aggregated) properties and activities of individual organisms that determined their success in the struggle, and hence that *caused* adaptation.

48. Lewontin 1970. Spelling out the causes of natural selection in this manner makes it clear that these processes can be intertwined.

49. Sober 1984, 50. Sober uses the term "force" rather than the terms "cause" or "process," which might be more common today. We note too that use of the term "law" may also seem dated. Here our usage attempts to update Sober's terminology, without changing his meaning.

50. Sober 1984, 59. Here we acknowledge that Sober's claim was made around 40 years ago, and in that intervening period regularities potentially qualifying as "source laws" may have emerged.

51. D. Walsh 2019.

52. Models of evolution that emphasize individual interactions, and the interactions between individuals and their ecology, can be found in the literature on evolutionary game theory and adaptive dynamics (Vincent and Brown 2005; Metz 2011; McNamara and Leimar 2020; Avila and Mullon 2023). Metz (2011) provides an engaging argument for the application of the adaptive dynamics modelling framework to also understand the relationship between development and evolution.

53. Sober 1984.

54. An important corollary here involves forms of selection that are frequency dependent, where the source of selection is another evolving entity.

55. M. Lynch 2007, 8597.

56. Such as claims made by: S. Gould 2002; West-Eberhard 2003; Jablonka and Lamb 2005; Pigliucci and Müller 2010; Laland et al. 2014, 2015; Love 2015; Sultan 2015; Huneman and Walsh 2017; Uller and Laland 2019.

57. Plastic responses: West-Eberhard 2003. Evolutionary novelties: Moczek 2020; Uller et al. 2020.

58. Jablonka and Lamb 2005; Uller and Helanterä 2017; Bonduriansky and Day 2018.

59. D. Walsh 2019. This has been demonstrated mathematically by González-Forero (2023, 562), who writes: "Whilst selection pushes genotypic and phenotypic evolution up the fitness landscape, development determines the admissible evolutionary pathway, such that evolutionary outcomes occur at path peaks rather than landscape peaks. Changes in development can generate path peaks, triggering genotypic or phenotypic diversification, even on constant,

single-peak landscapes. Phenotypic plasticity, niche construction, extra-genetic inheritance, and developmental bias alter the evolutionary path and hence the outcome." See also Fogarty and Wade 2022 and González-Forero 2024.

60. Sober 1984. M. Lynch (2007, 8598) writes: "No principle of population genetics has been overturned by an observation in molecular, cellular, or developmental biology."

61. As C. Hughes and Kaufman (2002) concluded in their discussion of arthropod adaptations: "The fangs of a centipede . . . and the claws of a lobster accord these organisms a fitness advantage. However, the crux of the mystery is this: From what developmental genetic changes did these novelties arise in the first place?"

62. Bilandžija et al. 2020.

63. Sifuentes-Romero et al. 2020; Kowalko et al. 2020.

64. West-Eberhard 2003.

65. Kavanagh et al. 2007, 2013.

66. Schwab et al. 2016; Hu et al. 2020; P. Rohner et al. 2022, 2024.

67.

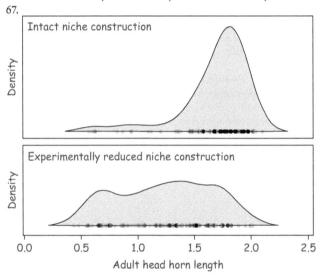

The data in this figure (courtesy of Patrick Rohner, personal communication) clearly illustrate how maternal and larval niche construction change the mean and phenotypic distribution of dung beetle horn size. This has some dramatic consequences, including strongly impacting the distribution of life-history traits (such as the proportions of major and minor males, and of their associated behaviors), affecting the intensity of sexual selection, and facilitating the accumulation of cryptic genetic variation (P. Rohner et al. 2024). For instance, P. Rohner and Moczek (2023, 2024) found that experimentally preventing larvae from modifying their environment increased additive genetic variance as well as heritability and evolvability for the trait of development time. Likewise, P. Rohner and Moczek (2023) found that niche construction shapes sexual dimorphism in overall body size, head horns, fore-tibia length and width, and elytron length and width. This study, and others from the same laboratory (Schwab et al. 2016, 2017; Hu et al. 2020; P. Rohner et al. 2021, 2022, 2024; P. Rohner and Moczek 2023, 2024), collectively imply that niche construction reduces selection on some traits (e.g., developmental time) but may enhance selection for genotypes best placed to exploit carbon nutrients made available by the niche construction (e.g., body size, horn size). More generally, the work illustrates niche construction's potential to shape additive variation, heritability, and evolvability.

68. Odling-Smee et al. 2003; Sultan 2015.

69. Following Leigh Van Valen's (1973, 488) dictum that "evolution is the control of development by ecology," developmental biology can determine how variations are created and how environmental agents not only select phenotypes but can also help construct them.

70. Lewontin 1982, 1983, 2002; Levins and Lewontin 1985; Odling-Smee et al. 2003; Sultan 2015.

71. Sultan 2015.

72. C. Gilbert et al. 2007.

73. Laland et al. 2008.

74. Our focus in this chapter is on natural selection; however, similar issues may arise with respect to other evolutionary processes, including genetic drift and mutational biases. For instance, Hartl and Taubes (1998, 525) write: "Almost every theoretical model in population genetics can be classified into one of two major types. In one type of model, mutations with stipulated selective effects are assumed to be present in the population as an initial condition. . . . The second major type of models does allow mutations to occur at random intervals of time, but the mutations are assumed to be selectively neutral or nearly neutral." Stoltzfus (2019) describes how these assumptions have masked recognition of the role of mutation biases in evolution, which, according to Stoltzfus, population genetics has failed to consider adequately.

75. Lewontin 1970.

76. B. Campbell 1974; Odling-Smee 1983; Dennett 1995; Hull et al. 2001. Strictly speaking, the same process can also erase information; for instance, when genes are lost. More specifically, we refer here to "Crick information" (Griffiths and Stotz, 2013).

77. Leimar et al. 2006; Frank 2009; Shea et al. 2011.

78. Slobodkin and Rapoport 1974; Odling-Smee 1988.

79. While it has been recognized at least since Hume ([1748] 1910) that past observations themselves do not establish the validity of inductive reasoning (see Howson [2000] for a contemporary evaluation), the effectiveness of that reasoning (i.e., the likelihood that such reasoning will lead to accurate predictions) is nonetheless a function of the temporal consistency of environmental conditions.

80. Odling-Smee et al. 2003. African American biologist Ernest Everett Just was among the first to appreciate the reciprocity between organism and environment. In 1933 he wrote; "We should not speak of the fitness of the environment or the fitness of the organism; rather we should regard organism and environment as one reacting system" (Byrnes and Newman 2014).

81. Laland et al. 2012.

82. Hansell 1984, 2000; Odling-Smee et al. 2003; Sultan 2015.

83. Constructions: Badyaev and Uller 2009. Choices: Bogert 1949; Huey et al. 2003.

84. Odling-Smee et al. 2003.

85. E.g., Callahan et al. 2014.

86. Odling-Smee et al. 2003.

87. West-Eberhard 2003.

88. West-Eberhard 2003.

89. Gerhart and Kirschner 1997; West-Eberhard 2003; M. Kirschner and Gerhart 2005.

90. Plotkin and Odling-Smee 1981; Plotkin 1994, 2010; Shea et al. 2011.

91. Odling-Smee et al. 2003; West-Eberhard 2003; Hoppitt and Laland 2013.

92. Badyaev and Uller 2009; Jablonka and Raz 2009; Bonduriansky and Day 2018.

93. Easier to control: Turner 2000; Laland et al. 2014a. Symbionts: S. Gilbert 2019, 2020a,b.

94. Turner 2000; Laland et al. 2014a.

95. Enhance fitness: McNamara and Dall 2010. Track the environment: Odling-Smee et al. 2003.

96. Sober 1984, 99.

97. While not used in this way by Sober, the toy could also be deployed to distinguish between selection and drift: unlike correlated selection (which is selection *of* a trait directly

resulting from selection *for* another trait), drift arises when chance events result in selection *of* a trait in the absence of selection *for* another trait (Godfrey-Smith 2009).

98. For instance, bird's nest size is well predicted by adult bird size (Vanadzina et al. 2023).

99. At the time of writing, this suggestion is perhaps best described as a highly plausible hypothesis, and subject to extensive ongoing investigations.

100. Powers et al. 2018, 2020a.

101. Levins and Lewontin 1985, 106.

102. Sometimes allele or genotype frequencies are plotted on the horizontal axes. For a comprehensive overview of the use of adaptive landscapes in contemporary evolutionary biology, see Svensson and Calsbeek (2012).

103. N. Barton and Turelli 1987. Perhaps ironically, Sewell Wright's last published paper (S. Wright 1988) is a kind of apology for adaptive topography, as he never intended it would be used that way. See also P. Moran 1964.

104. This equates directly to evolutionary biology's focus on consequence laws. In contrast, the shape of the fitness surface is determined by source laws (Sober 1984; Frank 2011).

105. Niche construction overlaps with several other concepts in ecology and evolution—notably, *ecosystem engineering* and *eco-evolutionary dynamics* (C. Jones et al., 1994; Matthews et al. 2014; Hendry 2020). What's distinctive about niche construction theory, however, is the claim that environmental modification by organisms (niche construction) and its legacy over time (ecological inheritance) are evolutionary processes—they *cause* evolutionary change (Odling-Smee et al. 2003; Laland et al. 2019; Odling-Smee 2024).

106. González-Forero 2023, 2024. To represent González-Forero's analyses accurately, in figure 11 one horizontal axis would be labeled 'genotype' and the other 'phenotype'. See also Tanaka et al. 2020; Fogarty and Wade 2022.

107. Tanaka et al. 2020; Fogarty and Wade 2022.

108. Tanaka et al. 2020.

109. Fisher 1930.

110. Tanaka et al. 2020. See also Fogarty and Wade 2022.

111. West-Eberhard 2003.

112. Laland et al. 2011.

113. Lewens 2019.

114. Lande and Arnold 1983.

115. Endler 1986; Kingsolver et al. 2001, 2012; Siepielski et al. 2009, 2013.

116. Lande and Arnold 1983.

117. Endler 1986; Hoekstra et al. 2001; Kingsolver et al. 2001, 2012; Siepielski et al. 2009, 2013.

118. Returning to Sober's terminology, the method shows that there has been *selection of* the character but does not establish that there has been *selection for* the character (Sober 1984).

119. Trut et al. 2009; Wilkins et al. 2014; Wilkins 2020; Powers et al. 2018, 2020a; Sifuentes-Romero et al. 2020.

120. Interestingly, this evolution probably occurred in the Canadian Arctic and the storage was for food in the Arctic winter (Perkins 2013).

121. Wade and Kalisz 1990.

122. We might, for instance, predict that camel hump size will be positively correlated with fitness when food is scarce but not when it is plentiful. Where the predicted pattern of covariation between fitness and the manipulated environmental agent arises, such experiments provide evidence that there has been *selection for* the character and that the character is the *cause* of fitness differences. Such experiments can also establish that the manipulated environmental agent is the source of selection.

123. These methods include *path analysis* and *causal* or *directed graphs* (Pearl 2009; Shipley 2016). See Otsuka (2016, 2019), Henshaw et al. (2020) and Edelaar et al. (2023) for examples and discussion. The methods allow researchers to translate a causal hypothesis into a

corresponding model and thus to distinguish between competing causal hypotheses by using observational data.

124. In principle, the incorporation of relevant environmental data would allow such methods to specify the source of selection, which would deepen the causal understanding.

125. To some extent this can be addressed by researchers looking for costs to alternatives that translates into fitness, but this is not the same as identifying the causes of those fitness differences or the reason why the realized alternative had highest fitness. Researchers also commonly consider constraints operating on the system, which explain the absence of a particular adaptive fit but not the absence of an alternative adaptive fit.

126. Schwab et al. 2017. See also P. Rohner et al. 2022, 2024.

127. Casasa and Moczek 2018.

128. Note, neither earlier natural selection nor phylogenetic constraints provide alternative explanations here. For instance, it might be reasoned that perhaps males with large horns have an advantage in sexual competition, and selection that favored heightened sensitivity of horn development to nutrients would enable individuals to enhance their fitness, an explanation that appears to situate the explanatory burden back with earlier selection. However, whether selection for horn size preceded the developmental bias arising through the co-option of the insulin/insulin-like pathway into head horn development and the recruitment of the doublesex gene into the horn regulatory network, or vice versa, is not the point. Dung beetle horns are under sexual selection because they either already were, or through selection could readily become, sensitive to the availability of carbon nutrients (i.e., they either were or could be made a sexually dimorphic plastic trait with the potential to enhance fitness). Not all dung beetle traits have that potential. Nor is the suggestion that the development of horns in *O. taurus* and the sensitivity of horn development to nutrition are conserved traits in this group of insects an alternative explanation. That account merely pushes back our explanation to earlier in evolutionary time (Uller and Helanterä 2019).

129. Quantitative genetic theory confirms that niche construction and ecological inheritance modify selection, alter the fitness landscape, and codetermine the heritability of niche-constructing traits (Fogarty and Wade 2022). Other mathematical approaches reach the same conclusion (Tanaka et al. 2020; González-Forero 2023, 2024). See also Edelaar et al. 2023.

130. Schwab et al. 2017; Dury et al. 2020; P. Rohner et al. 2024.

131. P. Rohner et al. 2022. A reaction norm describes the pattern of phenotypic expression of a single genotype across a range of environments (Schlichting and Pigliucci 1998). Evolutionary biologists have become used to thinking of reaction norms as genetically determined, but this is a convenient heuristic and not a biological fact. In reality, reaction norms are the product of developmental systems (Sultan 2019). In this instance, differential developmental niche construction was what caused the populations' evolutionary divergence.

132. As described in chapter 8, parallel work has established that maternal niche construction, including constructing the brood ball, incorporating a microbiome-laden pedestal, and choosing the depth at which the ball is buried, also dramatically affects offspring growth and fitness, and again creates dimensions along which populations diverge (Macagno et al. 2016, 2018).

133. Lewontin (1955) and Lewontin and Matsuo (1963) use the term "facilitation" rather than niche construction, but their focus is on how fruit flies work the medium to affect their reaction norm.

134. Schwab et al. 2016, 2017; Hu et al. 2020; P. Rohner et al. 2021, 2022, 2024; P. Rohner and Moczek 2023, 2024.

135. Odling-Smee et al. 2003; Schwab et al. 2016, 2017; Sultan 2015, 2019; D'Aguillo et al. 2021; P. Rohner et al. 2024. The same point holds for other sources of phenotypic variation, including extragenetic inheritance (Sultan 2015, 2019).

136. There is now extensive evidence that plastic and environment-altering traits of organisms can influence genetic variances and covariances of traits under selection, modifying the

rates and direction of evolution by changing how traits contribute to fitness, and thus influencing evolutionary dynamics (Odling-Smee et al. 2003; Sultan 2015; D'Aguillo et al. 2021). Animals choose between sun or shade, and select nest and oviposition sites, while plants preferentially grow roots in nutrient-rich soil (Bazzaz 1991; Odling-Smee et al. 2003; Sultan 2015; Munoz and Losos 2018; D'Aguillo et al. 2021), thereby buffering or accentuating selection.

137. Saastamoinen et al. 2010. The butterflies in question were *Bicyclus anynana*.

138. Duckworth et al. 2018; Duckworth 2019.

139. Sultan 2015. Other examples include some cichlids, which switch between brightly colored dominant morphs and cryptic "sneaker" forms, thereby adjusting the relationships between aggression, territoriality, coloration, and fitness (Hofmann et al. 1999; Burmeister et al. 2005). Dung beetles produce "sneaker" morphs too, when food supplies are limited, shifting selection toward sneaky mating behavior (Moczek and Emlen 2000). In such cases, males with short or absent horns would dig tunnels that intersect with the tunnels of the females. They would mate with the females while the long-horned males are guarding the entrance to the females' tunnels (Moczek and Emlen 2000).

140. Oh and Badyaev 2010.

141. Other studies show the maintenance of genetic variation in plumage traits can arise through the behavioral buffering of selection (Reinhold 2000; Gorelick and Bertram 2003). Indeed, behaviors that enable organisms to choose, or modify, their environment can have profound evolutionary consequences (Bogert 1949; Plotkin 1988; Huey et al. 2003; Odling-Smee et al. 2003; Duckworth 2009, 2019), because searching, sampling, and building are typically costly behaviors (Rosenzweig 1981; Hansell 1984, 1993; Stamps et al. 2005), which means the expression of these behaviors should differ among individuals, such that, all else being equal, the greatest investment is expected by individuals that stand to benefit the most from a change in their selective environment (Stearns 1992; Reznick et al. 2000; Badyaev and Qvarnström 2002; Kylafis and Loreau 2008, 2011; Oh and Badyaev 2010).

142. Odling-Smee et al. 2003; Sultan 2015; D'Aguillo et al. 2021.

143. J. Baldwin 1896; Bogert 1949; Wyles et al. 1983; M. West and King 1987; P. Bateson 1988; Plotkin 1988; Wcislo 1989; Huey et al. 2003; Odling-Smee et al. 2003; West-Eberhard 2003; Losos et al. 2004; Badyaev 2005, 2009; Jablonka and Lamb 2005; Sol et al. 2005a,c,d; Pigliucci et al. 2006; Duckworth 2008; Piersma and Van Gils 2011; Sultan 2015; Badyaev et al. 2017; Stellatelli et al. 2018; Edelaar and Bolnick 2019; Clark et al. 2020.

144. Laland et al. 2017; Clark et al. 2020.

145. Odling-Smee 1988; Odling-Smee et al. 2003.

146. Møller and Jennions 2002; Pujol et al. 2018. For instance, Møller and Jennions (2002, 498) write: "The merits of different avenues of research should be evaluated in the light of these findings." A related concern is a lack of sensitivity to meta-analyses. For illustration, an admirable recent large-scale analysis of how climate variation influenced selection reported that precipitation predicted variation in selection across plant and animal populations across many terrestrial biomes, but failed to find an effect of temperature (Siepielski et al. 2017). Perhaps with time this surprising finding will prove robust, yet it would surely be premature to infer that rising temperatures due to climate change have little effect on natural selection. Researchers have noted that natural and sexual selection can be highly variable, dynamic, and sensitive to plasticity and behavioral responses (Price et al. 2003; Price 2006; Cornwallis and Uller 2009). To the extent that such variables mediate the relationship between traits and fitness, incorporating relevant measures of developmental plasticity and niche construction into statistical models predicting fitness has the potential to boost the predictive power of evolutionary analyses (Cornwallis and Uller 2009; Pujol et al. 2018).

147. Sober 1984.

148. N. Barton 1990; Kirkpatrick 2009; B. Walsh and Blows 2009.

149. Finches: Badyaev 2009. Bluebirds: Duckworth 2019. Pleiotropic effects can also be mediated by developmental niche construction, as seen in *Arabidopsis* (Donohue 2014) and hermit crabs (Laidre 2012).

150. Domestication syndrome: Wilkins et al. 2014; Wilkins 2020. Cave fish: Bilandžija et al. 2020.

151. S. Gilbert 2019.

152. Durham 1991; Holden and Mace 1997; Burger et al. 2007; Gerbault et al. 2011; Evershed et al. 2022.

153. S. Gilbert 2018.

Chapter 10

1. Jablonka and Lamb (2014) make similar points. While research on extragenetic inheritance has accumulated rapidly since its publication, Jablonka and Lamb's book remains an essential source on mechanisms and evolutionary implications of extragenetic inheritance, as well as the historical development of the field.

2. This is not to suggest that we consider all components of extragenetic inheritance to be adaptations that evolved for the purpose of transmitting information across generations. While it is plausible that some aspects of extragenetic inheritance are such adaptations, much of inheritance is a by-product of the manufacture of a responsive offspring phenotype by parents. Theoretical studies show that such adaptive responsiveness can evolve even without selection *for* it (Rago et al. 2019), and, from the life-cycle perspective that we embrace, for a developing organism there is nothing special about inherited materials—some of the cues and materials to which the developing organism responds just happen to stem from the parents. Moreover, for the reasons we give in the text, for extragenetic inheritance to be an effective tool for short-term, rapid-response adaptation, it is necessary for developing organisms to cede control of the content of the inheritance channel. Once extragenetic inheritance pathways have evolved, they are vulnerable to exploitation by both parasitic organisms and parasitic information. As a consequence, not all inheritance is adaptive.

3. Waddington 1969b, 122.

4. Here we acknowledge our debt to Oyama (2000).

5. Wray et al. 2014.

6. We use the term "extragenetic inheritance" rather than the perhaps more commonly deployed "nongenetic inheritance" because, strictly speaking, no form of biological inheritance can be entirely nongenetic, since all require gene expression. This does not imply that genes specify the content of all extragenetic inheritance, a point that we stress in the text. See Griffiths and Stotz (2013) for discussion.

7. This does not necessarily imply that parents do a lot of forecasting, or that they constantly adjust what they transmit in response to encountered conditions. In part, the match between the organism's traits and the properties of its environment arises because parents help to build an offspring that is responsive to encountered conditions. Nonetheless, the general point remains that much of inheritance is concerned with making relevant developmental resources available to offspring.

8. Moczek 2012; Uller 2013; Uller and Helanterä 2017.

9. Quotes, respectively, from: National Human Genome Research Institute, Talking Glossary of Genomic and Genetic Terms, s.v. "inherited," accessed August 2023, https://www.genome.gov/genetics-glossary/Inherited; Wikipedia contributors, "Heredity," *Wikpedia, The Free Encyclopedia*, last updated August 24, 2023, https://en.wikipedia.org/wiki/Heredity; yourgenome, "What Is Inheritance?," last updated July 21, 2021, https://www.yourgenome.org/facts/what-is-inheritance.

10. Futuyma 2013, 11. Other authoritative definitions of heredity include: "The sum of the characteristics and potentialities genetically derived from one's ancestors" (*Merriam-Webster*, s.v. "heredity," accessed November 21, 2023, https://www.merriam-webster.com/dictionary/heredity); "The passing on of physical or mental characteristics genetically from one generation to another" (*Oxford Dictionaries*, s.v. "heredity," accessed November 21, 2023, https://premium.oxforddictionaries.com/definition/english/heredity); "The genetic transmission of characteristics from parent to offspring" (*American Heritage Dictionary of the English Language*, 5th ed., 2016, Houghton Mifflin Harcourt, quoted in *The Free Dictionary*, s.v. "heredity," accessed November 21, 2023, https://www.thefreedictionary.com/heredity); "The transmission of characteristics from parents to offspring via the chromosomes" (*Oxford Dictionary of Biology*, 5th ed., 2005); "Organisms carry instructions for how to build the proteins they use, as well as for when, where and in what quantities to make them, in their genetic material" (Herron and Freeman, 2014, 149); "The DNA carries the information used to build a new body, and to differentiate its various body parts" (Ridley 2004, 22); "The hereditary basis of every living organism is its genome, a long sequence of DNA that provides the complete set of hereditary information carried by the organism" (Krebs et al. 2008, 3); "A genome contains all the information needed for an individual to develop and function" (National Human Genome Research Institute, Talking Glossary of Genomic and Genetic Terms, s.v. "genome," last updated January 16, 2024, https://www.genome.gov/genetics-glossary/Genome#:~:text=In%20humans%2C%20the%20genome%20consists,individual%20to%20develop%20and%20function). Collectively, these quotes reflect the dominant, and perhaps consensual, view that heredity comes down to genetic transmission.

11. Jablonka and Lamb 1995, 2005; Jablonka and Raz 2009; J. Herman et al. 2014; Danchin et al. 2011, 2019; Uller 2019; Adrian-Kalchhauser et al. 2020. The separation of germline and soma may be more permeable anatomically than previously thought. In mammals, exosomes—small vesicles containing proteins and RNAs—are produced and exported by the epididymis, uterus, and oviducts. These vesicles may fuse with sperm cell membranes to give the sperm new properties (Martin-DeLeon 2016; U. Sharma et al. 2018). Some microRNAs and other small RNAs that sperm acquire in the epididymis may be critical for normal embryonic development (Conine et al. 2018; Sharma et al. 2018), while other exosomes' RNAs and proteins appear to affect brain behaviors (J. Chan et al. 2020; Y. Wang et al. 2021). This implies that somatic cells—those of the epididymis—can provide the sperm with molecules that are important for normal development and phenotype production. In addition, in some corals, approximately half of somatic mutations enter the germline and are passed on to the next generation (Vasquez Kuntz et al. 2022).

12. Extragenetic inheritance: Odling-Smee et al. 2003; Jablonka and Lamb 2005; Uller 2013; Bonduriansky and Day 2018. Parental effects: Mousseau and Fox 1998a,b; Badyaev and Uller 2009; J. Wolf and Wade 2009. Inherited microbiome: S. Gilbert et al. 2012, 2015; P. Gould et al. 2018; Osmanovic et al. 2018; Rampelli et al. 2021. Animal cultures: Avital and Jablonka 2000; Hoppitt and Laland 2013; Aplin et al. 2015; Whitehead and Rendell 2015; Laland 2017a; Danchin et al. 2011, 2018; R. Kendal et al. 2018; Whiten 2019, 2021.

13. Jablonka and Lamb 2005, 2014. Quote taken from back cover of Jablonka and Lamb (2005).

14. Bonduriansky and Day 2018. Ernst Mayr described soft inheritance as a "misconception," the refutation of which led to major advances in evolutionary thinking (Mayr 1982, 570).

15. While much extragenetic inheritance can affect short-term adjustment to environmental conditions, not all extragenetic inheritance operates in this way. Some aspects of extragenetic inheritance (e.g., various components of eggs) are very robustly and reliably constructed over long periods of time, in part because of their association with genetic variation, and in such instances the time scale of evolution does not clearly separate the genetic and nongenetic components of inheritance.

16. Jablonka and Lamb 2005, 2014; Bonduriansky and Day 2009, 2018; Danchin et al. 2011, 2018; Badyaev and Uller 2009; Uller 2013, 2019; Uller and Helanterä 2017. See Zimmer (2018) for a readable overview.

17. Bonduriansky and Day 2009, 2018; T. Day and Bonduriansky 2011; Bonduriansky 2012; Danchin et al. 2011, 2018; Jablonka and Lamb 2005, 2014.

18. Badyaev and Uller 2009.

19. Badyaev and Uller 2009; Bonduriansky and Day 2009, 2018; T. Day and Bonduriansky 2011; Bonduriansky 2012; Danchin et al. 2011, 2018; Jablonka and Lamb 2005, 2014.

20. Liebman and Chernoff 2012.

21. Bonduriansky and Day 2018.

22. Immler 2018; Uller 2012, 2013.

23. In numerous vertebrates, invertebrates, and plants, zygotic gene activity is not required until the blastula stage. Maternally derived proteins and mRNAs placed into the egg govern embryogenesis through the early stages of cleavage, after which they are degraded, and zygotic gene transcription is activated (Tadros and Lipshitz. 2009; Baroux and Grossniklaus 2015).

24. Robinson and Méndez-Gallardo 2011.

25. Clutton-Brock 1991; Mousseau and Fox 1998a,b; Staszewski et al. 2007; Groothuis and Schwabl 2008; Badyaev and Uller 2009; Hasselquist and Nilsson 2009; Royle et al. 2012; Bonduriansky and Day 2018.

26. Grun 1976; Mousseau and Fox 1998a,b.

27. Roach and Wulf 1987.

28. Pelegri 2003; S. Gilbert and Epel 2015.

29. Avital and Jablonka 2000; Gottlieb 2002; Maestripieri 2009; Rosenblatt 2010; Champagne and Curley 2012; Royle et al. 2012. A good example is the mother's influence on the birth weight of the fetus, which is critically important in dog and horse breeds, where the size can vary considerably (see S. Gilbert and Epel 2015, 281–285).

30. Uller 2008, 2013; J. Wolf and Wade 2009. We recognize that our use of such a broad definition generates overlaps with the other categories of *epigenetic inheritance, the inherited microbiome,* and *animal culture* we discuss. In this section, we focus on those parental effects not covered by other sections. Strictly speaking, by our broad definition, all our categories of extragenetic inheritance are parental effects. However, it is useful here to separate out and emphasize some subcategories of parental effects in order to bring out their evolutionary implications. For instance, both epigenetic and cultural inheritance can be construed as parental effects, but they are also processes that acquire and store information over multiple generations, sometimes with cumulative effects manifest across generations. Such processes might equally be construed as "grandparental effects," "great-grandparental effects," and so forth, and there is strong theoretical evidence that carry-over effects spanning multiple generations can strongly affect evolutionary dynamics (Cavalli-Sforza and Feldman 1981; Kirkpatrick and Lande 1989; Laland et al. 1996, 1999; Lehmann 2007, 2008).

31. Mousseau and Fox 1998a,b.

32. Oyama 2000.

33. Turner 2000; Odling-Smee et al. 2003; Erwin 2008; Sultan 2015. Ecological legacies also include some factors that negatively affect fitness. For instance, prions can remain in the environment for a long time, and chronic wasting disease—a prion disease that affects deer, elk, reindeer, sika deer, and moose—is thought to spread between animals through body fluids like feces, saliva, blood, or urine, either through direct contact or indirectly via contamination of soil, food, or water (Belay et al. 2004).

34. Turner 2000; Odling-Smee et al. 2003; Erwin 2008; Sultan 2015.

35. Odling-Smee et al. 2003; Sultan 2015.

36. Badyaev and Uller 2009.

37. Kirkpatrick and Lande 1989; Lachmann and Jablonka 1996; Laland et al. 1996, 1999; Wade 1998; Kerr et al. 1999; Lehmann 2007, 2008; Kylafis and Loreau 2008, 2011; Shea et al. 2011; Van Dyken and Wade 2012; Kuijper and Hoyle 2015; Leimar and McNamara 2015; McNamara et al. 2016; Dury and Wade 2020.

38. Mousseau and Fox 1998a,b; Groothuis and Schwabl 2008; Badyaev and Uller 2009; Uller 2008, 2012; Royle et al. 2012.

39. Marshall and Uller 2007; Uller 2008; Leimar and McNamara 2015.

40. Mousseau and Fox 1998a,b; Groothuis and Schwabl 2008; Badyaev and Uller 2009; Hasselquist and Nilsson 2009; Duckworth et al. 2015; Duckworth 2019.

41. Duckworth et al. 2018; Duckworth 2018.

42. Burgess and Marshall 2014; Uller et al. 2015; Lind et al. 2020.

43. Lachmann and Jablonka 1996; Kuijper and Hoyle 2015; Leimar and McNamara 2015; Dury and Wade 2020.

44. Seger and Brockman 1987; Simons 2011.

45. The nematode in question was *C. elegans.*

46. Dey et al. (2016) performed 60 generations of *C. elegans* experimental evolution in three environmental regimes that varied in the degree of mother-offspring normoxia-anoxia correlation. In strictly alternating environments, they found that the hermaphrodites evolved the ability to increase embryo glycogen provisioning when they experienced normoxia and to decrease embryo glycogen provisioning when they experienced anoxia. The survival of *C. elegans* in anoxia is constrained by the availability of glycerol during larval stages and enhanced by the ability of parents to provision their broods with glycogen, which means that in this condition a negative parental effect enhanced the survival of their offspring.

47. Proulx et al. 2019.

48. Nematodes/lizards: Feiner et al. 2020. Ecological inheritance: Belay et al. 2004. For instance, disease transmission can occur via the horizontal transfer of prions in multicellular organisms, with transmission often occurring through materials left by organisms in the ecological environment, such as in chronic wasting disease of deer (see earlier note).

49. Francis 2011.

50. Anastasiadi et al. 2021. Those specified by the DNA sequence are sometimes known as *obligatory* epigenetic variation (Anastasiadi et al. 2021). Environmentally sensitive epigenetic effects may or may not be genotype dependent (known, respectively, as *facilitated* and *pure* variants).

51. Jablonka and Lamb 1995, 2014; Jablonka and Raz 2009; Francis 2011; Heard and Martienssen 2014; Klosin and Lehner 2016; O'Dea et al. 2016; Adrian-Kalchhauser et al. 2021; Anastasiadi et al. 2021; Stajic and Jansen 2021.

52. Following O'Dea et al. (2016) and Anastasiadi et al. (2021).

53. DNA comprises four nucleotide bases—cytosine, guanine, adenine, and thymine. DNA methylation typically refers to the addition of a methyl group ($-CH_3$) to the 5′ carbon of cytosine nucleotides, although there are other forms such as 5-hydroxymethylation. While methylation usually results in the suppression of transcription, it can be associated with the activation of transcription—for instance, where suppressing regulatory elements become methylated.

54. Histone modifications include acetylation, phosphorylation, and methylation, which influence chromatin structure and the transcriptional activity of proximal genes. Histone acetylation and phosphorylation typically reduce chromatin compaction, easing access to the DNA for transcription, while histone methylation can result in either transcriptional activation or repression, depending on where it occurs (reviewed in Anastasiadi et al. [2021]).

55. O'Dea et al. 2016; Anastasiadi et al. 2021.

56. Jablonka and Lamb 1995, 2014; Jablonka and Raz 2009; Francis 2011; Heard and Martienssen 2014; Klosin and Lehner 2016; O'Dea et al. 2016; Adrian-Kalchhauser et al. 2021; Anastasiadi et al. 2021; Stajic and Jansen 2021.

57. Francis 2011; Anastasiadi et al. 2021. Strictly, this is in viviparous mammals.

58. Francis 2011; Anastasiadi et al. 2021.

59. Pray 2004; R. Painter et al. 2006a,b, 2008; de Rooij et al. 2006, 2007; Roseboom et al. 2006, 2011.

60. Tobi et al. 2018.

61. Heijmans et al. 2008.

62. Jirtle and Skinner 2007; Bohacek and Mansuy 2013. See Feldman and Ramachandran (2018) for a discussion of some the problems associated with estimates of heritability.

63. Cropley et al. 2012. Radford et al. (2014) found that starving a pregnant mouse caused changes in the sperm of her sons that affected the health of her grandchildren. The team, led by geneticist Anne Ferguson-Smith of the University of Cambridge and diabetes researcher Mary-Elizabeth Patti of Harvard Medical School studied the DNA of two generations of mice descended from an undernourished mother. The researchers gave pregnant mice food containing only half the calories they needed during the last week of gestation—a time when the epigenetic patterns in a male embryo's primordial sperm are erased, then reset. As Patti's group had previously shown, this treatment resulted in offspring and grandchildren that were underweight and prone to diabetes. The group examined DNA from the sperm of the males born to the starved moms. Compared with sons of control mice, their sperm had lower methylation on about 110 stretches of DNA. Often, the methyl groups were missing near genes involved in metabolism that may play a role in obesity and diabetes, and the expression of these genes was altered in relevant body tissues.

Epigenetic effects are reported in many other studies of humans. Recently, analyses of the Överkalix cohorts in northern Sweden has provided evidence for transgenerational epigenetic inheritance. These studies describe how food shortages induced by harvest failures, or food abundance after a rich harvest, affect mortality outcomes in two subsequent generations. The studies suggest that an epigenetic pathway, carrying information across generations, opens up just before puberty. Prepuberty may be one of several "windows" for germline reprogramming in response to nutritional signals. The findings imply that grandparental access to food during their slow growth period can modify diabetes and all-cause mortality in grandchildren. Vågerö et al. (2018) tested this hypothesis on a significantly larger dataset, collecting harvest data during the prepubertal period of grandparents to examine its potential association with mortality in children and grandchildren in the Uppsala Multigeneration Study. They found support for the main Överkalix finding: paternal grandfather's food access in prepuberty predicted his male, but not female, grandchildren's mortality (Bygren et al. 2001; Kaati et al. 2002, 2007; Pembrey et al. 2006; Soubry et al. 2014; Wu et al. 2015; Vågerö et al. 2018).

64. Many hundreds of published studies now attest to the importance of epigenetic inheritance in a wide variety of organisms. See Jablonka and Raz (2009), Heard and Martienssen (2014), Bonduriansky and Day (2018), and Anastasiadi et al. (2021) for reviews.

65. Anastasiadi et al. 2021; Hollwey et al. 2023.

66. Dias and Ressler 2014.

67. In the yellow toadflax—*Linaria vulgaris.* Tollefsbol 2019.

68. Anastasiadi et al. 2021.

69. Twenty generations: Houri-Ze'evi et al. 2021. Three to five: Anastasiadi et al. 2021. This estimate may change with further research. For instance, a recent study of epigenetic inheritance in water fleas (*Daphnia magna*) reported no attenuation in the magnitude of methylation after four generations of inheritance (Feiner et al. 2021).

70. Smoking: Tehranifar et al. 2018. Insecticides: Brevik et al. 2021. Heavy metals: Cong et al. 2019.

71. O'Dea et al. 2016; Anastasiadi et al. 2021.

72. Yehuda et al. 2000, 2005, 2009; Sarapas et al. 2011.

73. Francis 2011. While we raise the possibility that the Dutch Hunger Winter may consti-tute an example of adaptive parental adjustment of epigenetic variation, we also stress that care is required in the interpretation of this and other such claims of adaptive parental epigenetic adjustment. At the time of writing there is little clear evidence that these epigenetic effects are adaptive, or would have been adaptive under unchanged conditions. In the case of the Dutch Hunger Winter, a plausible alternative hypothesis is that the modifications to epigenetic attach-ments were largely disruptive and the observed reduced methylation is a consequence of se-lection acting on such variation (Tobi et al. 2018).

74. P. Bateson and Martin 1999; Uller 2008; Crean and Marshall 2009; P. Bateson et al. 2014. Possible examples of predictive adaptive responses include greater helmet development in *Daphnia cucullata* in response to maternal exposure to predator pheromones (Agrawal et al. 1999), rats' exposure to glucocorticoid during late gestation leading to an intolerance to glucose as adults (Nyirenda et al. 1998), and coat thickness determination in vole pups by the photo-period length experienced by the mother (T. Lee and Zucker 1988).

75. Not all effects of such epigenetic reprogramming are likely to be adaptive, even if the environment remains constant. The increased rates of cancer observed in descendants of mothers experiencing the Dutch Hunger Winter, for instance, may be a maladaptive side-effect of the changes in methylation. However, such effects appear later in the lives of offspring and would impact fitness only weakly, as the strength of selection on an individual decreases as it ages and passes sexual maturity, resulting in a time later in life where selection is weak, known as the "selection shadow" (Medawar 1952).

76. Eriksson et al. 2001; Gluckman and Hanson 2006, 2007.

77. Anastasiadi et al. 2021.

78. Jablonka and Lamb 2005; Burggren 2016; O'Dea et al. 2016; Houri-Ze'evi and Rechavi 2017; Jablonka 2017; Stajic and Jansen 2021; Anastasiadi et al. 2021.

79. Baker et al. 2019. The species was *Polygonum persicaria*.

80. Luna et al. 2012; Z. Fu and Dong 2013; Espinas et al. 2016; Y. He and Li 2018; O. Roth et al. 2018; Feiner et al. 2022. It has also been called *transgenerational immune priming* (Roth et al. 2018). Similar effects are observed in coral, where transplanted reef-building corals that modified their methylation to match local, established corals exhibited higher fitness than those that failed to do so (Dixon et al. 2018).

81. The presence of such mechanisms need not imply that they evolved to fulfil that func-tion. The inheritance of RNA may simply tap into a responsive physiological system, which could produce seemingly adaptive relationships between exposure and persistence without selection on variable RNA transfer between generations.

82. Houri-Ze'evi et al. 2016.

83. Houri-Ze'evi et al. 2021.

84. J. Herman and Sultan 2016. Adaptive transgenerational epigenetic memory of ancestral temperature regime has also been reported in *Arabidopsis thaliana* (Whittle et al. 2009).

85. Houri-Ze'evi et al. 2016.

86. Rine and Herskowwitz 1987; Gottschling et al. 1990; Holmes and Broach 1996; Stajic and Jansen 2021. The yeast was *Saccharomycetales*.

87. Seong et al. 2011; McCleary and Rine 2017; Stajic and Jansen 2021.

88. D. Charlesworth et al. (2017) claim that "a large body of genetic experiments has estab-lished the ineffectiveness of selection on homozygous lines, which lack genetic variation but still show phenotypic variation." However, this reasoning is problematic. Inbred lines are typi-cally generated through either self-fertilization or brother-sister mating, leaving both genetic *and epigenetic* variation greatly reduced. The experimental quantification of genetic and

epigenetic variation, as in the work on *Arabidopsis* described in the text (Bossdorf et al. 2010; Cortijo et al. 2014; Schmid et al. 2018; Stajic and Jansen 2021), provides a more reliable estimate of the contribution of epigenetic variation to selection.

89. Bossdorf et al. 2010; Cortijo et al. 2014; Stajic and Jansen 2021. Epigenetic variation explained more than 50% of the observed variance in those traits, and accounted for up to 90% of the broad-sense heritability—which is the proportion of phenotypic variance attributable to genetic causes, here including epigenetic factors (Bossdorf et al. 2010; Cortijo et al. 2014; Stajic and Jansen 2021). Genome-wide studies in this species have revealed extensive variation in DNA methylation patterns between populations (Dubin et al. 2015; Kawakatsu et al. 2016), potentially providing many targets for selection.

90. Schmid et al. 2018.

91. Offspring of ancestral and selected populations grown together in the same controlled environment exhibited significant inherited phenotypic differences even two generations after selection ceased, with these differences correlated with epigenetic but not genetic variation (Schmid et al. 2018).

92. The widespread belief that "without genetic variation, there can be no evolution" (Stearns and Hoekstra 2005, 99) must now be viewed as open to question.

93. O'Dea et al. 2016; Stajic and Jansen 2021; Anastasiadi et al. 2021.

94. O'Dea et al. 2016; Stajic and Jansen 2021; Anastasiadi et al. 2021.

95. Genetically identical bacterial cells differ in their tolerance; therefore, nongenetic factors, including epigenetic mechanisms, are implicated (e.g., Calo et al. 2014). Interestingly, when persisters are relieved of antibiotic stress they become active again and rapidly reestablish a population that can also be susceptible to antibiotics (T. Day 2016; Riber and Hansen 2021). This implies that the ability to persist is a transient phenotypic property rather than genetically specified, leading microbiologists to infer that epigenetic mechanisms may underly bacterial persistence (Riber and Hansen 2021). Mathematical modelling has reinforced the plausibility of this conclusion (T. Day 2016). The suggestion that pathogens, including bacteria, carry "epigenetic memories" of recent hosts also explains some curious features of disease virulence—for instance, that girls infected by boys, or boys infected by girls, are more likely to die of some diseases, including measles, chickenpox, and polio, than children infected by same-sex individuals (McLeod et al. 2021).

96. Lachmann and Jablonka 1996; Carja et al. 2014; Furrow and Feldman 2014.

97. Acar et al. 2008; Proulx et al. 2019. The study by Proulx and colleagues also provides evidence that transgenerational epigenetic inheritance helps populations of *C. elegans* to adapt to challenging oxygen-level fluctuations, predicting the current environment more effectively than maternal effects alone.

98. O'Dea et al. 2016.

99. Stajic and Jansen 2021.

100. O'Dea et al. 2016; Stajic and Jansen 2021; Anastasiadi et al. 2021.

101. Model-based estimates for the forward and backward CG epimutation rates were 2.56×10^{-4} and 6.30×10^{-4}, respectively (Van der Graaf et al. 2014). The authors suggest that the estimated CG forward-backward epimutation rates, which determine the CG methylation content of the genome, imply that about 30% of all CG sites in *A. thaliana* should be methylated, which is consistent with actual measurements (Cokus et al. 2008; Lister et al. 2008).

102. Slatkin 2009; D. Charlesworth et al. 2017.

103. Shea et al. 2011

104. Danchin et al. 2019; Stajic and Jansen 2021; Anastasiadi et al. 2021.

105. Stajic et al. 2019. The yeast was *Saccharomyces cerevisiae*.

106. Interestingly, intermediate levels of heritable silencing led to the most efficient adaptation; populations with low levels of silencing are vulnerable to extinction, but too high levels of

silencing also delay the fixation of adaptive mutations by reducing their adaptive advantage (Stajic et al. 2019).

107. Evolutionary rescue may also be achieved through niche construction and ecological inheritance (Longcamp and Draghi, 2023).

108. Ehrlich et al. 1986

109. Shen et al. 1994. Rate estimations from X. He et al. (2015) and Danchin et al. (2019).

110. Elevated point mutation from C to A and G in methylated genes is reported (Tomkova and Schuster-Böckler 2018), and histone modifications and noncoding RNAs also contribute to increased mutagenesis (Danchin et al. 2019). Epigenetic marks may also impact copy number variations (Vogt 2015, 2021; Pértille et al. 2019).

111. Schuster-Böckler and Lehner 2012.

112. Danchin et al. 2019; Stajic and Jansen 2021; Anastasiadi et al. 2021.

113. Methylation: Anastasiadi and Piferrer 2019; Pértille et al. 2019. Histone modifications: Stajic et al. 2019; Torres-Garcia et al. 2020. Noncoding RNAs: Calo et al. 2014. See Danchin et al. (2019) and Anastasiadi et al. (2021) for reviews.

114. Pértille et al. 2019, 685.

115. Kronholm et al. 2017. The alga was *Chlamydomonas*.

116. Bódi et al. 2017.

117. Calvo et al. 2009.

118. Similar findings are reported in a fungus (Calo et al. 2014).

119. Other studies show that experimental disruption of the gamete epigenome leads to developmental defects in offspring (reviewed in Klosin and Lehner 2016).

120. Snails: Thorson et al. 2017. Similar findings are also reported in three-spined sticklebacks (Heckwolf et al. 2020). Knotweed: C. Richards et al. 2012

121. These correlational data provide only circumstantial support, as adaptive differences between populations may simply be due to plasticity, rather than epigenetic inheritance. Salt marsh perennials: Foust et al. 2016. Interestingly, chemical disruption of DNA methylation in blind cave fish partially recovered eye size, implying that eye degeneration in this species arises in part through epigenetic gene silencing (Gore et al. 2018).

122. For instance, Day and Bonduriansky developed Price equation models combining genetic and nongenetic factors, demonstrating that epigenetic variation can influence the evolution of genetic variation and lead to "otherwise unexpected outcomes" (Bonduriansky and Day 2009, 2018; see also T. Day and Bonduriansky 2011; T. Day 2016). Likewise, Geoghegan and Spencer's "population-epigenetic" analyses explored the long-term evolutionary dynamics of epigenetic modifications, finding that there were potentially more stable equilibria for epigenetic variation than for analogous models of genetic inheritance, and that "epigenetic modifications to standard population-genetic models of selection can have major consequences" (Geoghegan and Spencer 2012, 2013). Furrow and Feldman's (2014) analysis of the coevolution of genetic and epigenetic regulation suggests that variation in the extent to which epigenetic inheritance impacts evolutionary adaptation is contingent on rates of environmental change. In rapidly changing environments, epigenetic regulation may evolve, but with low rates of transgenerational inheritance of epigenetic states, while in slowly changing environments genetic specification of epigenetic states is favored. In between, a variety of other epigenetic regulatory regimes are favored, including environment-responsive ("pure") epigenetic changes and more random "stochastic switching." See also Shea et al. 2011; Rivoire and Leibler 2014; Uller et al. 2015; McNamara et al. 2016.

123. Kohl et al. 2014.

124. Kamra 2005; S. Gilbert et al. 2012; Roughgarden et al. 2018; S. Gilbert 2019; N. Moran et al. 2019.

125. S. Gilbert et al. 2012; Funkhouser and Bordenstein 2013; Roughgarden et al. 2018.

126. Funkhouser and Bordenstein 2013.

127. Funkhouser and Bordenstein 2013; Roughgarden et al. 2018.

128. For illustration, in humans where microbiome inheritance occurs largely through intimate neighbor transfer, more than 72% of marker alleles present in mothers' microbiomes were also found in their offspring (Nayfach et al. 2016).

129. Funkhouser and Bordenstein 2013.

130. Kohl et al. 2014.

131. Dearing et al. 2005, 2022.

132. Kohl et al. 2014.

133. Kohl et al. 2014. The desert woodrats' habit of collecting and feeding on feces may be another example of genetic accommodation.

134. Rampelli et al. 2021.

135. Most modern dogs exhibit extreme copy number expansion of the amylase gene (AMY2B), an adaptive response to the shift from a carnivorous wolf diet to the starch-rich omnivorous diet of modern domesticated dogs (Botigué et al. 2017).

136. Rampelli et al. 2021.

137. Rampelli et al. 2021.

138. Cordaux and Gilbert 2017. Comparisons of the DNA sequences encoding pectin depolymerases and cellulases in several species of these beetles indicate that the these plant-digesting enzymes came to the beetles by horizontal gene transfer (HGT) from the plant-digesting microbes to the host beetle DNA (Kirsch et al. 2014; Pauchet, et al. 2014). HGT has been seen as the key phenomenon promoting the capacity of these beetles to use plants as their primary source of nutrition.

139. Alberdi et al. 2016; E. Rosenberg and Zilber-Rosenberg 2016; Henry et al. 2021.

140. Redman et al. 2002; Rodriguez et al. 2008.

141. Rodriguez et al. 2008.

142. Bean/soy bugs: Kikuchi et al. 2012. Coral: N. Webster and Reusch 2017. Bears (*Ursus arctos*): Sommer et al. 2016.

143. Yan et al. 2005. See also S. Gilbert 2020a.

144. Roughgarden et al. 2018.

145. Henry et al. 2021.

146. Rampelli et al. 2021.

147. Niche expansion: Kohl et al. 2014; Egan et al. 2020; Rampelli et al. 2021 Parasitoid resistance: Oliver et al. 2003. Heat stress: Montllor et al. 2002. Speciation: Tsuchida et al. 2004; Brucker and Bordenstein 2013; E. Rosenberg and Zilber-Rosenberg 2016. See E. Rosenberg and Zilber-Rosenberg (2016) for a review.

148. While we describe the experimental work of Sharon et al. (2010) with *Drosophila melanogaster*, their experiment is based on earlier work by Dodd (1989), who reported the same rapid appearance of reproductive isolation through rearing on different media in *Drosophila pseudoobscura*.

149. Sharon et al. 2010.

150. Sharon et al. 2010.

151. *Drosophila* species typically have a generation time of 10–30 days, while the generation time for bacteria ranges from 15 minutes to 24 hours.

152. Sharon et al. 2010.

153. E. Rosenberg and Zilber-Rosenberg 2016.

154. N. Moran and Yun 2015.

155. Franzenburg et al. 2013.

156. T. Suzuki et al. 2022. These empirical findings are also supported by mathematical models showing that gut mutualists can persist in host populations despite low fidelity of vertical transmission (Xiong et al. 2022).

157. Co-speciation: Groussin et al. 2017. Muegge et al. (2011) report a mismatch between microbial function in the gut and host phylogeny for mammalian hosts. Often, the functions that a microbe provides are more important than the species providing it, and the diet of the organism has an important effect on a host's microbia (Ley at al. 2008).

158. Concordance: Ley at al. 2008. Vertical inheritance: Muegge et al. 2011; Groussin et al. 2017. Interestingly, in humans and other apes there is evidence of co-speciation with prominent bacterial taxa (Moeller et al. 2016).

159. Muegge et al. 2011; Groussin et al. 2017. However, we note a species-specific core microbiome may be present.

160. Kohl et al. 2014. The collective microbial community can also alter environmental biogeochemical processes, as has been observed in hippos, who dump microbes into shared pools each time they excrete, creating a "meta-gut" (Dutton et al. 2021).

161. Kolodny et al. 2019. The bats were *Rousettus aegyptiacus*.

162. Archie and Theis 2011; Degnan et al. 2012; Ezenwa et al. 2012; Theis et al. 2012; Lax et al. 2014; Perofsky et al. 2017.

163. M. Lombardo 2008; Ezenwa et al. 2012.

164. Koch and Schmid-Hempel 2011. The bumblebees were *Bombus terrestris*.

165. Osmanovic et al. 2018.

166. Roughgarden et al. 2018. See also N. Moran 2007.

167. N. Moran and Sloan 2015; Douglas and Werren 2016; Roughgarden et al. 2018.

168. N. Moran and Sloan 2015.

169. Douglas and Werren 2016.

170. N. Moran 2007; Roughgarden et al. 2018.

171. Hoppitt and Laland 2013.

172. Hoppitt and Laland 2013; Whiten et al. 2017; Oudman et al. 2020.

173. Hoppitt and Laland 2013; Laland 2017a; Whiten et al. 2017

174. Slagsvold and Wiebe 2007, 2011; Slagsvold et al. 2013.

175. Helfman and Schultz 1984; Warner 1988, 199.

176. Cultural traditions are experimentally demonstrated in fruit flies, honeybees, wood crickets and many other insects (Leadbeater and Chittka 2007; Leadbeater and Dawson 2017; Laland 2017a; Danchin et al. 2018).

177. Leadbeater and Chitka 2007.

178. Leadbeater and Chittka 2007; Hoppitt and Laland 2013; Laland 2017a; Leadbeater and Dawson 2017;Whiten 2017; Danchin et al. 2018; Aplin 2019.

179. Rendell et al. 2010; Laland 2017a.

180. Richerson and Boyd 2005; Rendell et al. 2010.

181. Seeley 1995; Bonabeau, et al. 1999; Camazine et al. 2001; Krause et al. 2010; Sumpter 2010; King and Sueur 2011; Morand-Ferron and Quinn 2011; Sasaki and Biro 2017.

182. Laland 2017a. This statement implies that human culture is different from the culture of other animals in key respects, which sometimes allow maladaptive cultural information to spread among human populations (Richerson and Boyd 2005; Henrich 2016; Laland 2017a). Why this should be is discussed in chapter 13.

183. Whiten et al. 1999, 2001.

184. Routes to food sites: Laland and Williams 1997, 1998. Food patch preferences: R. Day et al. 2001; R. Kendal et al. 2004. Feeding: Stanley et al. 2008. Mating preferences: Dugatkin 1992; Dugatkin and Godin 1992, 1993; Briggs et al. 1996. Evasion: Sugita 1980; C. Brown and Laland 2002b; Kelley and Magurran 2003; Reader et al. 2003. Predator recognition: C. Brown and Laland 2002a. Inspection tactics: Kelley and Magurran 2003. The guppy is *Poecilia reticulata*.

185. Slagsvold and Wiebe 2007, 2011; Slagsvold et al. 2013; Laland 2017a; Whiten 2017.

186. Laland 2017a.

187. Laland 1990, 2017a.

188. Von Steiniger 1950; Galef and Clark 1971a,b; Galef and Beck 1985; Galef 1988.

189. Readers unfamiliar with the literature on attempts to exterminate wild rats using warfarin might think that colony-wide traditions of seeking out poisonous food would be difficult to detect as they would rapidly lead to colony extinction, and that this might leave open the possibility that rats transmit "bad" information socially. In fact, extensive field and experimental studies suggests that does not happen, and that rats avoid poison solely by transmitting "good" information about what to eat (Galef and Beck 1985; Galef 1988; Laland and Plotkin 1991, 1993; Galef 2003); see Laland (2017a) for a review of rat social learning.

190. Whiten et al. 1999, 2001; Fragaszy and Perry 2003; Hoppitt and Laland 2013; Laland 2017a; Whiten 2017.

191. Fragaszy and Perry 2003; Laland 2017a.

192. This conclusion contrasts with early studies of observational fear conditioning in monkeys, which appeared to suggest tight genetic constraints on what animals learn. In such studies, individuals were conditioned to acquire a fear of snakes through observing companions behaving fearfully in the presence of a snake but did not learn to become frightened of flowers through this mechanism (Mineka and Cook 1988).

193. Olsson and Phelps 2007. Blackbirds learn to recognize and mob predators through social transmission but can also acquire a fear of arbitrary objects such as plastic bottles in this manner (e.g., Curio 1988). Studies have established that rhesus monkeys can also be socially conditioned to acquire a fear of novel arbitrary objects, including kitchen utensils (e.g., Stephenson 1967).

194. Le et al. 2013.

195. Cultural variation can cause frequency changes through the differential survival and reproduction of the animal (natural selection) or through the differential adoption or discarding of cultural knowledge (cultural selection) (Cavalli-Sforza and Feldman 1981).

196. Foote et al. 2016; Hoelzel and Moura 2016.

197. West-Eberhard 2003.

198. There is evidence that animal cultures can be long-lived. For instance, archaeological remains show that chimpanzees have used stone tools to crack open nuts for at least 4,300 years (J. Mercader et al. 2007).

199. Insects: Mery et al. 2009; Danchin et al. 2018. Fishes: Dugatkin 1992; Witte and Massmann 2003; Godin et al. 2005. Birds: White and Galef 2000; White 2004; Swaddle et al. 2005. Mammals: Little et al. 2008; Galef 2009; A. Davies et al. 2020. Mathematical models: Gibson et al. 1991; Kirkpatrick and Dugatkin 1994; see also Sibly and Curnow 2022. Likewise, theory suggests that when young male birds learn their songs, usually from nearby adult males, they modify the selection of genes that affect how songs are acquired (in males) and which songs are preferred (in females) (Lachlan and Slater 1999) and can facilitate the evolution of brood parasitism (Beltman et al. 2003). The learned foraging traditions of killer whales provide another example (see chapters 1 and 4).

200. Brood parasitism: N. Davies and Welbergen 2009; Thorogood and Davies 2012. Bias evolution of signals: Ten Cate and Rowe 2007; Verzijden et al. 2012. Population divergence etc.: Laland 1994a,b; Beltman et al. 2004; Verzijden and Ten Cate 2007; Verzijden et al. 2012.

201. J. Allen et al. 2013.

202. Laland 2017a.

203. Snell-Rood et al. 2018; Laland et al. 2019. Behavioral innovations are discussed in Lefebvre et al. (1997), Reader and Laland (2003), and Reader et al. (2016).

204. Russon 2003. Orangutans are *Pongo pygmaeus*.

205. Laland 2004; Rendell et al. 2011; Laland et al. 2019, 2022.

206. Birds: Mason and Reidinger 1982. Fish: Coolen et al. 2003; M. Webster et al. 2019. Nest-site decisions: Sarin and Dukas 2009; Forsman and Seppänen 2011; Seppänen et al. 2011.

207. Trade-offs have been identified between sharing information and sharing infectious disease, since the same behavior can promote both. For instance, in primates, those species most reliant on social learning have the greatest burden of infectious disease (McCabe et al. 2015).

208. Brett Jesmer at the University of Wyoming and his colleagues describe how translocated bighorn sheep (*Ovis canadensis*) gradually adjust to novel environments through learning and cultural transmission (Jesmer et al. 2018). These animals migrate by exploiting the high-quality forage manifest in "green wave surfing," which requires possessing the requisite learned knowledge of where to find the quality food. The study reports iterative improvements over generations in the fraction of translocated populations that possess the relevant foraging knowledge.

209. The extent to which the newly exploited environments can be regarded as novel requires comment. There are, of course, features of the novel habitats to which the sheep are preadapted through ancestral selection (for instance, they are adapted to seek out and feed on the same high-quality forage in both ancestral and novel habitats). However, the ability of the sheep to surf the green wave was not automatically conferred by ancestral selection, because the first generation of transplanted sheep did not all find the high-quality food. The first generation of migrants perform considerably worse than their descendants. The point is that the sheep have to know where to find high-quality forage in order to eat it. The habitat into which they have been translocated is novel in the sense that they don't know where to find the food and past selection has not told them where to find it. That knowledge has to be newly acquired and gradually accumulates in the population over multiple generations through learning, in the process incrementing mean fitness.

210. Laland and Williams 1998; Richerson and Boyd 2005; Jablonka and Lamb 2014; Laland 2017a; Toyokawa et al. 2019.

211. Laland et al. 2000, 2010; Boivin et al. 2016.

212. Gerbault et al. 2011; but see Evershed et al. 2022.

213. Laland et al. 2010. These cases in our own species, like the evolutionary responses to cultural differences in diet observed in the killer whales, or the enhanced sensitivity to snakes in the monkeys, and other instances of gene-culture coevolution in animals (Whitehead et al. 2019), are functionally equivalent to how epigenetic inheritance can elicit genetic assimilation.

214. Feldman and Cavalli-Sforza 1976; Cavalli-Sforza and Feldman 1981; Lumsden and Wilson 1981; Boyd and Richerson 1985; Feldman and Laland 1996; Laland et al. 2001; Enquist et al. 2007; Laland and O'Brien 2010; Mesoudi 2011; Rendell et al. 2011; Henrich 2016; Creanza et al. 2017; Laland 2017a; Behar and Feldman 2018; Boyd 2018; Ram et al. 2019.

215. Learning typically has a beneficial effect on adaptation in relatively quickly changing environments, allowing individuals to adjust to changes occurring too fast to be tracked by the natural selection of genetic variation (Cavalli-Sforza and Feldman 1981; Boyd and Richerson 1985; Wakano et al. 2004; Aoki et al. 2005; Aoki and Feldman 2014). In unchanging or slowly changing environments the benefits of learning are more complex, with learning sometimes accelerating evolution by helping genotypes to locate otherwise difficult-to-find solutions (e.g., Hinton and Nowlan 1987), and other times weakening selection by reducing phenotypic differences between genotypes (Ancel 2000; Frank 2011). Nonetheless, these seemingly conflicting results are relatively well understood (Borenstein et al. 2006; Paenke et al. 2007; Frank 2011). The emerging consensus is that individual learning typically slows evolution in static unimodal fitness landscapes, but accelerates evolution in dynamic or static multimodal fitness landscapes. The existence of multiple optima usually slows down evolution as populations get trapped on suboptimal fitness peaks, but, by generating adaptive variation and thereby

smoothing the fitness landscape, learning increases the likelihood of a directly increasing path of fitness to the global optimum (Borenstein et al., 2006; Mills and Watson, 2006; Frank, 2011). Similar findings are reported for phenotypic plasticity (Price et al. 2003; Edelaar et al. 2017).

216. Theory suggests that this may be an oversimplification, and that genetic information commonly also provides a selection-based cue, as genetic variation becomes correlated with environmental state through the action of natural selection (Shea et al. 2011; McNamara et al. 2016). For discussion of genetic information, see Griffiths and Stotz (2013).

217. Y. He and Li 2018.

218. Jablonka and Lamb 2005.

219. Yeast: Stajic et al. 2019. Dogs: Rampelli et al. 2021. Rats: Galef and Beck 1985; Galef 1988, 2003; Laland and Plotkin 1991, 1993. See Laland (2017a) for a review of rat social learning.

220. *Polygonum*: Baker et al. 2019. Defense priming: Luna et al. 2012; Z. Fu and Dong 2013; Espinas et al. 2016; Y. He and Li 2018. Similar effects are observed in coral, where transplanted reef-building corals that modified their methylation to match local, established corals exhibited higher fitness than those that failed to do so (Dixon et al. 2018). The nematodes were *C. elegans*.

221. Proulx et al. 2019. See also related work in *Daphnia* by Graeve et al. (2021b).

222. English et al. 2015; McNamara et al. 2016.

223. Shea et al. 2011; Dall et al. 2015; Uller et al. 2015; McNamara et al. 2016.

224. Osmanovic et al. 2018.

225. Small RNA inheritance: Houri-Ze'evi et al. 2021. Cultural knowledge: Rendell et al. 2010; Hoppitt and Laland 2013; Laland 2017a.

226. O'Dea et al. 2016; Anastasiadi et al. 2021.

227. Slatkin 2009; D. Charlesworth et al. 2017.

228. Jablonka and Lamb 2005; Sultan 2015.

229. Jablonka and Lamb 2005; Sultan 2015.

230. Jablonka 2017.

231. Jablonka 2017.

232. Danchin et al. 2019; Stajic and Jansen 2021; Anastasiadi et al. 2021.

233. Woodrats: Kohl et al. 2014. Killer whales: Morin et al. 2010; Riesch et al. 2012; Foote et al. 2016.

234. Berdahl et al. 2018.

235. Houri-Ze'evi and Rechavi 2017.

236. Heard and Martienssen 2014.

237. Houri-Ze'evi and Rechavi 2017.

238. Anastasiadi et al. 2021.

239. Adrian-Kalchhauser et al. 2021.

240. Woodrats: Kohl et al. 2014. Monkeys: Le et al. 2013. The yeast were *Saccharomyces cerevisiae*.

241. Jablonka and Lamb 2005; Jablonka and Raz 2009; Heard and Martienssen 2014; Bonduriansky and Day 2018; Anastasiadi et al. 2021.

242. Durham 1991; Whiten et al. 1999; Richerson and Boyd 2005; Hoppitt and Laland 2013; Laland 2017a.

243. Some researchers have argued that there must be strong genetic control of extragenetic inheritance for evolution to produce adaptation (e.g., Dickins and Rahman 2012; Futuyma 2017; Dickins and Dickins 2023). However, only a small fraction of the data are consistent with this (Jablonka and Lamb 2005; Richerson and Boyd 2005; Bonduriansky and Day 2018; Anastasiadi et al. 2021). A degree of independence from genetic variation is necessary for extragenetic inheritance to play a role in short-term adaptation.

244. Oyama 2000.

245. Odling-Smee et al. 2003; West-Eberhard 2003; Badyaev and Uller 2009; Laland et al. 2015, 2022. We have already seen that different epigenetic mechanisms can enhance the stability of transgenerational epigenetic inheritance in yeast (Volpe et al. 2002; Holoch and Moazed 2015; Klosin and Lehner 2016), that epigenetic mechanisms and parental effects reinforce each other in nematodes (Proulx et al. 2019), and that without exposure to the ecological legacy of bacteria-rich feces and other environmental traces of conspecifics, transmission of the woodrats' capacity to digest creosote would rapidly break down (Kohl et al. 2014).

246. Fragaszy 2011. There are many examples. Black-capped chickadees (*Parus atricapillus*) learned to open the foil tops of milk bottles through exposure to the milk bottles opened by other birds (Sherry and Galef 1984). When leaving a feeding site, adult rats deposit scent trails that direct young rats seeking food to locations where food was ingested (Galef and Buckley 1996), while feeding adults deposit residual urine marks and feces, both in the vicinity of a food source and on foods they are eating, via which young rats acquire dietary preferences (Galef and Beck 1985; Galef and Heiber 1976; Laland and Plotkin 1991), and probably also acquire components of their microbiome that way. Fish secrete food cues in their mucus, as well as in their urine, to which other fish attend (Atton 2013). If a recently fed fish emits chemical cues of stress at the same time as these food cues, other fish seemingly learn that the new food is one to be avoided. Conversely, when there are no such stress chemicals in the water, the mucus cues are acted upon and observing fish rapidly develop a preference for the newly consumed diet (Atton 2013). Likewise, among bumblebees, when successful foragers bring nectar to the nest, they deposit the scented solution in honeypots, where other colony members sample it and thereby acquire a preference for the oral scent (Dornhaus and Chittka 1999; Leadbeater and Chittka 2007).

247. Musgrave et al. 2016. Other examples include young capuchin monkeys (*Sapajus apella*) that learn to find and retrieve beetle larvae hidden inside bamboo stalks by exposure to canes already opened by adults (Gunst 2008, 2010), and black rats (*Rattus rattus*) that learn to strip pinecones for their seeds by access to pinecones partially opened by their mothers (Terkel 1996). In addition, many fish learn to recognize novel predators through associating predator cues with an alarm substance that other fearful fish have released into the water (C. Brown and Laland 2003; Suboski et al. 1990), and countless animals learn safe routes through their environment by following the tracks and trails of conspecifics (Hoppitt and Laland 2013).

248. Meaney and Szyf 2005; Champagne and Meaney 2006, 2007; Champagne et al. 2007; Champagne 2008.

249. The adult offspring of good lickers have more glucocorticoid receptors in the brain than the offspring of poor lickers, which results in reduced stress. Good mothering promotes the demethylation of the gene that produces the glucocorticoid-receptor protein, while inadequate mothering leads to its continued methylation, leaving the stress axis hyperactive, and offspring predisposed to anxiety (Champagne and Meaney 2006, 2007; Champagne et al. 2007; Champagne 2008). In addition, exposing a pregnant mammalian laboratory animal to high stressors has been found to cause lifelong changes in her offspring's physiology, including by changing glucocorticoid secretion in the fetus (see Sarto-Jackson [2022] for an up-to-date review of this literature). For instance, if a pregnant rat encounters severe stressors her offspring will typically grow up to be more anxious. In this manner, mammalian offspring can learn about the outside world while still fetuses.

250. Bonduriansky and Day 2009, 2018; T. Day and Bonduriansky 2011; Shea et al. 2011; Geoghegan and Spencer 2012, 2013; Furrow and Feldman 2014; Rivoire and Leibler 2014; English et al 2015; Uller et al. 2015; T. Day 2016; McNamara et al. 2016; Adrian-Kalchhauser et al. 2021.

251. Oyama 2000.

252. Oyama 2000.

253. Waddington 1969b, 122.

254. There are several important implications of this change of perspective. One is that there is no longer good reason to assume that genes must be the sole or (for particular traits) even primary source of parent-offspring similarity. Data presented in this chapter make clear that the norm of reaction, for instance, is not a determinate property of the genotype but is conditioned by diverse forms of inherited environmental information (Sultan 2015, 2019). A second implication is that any lingering suggestion that genes alone contain "instructions" is undermined, as the information to build bodies is distributed across diverse inherited resources. For a century, genes have had a "privileged" status as the fundamental cause of development and heredity. A third implication is that, unlike Mayr (1961, 1980), contemporary researchers can no longer cleanly separate out the evolutionary processes by which hereditary information accumulates (i.e., "writing" the genome) from the developmental processes by which hereditary information is expressed (i.e., "reading" the genome), since the accrual of hereditary information occurs during development (Stotz 2019). Heredity is itself a developmental process (Oyama 2000). While regarding heredity as a developmental process may facilitate a deeper understanding of evolutionary adaptation, it raises some nontrivial terminological issues. For instance, it would seem to undermine a traditional understanding of Wilhelm Johannsen's (1911) genotype-phenotype distinction. For the "genotype" to include an organism's full hereditary information, researchers would need to recognize that it extends well beyond information in or arising from the genome sequence (for instance, encompassing epigenetic and learned information). Conversely, were the "phenotype" to remain an organism's actual observed properties, such as morphology, development, or behavior, some elements of which are inherited by nongenetic means, then some traits could simultaneously qualify as both genotype and phenotype. Fourth, perhaps the most profound implication, is that the much-maligned concept of "inheritance of acquired characteristics" (in the senses described here) must now be regarded as empirically validated, as well as taxonomically widespread. Arguments over whether the processes involved are Darwinian or Lamarckian, or whether Lamarck has been exonerated, are neither here nor there. The point is that mechanisms previously regarded as impossible are now known to be possible (e.g., the inheritance of learned knowledge through the germline in mice; Dias and Ressler 2014), that heritable variation can be generated in response to environmental conditions rather than at random (e.g., lobtail feeding in humpback whales; J. Allen et al. 2013), and that evolutionarily significant adaptations can arise in just a single generation and still be reliably inherited (e.g., media-based reproductive isolation in *Drosophila*; Sharon et al. 2010). These effects on the mechanics of evolutionary adaptation, undermine assumptions that have been central to the field for a century. The recognition that soft inheritance plays a vital role in evolutionary adaptation—one that complements, rather than is at odds with, the role played by genes—is likely to require new models or even new approaches to understanding evolutionary dynamics (Lachmann and Jablonka 1996; T. Day and Bonduriansky 2011; Shea et al. 2011; Dall et al. 2015; Laland et al. 2015; Bonduriansky and Day 2018; Adrian-Kalchhauser et al. 2021). A final implication of viewing inheritance as a developmental process is that it undermines the independence of Lewontin's (1970) three conditions for evolution by natural selection (D. Walsh 2015; Uller and Helanterä 2019; see chapters 1, 4, and 14 for discussion). Researchers need to build this causal interdependence into evolutionary explanations (D. Walsh 2015; Uller and Helanterä, 2019; D. Walsh and Rupik 2023).

Chapter 11

1. Lyons 1995.

2. Mivart 1871. Evo-devo: G. Müller 2007, 2021; Moczek 2008; Shubin et al. 2009; A. Wagner 2014. Evolutionary explanations: S. Gould 2002. The reputation of William Bateson, who coined the term "genetics" and championed Mendel's Laws, was damaged by his emphasis on discontinuities in inheritance that might precede formation of a new species (see W. Bateson 1894; P. Bateson 2002).

3. G. Müller and Wagner 1991; G. Wagner et al. 2000; Moczek 2017.

4. Erwin 2020.

5. A. Burke 1989; Nagashima et al. 2009.

6. Stephen J. Gould (1977, 104) famously questioned the adequacy of mainstream gradualism to account for evolutionary innovations, writing: "But how can a series of reasonable intermediate forms be constructed? Of what value could the first tiny step toward an eye be to its possessor? The dung-mimicking insect is well protected, but can there be any edge in looking only 5 percent like a turd?"

7. Love 2008; Moczek 2008, 2017; Hallgrímsson et al. 2012; G. Wagner 2014; T. Peterson and Müller 2016; Erwin 2020; G. Müller 2021.

8. Novelty has been defined as a process leading to the discontinuous origin of new homologous characters (G. Wagner 1989a,b, 1996, 2014; G. Müller and Wagner 1991). Here we build on the work of Günter Wagner (2014), who has stressed how evolutionary innovation can arise through the reorganization of "character identity networks," which resemble what we call "regulatory networks." Developmental reorganization and small-effect mutations are not mutually exclusive for at least two reasons: (1) both obviously can happen; (2) small mutational effects (e.g., a modest change in gene product) can have large consequences for the phenotype.

9. T. Peterson and Müller 2016.

10. Hajheidari et al. 2019.

11. Moczek 2017.

12. Moczek 2017.

13. Moczek 2017.

14. G. Wagner et al. 2000; Moczek 2017.

15. Although researchers have started to identify genes and gene regulatory networks (GRNs) that have been co-opted, how this actually translates into phenotypic structure often remains poorly understood (i.e., much of evo-devo is not very "devo," and the link between genetic change and phenotypic change is often unclear). Shedding light on this relationship is a clear target for future research, and the reuse of co-opted elements constitutes a potentially productive focus.

16. Moczek 2008, 2017; Hallgrímsson et al. 2012; G. Wagner 2014; T. Peterson and Müller 2016; G. Müller 2021.

17. Bruce and Patel 2020.

18. Bruce and Patel's (2020) study used gene expression patterns and gene knockout phenotypes to show that both hypotheses have claims to be correct. The experimental data suggest that two leg segments in the common ancestor of contemporary insects and crustaceans were incorporated into the body wall of the newly evolved insect. Later, the proximal exite of the leg was moved dorsally, to form wings.

19. Bruce and Patel 2020, 1710

20. Hu et al. 2019.

21. Hu et al. (2019) used RNA interference to inhibit gene expression. see also Nijhout 2019.

22. Hu et al. 2019; See also Nijhout 2019.

23. Hu et al. 2019.

24. Busey et al. 2016; Zattara et al. 2016.

25. Zattara et al. 2016.

26. Hu et al. 2019.

27. Busey et al. 2016. These authors found that paired posterior horns in *Onthophagus taurus* and *O. gazella*, two relatively distantly related species, derive from larval head regions positioned along a boundary that corresponds to the embryonic segment boundary between the CL and OC (clypeolabral and ocular) segments. Their results raise the possibility that some of the same developmental mechanisms that establish the CL-OC boundary during embryogenesis have become repurposed to also position horn formation during late postembryonic development.

28. Emlen et al. 2005.

29. Busey et al. 2016.

30. Nijhout 2019.

31. Zattara et al. 2017.

32. Targeted overexpression of eyeless/*Pax6* is known to induce the development of ectopic "eyelike" structures in both fruit flies and frogs, but while structurally impressive, these organs are not functionally operational (Halder et al. 1995; Chow et al. 1999).

33. Shubin et al. 2009; Held 2017; Uller et al. 2018. Likewise, the appearance of novel behaviors in some species may reflect the co-option of latent neural circuit properties (Ding et al. 2019).

34. Eyes: N. Mercader et al. 1999; Kozmik 2005; Kozmik et al. 2008. Hearts: Olson 2006; Xavier-Neto et al. 2007. Outgrowths: Panganiban et al. 1997; N. Mercader et al. 1999.

35. The same point is made by mathematical theory, which shows how developmental processes create admissible trajectories across the adaptive landscape, and that populations converge on path peaks rather than landscape peaks (González-Forero, 2023, 2024).

36. S. Gilbert and Epel 2015.

37. Newman 2010, 2019.

38. Barresi and Gilbert 2020.

39. G. Wagner et al. 2019.

40. Newman 2019.

41. Sagner and Briscoe 2017.

42. Newman 2010, 2019.

43. R. Watson and Thies 2019. A case can also be made that niche construction, too, may have played a critical role in the origin of life (Torday 2016; Damer 2019; Damer and Deamer 2020; Deamer 2019; R. Watson and Thies 2019; Odling-Smee 2024).

44. Newman 2010, 2019.

45. Newman and Bhat 2009; G. Müller 2007, 2010, 2021.

46. Hu et al. 2019.

47. Hu et al. 2019.

48. Hu et al. 2019.

49. Kauffman and Roli 2021.

50. Favé et al. 2015.

51. Hanna and Abouheif 2021.

52. R. Keller et al. 2014; Hanna and Abouheif 2021.

53. West-Eberhard 2003; J. Hunt 2012; Quiñones and Pen 2017; Kapheim et al. 2020. While our focus here is on the role of developmental plasticity, there is a potentially important role for large-scale mutations, such as inversions, that create supergenes, which appear to be critical in the appearance of some complex traits, such as sociality in fire ants (J. Wang et al. 2013).

54. Rajakumar et al. 2012.

55. Rajakumar et al. 2012; Parsons et al. 2019.

56. Parsons et al. 2019.

57. Another illustrative example of this coordinated innovation is the mechanotransductive origin of cichlid pharyngeal jaws, potentially as a consequence of a behavioral shift to processing harder food types (T. Peterson and Müller 2018).

58. *Polypterus senegalus.*

59. Standen et al. 2014.

60. Standen et al. 2014, 54.

61. G. Bennett and Moran 2015. Likewise, the acquisition of new symbionts has enabled the red turpentine beetle to become a major killer of Chinese pine trees (Sun et al. 2013; Taerum et al. 2013).

62. Chiu and Gilbert 2020; Mizrahi and Jami 2021.

63. International Human Genome Sequencing Consortium 2001; Ryan 2004.

64. V. Lynch et al. 2011.

65. G. Müller and Wagner 1991.

66. Barresi and Gilbert 2020.

67. Lettice et al. 2008. In fact, several mutations are now known to be associated with the appearance of polydactyly in vertebrates (Lange et al. 2014, 2018; Lange and Müller 2017; Lange, 2020).

68. In general, the number of fingers or toes that develop is related to the availability of mesenchymal cells (Lange et al. 2018). This explains why big dogs that have larger limb buds, such as Saint Bernards, appear to be particularly prone to polydactyly (Alberch 1985; Barresi and Gilbert 2020).

69. Turing 1952. Contemporary Turing models have been extended to incorporate spatial transport mechanisms and "local autoactivation–lateral inhibition" processes, in addition to the reaction-diffusion processes that Turing studied (Newman and Frisch 1979; Lange et al. 2014).

70. Generally, other digit elements.

71. Lange et al. 2014, 2018; Lange and Müller 2017; Lange 2020.

72. Newman et al. 2018. While we favor the self-organization explanation provided by Turing models, which we judge to be more consistent with the empirical data, some developmental biologists characterize the position of digits as genetically specified (see Newman [2007] for discussion).

73. Sheth et al. 2012.

74. Raspopovic et al. 2014.

75. Lange et al. 2018. The authors name their model the "Hemingway model" after the American novelist Ernest Hemingway, who kept and bred polydactylous cats.

76. In their cellular automaton model of data from polydactylous cats, Lange et al. (2018) identified a theoretical cell variable a critical threshold value of which would generate a discrete new stripe corresponding to an additional digit. There exists a natural threshold of pre-chondrogenic cell numbers that must aggregate in order to activate the cartilage differentiation pathway that produces skeletal elements. Increasing the variable further would eventually lead to a new threshold, and another stripe. The authors called the variable that determines the number of toes the "'reaction rate." It represents the sensitivity of cells to changing state that advances cartilage formation.

77. Interestingly, digit loss, as opposed to polydactyly, has a different developmental basis. Tetrapods that have fewer than five digits as adults start out with developing anlagen (chondrogenic domains) for five digits, but not all continue with differentiation and growth to form a digit.

78. Sheth et al. 2012; Raspopovic et al. 2014; Lange et al. 2018.

79. Bard 1981; K. Painter et al. 1999; Salazar-Ciudad and Jernvall 2002; Harjunmaa et al. 2012.

80. Glover et al. 2023.

81. Of course, laying down the cartilage is just the first step in determining the number and position of digits. Thereafter exploratory mechanisms kick in, with muscles, tendons, nerves, and blood vessels all aggregating around the cartilage. However, the broad patterning results from a self-organizing Turing process.

82. The observation that polydactyly comprises large discrete changes in phenotype, rather than small continuous ones, is one reason why William Bateson (1894) was particularly interested in this topic, and viewed polydactyly as a challenge to Darwin's emphasis on gradualism (Lange and Müller 2017).

83. The genes/pathways include *FGT*, *WNT*, *BMP*, and *NOTCH* signalling (Lange et al. 2014, 2018; Lange and Müller 2017). As an example of environmental factors, following an

investigation of polydactyly in guinea pigs, Sewell Wright pointed out that the number of toes increased with the age of the mother and was greater in pups born during winter compared to summer months (Lange and Müller 2017).

Were CRISPR technology deployed to extract the vertebrate point mutation and insert it in a worm or fly, no fingers or toes would be seen, because the mutation does not manufacture these organs. In marked contrast, the same mutation could plausibly elicit ectopic organs in related organisms that share the same conserved developmental circuitry underlying vertebrate digit formation.

84. Indeed, the word "allele" comes from the Greek word meaning "difference." The fact that a small change (e.g., a point mutation) in a regulatory gene can have a large effect means that there need not be a strong correspondence between the magnitude of genetic and phenotypic effects.

85. Lange et al. 2014; Lange and Müller 2017.

86. Another observed bias is that extra thumbs or big toes are more likely than extra little fingers. This arises because the former (preaxial digits) are later developing than the latter (postaxial digits) and hence more likely to encounter any left-over cell tissue.

87. Lange et al. 2014.

88. A. Jones et al. 2007, 2014; Pavlíčev et al. 2011; R. Watson et al. 2014. While odd numbers of toes are indeed rare, studies of polydactylous cats report a slight left-right asymmetry, with extra digits being marginally more likely to be found on the left side of the body (Lange et al. 2014). A sagittal asymmetry is also observed, with extra digits more common in the forelimbs than hind limbs (Lange et al. 2014).

89. Greer 1987, 1990. G. Müller and Alberch (1990) show the ontogenetic sequences of reptilian limb development.

90. De Bakker et al. 2013.

91. In fact, Turing-like reaction diffusion models can reproduce major aspects of skeletal patterning in vertebrate limbs (Zhu et al. 2010).

92. Whether this diversification took place in the Cambrian and whether it is correctly characterized as an explosion are moot points (Valentine et al. 1999; Erwin et al. 2011; Budd 2013).

93. McShea 1994; Erwin 2017; D. Jablonski 2017.

94. Davidson and Erwin 2006.

95. Coyne 2006.

96. Borenstein and Krakauer 2008.

97. Salazar-Ciudad and Jernvall 2005; Borenstein and Krakauer 2008. See also Wimsatt 2013.

98. Tetrapods and arthropods: Davidson and Erwin 2006; M. Hughes et al. 2013; more complex patterns have also been described (e.g., D. F. Wright 2017). Plants: Oyston et al. 2016.

99. G. Wagner 2014.

100. Uller et al. 2018.

101. Borenstein and Krakauer 2008. This is not to suggest that it is impossible for major innovations to occur later in evolutionary time, as for instance the evolution of mammals attests.

102. Cooney et al. 2017.

103. Campàs et al. 2010; Mallarino et al. 2011; Fritz et al. 2014.

104. Uller et al. 2018.

105. Erwin 2020.

106. Odling-Smee et al. 2003; Erwin 2008, 2020; Post and Palkovacs 2009; Bråthen and Ravolainen 2015; San Roman and Wagner 2018; Xue et al. 2020.

107. Erwin 2008, 2020. Sometimes described as *inceptive* niche construction (Odling-Smee et al. 2003).

108. Callahan et al. 2014.

109. San Roman and Wagner 2018. See also Flohr et al. 2013.

110. Erwin 2005.

111. Laland et al. 1999; Silver and Di Paolo 2006; Kylafis and Loreau 2008; Krakauer et al. 2009; Van Dyken and Wade 2012; Creanza and Feldman 2014; Chisholm et al. 2018; Tanaka et al. 2020; Xue et al. 2020; Fogarty and Wade 2022. See also Hochberg et al. (2017) for a discussion of the role of niche construction in behavioral innovation.

112. S. Gould and Lewontin 1979; S. Gould 1980b, 2002; B. Charlesworth et al. 1982; Davidson and Erwin 2006; Coyne 2006; Sepkoski 2012.

113. Crombach and Hogeweg 2008; A. Wagner 2008, 2011a,b.

114. A. Wagner 2011a,b.

115. Directional selection, say for the size of a character, can also lead to thresholds in development (e.g., cell number), which may automatically result in the generation of a novel trait. As directional selection affects an entire population, such thresholds are not limited to a single individual.

116. Crombach and Hogeweg 2008; A. Wagner 2008, 2011a,b.

117. G. Müller and Wagner 1991; S. Gilbert et al. 1996; G. Müller 2021.

118. B. Charlesworth et al. 1982, 477–478.

119. Wagner 1989a,b; G. Müller 2003b; Moczek 2017.

120. Wagner 1989a,b; G. Müller 2003b; Moczek 2017.

Chapter 12

1. The concept of "evolvability" occupies a confusing position in modern evolutionary theory. On the one hand, it is central to mainstream quantitative and population genetics (Hansen 2006, 2021; Hansen and Houle 2008; G. Wagner and Draghi 2010; Nuño de la Rosa 2017; Crother and Murray 2018; Brigandt et al. 2023). On the other, it is championed as a core, and perhaps defining (Hendrikse et al. 2007), concept for evolutionary developmental biology (M. Kirschner and Gerhart 1998; Hendrikse et al. 2007; G. Müller 2007, 2021; R. Brown 2014; R. Watson 2020) and the extended evolutionary synthesis (G. Müller 2007, 2021; Pigliucci 2008; Pigliucci and Müller 2010; Laland et al. 2015). The suggestion that evolvability is—or should be—a defining feature of evolutionary developmental biology's research agenda hangs critically on what evolvability means (G. Müller 2021). Clearly, evo-devo does not reduce to measuring genetic or phenotypic variability (G. Müller 2021). However, to the extent that a broader conception of "evolvability" is embraced (e.g., "evolvability is an organism's capacity to generate heritable phenotypic variation"; M. Kirschner and Gerhart 1998, 8420), Hendrikse et al.'s (2007) argument is in line with the suggestion that "devo evo" (i.e., the role of development in evolution) is a core focus of evo-devo research (e.g., G. Müller 2007).

2. Villegas et al. 2023.

3. Nuño de la Rosa 2017. In recent discussions of evolutionary biology, evolvability has achieved considerable prominence, albeit with researchers sometimes reliant on different frameworks (Nuño de la Rosa 2017; Villegas et al. 2023). Evolvability nonetheless appears to promote a degree of synthesis: topics such as modularity, robustness, bias and constraint, mutational biases, and the genotype-phenotype map have all been addressed under its rubric, and the concept is credited with playing a unifying role through such "trading zones" (Draghi and Wagner 2008; Nuño de la Rosa 2017).

4. Houle 1992; Hansen 2006; D. Charlesworth et al. 2017; Hansen and Pélabon 2021. This idea underlies R. Fisher's (1930) fundamental theorem of natural selection, which states that the change in mean fitness of a population is a function of the additive genetic variance.

5. Futuyma 2013.

6. Stoltzfus 2019. Ledyard Stebbins (1959, 305), for example, wrote: "Mutation neither directs evolution . . . nor even serves as the immediate source of variability on which selection may act," while Theodosius Dobzhansky asserted that "rates of evolution are not likely to be closely correlated with rates of mutation" (Dobzhansky et al. 1977, 72). These views draw on the long-standing recognition that quantitative variation in phenotypic traits is the product of many discrete genes and that recombination among these can produce substantial genetic variation that is unlikely to be eroded through selection.

7. Maynard Smith et al. 1985; Brakefield 2006; Uller et al. 2018. This conclusion is supported by analyses of simulated and real regulatory networks, which demonstrate that only a small part of the phenotypic space can be reached by a given developmental system, while the remainder is inaccessible (Kauffman 1983; Borenstein and Krakauer 2008; see also K. Dingle et al. 2015).

8. In earlier chapters we described experiments showing that eyespot color combinations in *Bicyclus* butterflies fail to respond to antagonistic selection (C. Allen et al. 2008) and show limited diversity across species commensurate with this (Brakefield 2010). Conversely, eyespot size patterns, which respond readily to selection, exhibit extensive diversity across species (Beldade et al. 2002; Brakefield and Roskam, 2006).

9. Hansen 2006; Hansen and Houle 2008; Hansen and Pélabon 2021.

10. Strictly, the additive genetic variances and covariances of the traits.

11. Lande 1979; Arnold 1992; Hansen and Pélabon 2021.

12. Pélabon et al. 2023.

13. Inbreeding: Phillips et al. 2001. Migration: Guillaume and Whitlock 2007. Mutation: A. Jones et al. 2007. Recombination: Draghi and Whitlock 2012. Selection: G. Wagner and Draghi 2010.

14. Draghi and Whitlock 2012; Uller et al. 2018; Hallgrímsson et al. 2023; Pavličev et al. 2023.

15. Houle 1991; Gromko 1995.

16. G. Wagner and Draghi 2010.

17. As well as in mutation accumulation lines (Scharloo 1970; Mezey and Houle 2005; Houle and Fierst 2013).

18. Klingenberg and Zaklan 2000; Houle et al. 2017.

19. The degree of mutational variance differed by more than two orders of magnitude across different dimensions of morphospace that showed significant mutational variation.

20. Houle et al. 2017, 449. P. Rohner and Berger (2023) show that the relationship between developmental bias and macroevolutionary diversification is also present in a sepsid fly (*Sepsis punctum*), stating that their data "challenges the classic view of strong relationships between genetic variation as solely reflecting constraints on adaptation" (1). That mutational bias can influence molecular evolution (e.g., Yampolsky and Stoltzfus 2001; Nei 2013; Stoltzfus and McCandlish 2017; Cano et al. 2023) and that selection can favor mechanisms that influence the rate at which mutation arises (Cox and Gibson 1974; Charlesworth 1976; Feldman and Liberman 1986) are well established. However, molecular evolutionists have generally assumed that mutational biases shape only selectively neutral features (Stoltzfus 2021). The *Drosophila* wing shape study goes beyond this to show that natural variation in traits that affect fitness can be predicted with knowledge of mutation rates. Similar results are reported in at least one other experimental study (Braendle et al. 2010) and in mathematical theory (Yampolsky and Stoltzfus 2001; Stoltzfus and McCandlish 2017; Stoltzfus 2019). Braendle et al. (2010) quantified phenotypic variability of the vulval developmental system across mutation accumulation lines of two species of nematodes. The study provided strong evidence for developmental bias, and once again the features that were most affected by mutation were those that varied the most in the stock population (and hence probably in nature). Recent theory establishes that the

effectiveness of mutational biases in shaping phenotypes does not depend on neutrality (Yampolsky and Stoltzfus 2001; Stoltzfus and McCandlish 2017; Stoltzfus 2019; reviewed in Stoltzfus 2021).

21. e.g., A. Jones et al. 2007; Chebib and Guillaume 2017.

22. G. Wagner and Altenberg 1996.

23. Pavlicev et al. 2011; R. Watson et al. 2014; but also see Hansen 2006.

24. Cheverud 1982, 1996.

25. Mutations can also have different effects because of their downstream consequences. Developmental modules are commonly organized hierarchically, in a sequential progression, which means that the phenotypic consequences of a mutation depend strongly on its position in the sequence (Davidson and Erwin 2006). Broadly speaking, mutations in transcription factors acting early in the process may be more likely to affect multiple tissues, while mutations in late-acting regulatory elements will have less profound and more localized effects. Plausibly, this might imply that early-acting mutations will be locked into the correlated evolution of multiple traits, bound to the major axes of genetic variation and selection (Kirkpatrick 2009; Walsh and Blows 2009), whereas late-acting mutations will be more likely to evolve independently of these axes. If correct, this reasoning suggests that the timing in the development of a trait may shed light on its evolvability, with, other factors being equal, late-developing traits more evolvable than early-developing traits.

26. One might imagine that novel mutations could easily break the symmetry, but to the extent that this occurs, further selection will favor additional epistasis, leading to robustness in response to genetic mutation among symmetrical phenotypes, and reduced evolvability in asymmetrical forms. Theoretical models that allow the interaction between genes controlling two traits to evolve find that stabilizing or directional correlational selection causes the mutational effects to become biased toward phenotypes that are aligned with the fitness landscape (A. Jones et al. 2007, 2014; Pavlicev et al. 2011; R. Watson et al. 2014). Selection strengthens the interactions between genes (i.e., epistatic effects) that produce the favorable correlation among characters, while it reduces the strength of interactions among genes that produce unfavorable correlation. If trait correlations evolve more slowly than the quantitative traits themselves, the combination of trait values arising through new mutations will be biased toward those that have been favored in the past.

27. Crombach and Hogeweg 2008.

28. Wilke and Adami 2003; Crombach and Hogeweg 2008; R. Watson et al. 2014.

29. For instance, drawing on Cheverud's (1982, 1996) thesis that selection will produce developmental processes with concordant variation among functionally related traits, Hendrikse et al. (2007) predict that mammalian species that use fore and rear limbs for different functions (e.g., bats) will exhibit lower interlimb covariation than mammalian species that use all four limbs for walking and running. N. Young and Hallgrímsson (2005) have shown that this is correct.

30. *Internal selection*, which maintains the functionality of integrated structures in organisms, further reduces the probability of mutations with deleterious phenotypic effects, and focuses variation in useful directions (Schwenk and Wagner 2004; Pavlicev et al. 2023).

31. M. Kirschner and Gerhart 2005.

32. If the environment changes in a way that shifts the fitness landscape and variation in a different dimension is required, then ancestral selection will have reduced evolvability.

33. Abzhanov et al. 2004.

34. Abzhanov et al. 2006.

35. Grant and Grant 2002.

36. Hansen 2006; Crombach and Hogeweg 2008; Draghi and Wagner 2008; Jiménez et al. 2015; Payne and Wagner 2019; Pavlicev et al. 2023. See Pavlicev et al. (2023) for a discussion

of different forms of explicit G-P map models. See also S. Rice (2008a,b). This higher-level property has been labelled "genetic architecture" (e.g., Hansen 2006) and "genotype-phenotype map" (e.g., Hallgrímsson et al. 2023; Pavličev et al. 2023), but we prefer the term "regulatory network." That is because regulatory networks operate at the level of patterns of interactions among genes and involve many nongenetic constituents, so (recognizing that "epigenetic" is commonly deployed in a narrower sense to refer to gene regulatory mechanisms) "regulatory network" appears apt. We regard this higher-level property to be an "epigenetic" (*sensu* Waddington 1940, 1959) rather than genetic phenomenon, so avoid the term "genetic architecture," but nonetheless there are parallels with Hansen's (2006, 2011) analysis. G. Müller (2007) makes a related argument. Similarly, the term "genotype-phenotype map" or (G-P map) is used with different meanings in the biological literature, some of which appear to assume a simple mapping of genotype to phenotype (when in fact the relationship is highly nonlinear), or a genetically determined phenotype (Hallgrímsson et al. 2023). Pavličev et al. (2023) propose a mechanistic G-P map, which by including mechanistic detail of regulatory interactions may be closer to our intended meaning of "network structure." Thinking of the relationship between inputs and phenotypic outputs in terms of "maps" has been extended to also include environmental and parental sources of inputs (Brun-Usan et al. 2022).

37. Draghi and Wagner (2008) label this property "network excitation."

38. Pavličev et al. (2020) make the same point with respect to G-P map structures.

39. E.g., Draghi and Whitlock 2012; R. Watson et al. 2014.

40. Kashtan et al. 2007; R. Watson et al. 2014.

41. Parter et al. 2008; R. Watson et al. 2014; see also Crombach and Hogeweg 2008; Draghi and Wagner 2009; Kouvaris et al. 2017.

42. A. Wagner 2011a,b; Uller et al. 2018.

43. Brattström et al. 2020.

44. Deletion of a *Pitx1* enhancer; Y. Chan et al. 2010.

45. Network models also potentially explain the discordance in evolvabilities observed on micro and macroevolutionary time scales (Hansen 2021). The modular structure and nonlinear interactions of regulatory networks allow developmental systems to exhibit both robustness and innovation (A. Wagner 2011a,b, 2013; Uller et al. 2018; Payne and Wagner 2019). Disruptive and fluctuating selection will shift evolving populations to the boundaries of genotype space, where rapid phenotypic change can ensue (A. Wagner 2011a,b, 2013), generating high microevolutionary evolvabilities, albeit channeled in specific dimensions. However, such populations are also expected to oscillate in the vicinity of those boundaries, while populations subject to stabilizing selection will remain distant from them for long periods. In both cases, on longer time spans little directional phenotypic change, stasis, and low macroevolutionary evolvabilities ensue.

46. Mathematical theory suggests that the effects of developmental processes on evolution can be stable rather than transient, and influence evolutionary outcomes. For instance, by structuring variation, developmental processes create pathways through multidimensional morphospace along which populations evolve though selection. Populations converge on the points on paths that have highest fitness ("path peaks") rather than fitness surface peaks (González-Forero 2023, 2024).

47. Kirkpatrick 2009; B. Walsh and Blows 2009. Kirkpatrick (2009) estimated the dimensionality of genetic variation in dairy and beef cattle, mice, fish, and fly populations and found that selectable genetic variation lies in a small number of dimensions of correlated trait variation. Kirkpatrick (272) concludes that "genetic correlations between quantitative traits may be at least as important as the total amount of genetic variation in limiting a population's response to selection." For morphological traits ranging from vertebrate limb dimensions to the shapes of fly wings and human faces, "the vast majority of phenotypic variation is captured by a few underlying factors" (Hallgrímsson et al. 2023, 183). Subsequent studies have found mixed levels of genetic correlation between traits, yet a recent review concluded that "it remains possible

that much of the evolvability . . . is bound up in a few underlying dimensions" (Hansen and Pélabon 2021, 164–165).

48. Draghi and Wagner 2008; Uller et al. 2018.

49. Distin 2023.

50. Uller et al., forthcoming.

51. Waddington 1957; West-Eberhard 2003; Gerhart and Kirschner 2007. Hansen (2006, 129) stresses that "not only the production of variation . . . but also the ability to maintain variation" is important to evolvability. Hansen and Pélabon (2021) discuss explanations for the existence of standing genetic variation, the dominant position being a mutation-stabilizing selection balance. However, the data remain ambiguous, and an important role for plasticity in preserving cryptic genetic variation cannot be ruled out. Distin (2023) emphasizes how evolvability research has often placed emphasis on developmental plasticity and other features of organisms that facilitate short-term adaptability.

52. Wund et al. 2008.

53. Pilakouta et al. 2023.

54. Sex determination provides another example of the ability of developmental circuitry to process both genetic and environmentally instigated inputs. In some vertebrates (e.g., fish and reptiles) sex is determined environmentally (e.g., depending on temperature, or social interactions), but in others (e.g., birds, mammals) it depends on sex chromosomes. However, the underlying mechanisms for sex determination are similar in all vertebrates, it is just that the trigger (environmental or genetic) may vary. Both provide little information about the outcome but just act on an evolutionarily conserved "switch" (Gerhart and Kirschner 2007).

55. Laland et al. 2014b, 2015.

56. Muschick et al. 2011; Navon et al. 2020. A reuse of signaling pathways is also a distinctive feature of morphological variation in the skull and facial bones and associated soft tissue of vertebrates. The extensive reliance on a small number of developmental pathways substantially reduces the degrees of freedom for phenotypic development and helps to explain why there often appears to be a restricted number of phenotypic responses to environmental challenges. In cichlids, jaws undergo greater change than other regions of morphology in response to variation in diet (Parsons et al. 2016), a finding that is concordant with other evidence suggestive of a separate developmental module for jaws distinct from the rest of the head (Parsons and Albertson 2009; Kuratani 2021). Thus, while plasticity arises in response to diet, the variation generated seems to be channeled toward robust patterns of trait covariance that create a nonrandom localized response (Parsons et al. 2019). Such findings imply that developmental systems are biased to respond to environmental challenges in characteristic ways (Uller et al. 2018; Parsons et al. 2019).

57. Bilandžija et al. 2020.

58. Bilandžija et al. 2020.

59. N. Rohner et al. 2013; Bilandžija et al. 2020.

60. Feiner et al. 2020. Plasticity can facilitate survival in new or challenging stressful environments, including through forms of "bet hedging." See West-Eberhard 2003, Pfennig et al. 2010, Scoville and Pfrender 2010; Moczek et al. 2011, Simons 2014; Furness et al. 2015.

61. Bilandžija et al. 2020.

62. Kowalko et al. 2020; Sifuentes-Romero et al. 2020.

63. Yamamoto et al. 2004.

64. Jeffery 2008.

65. Darwin's finches appear to have diversified in a similar way, consistent with the flexible stem hypothesis (Tebbich et al. 2010).

66. Espinosa-Soto et al. 2011; Fierst 2011; A. Wagner 2011a,b; Draghi and Whitlock 2012; Van Gestel and Weissing 2016.

67. Espinosa-Soto et al. 2011; Fierst 2011; A. Wagner 2011a,b; Draghi and Whitlock 2012; Van Gestel and Weissing 2016.

68. Draghi and Whitlock 2012.

69. Badyaev and Foresman 2000; Wund et al. 2008; Ledón-Rettig et al. 2008; Muschick et al. 2011; see also Pfennig et al. 2010, and references therein.

70. Noble et al. 2019.

71. Selection for plasticity can increase genetic variation along the dimensions of the phenotype that are plastic (Ancel and Fontana, 2000; Hansen 2006; Draghi and Whitlock 2012), but selection can also favor the capacity to evolve efficiently (G. Wagner and Altenberg 1996; G. Wagner and Draghi 2010; Crombach and Hogeweg 2008; R. Watson et al. 2014), including how organisms respond to environmental variation. So, should we expect plasticity-led evolution or selection-led plasticity? The answer is *both*. An analysis by Miguel Brun-Usan and colleagues (2021) used network models to address this issue, finding that selection for genetic evolvability favors adaptive plasticity, while selection for adaptive plasticity increases genetic evolvability. However, selection for plasticity is expected to dominate because selection on responses to environmental variation is generally more efficient than selection on responses to mutation. Adaptive phenotypes will evolve more readily under selection for plasticity than selection for evolvability, making adaptive plasticity a potential source of genetic evolvability (see also Uller et al., forthcoming).

72. M. Kirschner and Gerhart 1998, 2005; West-Eberhard 2003; Snell-Rood et al. 2018; Feiner et al. 2020.

73. Standen et al. 2014

74. There are many types of learning, but here we focus primarily on *operant* (*instrumental*) conditioning, which is thought to be the primary means by which animals acquire behavior (Staddon 2016).

75. Laland et al. 2020, 2022.

76. Cavalli-Sforza and Feldman 1981; Boyd and Richerson 1985; Todd 1991. In stationary or slowly changing environments, learning typically reduces fitness differences between genotypes and thereby slows the evolutionary approach to a single fitness peak, while accelerating evolution in situations where there are multiple fitness peaks (Borenstein et al. 2006; Paenke et al. 2007; Frank 2011). In the latter case, the existence of multiple optima usually slows down the evolutionary process, as populations become trapped on suboptimal fitness peaks. However, by reducing the differences between genotypes, learning increases the likelihood of a population evolving to the global optimum (Borenstein et al. 2006; Mills and Watson 2006; Frank 2011). These findings parallel analyses using GRNs, which, in contrast to more traditional reaction-norm modelling frameworks, also found that adaptive plasticity can reduce the likelihood of getting stuck on local fitness peaks (Van Gestel and Weissing 2016; Kounios et al. 2016).

77. Gerhart and Kirschner 2007; Snell-Rood 2012.

78. Lefebvre et al. 1997; Reader and Laland 2003; Reader et al. 2016.

79. Sol and Lefebvre 2000; Sol et al. 2002, 2005a; Sol 2003. Also consistent with the suggestion that innovation is adaptive is the finding that migratory birds are less innovative than nonmigrants, with innovation in the latter helping them survive the harsh winter months (Sol et al. 2005b).

80. Nicolakakis et al. 2003; Sol 2003; Sol et al. 2005c. Also, many animals have been able to colonize urban environments by learning new behaviors (Sol et al. 2013), which can trigger subsequent genetic adaptation (McDonnell and Hahs 2015). That birds have evolved to sing longer and louder, frogs to call at a higher pitch, and mammals and birds to be bolder in urban environments (McDonnell and Hahs 2015) emphasizes how learning and other forms of plasticity may be highly relevant to the evolvability of animals. Moreover, as described in chapter 5,

many developmental systems operate by iteratively generating variation at random, testing its functionality, and selecting successful solutions. These "exploratory mechanisms" mimic adaptation by natural selection, except that they allow for information gain by individual organisms within their lifetimes, rather than the acquisition of genetic information over multiple generations in a population. Where exploratory systems are operating, an "individual-level evolvability" might be said to be manifest; antibodies, for instance, "evolve" over developmental time in response to foreign antigens, and the adaptive immune system of a single individual may exhibit different capacities to evolve antibodies in response to distinct threats. This individual-level evolvability generates plastic responses that contribute to the more familiar population-level evolvability that is the focus of this chapter, and genes involved in the vertebrate immune system tend to have higher rates of adaptive evolution than other genes in the genome (Shultz and Sackton 2019; Kosiol et al. 2008). Interestingly, the antibody diversity generated by the vertebrate immune system is greater in mammals than in amphibians, where it is greater than in fish, implying that the evolvability of the immune system has itself evolved over macroevolutionary time (Gerhart and Kirschner 1997).

81. Macagno et al. 2016, 2018; Schwab et al. 2017.

82. Laland et al. 1996, 1999; Odling-Smee et al. 2003; Saltz and Nuzhdin 2014.

83. Schwab et al. 2017; Hu et al. 2020; P. Rohner and Moczek 2020, Rohner et al. 2024.

84. Reaction norms are the product of developmental systems (Sultan 2019).

85. Cheverud 1982, 1996; Pavličev et al. 2011; R. Watson et al. 2014; but also see Hansen 2006.

86. Laland et al. 2015.

87. With niche construction, epistasis is mediated by shared association with a constructed environmental resource rather than by linkage disequilibrium or pleiotropy. Note, some of these interactions can be understood as maternal or indirect genetic effects (Bijma, 2014), while others cannot (e.g., where the larva modifies its own environment, thereby modifying its own development, growth, and fitness).

88. Odling-Smee et al. 2003; Saltz and Nuzhdin 2014.

89. Odling-Smee et al. 2003; Laland et al. 2017.

90. In *Arabidopsis thaliana*. Chiang et al. 2013; Saltz and Nuzhdin 2014; Dong 2022.

91. Via 1999. Similarly, in cliff swallows (*Hirundo pyrrhonota*) there is evidence that colony size preference, predator surveillance, and infection status covary (C. Brown and Brown, 1987, 2000; C. Brown et al. 2001; Saltz and Nuzhdin 2014).

92. Odling-Smee et al. 2003; S. Gilbert et al. 2012; Chiu and Gilbert 2015; S. Gilbert 2019, 2020a,b.

93. Such statistical associations are generally known as "linkage disequilibrium," which refers to the degree to which the frequencies of alleles of one genetic locus are correlated with alleles of another genetic locus within a population.

94. Odling-Smee et al. 2003; Riederer et al. 2022. To the extent that evolvability is viewed as a dispositional property of the organism (e.g., Love 2003; R. Brown 2014; Brigandt et al. 2023), and is contrasted with selection, then this dual role for niche construction as a source of both covariation and correlated selection may present difficulties. At least one dispositional property of organisms (niche construction) muddies the water by creating clusters of selectable variation and simultaneously creating fitness differences among those clusters. More generally, the fuzzy, dynamic, and reciprocally caused boundary between organism and environment means it may not always be possible to separate their properties into neat boxes.

95. Clark et al. 2020.

96. West-Eberhard 2003.

97. Laland et al. 2017.

98. Odling-Smee et al. 2003; Clark et al. 2020.

99. Erwin 2008; Bråthen and Ravolainen 2015; San Roman and Wagner 2018; Xue et al. 2020.

100. San Roman and Wagner 2018.

101. Bråthen and Ravolainen 2015.

102. Erwin 2005.

103. Xue et al. 2020.

104. Jablonka and Lamb 2005. Distin (2023) argues that different mechanisms may underly short-term and long-term evolvability.

105. Epigenetic: Jablonka and Lamb 2005; Dubin et al. 2015; Kawakatsu et al. 2016; Proulx et al. 2019; Houri-Ze'evi et al. 2021; Anastasiadi et al. 2021. Symbiotic: S. Gilbert et al. 2012; Roughgarden et al. 2018; S. Gilbert 2019; N. Moran et al. 2019. Cultural: Jablonka and Lamb 2005; Sultan 2015; Laland 2017a; Whitehead et al. 2019; Whiten 2019, 2021. Payne and Wagner (2019) review evidence that selection on phenotypic diversity (arising through stochastic gene expression; errors in protein synthesis; epigenetic modifications, including prions; and protein promiscuity) can contribute to evolvability.

106. Selected populations showed a reduction in epigenetic diversity, including changes in methylation state, that were associated with the phenotypic changes, but no evidence for genetic changes. Offspring of ancestral and selected populations grown together in the same controlled environment exhibited significant inherited phenotypic differences even two generations after selection ceased, with these differences correlated with epigenetic but not genetic variation (Schmid et al. 2018).

107. These studies were with *Saccharomyces cerevisiae.*

108. Stajic et al. 2019.

109. Microbiome: Dearing et al. 2005, 2022. Cultural knowledge: J. Allen et al. 2013.

110. Hansen and Pélabon 2021, 153.

111. For instance, Hansen (2006, 130) writes: "Mutation is the ultimate source of variation and thus sets a fundamental upper limit to evolvability." Nonetheless, a small number of prescient commentators have envisaged an important role for extragenetic inheritance in evolutionary adaptation and evolvability (Jablonka and Lamb 2005; O'Dea et al. 2016; Payne and Wagner 2019; Riederer et al. 2022).

112. The instability of extragenetic inheritance, where variants may persist for only a handful of generations, has frequently been viewed as undermining any important role for it in adaptive evolution (Slatkin 2009; D. Charlesworth et al. 2017). Yet, many biologists are seemingly comfortable relying on the **G** matrix in quantitative genetics to make short-term predictions as to the direction of adaptive evolution (Schluter 1996; Hansen and Houle 2008; Bolstad et al. 2014; McGlothlin et al. 2018), despite instability in its structure (Draghi and Whitlock 2012). While the assumption of "missing heritability" is itself a problematic concept (Feldman and Ramachandran 2018), many researchers have inferred from the common finding of genome-wide association studies that genetic polymorphisms explain little heritable phenotypic variation that therefore quantitative traits must be influenced by many genes of small effect, which supports an infinitesimal model of inheritance and the relative stability of the **G** matrix (Hansen and Pélabon 2021). By comparison, other researchers have interpreted the findings of GWAS as suggesting extragenetic forms of inheritance are more important than currently widely recognized by geneticists (Jablonka and Lamb 2005; Sultan 2015; Laland 2017a; Feldman and Ramachandran 2018).

113. Perhaps because of these challenges, historically many researchers have been reticent to regard the selection of extragenetic variation as "proper" evolution. Despite the preponderance of evidence for the significant role of extragenetic inheritance in adaptation, there have been repeated claims that this is not evolution, except in a metaphorical sense (S. Gould 1991; Pinker 1995; Bamforth 2002; Gabora 2006; Dickins and Rahman 2012). On this view, cultural evolution, for instance, is analogous to genetic evolution, but not actually biological evolution. We reject this position.

114. Whiten et al. 2017.

115. Jablonka and Lamb 2005, 2006; Klosin and Lehner 2016; O'Dea et al. 2016; Jablonka 2017. As well as transposable elements, it may originally have had a function controlling repetitive DNA.

116. Maynard-Smith and Szathmary 1995; Odling-Smee et al. 2003; Jablonka and Lamb 2005, 2006; Newman 2010; E. Rosenberg and Zilber-Rosenberg 2016; Laland 2017a; S. Gilbert 2019; Helanterä and Uller 2019; R. Watson and Thies 2019.

117. Dyall et al. 2004.

118. Koonin 2016; E. Rosenberg and Zilber-Rosenberg 2016. While we have not dwelt on it, a case can be made that horizontal gene transfer is an important contributor to evolvability, and that factors that determine gene-exchange communities (e.g., genome size, G/C composition, carbon utilization, oxygen tolerance) may account for significant variation in evolvability across taxa (Jain et al. 2003; Koonin 2016). For example, the loss of the tails of apes came from the insertion of a viral element into a gene responsible for mesoderm growth (Xia et al., 2024).

119. E. Rosenberg and Zilber-Rosenberg 2016; S. Gilbert 2019.

120. S. Gilbert 2019. The bacteria actively induce the formation of the rumen that will house them, thereby providing both developmental and nutritional symbiosis.

121. Judson 2017.

122. Woznica et al. 2017.

123. E.g., Pavličev et al. 2011; Hansen and Pélabon 2021.

124. Pavličev et al. 2011; R. Watson et al. 2014.

125. But see Livnat and Melamed 2023.

126. M. Kirschner and Gerhart 1998, 2005; Gerhart and Kirschner 2007.

127. Hallgrímsson et al. (2023) refer to a "conceptual divide" between these fields. There is an important mechanistic distinction between selection among genetic variants and selection among packages of developmental modules, even if at the genetic level both involve point mutations. This distinction also helps explain how the concept of evolvability can be embraced simultaneously by scientists who differ in what they believe are key evolutionary processes.

128. Greer 1987, 1990; De Bakker et al. 2013; Lange et al. 2014; Lange 2020. Robustness is built into digit production by the developmental mechanism, because its reliance on a Turing process means the system effectively has virtually a whole wavelength of leeway to play with. A modest increase in the number of limb bud cells, or changes in the concentration of the interacting morphogens that result in a small decrease in the wavelength of the Turing pattern, will not change the number of digits unless sufficiently extreme to cross the threshold. The reason why five digits should be the vertebrate norm remains unclear, but one plausible hypothesis is that five reflects vertebrate limb bud size, which is remarkably consistent across taxa despite vast differences in adult size (G. Müller, personal communication).

129. Pavličev et al. 2023. See Pavličev et al. (2023) for a discussion of different forms of explicit G-P map models. See also S. Rice (2008b). These considerations suggest that gene- or single-nucleotide-polymorphism-level analyses will often fail to pick up important phenotypic regularities (Pavličev and Wagner 2012; Hallgrímsson et al. 2023; Pavličev et al. 2023).

130. Parter et al. 2008; R. Watson et al. 2014. See also Crombach and Hogeweg 2008; Draghi and Wagner 2009; Kouvaris et al. 2017.

131. R. Fisher 1930.

132. The laws of physics and chemistry impose structure on the developmental system, as when the diffusion of interacting morphogens generates Turing-like stripes of cell types that develop into distinct digits (Sheth et al. 2012; Raspopovic et al. 2014; Lange et al. 2018).

133. Gerhart and Kirschner 2007.

134. Gerhart and Kirschner 2007.

135. This assertion alludes to Richard Goldschmidt's (1940) "hopeful monsters." However, it is not chance but developmental mechanisms that create the alternative phenotype trajectories in response to perturbation. This argument builds on a long-standing tradition within developmental biology (e.g., Waddington 1957; Zeeman 1977; Alberch 1989; Thom 1989).

Chapter 13

1. Darwin made this statement in a letter to Alfred Russell Wallace, written on December 22, 1857 ("To A. R. Wallace 22 December 1857," Darwin Correspondence Project, letter no 2192, https://www.darwinproject.ac.uk/letter/?docId=letters/DCP-LETT-2192.xml).

2. Obviously, the major arguments against human evolution come from creationists, proponents of intelligent design, and other nonscientific sources. Darwin's only mention of human evolution in the *Origin of Species* is to state that "light will be thrown on the origin of man and his history" (Darwin [1859] 1968, 458). He delayed his treatment of the evolution of our species for over a decade, until 1871 with *The Descent of Man*, which was closely followed by *Expression of the Emotions in Man and Animals* in 1872.

3. Here we refer to the mainstream evolutionary biology literature. For instance, in what is perhaps the most successful undergraduate textbook, Futuyma (2013) devotes just three pages to our species, stating that "no topic in evolutionary biology is more ... controversial" and referring to "the immense difference between human mental capacities and those of any other mammals" (652).

4. Allmon 2011.

5. Whitehead et al. 2019.

6. Indeed, many human evolution researchers have already embraced these tools, and our principal role here is to draw attention to how they are being deployed.

7. Bramble and Lieberman 2004.

8. Bipedalism also required modifications to the feet and toes. Stringer and Andrews 2012; Boyd and Silk 2017; Newson and Richerson 2021.

9. Bichir fish: Standen et al. 2014. Goats and dogs: West-Eberhard 2003.

10. M. Kirschner and Gerhart 1998, 2005; Gerhart and Kirschner 2007; Richerson 2019.

11. Cheverud 1982, 1996; A. Jones et al. 2007, 2014; Pavličev et al. 2011; R. Watson et al. 2014.

12. J. Young et al. 2010; B. Fischer and Mitteroecker 2015.

13. Bramble and Lieberman 2004. Endurance running is defined as "running many kilometres over extended time periods using aerobic metabolism" (Bramble and Lieberman 2004, 345).

14. Bramble and Lieberman 2004; Henrich 2016. In the Kalahari, where human use of endurance running as a hunting strategy has been best studied, hunters tend to pursue game at midday, when temperatures are hottest, so that prey will overheat more quickly (see Henrich 2016).

15. Bramble and Lieberman 2004.

16. Ruxton and Wilkinson 2011; Lu et al. 2016.

17. Henrich 2016.

18. Cheverud 1982, 1996; A. Jones et al. 2007, 2014; Pavličev et al. 2011; R. Watson et al. 2014. "Tinkering": Lieberman et al. 2009, 89.

19. Grunstra et al. 2019.

20. Washburn 1960. An alternative suggestion to the obstetrical dilemma, known as the "pelvic floor hypothesis," suggests that a narrow pelvis is advantageous for supporting the stomach's organs when individuals are standing bipedally (Abitbol, 1988). See Grunstra et al. (2019) and DeSilva (2021) for reviews of this literature and competing hypotheses.

21. This generates a fitness function that Mitteroecker et al. (2016) characterize as resembling a "cliff edge." Medical data show a decrease of morbidity and mortality (i.e., increased

evolutionary fitness) with increasing neonatal weight and decreasing maternal pelvic width, up to the point where the baby no longer fits through the birth canal, at which juncture fitness drops sharply. see also Pavličev et al. 2020.

22. Learned and socially transmitted information about childbirth and the practice of midwifery constitutes verbally inherited knowledge that dates back to prehistory, perhaps at least to the Paleolithic (Towler and Bramall 1986). Assisting with childbirth was one of many contributions that grandmothers and other experienced family matriarchs are thought to have contributed to the survival and well-being of offspring among our ancestors (K. Hawkes et al. 1998; K. Hawkes 2003; Hrdy 2009; Trevathan 2011; Sarto-Jackson et al. 2017).

23. B. Fischer and Mitteroecker 2015.

24. Cheverud 1982, 1996; A. Jones et al. 2007, 2014; Pavličev et al. 2011; R. Watson et al. 2014.

25. We are not the first to note that, while literally thousands of scientific articles have been written accounting for sex differences in human stature in terms of distinctive foraging roles, divergent mating preferences, or male-male competition, a more parsimonious explanation is that sex differences in human height are at least partly a side effect of hormone differences and timing of development related to childbirth (see Dunsworth 2020). When estrogen peaks during female development, which is crucial for the initiation of menstrual cycles, the growth plates fuse and teenage girls stop growing taller. Typically, teenage boys don't have ovaries or menstrual cycles, and so don't have the levels of hormones that cause bones to stop growing until later. Even if there were no other natural or sexual selection operating, a sex difference in stature is to be expected.

26. B. Fischer et al. 2021.

27. Grunstra et al. 2019; B. Fischer et al. 2021.

28. Hormonally regulated shifts in suites of morphological and behavioral traits during human evolution are likely to have influenced human males too (Gettler et al. 2012).

29. B. Fischer et al. 2021, 628.

30. Several other traits that differentiate our species from most other mammals characterize domesticated species, including juvenile facial features, frequent female sexual cycles, and elevated sociability and social tolerance. Other features of domesticated species, such as floppy ears, piebald coloration, or smaller brains, are not seen in humans—in fact, human brains have grown larger during the course of human evolution. However, it is a common feature of all domesticated species that they have only a subset of the traits associated with domestication syndrome (Wilkins et al. 2014; Wilkins 2017; Sánchez-Villagra and Van Schaik 2019).

31. Wilkins et al. 2014; Theofanopoulou et al. 2017; Wilkins 2017, 2019.

32. Trait associations: Theofanopoulou et al. 2017. Correlational selection: Cheverud 1982, 1996; A. Jones et al. 2007, 2014; Pavličev et al. 2011; R. Watson et al. 2014.

33. Parsons et al. 2020.

34. Salazar-Ciudad and Jernvall 2010; Harjunmaa et al. 2014. See also Kavanagh et al. 2007; Evans et al. 2016.

35. Roberts 2021.

36. Kahn and Ehrlich 2018.

37. Wrangham 2009. See also Henrich 2016.

38. Henrich 2016.

39. Wrangham 2009.

40. Wrangham 2009.

41. Wrangham 2009.

42. There is a lot of controversy over when humans first controlled fire, with some evidence for much later dates (Sandgathe et al. 2011).

43. M. Richards et al. 2003; Stedman et al. 2004; Laland et al. 2010. Such variants include *MYH16* and *ENAM*.

44. Recent discoveries that two relatively small-brained species of *Homo* appear to have exhibited complex behavior has elicited some discussion as to how critical brain size really was to human cognitive and behavioral evolution. *Homo naledi* is thought to have buried its dead and produced cave art (Fuentes et al. 2023), while *Homo florescensis* is thought to have used tools and fire (P. Brown et al. 2004). To what extent these findings constitute a challenge to the account of human brain evolution given here is currently difficult to gauge. The relationship between brain size and intelligence is complex, and factors such as relative brain size, neuron density, brain structure, brain organization, and brain connectivity are all thought to be important (Striedter 2005; Sherwood and Gómez-Robles 2017).

45. Hominin brain evolution is not just about size or number of neurons, but also includes changes in organizational structure critical to enhanced cognition (Sherwood and Gómez-Robles 2017; Sousa et al. 2017). These include a higher proportion of neocortex (especially the prefrontal and temporal cortices), increased connections within the prefrontal cortex, increased dendritic arbor, increased glia-to-neuron ratio, prolonged development, higher activity of metabolic pathways, and many other differences (Somel et al. 2014; Sherwood and Gómez-Robles 2017; Sherwood 2018). The enlarged and better-connected human prefrontal cortex, in particular, is thought to be important to human cognition by enhancing executive control (Striedter 2005), while the correlated expansion of the cerebellum is thought to play a role in sensory-motor control and action planning (R. Barton and Venditti 2014). There has also been a reorganization of neural circuits in the human brain over recent evolution, with evolutionary modifications in gene expression and the distribution of neurons associated with neuromodulatory systems, which may underlie some cognitive and behavioral differences between species (Sousa et al. 2017).

Humans experience a prolonged period of infant development and brain growth. While the brains of other apes cease growing shortly after birth, the human brain keeps growing at the fetal rate of neuron proliferation for least two years (Leigh 2004). During this early postnatal time, we add about 250,000 neurons per minute and create no fewer than 30,000 synapses per square centimeter of cerebral cortex each second (Purves and Lichtman 1985; Rose 1998; Barinaga 2003).

46. Numerous genes appear to be responsible for this growth. Some human-specific mutations in our genome (i.e., those alleles not found in chimpanzees or bonobos) are those that lower the activity of mitosis repressors (Dennis et al. 2012; McLean et al. 2011). Other human-specific sequences are in noncoding RNAs that regulate neural growth (K. Pollard et al. 2006; Prabhakar et al. 2006).

47. Striedter 2005; Sherwood and Gómez-Robles 2017; Herculano-Houzel 2020, 2021.

48. Pontzer et al. 2016; Sherwood and Gómez-Robles 2017. Calculation of the percentage of resting metabolic rate required to support the brain shows that in human men and women the adult brain takes up 19.1% and 24.0%, respectively, of the body's total energetic budget; the relative cost is even higher in newborns (52.5% in males and 59.8% in females) and children (66.3% in males and 65.0% in females) (Kuzawa et al. 2014; Sherwood and Gómez-Robles 2017). These costs reflect not only brain size but also our altricial development, an enlarged and high-energy-consuming neocortex, and increased synaptic activity, which are all metabolically expensive traits (Sherwood and Gómez-Robles 2017).

49. Heldstab et al. 2022.

50. The trend of increases in human brain size and cultural complexity begins with the transition to the Pleistocene, with its substantial increase in climate variation and exaggerated pole-to-equator temperature gradient, conditions thought to favor learning and cultural transmission (Richerson and Boyd 2013, 2020, 2022). More generally, substantial evidence connects hominin evolution to changes or variability in climate (DeMenocal 2011; Anton et al. 2014).

Many genes expressed in energy production have been upregulated during recent human evolution (Uddin et al. 2004; Somel et al. 2014). The costs of energy production may have led to the extinction of the Neanderthals. According to Wragg Sykes (2020), Neanderthals' large bodies and large brains demanded nearly twice the nutrient intake compared with the ancestors of contemporary *Homo sapiens*. This intake was possible until the ice age that began 42,000 years ago, when fatty animals migrated and vegetation failed.

51. Whiten et al. 1999; S. Perry et al. 2003; Van Schaik et al. 2003. Izar et al. (2022) provide evidence that the use of stone tools to forage—a culturally learned trait—improves diet quality in wild monkeys.

52. Whiten et al. 1999; S. Perry et al. 2003; Van Schaik et al. 2003. This conclusion is also supported by theory showing that social learning increases the efficiency of learning (Rendell et al. 2010).

53. Heldstab et al. 2022.

54. Kaplan et al. 2000; Tomasello 2009; Dunbar and Shultz 2017; Laland 2017a}; Street et al. 2017. Mechanistic models of brain evolution have reached the same conclusion (González-Forero and Gardner 2018; McElreath 2018).

55. Reader and Laland 2002; Lefebvre et al. 2004; Reader et al. 2011; Navarrete et al. 2016; Street et al. 2017. While the associations between cultural variables and brain size may not be causal, as many other factors covary with brain size (Dunbar and Shultz 2017), cultural processes may have contributed to brain evolution and intelligence (see Laland 2017a for review).

56. Kaplan et al. 2000; Kaplan and Robson 2002.

57. Kaplan et al. 2000; Kaplan and Robson 2002. See also Muthukrishna et al. (2018) and Muthukrishna (2023) for a related theoretical argument.

58. Anthropologists Leslie Aiello and Peter Wheeler reported a general trend observed across primates that species with smaller guts often have larger brains. They proposed an "expensive-tissue hypothesis" in which "a high-quality diet relaxes the metabolic constraints on encephalization by permitting a relatively smaller gut" (Aiello and Wheeler 1995, 208). Recent analyses suggest a more complex picture, but nonetheless "numerous studies converge on the conclusion that cognitive abilities can only drive brain size evolution in vertebrate lineages where they result in an improved energy balance" (Heldstab et al. 2022).

59. Reader and Laland 2002; Herrmann et al. 2007, 2010; Whiten and van Schaik 2007; K. Hill et al. 2009; Tomasello 2009; Whiten and Erdal 2012; Henrich 2016; Laland 2017a. Beginning about 35 million years ago, a major grade shift occurred with the evolution of anthropoids (monkeys and apes) that had large brains linked to binocular color vision and a diet shifted toward consumption of plant materials (Clutton-Brock and Harvey 1980; DeCasien et al. 2017; Dunbar and Shultz 2017). This was followed by further encephalization in the apes that specialized in extracting foods from protected substrates (Whiten et al. 1999; Kaplan et al. 2000; Van Schaik et al. 2003).

60. In Neanderthals there appears to have been a similar balance, also relying on cultural knowledge to support a large brain. Wragg Sykes (2020) estimated that the Neanderthals' diet demanded over 5,000 kcal a day—nearly twice that of our modern human ancestors—and suggested the demise of the Neanderthals resulted from a failure to meet these metabolic demands. While plausible, this explanation remains speculative, and is just one of several explanations for the disappearance of Neanderthals (see Kolodny and Feldman [2017] for a recent review of this literature and alternative explanation).

61. Cheverud 1982, 1996; A. Jones et al. 2007, 2014; Pavličev et al. 2011; R. Watson et al. 2014. In humans, strong associations between brain shape and face shape arise, which are thought to result because early in development the face and brain mutually shape each other (Naqvi et al. 2021). Genetic analyses identify shared loci influencing brain and face shape, which include transcription factors involved in craniofacial development, as well as contributors to signaling pathways involved in brain-face crosstalk. Note, no relationship was found between shared brain-face/genome-wide-association-study signals and variants affecting behavioral-cognitive traits.

The aforementioned interplay between brain and face appears to occur prior to the stages of brain development that affect cognitive function.

62. Finlay and Darlington 1995.

63. Finlay and Darlington 1995; Striedter 2005.

64. Finlay and Darlington 1995; M. Anderson and Finlay 2014; Finlay and Uchiyama 2017. Cognitive neuroscientists have tended to stress concerted brain evolution (Finlay and Darlington 1995; Finlay and Uchiyama 2015), in marked contrast to the dominant tradition within behavioral ecology, which emphasizes mosaic brain evolution (e.g., R. Barton and Harvey 2000). At least in mammals, this can be viewed as a partitioning of the workload, with cognitive neuroscience focused more on the "easy" mode of concerted brain evolution and behavioral ecology on the "hard" mode of mosaic evolution.

65. In fact, very similar patterns are also observed in birds, where again enlarged brains, enhanced cognition, and reliance on culture are strongly associated in taxa such as corvids and parrots (Bugnyar and Kortschal 2002; G. Hunt and Gray 2003; Emery and Clayton 2004; Emery 2016). Like apes, corvids and parrots possess large brains relative to their body size—the same size in relative terms as chimpanzees (Emery and Clayton, 2004).

66. Navarrete et al. 2016; Street et al. 2017. Comparative phylogenetic analyses of primates have found that a variety of cognitive measures (rates of social learning, behavioral innovation, tool use, extractive foraging, and deception) are strongly correlated, with the dominant principal component referred to as "G," analogous to general intelligence in humans (Reader and Laland 2002; Reader et al. 2011; Navarrete et al. 2016; Street et al. 2017). Further studies report that those measures of primate cognition that load highly on G evolve faster than those that do not (Fernandes et al. 2014).

67. M. Anderson 2010; M. Anderson and Finlay 2014.

68. Speech: Stout and Chaminade 2012. Predicting: J. Fischer et al. 2016.

69. Finlay and Darlington 1995; M. Anderson 2010; M. Anderson and Finlay 2014. Evolution of enhanced cognition through expansion of the whole brain can be economical because the costs of additional neural tissue are ameliorated by the demand-based mechanisms of neural plasticity (which remove unnecessary connections), and because much complex cognition is distributed over brain structures that are large and late developing (which means earlier developing structures are relatively unaffected).

70. This historically important idea is known as *localizationism* (see Sarto-Jackson [2022] for a review of this theory and how modern neuroscience has undermined it).

71. That is, few brain regions and few cognitive functions are modular in an encapsulated way (Buller and Hardcastle 2000; M. Anderson 2010; M. Anderson and Finlay 2014). Rather, there is extensive evidence for top-down effects, cross-domain integration, cross-modal neural plasticity, and overlapping neural circuitry for functionally distinct tasks. M. Anderson (2010) conducted a meta-analysis of data from 1,469 functional MRI studies, establishing that, independently of scale, a typical brain region is activated by tasks in multiple (mean = 9) different domains.

72. M. Anderson 2010. Another example of reuse is that the same brain circuits (including the hippocampus) are activated when healthy participants are asked to remember past and imagine future episodes (Schacter et al. 2007).

73. Further evidence for the reuse of brain circuitry during human evolution comes from the observation that more recently evolved brain functions exploit a greater number of widely scattered brain areas than do evolutionarily older functions (M. Anderson 2010). That makes sense, as recent functions are more likely than older functions to encounter useful preexisting neural circuits that can be incorporated into the developing complex. For illustration, language and reasoning are recently evolved capabilities and exhibit scattered circuitry that integrates spatially separate regions of the brain (M. Anderson 2010; M. Anderson and Finlay 2014).

Human cognitive sophistication may result, in part, because of the extensive reuse of neural circuitry, which has generated greater integration in human brains than in other mammals.

Innovation and evolvability depending on reuse: M. Kirschner and Gerhart 1998, 2005; Gerhart and Kirschner 2007; G. Müller 2007. In contrast, this co-option and reuse is prima facie more difficult to reconcile with notions of structural modularity or with encapsulated conceptions of functional modularity, as, for instance, stressed by many evolutionary psychologists (Cosmides and Tooby 1987; D. Buss 1999; H. Barrett 2015). The same point is well made by Sarto-Jackson (2022).

74. E.g., Healy and Krebs 1996; Shettleworth 2010.

75. Sarto-Jackson 2022.

76. Associative learning is distributed throughout the entire nervous system (Finlay and Uchiyama 2017), with a slow form in the isocortex contrasted with more rapid association in the hippocampus. While reinforcement and habit learning are linked to the basal ganglia and associated structures, the cerebellum plays a key role in error prediction and correction, and imitation is also linked to the motor cortex and parietal lobe.

77. Striedter 2005.

78. Deacon 1990; Finlay and Darlington 1995; Striedter 2005.

79. Deacon 1990.

80. Deacon's rule is based on two fundamental principles of brain development (Deacon 1990). First, developing axons often compete with one another to connect with target sites. Second, this competition is generally won by those axons that participate in "firing" the target cells (Rakic 1986; Purves 1988). When a brain region evolves to become proportionally larger, its axons are given a competitive advantage over those from other regions, because the more axons a region can send to a target, the more likely it is that excitation of the target cells will occur, allowing the connections to strengthen. These new connections can also "displace" older connections, particularly if the source of these old connections has decreased in proportional size.

81. Finlay and Darlington 1995; Striedter 2005.

82. The neocortex becomes an increasingly large proportion of the whole brain as brain size increases, and so would be expected to be comparatively easily changed through selection on whole brain (Finlay and Darlington 1995; R. Barton and Venditti 2014), which fits with the finding that much primate cognition evolves in a concerted fashion (Fernandes et al. 2014). Likewise, the cerebellum is known to play an important role in motor control, and to the extent that enlargement of the cerebellum is indicative of functional gains in the bodily movements necessary for tool use and extractive foraging, increases in this region might be subject to positive selection during human evolution, as neuroanatomical data suggest (R. Barton 2012).

83. Heffner and Masterton 1975, 1983; Striedter 2005.

84. Deacon 1990. This idea often goes under the term *embodied cognition* (L. Barrett 2011).

85. Finlay and Uchiyama 2017.

86. Finlay and Uchiyama 2017. See also Striedter 2005; Krubitzer and Stolzenberg 2014. These range from the presence of Von Economo cells in the cortex to the depth of layer IV in the visual cortex to the size of the prefrontal cortex (see Finlay and Uchiyama [2017] for review). We do not deny that such departures from primate or mammalian expectation may play important roles in human cognition. For instance, the disproportionately large cerebellum and prefrontal cortex are plausibly critical to human tool making and usage and behavioral innovation, respectively (Striedter 2005; R. Barton 2012). However, we do not accept that such departures are the only, or even principal, indicators of adaptation. Previous chapters review extensive empirical and theoretical work suggesting that much evolutionary change occurs along major dimensions of trait covariation, and it is highly plausible that such dimensions will influence,

and perhaps coincide with, taxon-wide regression lines in comparative statistical analyses of brain size.

87. Finlay and Uchiyama 2017; Laland and Seed 2021.

88. Finlay and Uchiyama 2017.

89. Finlay and Uchiyama 2017.

90. There are, in fact, other viable hypotheses for human altriciality. Dunsworth et al. (2012) provide an alternative, although perhaps complementary, metabolic explanation.

91. Finlay and Uchiyama 2017. The timing of birth might be uncorrelated with the rate of neural maturation (Finlay and Uchiyama 2017), but in mammals one can predict the staging of literally hundreds of developmental events, such as when an animal's brain will be fully developed, when it will walk, or even how long it will live, from knowledge of its adult brain mass (Workman et al. 2013; Finlay and Uchiyama 2017; Street et al. 2017). Despite this, plasticity does affect the timing of developmental events, and many aspects of brain organization are initiated and regulated by experience (Finlay and Uchiyama 2017). Again this suggests the existence of a major axis of developmentally biased covariation that sets the direction of evolutionary change. Note, while the general assumption has been that selection for large brains brought about changes in developmental events, one could just as well argue that a large brain is a necessary by-product of selection for extended development (Finlay and Uchiyama 2017).

92. K. Rosenberg 2023, 436. Rosenberg emphasizes how, while human newborns are undeveloped in terms of brain and body size, and motorically immature, we can be regarded as precocious with respect to social learning, which has profound social implications.

93. Sherwood 2018; Sherwood and Gómez-Robles 2017. For instance, human brains are especially plastic owing to shifts in the timing of neuroanatomical developmental changes relative to birth, prolonged myelination, and molecular mechanisms that enhance the formation and refinement of synapses.

94. Sherwood and Gómez-Robles 2017.

95. Peak expression of synaptic genes in the prefrontal cortex occurs approximately five years after birth in humans, compared with several months after birth in chimpanzees (Somel et al. 2013, 2014).

96. These include the timing of brain growth relative to birth, rates of synaptogenesis and myelination, and shifts in gene expression and epigenetic modifications (Sherwood and Gómez-Robles 2017).

97. Ancient DNA: Sherwood and Gómez-Robles 2017. Genetic studies have, for instance, identified hundreds of human genes subject to recent positive selection that show major changes in expression in the brain, including the upregulation of genes expressed in energy production and neuronal signaling in the neocortex as well as extended plasticity in brain development (Uddin et al. 2004; Voight et al. 2006; E. Wang et al. 2006; Sabeti et al. 2006, 2007; Nielsen et al. 2007; Somel et al. 2014).

98. Raghanti et al. 2018.

99. Raghanti et al. 2018.

100. Hrdy 2009; Sear 2016; Finlay and Uchiyama 2017.

101. Laland and Seed 2021.

102. Laland and Seed 2021. In chapter 10, we discussed inheritance as a series of anticipatory survival strategies. To survive, all organisms must develop phenotypes that successfully predict what the environment will be. Genetic inheritance selects for phenotypes that would persist in stable environments over long periods of time, whereas extragenetic forms of inheritance are mainly predictive over shorter periods, and phenotypic plasticity is predictive in the present generation. Humans possess a brain that can predict the environmental circumstances and frequently produce an appropriate response on a moment-to-moment basis. We are "planning animals" (S. Gilbert 2003b). As human geneticist Barton Childs (2003, 108) has written, this

neural developmental plasticity "allows us to escape the tyranny of our genes." Or as neurobiologists Dale Purves and Jeffery Lichtman (1985, 15) stated: "The interaction of individual animals and their world continues to shape the nervous system throughout life in ways that could never have been programmed. Modification of the nervous system by experience is thus the last and most subtle developmental strategy."

103. Ruiz and Santos 2013.

104. Conway and Christiansen 2001; Tomasello 2011.

105. Karmiloff-Smith 1995; Carey 2009; Tomasello 2018; Laland and Seed 2021.

106. Roche et al. 1999.

107. Toth 1987; Stout 2011; T.J.H. Morgan et al. 2015. Stone tool remains show systematic flake detachment, maintenance of flaking angles, and repair of damaged cores (Delagnes and Roche 2005).

108. Socially transmitted: Schick and Toth 2006; Braun et al. 2009; Stout 2011; Hovers 2012; T.J.H. Morgan et al. 2015. Cultural inheritance: Herrmann et al. 2007, 2010; K. Hill et al. 2009; Laland 2017a; Reader and Laland 2002; Whiten and Erdal 2012; Whiten and Van Schaik 2007; Henrich 2016; Tomasello 2009.

109. Stout 2011; Hovers 2012; T.J.H. Morgan et al. 2015.

110. Schick and Toth 2006; Stout 2011; T.J.H. Morgan et al. 2015.

111. T.J.H. Morgan et al. 2015. Initially, gesture may have been more important than speech in teaching tool manufacture and usage (Cataldo et al. 2018).

112. Uomini and Meyer 2013.

113. Stout 2011; Uomini and Meyer 2013; T.J.H. Morgan et al. 2015.

114. Schick and Toth 2006; Tomasello 2009; Stout 2011; Hovers 2012; Whiten and Erdal 2012; Henrich 2016; Fuentes 2017.

115. Reader and Laland 2002; Whiten and Van Schaik 2007; Herrmann et al. 2007, 2010; K. Hill et al. 2009; Whiten and Erdal 2012; Henrich 2016; Laland 2017a; Creanza et al. 2017.

116. McGuigan and Whiten 2009; Spelke 2009; Ruiz and Santos 2013; Laland and Seed 2021. Observational learning studies show that humans, but not chimpanzees, copy causally irrelevant actions such as tapping a box before taking out a reward (Horner and Whiten 2005), an observation labeled "overimitation" because of its seeming irrationality (Hoehl et al. 2019). From an early age, children infer the existence of a pedagogical intention, which makes it rational to privilege information given by a teacher even when this conflicts with functional information (Buchsbaum et al. 2012). Children are extremely competent tool users who think about physical events in terms of their causal underpinnings from an early age (A. Brown 1990; Gopnik et al. 2017). Some researchers suggest that enhanced human cognition derives primarily from our prowess in tool use and physical cognition, although this suggestion is contentious (e.g., Povinelli 2000). In contrast, other studies report that adult humans also have an impoverished understanding of the causal bases of even simple tools (Lawson 2006) and can invent mythical causal stories about technology (Henrich 2016; Derex et al. 2019).

117. Call 2013; Visalberghi et al. 2017. E.g., chimpanzees, capuchins. Flexible tool use is also seen in some birds, including the Galapagos woodpecker finch and New Caledonian crow (Rutz et al. 2010; Biro et al. 2013; Visalberghi et al. 2017).

118. Visalberghi et al. 2017.

119. Overington et al. 2009; Navarrete et al. 2016.

120. Kaplan et al. 2000; Reader and Laland 2002; Reader et al. 2011; Navarrete et al. 2016; Street et al. 2017; Dunbar and Schultz 2017.

121. Fuentes 2017.

122. Browne 2021; DeSilva and Browne 2021.

123. Hrdy 2009; Dyble et al. 2015; Fuentes 2017; Hedenstierna-Jonson et al. 2017; Haas et al. 2020.

124. Hrdy 2009; Dyble et al. 2015; Fuentes 2017; Hedenstierna-Jonson et al. 2017; Toler 2019; Haas et al. 2020; Saini 2023.

125. Henrich 2016; Laland 2017a; Boyd 2018; Newson and Richerson 2021.

126. Sabeti et al. 2006, 2007; Voight et al. 2006; E. Wang et al. 2006; Nielsen et al. 2007; Laland et al. 2010.

127. Variants favored: Sabeti et al. 2006, 2007; Voight et al. 2006; E. Wang et al. 2006; Nielsen et al. 2007; Laland et al. 2010; Gerbault et al. 2011. Disease or stress: Evershed et al. 2022.

128. Phylogenetic comparative and other evolutionary analyses provide clear evidence of niche construction, in which culture acted as a selective pressure and the human genome responded (Aoki 1986; Feldman and Cavalli-Sforza 1989; Holden and Mace 1997; Gerbault et al. 2011). For instance, there is no doubt that dairy farming came first, and only subsequently were alleles for adult lactase persistence favored. The *LP* allele was absent in ancient DNA extracted from early Neolithic Europeans, suggesting that it was absent or at low frequency 7,000–8,000 years ago (Burger et al. 2007). It is possible that the consumption of dairy products conferred a significant fitness advantage only during times of famine or disease (Evershed et al. 2022). Another compelling example of gene-culture coevolution and plasticity-led selection concerns the cultural practice of diving, which has led to natural selection increasing the size of the spleen in some human populations (Ilardo et al. 2018).

129. Laland et al. 2010; O'Brien and Laland 2012.

130. Durham 1991; O'Brien and Laland 2012.

131. Durham 1991; O'Brien and Laland 2012.

132. Barnes et al. 2011.

133. S. Burke and Duffy 2022.

134. Earlier chapters described similar examples, such as the Dutch Hunger Winter, that may involve epigenetic inheritance (Tobi et al. 2018).

135. Odling-Smee et al. 2003; Lansing et al. 2009; J. Kendal et al. 2011; Lansing and Fox 2011; McClure 2015; Boivin et al. 2016; Laland 2017a; Zeder 2017, 2018; Quintus and Cochrane 2018.

136. Odling-Smee et al. 2003; Lansing et al. 2009; J. Kendal et al. 2011; Lansing and Fox 2011; McClure 2015; Boivin et al. 2016; Laland 2017a; Zeder 2017, 2018; Quintus and Cochrane 2018.

137. Mesoudi et al. 2013. Note, technological niches can feed back in several ways, including by creating demand for products.

138. Laland et al. 2014a; Laubichler and Renn 2015; Riahi 2023.

139. Chudek and Henrich 2011; Laubichler and Renn 2015; Henrich 2016; Muthukrishna and Henrich 2016; Laland 2017a; Boyd 2018; Henrich and Muthukrishna 2021; Newson and Richerson 2021; Riahi 2023.

140. B. Smith 2007a,b, 2011, 2016; Zeder and Smith 2009; Ren et al. 2016; Piperno 2017; Zeder 2017, 2018; U. Lombardo et al. 2020; H. Barrett and Armstrong 2023; Dorninger et al. 2023.

141. Boivin et al. 2016; Ellis 2018; M. Kemp et al. 2020; J. Thompson et al. 2021. See also Ilardo et al. 2018.

142. Wrangham 2009.

143. Judson 2017.

144. Odling-Smee 2024. We are indebted to John Odling-Smee for drawing this point to our attention. See also Lala and O'Brien 2023.

145. Judson 2017.

146. O'Brien and Laland 2012; Wells and Stock 2020.

147. Wrangham 2009; Boivin et al. 2016; J. Thompson et al. 2021. For example, the transport of fire from one region to another has allowed Australian hunter-gather communities to increase resource productivity by promoting landscapes with diverse successional stages in which small

game flourish, in the process triggering ecological cascades (Bliege Bird et al. 2020; J. Thompson et al. 2021).

148. Laland and Brown 2006; Lala and O'Brien 2023.

149. Hawks et al. 2007.

150. Rendell et al. 2010.

151. Boyd and Richerson 1985; Feldman and Laland 1996; Richerson and Boyd 2005; Henrich 2016; Laland 2017a; Boyd 2018. Note, by generating correlated variation in genotypes and phenotypes, gene-culture coevolution is itself a form of developmental bias.

152. Richerson and Boyd 2005; Henrich 2016; Laland 2017a; Boyd 2018.

153. Cochran and Harpending 2009; Hawks et al. 2007.

154. Laland et al. 2000; Richerson and Boyd 2005; Laland and Brown 2006.

155. Lala and O'Brien 2023.

156. Darwin (1871) 1981, 84.

157. Darwin (1871) 1981; Radick 2007.

158. Hauser 1996; Hauser et al. 2002; Radick 2007; Fitch 2005, 2010. Of Darwin's contemporaries, for example, in a series of influential lectures, University of Oxford professor of linguistics Max Müller asserted that language was the "one great barrier between the brute and man" and "no process of natural selection will ever distil significant words out of the notes of birds and the cries of beasts" (M. Müller 1861, 22–23, 354).

159. Hauser et al. 2002; Christiansen and Kirby 2003; Bolhuis et al. 2010; Fitch 2010; B. Wheeler and Fischer 2012; Corballis 2017.

160. Seyfarth et al. 1980.

161. The claim of functionally referential communication has been made for the social, food, and alarm calls of a range of primates, meerkats, prairie dogs, and chickens (Seyfarth and Cheney 2000; Hauser et al. 2002; B. Wheeler and Fischer 2012).

162. B. Wheeler and Fischer 2012, 2015. Acoustically distinctive calls may represent different emotional states rather than specific predators (Owren and Rendall 1997), with referentiality restricted to the receiver (Hauser et al. 2002).

Most primate species appear to have only limited ability to modify their natural vocalizations (Hauser et al. 2002; Pfenning et al. 2014). In contrast, vocal learning is widespread in bats, cetaceans, elephants, pinnipeds, songbirds, and parrots, and birdsong has long been studied as a model for human speech acquisition (Bolhuis et al. 2010). Researchers note several parallels between birdsong and language, including vocal learning, auditory feedback, error correction, sensitive periods, babbling (subsong), shared brain circuitry (e.g., mirror neurons), shared genes (e.g., *FOXP2*), and syntax. However, avian and cetacean songs lack a representational quality and seemingly do not convey complex messages.

163. Gardner and Gardner 1969; Pepperberg 2017; Terrace 1979. In the 1960s a young chimpanzee called Washoe was taught American Sign Language. The investigation triggered a series of studies in which apes, birds, and cetaceans were taught signs or symbols. Through such experiments, animals have been shown to be capable of using signs intentionally to make their needs clear, to comprehend the meaning of a large repertoire of signs, and to answer questions using signs (L. Herman et al. 1984; Pepperberg 1999, 2017; Hauser et al. 2002).

164. Hauser 1996; Hauser et al. 2002; Laland and Seed 2021. Such investigations demonstrate that many animals can achieve a rudimentary capability for symbolic communication. Yet these capabilities fall short of human language in important respects, including that they are typically relatively simple one- or two-symbol combinations, are tied to the present, and show neither recursion nor evidence for the comprehension of tense, syntax, or grammar. For example, the longest recorded "utterance" of chimpanzee Nim Chimpsky, who was taught sign language, was: "Give orange me give eat orange me eat orange give me eat orange give me you" (Terrace 1979). In contrast, two-year-old children produce a variety of complex sentences,

comprising verbs, nouns, prepositions, and determiners, in the correct grammatical relations and on a diverse range of topics; three-year-olds can communicate about the past and future and distant objects; and high-school graduates know up to 60,000 words (Hauser et al. 2002).

165. Pika et al. 2005; Hobaiter and Byrne 2011; Corballis 2017.

166. Organizing thought: Chomsky 1968, 1990; Everaert et al. 2015. Syntactic structure: Bolhuis et al. 2014. Researchers have distinguished two principal schools of contemporary linguistics, the *formalist* and *functionalist* schools (Kolodny and Edelman 2018). For the former, dominated by the views of Noam Chomsky, the primary use of language is taken to be the structuring of thought. For the latter, language is for communication, but grammar still plays a central role as a tool for encoding meanings into forms that can be easily transmitted and decoded. Either way, there remains a stark divide between the complexity of utterances (and hence thought) of humans and other animals, which leaves the evolution of language a challenge (see Fitch [2005, 2010] for an overview of these literatures).

167. Nakahara et al. 2001; Stout and Chaminade 2012; Jin et al. 2014; J. Fischer et al. 2016; Kolodny and Edelman 2018.

168. Kolodny and Edelman 2018. Both experiments and theory have shown that long sequences of actions are unlikely to be learned through trial and error and that social learning greatly enhances the likelihood of finding an optimal sequence (Whalen et al. 2015). Sequences in which the reward comes at the end of a long string of elements, where there are a large number of options at earlier stages, and where the final rewarding action must be preceded by a series of nonrewarding or aversive actions are particularly challenging to learn without the help of others. For acquisition of such tasks, social learning is critical.

169. Tomasello 2009; Stout 2011; Hovers 2012; Steele et al. 2012; Whiten and Erdal 2012; Uomini and Meyer 2013; T.J.H. Morgan et al. 2015; Henrich 2016; Laland 2017a,b; Kolodny and Edelman 2018.

170. Castro and Toro 2004; Fitch 2004, 2005, 2010; Gergely and Csibra 2005; Gergely et al. 2007; Tomasello 2008, 2009; Csibra and Gergely 2011; Laland 2017a,b; Kolodny and Edelman 2018. In these contexts, language likely provided our ancestors with both efficient means of teaching and tools for thought, connecting symbols and meanings, stabilizing and interweaving concepts, integrating experiences, and supporting inductive reasoning and theory-based inference. These cognitive capabilities may also have supported other forms of cooperation, including mutualism, trade, reciprocity, and cultural group selection (Sober and Wilson 1999; Tomasello 2009; Nowak and Highfield 2011; Pagel 2012; Henrich 2016).

171. M. Anderson 2010.

172. M. Anderson 2010; M. Anderson and Finlay 2014.

173. This might explain why language can be deployed flexibly to enhance other aspects of cognition, and perhaps why human language can be learned so rapidly (M. Anderson 2010; M. Anderson and Finlay 2014; Laland and Seed 2021).

174. Skeide et al. 2017; Heyes 2018.

175. Heyes 2018. Naturally, reading may result in a cognitive gadget because it enables "reuse" of preexisting circuitry (e.g., for speech) (M. Anderson 2010), and hence does not necessarily imply that our language capability, or cognition in general, is composed exclusively of such gadgets (Laland and Seed 2021).

176. Hecht et al. 2015.

177. D'Souza and Karmiloff-Smith 2011; Heyes 2018.

178. Kolodny and Edelman 2018.

179. Darwin (1871) 1981, 60–61.

180. Kirby et al. 2007, 2008, 2015; K. Smith and Kirby 2008. University of Edinburgh linguist Simon Kirby and his colleagues labelled this process "cultural selection for learnability" (Brighton et al. 2005).

181. E.g., Chomsky 1980; Pinker 1995.

182. Deacon 1997; Kirby et al. 2007, 2008, 2015; K. Smith and Kirby 2008. The cultural evolution of language has been studied through mathematical modeling, and researchers have established that key properties of language could evolve in this manner (Kirby et al. 2007; K. Smith and Kirby 2008). For instance, compositionality—the idea that the meaning of a complex signal is a function of the meaning of its parts—can arise through a process of cultural evolution, and typically more rapidly and at lower cost than through genetic evolution (K. Smith and Kirby 2008).

183. Kirby et al. 2007, 2008, 2015; K. Smith and Kirby 2008.

184. Kirby et al. 2007, 2008, 2015; K. Smith and Kirby 2008.

185. Natural selection on anatomy and cognition: e.g., Jablonka and Lamb 2006.

Children are known to hear some linguistic structures selectively and to ignore others, a phenomenon that may have generated selection for language structure that is child friendly (Falk 2004). Infant-directed speech is typically slower and higher in pitch than regular speech and uses shorter and simpler words. Important elements of infant-directed speech, such as children's sensitivity to its linguistic features, or adults' tendency to engage in behavior that elicits rewarding responses (e.g., smiles), may have been favored through a gene-culture coevolutionary process.

Language use is thought to have generated strong selection for cognitive adaptations that facilitated language learning and information transmission. For instance, compared with other primates, humans appear particularly adept at inferring the meaning of what others are saying (Gopnik and Meltzoff 1997; Buchsbaum et al. 2011, 2012). While this ability is partly attributable to the pedagogical activities of the transmitter, an enhanced capability of the receiver to extract meaning through observation of others is also evidenced.

Children partly construct their own cognition by seeking to understand their own thinking, which leads to representational flexibility, theory building, and mental simulation (Karmiloff-Smith 1995). This process of *representational redescription* is either not available to, or is impoverished in, other primates (Karmiloff-Smith 1995; Laland and Seed 2021). Young children change to exhibit more meta-cognitive awareness or explicit theorizing around ages four to five, and start to represent their implicit knowledge explicitly, making it available to other domains—for instance, tool use (Karmiloff-Smith 1995). This ability can also be deployed to understand or infer what others are thinking ("theory of mind").

Furless faces: Wilkins 2017. Eyes: Compared with other primates, humans have especially visible eyes, and our contrasting sclera are thought to enable gaze tracking (Tomasello et al. 2007). This feature of human eyes is thought to have evolved to make it easier for conspecifics to follow an individual's gaze direction in close-range joint attentional and communicative interactions. Experiments show that humans are especially reliant on eyes in gaze-following situations, and suggest that eyes may have evolved a new social function in human evolution, to support cooperative social interactions (Tomasello et al. 2007).

186. Fitch 2006, 2013; Patel 2006; Patel et al. 2009; S. Kirschner and Tomasello 2010; Dunbar et al. 2012; Tarr et al. 2014; Laland et al. 2016.

187. Boyd and Richerson 1985; Richerson et al. 2010, 2016; Nowak and Highfield 2011; Pagel 2012; Henrich 2016; Muthukrishna and Henrich 2016; Laland 2017a; Boyd 2018; Henrich and Muthukrishna 2021.

188. Boyd and Richerson 1985; Henrich 2016; Tomasello 2009.

189. Boyd and Richerson 1985; Chudek and Henrich 2011; Tomasello 2009.

190. Boyd and Richerson 1985; Chudek and Henrich 2011; Tomasello 2009.

191. Boyd and Richerson 1985; Wilson et al. 2008; Wilson 2010; Henrich 2016; Richerson et al. 2016; Boyd 2018.

192. Wilson et al. 2008; Waring and Wood 2021.

193. Klingenberg 2008; Kavanagh et al. 2013; Uller et al. 2018.

194. Uller et al. 2018

195. Wilkins 2017.

196. M. Kirschner and Gerhart 1998, 2005; Gerhart and Kirschner 2007.

197. Holden and Mace 1997; Burger et al. 2007; Gerbault et al. 2011; Evershed et al. 2022.

198. Laland et al. 2010.

199. Odling-Smee et al. 2003; J. Kendal et al. 2011; O'Brien and Laland 2012. Note, an important caveat here is that this emphasis on agency does not imply human niche construction is necessarily conscious or deliberate.

200. Cognition and language: Bickerton 2009; Fuentes 2017; Laland 2017a,b; Tomlinson 2018; Rouse 2023. Domestication and agriculture: B. Smith 2007a,b, 2011, 2016; Zeder and Smith 2009; Ren et al. 2016; Piperno 2017; Zeder 2017, 2018; U. Lombardo et al. 2020. Other aspects: Fuentes 2009, 2017; J. Kendal et al. 2011; Anton et al. 2014.

201. Ellis 2015; Boivin et al. 2016.

202. B. Smith 2007a,b; Alberti 2015; Zeder 2017; Otto 2018; Derryberry et al. 2020.

203. Contemporary biomedical science places increasing emphasis on the fact that medical phenotypes can be inherited in a DNA-independent way (e.g., Lacal and Ventura 2018; Senaldi and Smith-Raska 2020), suggesting that epigenetic variation may regularly impact fitness. Selection of inherited symbionts may also contribute to human adaptability. Since Darwin, the human appendix has been regarded as a vestigial trait with no current function. Recently, however, it has been found to act as a repository for beneficial gut bacteria that can recolonize the intestine after bouts of stomach disease and infection (Bollinger et al. 2007; Roberts 2021). The appendix, it would seem, is not an atrophied remnant of the past at all, but rather a refuge for vital symbionts. Human evolution has also involved horizontal gene transfer. An example is the ability of Japanese, but not Westerners, to digest agar because they have a bacterium in their gut that contains a gene that codes for agarase (Hehemann et al. 2010).

204. Hawks et al. 2007.

205. Cavalli-Sforza and Feldman 1981; Lumsden and Wilson 1981; Boyd and Richerson 1985; Feldman and Laland 1996; Henrich and McElreath 2003; Richerson and Boyd 2005; Boyd 2018; Newson and Richerson 2021.

206. Pinker 2010; Kurzban 2011.

207. Boyd and Richerson 1985; Odling-Smee et al. 2003; Henrich 2016; Laland 2017a; Creanza et al. 2017.

208. Rose et al. 1984; Levins and Lewontin 1985; Feldman and Ramachandran 2018; Uchiyama et al. 2022; Feldman and Riskin 2022.

209. To give Darwin his due, historians have suggested recently that Darwin anticipated many of the themes that characterize the broader view of evolutionary causation depicted here (Bradley 2020).

Chapter 14

1. Latin name *Elysia chlorotica*.

2. Rumpho et al. 2011; Sultan 2015. While the adult green sea slug can survive entirely on energy derived from photosynthesis, it nonetheless will engage in opportunistic feeding (Rumpho et al. 2011; Sultan 2015).

3. Sagan 1967. Chloroplasts are thought to have evolved from a primitive cyanobacterium, and hence are themselves modified symbionts (Sabater 2018).

4. Sultan 2015.

5. Sultan 2015.

6. These interdependencies go beyond the mere existence of covariance between fitness and phenotype, as for instance commonly modelled using the Price equation (see Uller and Helanterä 2017).

7. Boyd and Richerson 1985; Odling-Smee et al. 2003; Henrich 2016; Creanza et al. 2017; Laland 2017a.

8. For instance, Mayr (1980, 9–10) wrote: "The clarification of the biochemical mechanism by which the genetic program is translated into the phenotype tells us absolutely nothing about the steps by which natural selection has built up the particular genetic program."

9. Oyama 1985; Oyama et al. 2001; Laland et al. 2015.

10. Waddington 1953, 1959; Lewontin 1983; West-Eberhard 2003.

11. A more formal definition of reciprocal causation is that process A is a cause of process B and process B is a cause of process A. Reciprocal causation captures the idea that developing organisms are not just products, but are also causes, of evolution (Laland et al. 2015).

12. Forgacs and Newman 2005; Newman 2010, 2023.

13. Gerhart and Kirschner 1997; West-Eberhard 2003; M. Kirschner and Gerhart 2005.

14. Wray et al. 2014. See Lewens (2019) for a critique of the argument that only processes that directly change gene frequencies should be recognized as evolutionary processes.

15. For example, see Futuyma (2017).

16. Mayr 1961. Mayr depicted ultimate causes as providing historical evolutionary accounts for the existence of characters, and proximate causes as explaining how biological systems work. From Mayr's standpoint, evolutionary biologists are concerned with ultimate causes and developmental biologists with proximate causes. This standpoint remains prevalent (e.g., S. West et al. 2007, 2011; Scott-Phillips et al. 2011).

17. e.g., Laland et al. 2011, 2013; Calcott 2013; Watt 2013; Pigliucci 2019; Baedke 2021.

18. e.g., West-Eberhard 2003; Laland et al. 2011.

19. West-Eberhard 2003, 11. Several debates in contemporary evolutionary biology concern the extent to which causal effects of environment on organism (e.g., plasticity, epigenetic inheritance) and causal effects of organism on environment (e.g., niche construction) can also be evolutionary causes, in particular causes of adaptive evolution (Laland et al. 2011, 2015).

20. Laland et al. 2011, 2012, 2013; Baedke 2021.

21. Nonetheless, in most evolutionary biology textbooks *developmental bias* (and *plasticity, niche construction*, and *extragenetic inheritance too*) is barely, if at all, recognized explicitly as a process that contributes fundamentally to evolution by natural selection. In ten leading undergraduate evolution textbooks, which are many hundreds of pages long, the topics of developmental bias, developmental plasticity, inclusive inheritance, and niche construction each receive treatments of 10 pages or fewer (mean 3.2 pages, mode 0 pages; Futuyma 1998, 2013; Ridley 2004; Stearns and Hoekstra 2005; N. Barton et al. 2007; Arthur 2011; Bergstrom and Dugatkin 2012; Zimmer and Emlen 2013; Herron and Freeman 2014; Losos 2014). In contrast to most evolutionary biology undergraduate textbooks, Arthur (2011) and Bard (2022) incorporate a strong evo-devo component, and Arthur (2011) contains an extended treatment of developmental bias. These concepts are still sometimes regarded as peripheral ("add-ons"; Wray et al. 2014) rather than central to evolutionary explanation.

22. Amundson 2005. See also Laland et al. (2008) and Uller and Laland (2019).

23. Molar development: Kavanagh et al. 2007, 2013. Wing spots: Beldade et al. 2002; Brakefield and Roskam 2006; C. Allen et al. 2008; Brattström et al. 2020. Embryonic tissues: Abzhanov et al. 2004, 2006.

24. Provine 1971; S. Gould 2002; Amundson 2005; Stoltzfus 2019. Ledyard Stebbins (1959, 305), for example, wrote: "Mutation neither directs evolution . . . nor even serves as the immediate source of variability on which selection may act," while Theodosius Dobzhansky

asserted that "rates of evolution are not likely to be closely correlated with rates of mutation" (Dobzhansky et al. 1977, 72). This draws on a long-standing recognition that quantitative variation in phenotypic traits is the product of many discrete genes and that recombination can produce substantial genetic variation that is unlikely to be eroded through selection. Neglect of the possibility of taxonomically widespread developmental bias led Mayr to the view that eyes had evolved independently at least 40, and possibly up to 65, separate times (Salvini-Plawen and Mayr 1961). However, it is now clear that the specification of eye photoreceptors is based on a developmental pathway underlying all the types of eyes (Gehring 2005).

25. N. Barton 1990; Kirkpatrick 2009; B. Walsh and Blows 2009; A. Wagner 2014; Mitteroecker et al. 2016; Hallgrímsson et al. 2023.

26. González-Forero 2023, 2024.

27. Uller et al. 2018.

28. As we emphasize in chapters 2, 7, and 12, a categorical distinction between what is biologically possible and what is not neglects how development can probabilistically bias outcomes, while constraint fails to recognize the reciprocal nature of evolutionary causation.

29. Uller et al. 2018; see González-Forero (in press) for an example.

30. Oyama 1985, 5.

31. Sultan 2015, 2019. Developmental plasticity and developmental symbiosis had been seen as oddities and peripheral to evolution. With the discovery that both phenomena are widespread, if not universal, plasticity and symbiosis are becoming central to evolutionary studies.

32. Oyama 1985, 2000.

33. Griffiths and Stotz 2013.

34. Feldman and Cavalli-Sforza 1981; Boyd and Richerson 1985; Odling-Smee et al. 2003; McElreath and Boyd 2007; Bonduriansky and Day 2009, 2018; T. Day and Bonduriansky 2011; Shea et al. 2011; Geoghegan and Spencer 2012, 2013; Furrow and Feldman 2014; Rivoire and Leibler 2014; Uller et al. 2015; T. Day 2016; McNamara et al. 2016; Adrian-Kalchhauser et al. 2021.

35. Strictly, this should be "quasi-autonomous" (D. Walsh 2015).

36. Lewontin 1970.

37. Badyaev 2011; D. Walsh 2015; Uller and Helanterä 2019.

38. This claim might, at first sight, appear to be contradicted by the Lande equation of quantitative genetics, which specifies that $\mathbf{G}$ gives directionality to evolution even though it does not directly describe fitness differences. However, as any directionality in evolution arising from $\mathbf{G}$ (and $\mathbf{M}$) is commonly explained as the result of ancestral natural selection, developmental biases have not historically been afforded a significant explanatory role, over and above accounting for the absence of evolution or adaptation (Uller et al. 2018).

39. In practice, researchers might also need to consider additional processes, such as drift, which for simplicity we ignore here.

40. Stoltzfus 2019, 2021.

41. In this example, the parallels with mutation pressure are sufficiently close for researchers to accept bias as playing an evolutionary role, albeit typically through reconceptualizing the developmental bias as a more familiar mutational bias. Here a developmental perspective is delivering more than an understanding of the rate at which phenotypes arise to specify, in addition, why that particular phenotype results.

42. J. Allen et al. 2013.

43. O'Dea et al. 2016.

44. S. Gilbert et al. 2012.

45. Stoltzfus 2019.

46. Schwab et al. 2016, 2017; Hu et al. 2020; P. Rohner et al. 2021, 2024; P. Rohner and Moczek 2023, 2024.

47. Schwab et al. 2016, 2017; Hu et al. 2020; P. Rohner et al. 2021, 2024; P. Rohner and Moczek 2023, 2024.

48. Lewontin 1982; Sober 1984; Oyama 2000; Uller and Laland 2019.

49. Whether Lamarck's work is viewed as discredited or supported by recent findings of extensive extragenetic inheritance is open to debate. See Jablonka and Lamb (1995) for a treatment sympathetic to Lamarck.

50. Sober 1984.

51. Fogarty and Wade 2022; see also Wade and Sultan 2024.

52. Fogarty and Wade 2022, 8.

53. Fogarty and Wade 2022, 8.

54. Jablonka and Lamb 2005, 2014; Uller and Helanterä 2017.

55. González-Forero's (2023, 2024) analysis shows that evolution stops at path peaks, rather than peaks in the adaptive landscape, and therefore developmental processes can have long-term rather than just transient effects on evolutionary dynamics. Here, the paths are given by developmental processes, as long-term evolution requires following both genetic and phenotypic evolution, and these are linked by development.

56. S. Rice 2008a,b; Tanaka et al. 2020; Fogarty and Wade 2022; González-Forero 2023, 2024.

57. See also Osmanovic et al. 2018; Roughgarden 2020.

58. R. Watson and Thies 2019.

59. R. Watson and Thies 2019.

60. Lewontin 1983. Here we build on the conceptual revision of the "extended evolutionary synthesis" (Pigliucci and Müller 2010; Laland et al. 2015; G. Müller 2017, 2021) to recognize additional classes of evolutionary cause (see also Endler 1986), including those that generate novel variation, bias selection, and contribute to inheritance (see figure 2 in Laland et al. 2015).

61. González-Forero's (2023, 2024) analysis, which models genetic and phenotypic evolution simultaneously, provides a precise specification of the first two of these points: development and selection jointly define evolutionary outcomes, which are typically at path peaks. Note, our claims would not hold under standard assumptions derived from evolutionary models that model either genetic or phenotypic evolution (but typically not development), where outcomes are at landscape peaks because landscape peaks are predefined by selection alone. However, even in purely genetic analyses, with epistasis Wright's adaptive topography does not necessarily produce evolutionary equilibria.

62. For instance, following Jablonka and Lamb (2005, 2014; see also Pal 1998), we argued in chapter 10 that extragenetic inheritance can operate as a fast-response capability, allowing diverse organisms to survive and adapt to novel, challenging, or changeable conditions. A clear implication is that the selection of extragenetic inheritance should be recognized as organic evolution. A broader definition of evolution, say, *Evolution is a transgenerational change in the distribution of heritable traits of organisms in a population,* may be necessary. The concepts of "adaptation," "fitness," "genotype," and "phenotype" may also need modification (Oyama 1985; Griffiths and Gray 1994; Oyama et al. 2000; Jablonka and Lamb 2005, 2014; Laland et al. 2015; Bonduriansky and Day 2018). Definitions of adaptation might have to be adjusted to recognize the selection of diverse forms of heritable variation and perhaps a role for internally and externally expressed constructive processes. The concepts of "genotype" and "phenotype" might have to be revised to recognize that an organism's full hereditary information extends beyond information in the genome sequence, and perhaps to allow some traits to qualify as both genotype and phenotype (Laland et al. 2015; Pontarotti et al. 2022). These are nontrivial conceptual challenges, but we submit that they are warranted and will prove useful.

63. For instance, the recognition that heredity commonly involves the transgenerational inheritance of epigenetic marks and socially learned knowledge raises the possibility that

random change in these processes could influence the dynamics of phenotypic evolution. In the case of culture, some effects of "cultural drift" are already documented (Cavalli-Sforza and Feldman 1981; Mesoudi 2011). Neutral models have been applied to some instances of animal social learning, such as birdsong, where for species such as chaffinches (*Fringilla coelebs*) one performed song variant is seemingly as good as another, and random copying describes the observed patterns of change (Slater 1986; Slater et al. 1980). Researchers have also successfully applied neutral models borrowed from population genetics to a range of human cultural phenomena, from the popularity of dog breeds to the decorative patterns on pottery. That animals have socially learned migration routes, and influence one another's movements, has implications for gene flow, admixture, and genetic substructure (Bentley et al. 2004; Cantor et al. 2015; Berdahl et al. 2018; Jesmer et al. 2018; Brakes et al. 2019). Virtually all these issues are underexplored. There is also evidence that developmental processes can directly affect the frequency of mutation. For example, a mutation of the human *FGFR3* gene has been shown to be responsible for causing achondroplasia, the most common form of human dwarfism. The mutation produces an FGFR3 protein that is constitutively active and signals the premature cessation of bone growth through the precocious phosphorylation of a transcription factor. Interestingly, most, if not all, new mutations in the human *FGFR3* gene arise in the sperm. Since spermatogenesis is continuous, the chromosomes of sperm stem cells have undergone about 35 replications by age 15, 380 by age 30, and 840 by age 50. Copying errors can occur at each round of DNA replication, hence later replications would be expected to have more errors than earlier ones, and thus mutation is much more common in the sperm of older men (Conti and Eisenberg 2016; Wilkin et al. 1998). Moreover, the mutant FGFR3 protein also appears to give spermatogonial stem cells a higher rate of replication, such that clones having the mutant gene will be represented in the testes more frequently than expected. Indeed, mutant clones have a "selective advantage" inside the testes (Goriely and Wilkie 2012; Shinde et al. 2013).

64. Houle et al. 2017.

65. Kavanagh et al. 2007.

66. González-Forero 2023, 2024. Note, there is now good theoretical evidence that the effectiveness of mutational biases in shaping phenotypes does not depend on neutrality (e.g., Yampolsky and Stoltzfus 2001; Stoltzfus and McCandlish 2017; Stoltzfus 2019).

67. P. Rohner and Moczek 2023, 2024; P. Rohner et al. 2024.

68. Schwab et al. 2016, 2017; Hu et al. 2020; P. Rohner et al. 2021, 2024; Bilandžija et al. 2020; Powers et al. 2020a,b; Sifuentes-Romero et al. 2020; P. Rohner and Moczek 2023, 2024.

69. Gilbert and Epel 2009. González-Forero (2023, 2024) shows that knowledge of developmental mechanisms can be used to predict correlational selection and responses to selection.

70. Cavalli-Sforza and Feldman 1981; Boyd and Richerson 1985.

71. In fact, theory suggests that even under the assumption of mechanistic independence, causation is not legitimately traced back solely to fitness differences (González-Forero 2023, 2024).

72. R. Watson and Thies 2018; Uller and Helanterä 2019; Fogarty and Wade 2022; González-Forero 2023, 2024; Wade and Sultan 2024. The conditions under which the three subprocesses interact, the resultant evolution of the process of natural selection, and the contribution of that evolution to major transitions in evolution (e.g., origins of multicellularity, the potential for cultural group selection) are also ripe for investigation.

73. S. Gilbert et al. 1996.

74. Moritz (2002) emphasizes the need for conservation biology to focus on the details of evolutionary processes.

75. Laland et al. 2015.

76. B. Charlesworth 1996.

77. Waddington 1953, 1959; West-Eberhard 2003; Laland et al. 2014, 2015; G. Müller 2017.

78. West-Eberhard 2003; Wray et al. 2014; Laland et al. 2015; Futuyma 2017; Uller et al. 2020; Sultan 2019.

79. E.g., Lande 2009. Reaction norms: Schlicting and Pigliucci 1998.

80. The idealization: Uller et al. 2020. Add-on: Wray et al. 2014.

81. Waddington 1953, 1959; West-Eberhard 2003.

82. West-Eberhard 2003, chapter 3. See also Gerhart and Kirschner 1997; M. Kirschner and Gerhart 2005.

83. See, for instance, West-Eberhard 2003, chapter 1.

84. Sultan 2015, 2019; S. Gilbert et al. 2015.

85. Brun-Usan et al. 2022.

86. Lewontin 1983, 280.

87. Waddington 1959; Fabris 2021.

88. Cavalli-Sforza and Feldman 1981; Boyd and Richerson 1985; Jablonka and Lamb 2005, 2014; Mesoudi 2011; Bonduriansky and Day 2018.

REFERENCES

Abitbol, M. M. 1988. Evolution of the ischial spine and of the pelvic floor in the Hominoidea. *American Journal of Physical Anthropology* 75:53–67.

Abushama, F. T. 1974. Water-relations of the termites *Macrotermes bellicosus* (Smeathman) and *Trinervitermes geminatus* (Wasmann). *Zeitschrift für angewandte Entomologie* 75:124–134.

Abzhanov, A., W. P. Kuo, C. Hartmann, B. R. Grant, P. R. Grant, and C. J. Tabin. 2006. The calmodulin pathway and evolution of elongated beak morphology in Darwin's finches. *Nature* 442:563–567.

Abzhanov, A., M. Protas, B. R. Grant, P. R. Grant, and C. J. Tabin. 2004. Bmp4 and morphological variation of beaks in Darwin's finches. *Science* 305:1462–1465.

Acar, M., J. T. Mettetal, and A. van Oudenaarden. 2008. Stochastic switching as a survival strategy in fluctuating environments. *Nature Genetics* 40:471–475.

Adrian-Kalchhauser, I., S. E. Sultan, L.N.S. Shama, H. Spence-Jones, S. Tiso, C.I.K. Valsecchi, and F. J. Weissing. 2020. Understanding "non-genetic" inheritance: Insights from molecular-evolutionary crosstalk. *Trends in Ecology and Evolution* 35:1078–1089.

———. 2021. Inherited gene regulation unifies molecular approaches to nongenetic inheritance: Response to Edelaar et al. *Trends in Ecology and Evolution* 36:477.

Agrawal, A. A., C. Laforsch, and R. Tollrian. 1999. Transgenerational induction of defences in animals and plants. *Nature* 40:60–63.

Ahnert, S. E., and T.M.A. Fink. 2016. Form and function in gene regulatory networks: The structure of network motifs determines fundamental properties of their dynamical state space. *Journal of the Royal Society Interface* 13:20160179.

Aiello, L. C., and P. Wheeler. 1995. The expensive-tissue hypothesis: The brain and the digestive system in human and primate evolution. *Current Anthropology* 36:199–221.

Alberch, P. 1980. Ontogenesis and morphological diversification. *American Zoologist* 20:653–667.

———. 1985. Developmental constraints: Why St. Bernards often have an extra digit and poodles never do. *American Naturalist* 126:430–433.

———. 1989. The logic of monsters: Evidence for internal constraint in development and evolution. *Geobios* 12:21–57.

Alberch, P., and E. A. Gale. 1985. A developmental analysis of an evolutionary trend: Digital reduction in amphibians. *Evolution* 39:8–23.

Alberdi, A., O. Aizpurua, K. Bohmann, M. L. Zepeda-Mendoza, and M.T.P. Gilbert. 2016. Do vertebrate gut metagenomes confer rapid ecological adaptation? *Trends in Ecology and Evolution* 3:689–699.

Alberti, M. 2015. Eco-evolutionary dynamics in an urbanizing planet. *Trends in Ecology and Evolution* 30:114–126.

Allen, C. E., P. Beldade, B. J. Zwaan, and P. M. Brakefield. 2008. Differences in the selection response of serially repeated color pattern characters: Standing variation, development, and evolution. *BMC Evolutionary Biology* 8:94.

Allen, G. E. 1970. Biology and culture: Science and society in the eugenic thought of H. J. Muller. *BioScience* 20:346–353.

Allen, J., M. Weinrich, W.J.E. Hoppitt, and L. Rendell. 2013. Network-based diffusion analysis reveals cultural transmission of lobtail feeding in humpback whales. *Science* 340:485–488.

Allmon, W. D. 2011. Why don't people think evolution is true? Implications for teaching, in and out of the classroom. *Evolution: Education and Outreach* 4:648–665.

Alon, U. 2020. *An Introduction to Systems Biology*, 2nd ed. Boca Raton, FL: Chapman and Hall.

Amundson, R. 1994. Two concepts of constraint: Adaptationism and the challenge from developmental biology. *Philosophy of Science* 61:556–578.

———. 2005. *The Changing Role of the Embryo in Evolutionary Thought: Roots of Evo-Devo.* Cambridge: Cambridge University Press.

Anastasiadi, D., and F. Piferrer. 2019. Epimutations in developmental genes underlie the onset of domestication in farmed European sea bass. *Molecular Biology and Evolution* 36(10): 2252–2264.

Anastasiadi, D., C. J. Venney, L. Bernatchez, and M. Wellenreuther. 2021. Epigenetic inheritance and reproductive mode in plants and animals. *Trends in Ecology and Evolution* 36:1124–1140.

Ancel, L. W. 2000. Undermining the Baldwin expediting effect: Does phenotypic plasticity accelerate evolution? *Theoretical Population Biology* 58:307–319.

Ancel, L. W., and W. Fontana. 2000. Plasticity, evolvability, and modularity in RNA. *Journal of Experimental Zoology* 288:242–283.

Anderson, M. L. 2010. Neural reuse: A fundamental organizational principle of the brain. *Behavioral and Brain Sciences* 33:245–266.

Anderson, M. L., and B. L. Finlay. 2014. Allocating structure to function: The strong links between neuroplasticity and natural selection. *Frontiers in Human Science* 7:918.

Anderson, R. W. 1995. Learning and evolution: A quantitative genetics approach. *Journal of Theoretical Biology* 175:89–101.

Anton, S. C., R. Potts, and L. C. Aiello. 2014. Evolution of early *Homo*: An integrated biological perspective. *Science* 345:6192.

Antonovics, J., and P. H. van Tienderen. 1991. Ontoecogenophyloconstraints? The chaos of constraint terminology. *Trends in Ecology and Evolution* 6:166–168.

Aoki, K. 1986. A stochastic model of gene-culture coevolution suggested by the "culture historical hypothesis" for the evolution of adult lactose absorption in humans. *Proceedings of the National Academy of Sciences of the USA* 83:2929–2933.

Aoki, K., and M. W. Feldman. 2014. Evolution of learning strategies in temporally and spatially variable environments: A review of theory. *Theoretical Population Biology* 91:3–19.

Aoki, K., J. Y. Wakano, and M. W. Feldman. 2005. The emergence of social learning in a temporally changing environment: A theoretical model. *Current Anthropology* 46:334–340.

Aplin, L. M. 2019. Culture and cultural evolution in birds: A review of the evidence. *Animal Behaviour* 147:179–187.

Aplin, L. M., D. R. Farine, J. Morand-Ferron, A. Cockburn, A. Thornton, and B. C. Sheldon. 2015. Counting conformity: Evaluating the units of information in frequency-dependent social learning. *Animal Behaviour* 110:e5–e8.

Aragona, M., A. Sifrim, M. Malfait, Y. Song, J. van Herck, S. Dekoninck, S. Gargouri, et al. 2020. Mechanisms of stretch-mediated skin expansion at single-cell resolution. *Nature* 584:268–273.

Archie, E. A., and K. R. Theis. 2011. Animal behaviour meets microbial ecology. *Animal Behaviour* 82:425–436.

Arendt, D., J. M. Musser, C.V.H. Baker, A. Bergman, C. Cepko, D. H. Erwin, and M. Pavličev. 2016. The origin and evolution of cell types. *Nature Reviews Genetics* 17:744–757.

Armbruster, W. S., C. Pélabon, G. H. Bolstad, and T. F. Hansen. 2014. Integrated phenotypes: Understanding trait covariation in plants and animals. *Philosophical Transactions of the Royal Society B* 369:20130245.

Arnold, S. 1992. Constraints on phenotypic evolution. *American Naturalist* 140:S85–S107.

Arthur, W. 2004. *Biased Embryo and Evolution.* Cambridge: Cambridge University Press.

———. 2011. *Evolution. A Developmental Approach.* Hoboken, NJ: Wiley-Blackwell.

Athey, J., A. Alexaki, E. Osipova, A. Rostovtsev, L. V. Santana-Quintero, U. Katneni, V. Simonyan, et al. 2017. A new and updated resource for codon usage tables. *BMC Bioinformatics* 18:391.

Atton, N. 2013. Investigations into stickleback social learning. PhD dissertation, University of St Andrews.

Avila, P., and C. Mullon. 2023. Evolutionary game theory and the adaptive dynamics approach: Adaptation where individuals interact. *Philosophical Transactions of the Royal Society B* 378:20210502.

Avital, E., and E. Jablonka. 2000. *Animal Traditions: Behavioural Inheritance in Evolution.* Cambridge: Cambridge University Press.

Ayad, N.M.E., S. Kaushik, and V. M. Weaver. 2019. Tissue mechanics, an important regulator of development and disease. *Philosophical Transactions of the Royal Society B* 374:20180215.

Bäckhed, F., R. E. Ley, J. L. Sonnenburg, D. A. Peterson, and J. I. Gordon. 2005. Host-bacterial mutualism in the human intestine. *Science* 307:1915–1920.

Badyaev, A. V. 2005. Stress-induced variation in evolution: From behavioural plasticity to genetic assimilation. *Proceedings of the Royal Society B* 272:877–886.

———. 2009. Evolutionary significance of phenotypic accommodation in novel environments: An empirical test of the Baldwin effect. *Philosophical Transactions of the Royal Society B* 364:1125–1141.

———. 2011. Origin of the fittest: Link between emergent variation and evolutionary change as a critical question in evolutionary biology. *Proceedings of the Royal Society B* 278:1921–1929.

———. 2014. Epigenetic resolution of the "curse of complexity" in adaptive evolution of complex traits. *Journal of Physiology* 592:2251–2260.

Badyaev, A. V., and K. R. Foresman. 2000. Extreme environmental change and evolution: Stress-induced morphological variation is strongly concordant with patterns of evolutionary divergence in shrew mandibles. *Proceedings of the Royal Society of London B* 267:371–377.

Badyaev, A. V., and A. Qvarnström. 2002. Putting sexual traits into the context of an organism: A life-history perspective in studies of sexual selection. *Auk* 119:301–310.

Badyaev, A. V., and T. Uller. 2009. Parental effects in ecology and evolution: Mechanisms, processes and implications. *Philosophical Transactions of the Royal Society B* 364:1169–1177.

Badyaev, A. V., A. L. Potticary, and E. S. Morrison. 2017. Most colorful example of genetic assimilation? Exploring the evolutionary destiny of recurrent phenotypic accommodation. *American Naturalist* 190:266–280.

Baedke, J. 2019. O organism, where art thou? Old and new challenges for organism-centered biology. *Journal of the History of Biology* 52:293–324.

———. 2021. What's wrong with evolutionary causation? *Acta biotheoretica* 69:79–89.

Baedke, J., and A. Fábregas-Tejeda. 2023. The organism in evolutionary explanation: From early twentieth century to the extended evolutionary synthesis. In: *Evolutionary Biology: Contemporary and Historical Reflections upon Core Theory,* ed. T. E. Dickins and B.J.A Dickins, 121–150. Berlin: Springer.

Baedke, J., A. Fábregas-Tejeda, and F. Vergara-Silva. 2020. Does the extended evolutionary synthesis entail extended explanatory power? *Biology and Philosophy* 35:20.

Baker, B. H., S. E. Sultan, M. Lopez-Ichikawa, and R. Waterman. 2019. Transgenerational effects of parental light environment on progeny competitive performance and lifetime fitness. *Philosophical Transactions of the Royal Society B* 374:20180182.

Baldwin, J. M. 1896. A new factor in evolution. *American Naturalist* 30:536–553.

Baldwin, R. L., R. W. Li, Y. Jia, and C. J. Li. 2018. Transcriptomic impacts of rumen epithelium induced by butyrate infusion in dairy cattle in dry period. *Gene Regulation and Systems Biology* 12:1177625018774798.

Bamforth, D. B. 2002. Evidence and metaphor in evolutionary archaeology. *American Antiquity* 67:435–452.

Bard, J.B.L. 1981. A model for generating aspects of zebra and other mammalian coat patterns. *Journal of Theoretical Biology* 93:363–385.

———. 2022. *Evolution: The Origins and Mechanisms of Diversity*. Boca Raton, FL: CRC.

Barinaga, M. 2003. Newborn neurons search for meaning. *Science* 299:32–34.

Barnes, I., A. Duda, O. G. Pybus, and M. G. Thomas. 2011. Ancient urbanization predicts genetic resistance to tuberculosis. *Evolution* 65:842–848.

Baroux, C., and U. Grossniklaus. 2015. The maternal-to-zygotic transition in flowering plants: Evidence, mechanisms, and plasticity. *Current Topics in Developmental Biology* 113:351–371.

Barresi, M.J.F., and S. F. Gilbert. 2020. *Developmental Biology*, 12th ed. New York: Oxford University Press.

Barrett, H. C. 2015. *The Shape of Thought*. New York: Oxford University Press.

Barrett, H. C., and J. Armstrong. 2023. Climate-change adaptation and the back of the invisible hand. *Philosophical Transactions of the Royal Society B* 378:20220406.

Barrett, L. 2011. *Beyond the Brain*. Princeton, NJ: Princeton University Press.

Barton, N. H. 1990. Pleiotropic models of quantitative variation. *Genetics* 124:773–782.

Barton, N. H., and M. Turelli. 1987. Adaptive landscapes, genetic distance and the evolution of quantitative characters. *Genetical Research* 49:157–173.

Barton, N. H., D. E. Briggs, J. A. Eisen, D. B. Goldstein, and N. H. Patel. 2007. *Evolution*. Long Island, NY: Cold Spring Harbor Laboratory.

Barton, R. A. 2012. Embodied cognitive evolution and the cerebellum. *Philosophical Transactions of the Royal Society B* 367:2097–2107.

Barton, R. A., and P. H. Harvey. 2000. Mosaic evolution of brain structure in mammals. *Nature* 405:1055–1058.

Barton, R. A., and C. Venditti. 2014. Rapid evolution of the cerebellum in humans and other great apes. *Current Biology* 24:2440–2444.

Bateson, P. 1988. The active role of behaviour in evolution. In: *Evolutionary Processes and Metaphors*, ed. M. W. Fox and S. W. Fox, 191–207. New York: John Wiley.

———. 2002. William Bateson: A biologist ahead of his time. *Journal of Genetics* 81:49–58.

———. 2017. Robustness and plasticity in development. *WIREs Cognitive Science* 8:e1386.

Bateson, P., and P. Gluckman. 2011. *Plasticity, Robustness, Development and Evolution*. Cambridge: Cambridge University Press.

Bateson P., and P. Martin. 1999. *Design for a Life*. London: Cape.

Bateson, P., P. Gluckman, and M. Hanson. 2014. The biology of developmental plasticity and the predictive adaptive response hypothesis. *Journal of Physiology* 592:2357–2368.

Bateson, W. 1894. *Materials for the Study of Variation: Treated with Especial Regard to Discontinuity in the Origin of Species*. London: Macmillan.

Bazzaz, F. A. 1991. Habitat selection in plants. *American Naturalist* 137: S116–S130.

Beatty, J. 1988. Ecology and evolutionary biology in the war and post-war years: Questions and comments. *Journal of the History of Biology* 21:245–263.

———. 1991. Genetics in the atomic age: The atomic bomb casualty commission, 1947–1956. In: *The Expansion of American Biology*, ed. K. R. Benson, J. Maienschein, and R. Rainger, 284–324. New Brunswick, NJ: Rutgers University Press.

———. 1994a. Opportunities for genetics in the atomic age. Paper presented at the Fourth Mellon Symposium, Institutional and Disciplinary Contexts of the Life Sciences, MIT, Cambridge, MA, April 1994.

———. 1994b. The proximate/ultimate distinction in the multiple careers of Ernst Mayr. *Biology and Philosophy* 9:333–356.

Behar, H., and M. W. Feldman. 2018. Gene-culture coevolution under selection. *Theoretical Population Biology* 121:33–44.

Behrmann-Godel, J., A. W. Nolte, J. Kreiselmaier, R. Berka, and J. Freyhof. 2017. The first European cave fish. *Current Biology* 27:R257–R258.

Belay, E. D., R. A. Maddox, E. S. Williams, M. W. Miller, P. Gambetti, and L. B. Schonberger. 2004. Chronic wasting disease and potential transmission to humans. *Emerging Infectious Diseases* 10:977–984.

Beldade, P., K. Koops, and P. M. Brakefield. 2002. Developmental constraints versus flexibility in morphological evolution. *Nature* 416:844–847.

Bellwood, D. R. 2003. Origins and escalation of herbivory in fishes: A functional perspective. *Paleobiology* 29:71–83.

Beltman, J. B., P. Haccou, and C. ten Cate. 2003. The impact of learning foster species' song on the evolution of specialist avian brood parasitism. *Behavioral Ecology* 14:917–923.

———. 2004. Learning and colonization of new niches: A first step toward speciation. *Evolution* 58:35–46.

Benito, E., C. Kerimoglu, B. Ramachandran, T. Pena-Centeno, G. Jain, R. M. Stilling, and M. R. Islam, et al. 2018. RNA-dependent intergenerational inheritance of enhanced synaptic plasticity after environmental enrichment. *Cell Reports* 23:546–554.

Bennett, A. F., R. B. Huey, H. John-Alder, and K. A. Nagy. 1984. The parasol tail and thermoregulatory behavior of the cape ground squirrel *Xerus inauris*. *Physiological Zoology* 57:57–62.

Bennett, G. M., and N. A. Moran. 2015. Heritable symbiosis: The advantages and perils of an evolutionary rabbit hole. *Proceedings of the National Academy of Sciences of the USA* 112:10169–10176.

Bentley, R. A., M. W. Hahn, and S. J. Shennan. 2004. Random drift and culture change. *Proceedings of the Royal Society B* 271:1443–1450.

Berdahl, A. M., A. B. Kao, A. Flack, P.A.H. Westley, E. A. Codling, I. D. Couzin, A. I. Dell, et al. 2018. Collective animal navigation and migratory culture: From theoretical models to empirical evidence. *Philosophical Transactions of the Royal Society B* 373:20170009.

Bergstrom, C. T., and L. A. Dugatkin. 2012. *Evolution*. New York: W. W. Norton.

Bermúdez de Castro, J. M., M. Modesto-Mata, C. García-Campos, S. Sarmiento, L. Martín-Francés, M. Martínez de Pinillos, and M. Martinón-Torres. 2021. Testing the inhibitory cascade model in a recent human sample. *Journal of Anatomy* 239:1170–1181.

Bever, G. S., T. R. Lyson, D. J. Field, and B.-A. S. Bhullar. 2016. The amniote temporal roof and the diapsid origin of the turtle skull. *Zoology* 119:471–473.

Bickerton, A. 2009. *Adam's Tongue*. New York: Hill and Wang.

Bijma, P. 2014. The quantitative genetics of indirect genetic effects: A selective review of modelling issues. *Heredity* 112:61–69.

Bilandžija, H., L. Abraham, L. Ma, K. J. Renner, and W. R. Jeffery. 2018. Behavioural changes controlled by catecholaminergic systems explain recurrent loss of pigmentation in cavefish. *Proceedings of the Royal Society B* 285:20180243.

Bilandžija, H., B. Hollifield, M. Steck, G. Meng, M. Ng, A. D. Koch, R. Gračan, et al. 2020. Phenotypic plasticity as a mechanism of cave colonization and adaptation. *eLife* 9:e51830.

Biro, D., M. Haslam, and C. Rutz. 2013. Tool use as adaptation. *Philosophical Transactions of the Royal Society B* 368:20120408.

Bizzarri, M., D. E. Brash, J. Briscoe, V. A. Grieneisen, C. D. Stern, and M. Levin. 2019. A call for a better understanding of causation in cell biology. *Nature Reviews Molecular Cell Biology* 20:261–262.

Bliege Bird, R., C. McGuire, D. W. Bird, M. H. Price, D. Zeanah, and D. G. Nimmo. 2020. Fire mosaics and habitat choice in nomadic foragers. *Proceedings of the National Academy of Sciences of the USA* 117:12904–12914.

Bódi, Z., Z. Farkas, D. Nevozhay, D. Kalapis, V. Lázár, B. Csörgő, Á. Nyerges, et al. 2017. Phenotypic heterogeneity promotes adaptive evolution. *PLOS Biology* 15:e2000644.

Bogert, C. M. 1949. Thermoregulation in reptiles, a factor in evolution. *Evolution* 3:195–211.

Bohacek, J., and I. M. Mansuy. 2013. Epigenetic inheritance of disease and disease risk. *Neuropsychopharmacology* 38:220–236.

Boivin, N. L., M. A. Zeder, D. Q. Fuller, A. Crowther, G. Larson, J. M. Erlandson, T. Denham, et al. 2016. Ecological consequences of human niche construction: Examining long-term anthropogenic shaping of global species distributions. *Proceedings of the National Academy of Sciences of the USA* 113:6388–6396.

Bolhuis, J. J., L. Giraldeau, and J. A. Hogan. 2022. *The Behavior of Animals: Mechanisms, Function, and Evolution*, 2nd ed. Hoboken, NJ: Wiley Blackwell.

Bolhuis, J. J., K. Okanoya, and C. Scharff. 2010. Twitter evolution: Converging mechanisms in birdsong and human speech. *Nature Reviews Neuroscience* 11:747–759.

Bolhuis, J. J., I. Tattersall, N. Chomsky, and R. C. Berwick. 2014. How could language have evolved? *PLOS Biology* 12:e1001934.

Bollinger, R. R., A. S. Barbas, E. L. Bush, S. S. Lin, and W. Parker. 2007. Biofilms in the large bowel suggest an apparent function of the human vermiform appendix. *Journal of Theoretical Biology* 249:826–831.

Bolstad, G. H., T. F. Hansen, C. Pélabon, M. Falahati-Anbaran, R. Perez-Barrales, and W. S. Armbruster. 2014. Genetic constraints predict evolutionary divergence in *Dalechampia* blossoms. *Philosophical Transactions of the Royal Society B* 369:20130255.

Bonabeau, E., M. Dorigo, and G. Theraulaz. 1999. *Swarm Intelligence: From Natural to Artificial Systems*. Oxford: Oxford University Press.

Bonduriansky, R. 2012. Rethinking heredity again. *Trends in Ecology and Evolution* 27:330–336.

Bonduriansky, R., and T. Day. 2009. Nongenetic inheritance and its evolutionary implications. *Annual Review of Ecology, Evolution, and Systematics* 40:103–125.

———. 2018. *Extended Heredity: A New Understanding of Inheritance and Evolution*. Princeton, NJ: Princeton University Press.

Bonner, J. T. 1982. *Evolution and Development: Report of the Dahlem Workshop on Evolution and Development, Berlin 1981, May 10–15*. Berlin: Springer.

Borenstein, E., and D. C. Krakauer. 2008. An end to endless forms: Epistasis, phenotype distribution bias, and nonuniform evolution. *PLOS Computational Biology* 4:e1000202.

Borenstein, E., I. Meilijson, and E. Ruppin. 2006. The effect of phenotypic plasticity on evolution in multipeaked fitness landscapes. *Journal of Evolutionary Biology* 19:1555–1570.

Borowsky, R. 2023. Selection maintains the phenotypic divergence of cave and surface fish. *American Naturalist* 202:55–63.

Bossdorf, O., D. Arcuri, C. L. Richards, and M. Pigliucci. 2010. Experimental alteration of DNA methylation affects the phenotypic plasticity of ecologically relevant traits in *Arabidopsis thaliana*. *Evolutionary Ecology* 24:541–553.

Bosse, M., L. G. Spurgin, V. N. Laine, E. F. Cole, J. A. Firth, P. Gienapp, and A. G. Gosler. 2017. Recent natural selection causes adaptive evolution of an avian polygenic trait. *Science* 358:365–368.

Botigué, L. R., S. Song, A. Scheu, S. Gopalan, A. L. Pendleton, M. Oetjens, A. M. Taravella, et al. 2017. Ancient European dog genomes reveal continuity since the early Neolithic. *Nature Communications* 8:16082.

Boyd, R. 2018. *A Different Kind of Animal: How Culture Transformed Our Species*. Princeton, NJ: Princeton University Press.

Boyd, R., and P. J. Richerson. 1985. *Culture and the Evolutionary Process*. Chicago: University of Chicago Press.

Boyd, R., and J. B. Silk. 2017. *How Humans Evolved*, 8th ed. New York: W. W. Norton.

Bradley, B. 2020. *Darwin's Psychology*. New York: Oxford University Press.

Braendle, C., C. F. Baer, and M. A. Félix. 2010. Bias and evolution of the mutationally accessible phenotypic space in a developmental system. *PLOS Genetics* 6:e1000877.

Brakefield, P. M. 2006. Evo-devo and constraints on selection. *Trends in Ecology and Evolution* 21:362–368.

———. 2008. Why are forms the way they are? *Trends in Ecology and Evolution* 23:62–63.

———. 2010. Radiations of mycalesine butterflies and opening up their exploration of morphospace. *American Naturalist* 176:S77–S87.

———. 2011. Evo-devo and accounting for Darwin's endless forms. *Philosophical Transactions of the Royal Society B* 366: 2069–2075.

Brakefield, P. M., and N. Reitsma. 1991. Phenotypic plasticity, seasonal climate and the population biology of *Bicyclus* butterflies (Satyridae) in Malawi. *Ecological Entomology* 16:291–303.

Brakefield, P. M., and J. C. Roskam. 2006. Exploring evolutionary constraints is a task for an integrative evolutionary biology. *American Naturalist* 168:S4–S13.

Brakes, P., S.R.X. Dall, L. M. Aplin, S. Bearhop, E. L. Carroll, C. Ciucci, V. Fishlock, et al. 2019. Animal cultures matter for conservation. *Science* 363:1032–1034.

Bramble, D. M., and D. E. Lieberman. 2004. Endurance running and the evolution of *Homo*. *Nature* 432:345–352.

Bråthen, K. A., and V. T. Ravolainen. 2015. Niche construction by growth forms is as strong a predictor of species diversity as environmental gradients. *Journal of Ecology* 103:701–713.

Brattström, O., K. Aduse-Poku, E. van Bergen, V. French, and P. M. Brakefield. 2020. A release from developmental bias accelerates morphological diversification in butterfly eyespots. *Proceedings of the National Academy of Sciences of the USA* 117:27474–27480.

Braun, D. R., T. Plummer, P. W. Ditchfield, L. C. Bishop, and J. V. Ferraro. 2009. Oldowan technology and raw material variability at Kanjera South. In: *Interdisciplinary Approaches to Oldowan*, ed. E. Hovers and D. R. Braun, 99–110. New York: Springer.

Brawand, D., C. E. Wagner, Y. I. Li, M. Malinsky, I. Keller, S. Fan, O. Simakov, et al. 2014. The genomic substrate for adaptive radiation in African cichlid fish. *Nature* 513:375–381.

Brevik, K., E. M. Bueno, S. McKay, S. D. Schoville, and Y. H. Chen. 2021. Insecticide exposure affects intergenerational patterns of DNA methylation in the Colorado potato beetle, *Leptinotarsa decemlineata*. *Evolutionary Applications* 14:746–757.

Brigandt, I. 2015. From developmental constraint to evolvability: How concepts figure in explanation and disciplinary identity. In: *Conceptual Change in Biology: Scientific and Philosophical Perspectives on Evolution and Development*, ed. A. C. Love, 305–325. New York: Springer.

———. 2020. Historical and philosophical perspectives on the study of developmental bias. *Evolution and Development* 22:7–19.

Brigandt, I., C. Villegas, A. C. Love, L. Nuño de la Rosa. 2023. Evolvability as a disposition: Philosophical distinctions, scientific implications. Chapter 4 in *Evolvability: A Unifying Concept in Evolutionary Biology?*, ed. T. F. Hansen, D. Houle, M. Pavličev, and C. Pélabon, Vienna Series in Theoretical Biology. Cambridge, MA: MIT Press.

Briggs, S. E., J.G.J. Godin, and L. A. Dugatkin. 1996. Mate-choice copying under predation risk in the Trinidadian guppy (*Poecilia reticulata*). *Behavioral Ecology* 7:151–157.

Brighton, H., S. Kirby, and K. Smith. 2005. Cultural selection for learnability: Three principles underlying the view that language adapts to be learnable. In: *Language Origins: Perspectives on Evolution*, ed. M. Tallerman, 291–309. Oxford: Oxford University Press.

Brocklehurst, N. 2017. Rates of morphological evolution in Captorhinidae: An adaptive radiation of Permian herbivores. *PeerJ* 5:e3200.

Brown, A. L. 1990. Domain-specific principles affect learning and transfer in children. *Cognitive Science* 14:107–133.

Brown, C., and K. N. Laland. 2001. Social learning and life skills training for hatchery reared fish. *Journal of Fish Biology* 59:471–493.

———. 2002a. Social enhancement and social inhibition of foraging behaviour in hatchery-reared Atlantic salmon. *Journal of Fish Biology* 61:987–998.

———. 2002b. Social learning of a novel avoidance task in the guppy: Conformity and social release. *Animal Behaviour* 64:41–47.

———. 2003. Social learning in fishes: A review. *Fish and Fisheries* 4:280–288.

Brown, C. R., and M. B. Brown. 1987. Group-living in cliff swallows as an advantage in avoiding predators. *Behavioral Ecology and Sociobiology* 21:97–107.

———. 2000. Heritable basis for choice of group size in a colonial bird. *Proceedings of the National Academy of Sciences of the USA* 97:14825–14830.

Brown, C. R., N. Komar, S. B. Quick, R. A. Sethi, N. A. Panella, M. B. Brown, and M. Pfeffer. 2001. Arbovirus infection increases with group size. *Proceedings of the Royal Society of London B* 268:1833–1840.

Brown, P., T. Sutikna, M. J. Morwood, R. P. Soejono, Jatmiko, E. Wayhu Saptomo, and R. Awe Due. 2004. A new small-bodied hominin from the late Pleistocene of Flores, Indonesia. *Nature* 431:1055–1061.

Brown, R. L. 2014. What evolvability really is. *British Journal for the Philosophy of Science* 65:549–572.

Browne, J. 2021. Introduction to *A Most Interesting Problem*, ed. J. Desilva and J. Browne. Princeton, NJ: Princeton University Press.

Bruce, H. S., and N. H. Patel. 2020. Knockout of crustacean leg patterning genes suggests that insect wings and body walls evolved from ancient leg segments. *Nature Ecology and Evolution* 4:1703–1712.

Brucker, R. M., and S. R. Bordenstein. 2013. The hologenomic basis of speciation: Gut bacteria cause hybrid lethality in the genus *Nasonia*. *Science* 341:667–669.

Brun-Usan, M., A. Rago, C. Thies, T. Uller, and R. Watson. 2021. Development and selective grain make plasticity 'take the lead' in adaptive evolution. *BMC Ecology and Evolution* 21:1–17.

Brun-Usan, M., R. Zimm, and T. Uller. 2022. Beyond genotype-phenotype maps: Toward a phenotype-centered perspective on evolution. *BioEssays* 44:2100225.

Buchsbaum, D., S. Bridgers, D. S. Weisberg, and A. Gopnik. 2012. The power of possibility: Causal learning, counterfactual reasoning, and pretend play. *Philosophical Transactions of the Royal Society B* 367:2202–2212.

Buchsbaum, D., A. Gopnik, T. L. Griffiths, and P. Shafto. 2011. Children's imitation of causal action sequences is influenced by statistical and pedagogical evidence. *Cognition* 120:331–340.

Budd, G. E. 2013. At the origin of animals: The revolutionary Cambrian fossil record. *Current Genomics* 14:344–354.

Bugnyar, T., and K. Kortschal. 2002. Observational learning and the raiding of food caches in ravens (*Corvus corax*): Is it "tactical deception"? *Animal Behaviour* 64:185–195.

Buller, D. J., and V. G. Hardcastle. 2000. Evolutionary psychology, meet developmental neuro-biology: Against promiscuous modularity. *Brain Mind* 1:307–325.

Burger, J., M. Kirchner, B. Bramanti, W. Haak, and M. G. Thomas. 2007. Absence of the lactase-persistence-associated allele in early Neolithic Europeans. *Proceedings of the National Academy of Sciences of the USA* 104:3736.

Burgess, S. C., and D. J. Marshall. 2014. Adaptive parental effects: The importance of estimating environmental predictability and offspring fitness appropriately. *Oikos* 12:769–776.

Burggren, W. 2016. Epigenetic inheritance and its role in evolutionary biology: Re-evaluation and new perspectives. *Biology* 5:24.

Burke, A. C. 1989. Development of the turtle carapace: Implications for the evolution of a novel bauplan. *Journal of Morphology* 199:363–378.

———. 1991. The development and evolution of the turtle body plan: Inferring intrinsic aspects of the evolutionary process from experimental embryology. *American Zoologist* 31:616–627.

Burke, S., and T. Duffy. 2022. Famine, tea, and bread in Ireland: C282Y and modern human microevolution. *American Journal of Biological Anthropology* 178:211–229.

Burmeister, S. S., E. D. Jarvis, and R. D. Fernald. 2005. Rapid behavioral and genomic responses to social opportunity. *PLOS Biology* 3:e363.

Busey, H. A., E. E. Zattara, and A. P. Moczek. 2016. Conservation, innovation, and bias: Embryonic segment boundaries position posterior, but not anterior, head horns in adult beetles. *Journal of Experimental Zoology Part B* 326:271–279.

Buss, D. M. 1999. *Evolutionary Psychology: The New Science of the Mind.* London: Allyn and Bacon.

Buss, L. 1987. *The Evolution of Individuality.* Princeton, NJ: Princeton University Press.

Bygren, L. O., G. Kaati, and S. Edvinsson. 2001. Longevity determined by paternal ancestors' nutrition during their slow growth period. *Acta biotheoretica* 49:53–59.

Byrnes W. M., and S. A. Newman. 2014. Ernest Everett Just: Egg and embryo as excitable systems. *Journal of Experimental Biology Part B* 322:191–201.

Calcott, B. 2013. Why how and why aren't enough: More problems with Mayr's proximate-ultimate distinction. *Biology and Philosophy* 28:767–780.

Call, J. 2013. Three ingredients for becoming a creative tool user. In: *Tool Use in Animals: Cognition and Ecology*, ed. C. M. Sanz, J. Call, and C. Boesch, 3–20. Cambridge: Cambridge University Press.

Callahan, B. J., T. Fukami, and D. S. Fisher. 2014. Rapid evolution of adaptive niche construction in experimental microbial populations. *Evolution* 68:3307–3316.

Callebaut, W., and D. Rasskin-Gutman. 2005. *Modularity: Understanding the Development and Evolution of Natural Complex Systems.* Cambridge, MA: MIT Press.

Calo, S., C. Shertz-Wall, S. C. Lee, R. J. Bastidas, F. E. Nicolás, J. A. Granek, P. Mieczkowski, et al. 2014. Antifungal drug resistance evoked via RNAi-dependent epimutations. *Nature* 513:555–558.

Calvo, S. E., D. J. Pagliarini, and V. K. Mootha. 2009. Upstream open reading frames cause widespread reduction of protein expression and are polymorphic among humans. *Proceedings of the National Academy of Sciences of the USA* 106:7507–7512.

Camara, M. D., and M. Pigliucci. 1999. Mutational contributions to genetic variance-covariance matrices: An experimental approach using induced mutations in *Arabidopsis thaliana*. *Evolution* 53:1692–1703.

Camara, M. D., C. A. Ancell, and M. Pigliucci. 2000. Induced mutations: A novel tool to study phenotypic integration and evolutionary constraints in *Arabidopsis thaliana*. *Evolutionary Ecology Research* 2:1009–1029.

Camazine, S., J. L. Deneubourg, N. R. Franks, J. Sneyd, G. Theraula, and E. Bonabeau. 2001. *Self-Organization in Biological Systems.* Princeton, NJ: Princeton University Press.

Campàs, O., R. Mallarino, A. Herrel, A. Abzhanov, and M. P. Brenner. 2010. Scaling and shear transformations capture beak shape variation in Darwin's finches. *Proceedings of the National Academy of Sciences of the USA* 107:3356–3360.

Campbell, B. 1974. *In Search of Man: Some Questions and Answers in African Archaeology and Primatology*. São Francisco, Brazil: Leakey Foundation.

Campbell, D. T. 1960. Blind variation and selective retention in creative thought as in other knowledge processes. *Psychological Review* 67:380–400.

Cano, A. V., B. L. Gitschlag, H. Rozhoňová, A. Stoltzfus, D. M. McCandlish, and J. L. Payne. 2023. Mutation bias and the predictability of evolution. *Philosophical Transactions of the Royal Society B* 378:20220055.

Cantor, M., L. G. Shoemaker, R. B. Cabral, C. O. Flores, M. Varga, and H. Whitehead. 2015. Multilevel animal societies can emerge from cultural transmission. *Nature Communications* 6:8091.

Carey, S. 2009. *The Origin of Concepts*. New York, NY: Oxford University Press.

Carja, O., R. E. Furrow, and M. W. Feldman. 2014. The role of migration in the evolution of phenotypic switching. *Proceedings of the Royal Society B* 281:20141677.

Carroll, E. L., C. S. Baker, M. Watson, R. Alderman, J. Bannister, O. E. Gaggiotti, D. R. Gröcke, et al. 2015. Cultural traditions across a migratory network shape the genetic structure of southern right whales around Australia and New Zealand. *Scientific Reports* 5:16182.

Carroll, S. B. 2005. *Endless Forms Most Beautiful: The New Science of Evo-Devo and the Making of the Animal Kingdom*. New York: W. W. Norton.

Casasa, S., and A. P. Moczek. 2018. Insulin signalling's role in mediating tissue-specific nutritional plasticity and robustness in the horn-polyphenic beetle *Onthophagus taurus*. *Proceedings of the Royal Society B* 285:20181631.

Casasa, S., E. E. Zattara, and A. P. Moczek. 2020. Nutrition-responsive gene expression and the developmental evolution of insect polyphenism. *Nature Ecology and Evolution* 4:970–978.

Castro, L., and M. A. Toro. 2004. The evolution of culture: From primate social learning to human culture. *Proceedings of the National Academy of Sciences of the USA* 101:10235–10240.

Cataldo, D. M., A. B. Migliano, and L. Vinicius. 2018. Speech, stone tool-making and the evolution of language. *PLOS One* 13:e0191071.

Cavalli-Sforza, L. L., and M. W. Feldman. 1981. *Cultural Transmission and Evolution: A Quantitative Approach*. Princeton, NJ: Princeton University Press.

Champagne, F. A. 2008. Epigenetic mechanisms and the transgenerational effects of maternal care. *Frontiers in Neuroendocrinology* 29:386–397.

Champagne F. A., and J. M. Curley. 2012. Genetics and epigenetics of parental care. In: *The Evolution of Parental Care*, ed. N. J. Royle, P. T. Smiseth, and M. Kolliker, 304–324. Oxford: Oxford University Press.

Champagne, F. A., and M. J. Meaney. 2006. Stress during gestation alters postpartum maternal care and the development of the offspring in a rodent model. *Biological Psychiatry* 59:1227–1235.

———. 2007. Transgenerational effects of social environment on variations in maternal care and behavioral response to novelty. *Behavioral Neuroscience* 121:1353–1363.

Champagne, F. A., J. P. Curley, E. B. Keverne, and P.P.G. Bateson. 2007. Natural variations in postpartum maternal care in inbred and outbred mice. *Physiology and Behavior* 91:325–334.

Chan, J. C., C. P. Morgan, N. A. Leu, A. Shetty, Y. M. Cisse, B. M. Nugent, K. E. Morrison, et al. 2020. Reproductive tract extracellular vesicles are sufficient to transmit intergenerational stress and program neurodevelopment. *Nature Communications* 11:1499.

Chan, Y. F., M. E. Marks, F. C. Jones, G. Villarreal, M. D. Shapiro, S. D. Brady, A. M. Southwick, et al. 2010. Adaptive evolution of pelvic reduction in sticklebacks by recurrent deletion of a *Pitx1* enhancer. *Science* 327:302–305.

Charlesworth, B. 1976. Recombination modification in a fluctuating environment. *Genetics* 83:181–195.

———. 1996. The good fairy godmother of evolutionary genetics. *Current Biology* 6:220

Charlesworth, B., and C. H. Langley. 1989. The population genetics of *Drosophila* transposable elements. *Annual Review of Genetics* 23:251–287.

Charlesworth, B., R. Lande, and M. Slatkin. 1982. A Neo-Darwinian commentary on macroevolution. *Evolution* 36:474–498.

Charlesworth, D., N. H. Barton, and B. Charlesworth. 2017. The sources of adaptive variation. *Proceedings of the Royal Society B* 284:20162864.

Chebib, J., and F. Guillaume. 2017. What affects the predictability of evolutionary constraints using a G-matrix? The relative effects of modular pleiotropy and mutational correlation. *Evolution* 71:2298–2312.

Chen, Y.-C., A. Arnatkevičiūtė, E. McTavish, J. C. Pang, S. Chopra, C. Suo, A. Fornito, et al. 2022. The individuality of shape asymmetries of the human cerebral cortex. *eLife* 11:e75056.

Cheverud, J. M. 1982. Phenotypic, genetic, and environmental morphological integration in the cranium. *Evolution* 36:499–516.

———. 1984. Evolution by kin selection: A quantitative genetic model illustrated by maternal performance in mice. *Evolution* 38:766–777.

———. 1996. Quantitative genetic analysis of cranial morphology in the cottontop (*Sanguinus oedipus*) and saddle-back (*S. fuscicollis*) tamarins. *Journal of Evolutionary Biology* 9:5–42.

Chiang, G.C.K., D. Barua, E. Dittmar, E. M. Kramer, R. Rubio de Casas, and K. Donohue. 2013. Pleiotropy in the wild: The dormancy gene *Dog1* exerts cascading control on life cycles. *Evolution* 67:883–893.

Chiari, Y., V. Cahais, N. Galtier, and F. Delsuc. 2012. Phylogenomic analyses support the position of turtles as the sister group of birds and crocodiles (Archosauria). *BMC Biology* 10:65.

Childs, B. 2003. *Genetic Medicine: A Logic of Disease*. Baltimore, MD: Johns Hopkins University Press.

Chisholm, R. H., B. D. Connelly, B. Kerr, and M. M. Tanaka. 2018. The role of pleiotropy in the evolutionary maintenance of positive niche construction. *American Naturalist* 192:35–48.

Chiu, L., and S. F. Gilbert. 2015. The birth of the holobiont: Multi-species birthing through mutual scaffolding and niche construction. *Biosemiotics.* 8:191–210.

———. 2020. Niche construction and the transition to herbivory: Phenotype switching and the origination of new nutritional modes. In: *Phenotype Switching: Implications in Biology and Medicine*, ed. H. Levine, M. Jolly, P. Kulkarni, and V. Nanjundiah, 459–482. London: Elsevier.

Chomsky, N. 1968. *Language and Mind*. Cambridge: Cambridge University Press.

———. 1980. *Rules and Representations*. New York: Columbia University Press.

———. 1990. On the nature, acquisition and use of language. In: *Mind and Cognition: A Reader*, ed. W. G. Lycan, 627–645. Cambridge, MA: Blackwell.

Chow, R. L., C. R. Altmann, R. A. Lang, and A. Hemmati-Brivanlou. 1999. *Pax6* induces ectopic eyes in a vertebrate. *Development* 126:4213–4222.

Christensen, M. M., O. Hallikas, R. D. Roy, V. Väänänen, O. E. Stenberg, T. J. Häkkinen, J. C. François, et al. 2023. The developmental basis for scaling of mammalian tooth size. *Proceedings of the National Academy of Sciences of the USA* 120:e2300374120.

Christiansen, M. H., and S. Kirby. 2003. *Language Evolution*. New York: Oxford University Press.

Chu, D. 2011. Complexity: Against systems. *Theory in Biosciences* 130:229–245.

Chu, D., R. Strand, and R. Fjelland. 2003. Essays and commentaries: Theories of complexity. *Complex* 8:19–30.

Chudek, M., and J. Henrich. 2011. Culture-gene coevolution, norm-psychology and the emergence of human prosociality. *Trends in Cognitive Sciences* 15:218–226.

Ciliberti, S., O. C. Martin, and A. Wagner. 2007. Innovation and robustness in complex regulatory gene networks. *Proceedings of the National Academy of Sciences of the USA* 104:13591–13596.

Clark, A. D., D. Deffner, K. N. Laland, F. J. Odling-Smee, and J. Endler. 2020. Niche construction affects the variability and strength of natural selection. *American Naturalist* 195:16–30.

Clutton-Brock, T. H. 1991. *The Evolution of Parental Care.* Princeton, NJ: Princeton University Press.

Clutton- Brock, T. H., and P. H. Harvey. 1980. Primates, brain and ecology. *Journal of Zoology* 190:309–323.

Cochran. G., and H. Harpending. 2009. *10,000 Year Explosion: How Civilization Accelerated Human Evolution.* New York: Basic Books.

Cokus, S. J., S. Feng, X. Zhang, Z. Chen, B. Merriman, C. D. Haudenschild, S. Pradhan, et al. 2008. Shotgun bisulphite sequencing of the *Arabidopsis* genome reveals DNA methylation patterning. *Nature* 452:215–219.

Cong, W., Y. Miao, L. Xu, Y. Zhang, C. Yuan, J. Wang, T. Zhuang, et al. 2019. Transgenerational memory of gene expression changes induced by heavy metal stress in rice (*Oryza sativa* L.). *BMC Plant Biology* 19:282.

Conine, C. C., F. Sun, L. Song, J. A. Rivera-Pérez, and O. J. Rando. 2018. Small RNAs gained during epididymal transit of sperm are essential for embryonic development in mice. *Developmental Cell* 46:470–480.e3.

Conti, S., and M. Eisenberg. 2016. Paternal aging and increased risk of congenital disease, psychiatric disorders, and cancer. *Asian Journal of Andrology* 18:420–424.

Conway, C. M., and M. H. Christiansen. 2001. Sequential learning in non-human primates. *Trends in Cognitive Sciences* 5:539–546.

Coolen, I., Y. V. Bergen, R. L. Day, and K. N. Laland. 2003. Species difference in adaptive use of public information in sticklebacks. *Proceedings of the Royal Society B* 270:2413–2419.

Cooney, C. R., J. A. Bright, E.J.R. Capp, A. M. Chira, E. C. Hughes, C.J.A. Moody, L. O. Nouri, et al. 2017. Mega-evolutionary dynamics of the adaptive radiation of birds. *Nature* 542:344–347.

Cooper, A., C.S.M. Turney, J. Palmer, A. Hogg, M. McGlone, J. Wilmshurst, A. M. Lorrey, et al. 2021. A global environmental crisis 42,000 years ago. *Science* 371:811–818.

Corballis, M. C. 2017. Language evolution: A changing perspective. *Trends in Cognitive Sciences* 21:229–236.

Cordaux, R., and C. Gilbert. 2017. Evolutionary significance of *Wolbachia*-to-animal horizontal gene transfer: Female sex determination and the *f* element in the isopod *Armadillidium vulgare. Genes* 8:186.

Cornwallis, C. K., and T. Uller. 2010. Towards an evolutionary ecology of sexual traits. *Trends in Ecology and Evolution* 25:145–152.

Corral, R., D. Winer, and E. A. Windsor. 2019. Effects of temperature and water turbulence on vertebral number and body shape in *Astyanax mexicanus* (Teleostei: Characidae). *PLOS One* 14:e0219677.

Cortijo, S., R. Wardenaar, M. Colomé-Tatché, A. Gilly, M. Etcheverry, K. Labadie, E. Caillieux, et al. 2014. Mapping the epigenetic basis of complex traits. *Science* 343:1145–1148.

Cosmides, L., and J. Tooby. 1987. From evolution to behavior: Evolutionary psychology as the missing link. In: *The Latest on the Best: Essays on Evolution and Optimality*, ed. J. Dupre, 277–306. Cambridge, MA: MIT Press.

Cossetti, C., L. Lugini, L. Astrologo, I. Saggio, S. Fais, and C. Spadafora. 2014. Soma-to-germline transmission of RNA in mice xenografted with human tumour cells: Possible transport by exosomes. *PLOS One* 9:e101629.

Cote, J., J. Clobert, T. Brodin, S. Fogarty, and A. Sih. 2010. Personality-dependent dispersal: Characterization, ontogeny and consequences for spatially structured populations. *Philosophical Transactions of the Royal Society B* 365:4065–4076.

Couzens, A.M.C., K. E. Sears, and M, Rücklin. 2021. Developmental influence on evolutionary rates and the origin of placental mammal tooth complexity. *Proceedings of the National Academy of Sciences of the USA* 118:e2019294118.

Cox, E. C., and T. C. Gibson. 1974. Selection for high mutation rates in chemostats. *Genetics* 77:169–184.

Coyne, J. A. 2006. Comment on gene regulatory networks and the evolution of animal body plans. *Science* 313:761.

Crawford, E. C., Jr., and K. Schmidt-Nielsen. 1967. Temperature regulation and evaporative cooling in the ostrich. *American Journal of Physiology–Legacy Content* 212:347–353.

Crawford, N. G., B. C. Faircloth, J. E. McCormack, R. T. Brumfield, K. Winker, and T. C. Glenn. 2012. More than 1000 ultraconserved elements provide evidence that turtles are the sister group of archosaurs. *Biology Letters* 8:783–786.

Crean, A., and D. Marshall. 2009. Coping with environmental uncertainty: Dynamic bet hedging as a maternal effect. *Philosophical Transactions of the Royal Society B* 364:1087–1096.

Creanza, N., and M. W. Feldman. 2014. Complexity in models of cultural niche construction with selection and homophily. *Proceedings of the National Academy of Sciences of the USA* 111:10830–10837.

———. 2016. Worldwide genetic and cultural change in human evolution. *Current Opinion in Genetics and Development* 41:85–92.

Creanza, N., O. Kolodny, and M. W. Feldman. 2017. Cultural evolutionary theory: How culture evolves and why it matters. *Proceedings of the National Academy of Sciences of the USA* 114:7782.

Crick, F. H. 1958. On protein synthesis. *Symposia of the Society for Experimental Biology* 12:138–163.

Crombach, A., and P. Hogeweg. 2008. Evolution of evolvability in gene regulatory networks. *PLOS Computational Biology* 4:e1000112.

Cropley, J. E., T.H.Y. Dang, D.I.K. Martin, and C. M. Suter. 2012. The penetrance of an epigenetic trait in mice is progressively yet reversibly increased by selection and environment. *Proceedings of the Royal Society B* 279:2347–2353.

Crother, B. I., and C. M. Murray. 2018. Linking a biological mechanism to evolvability. *Journal of Phylogenetics and Evolutionary Biology* 6:192.

Csibra, G., and G. Gergely. 2011. Natural pedagogy as evolutionary adaptation. *Philosophical Transactions of the Royal Society B* 366:1149–1157.

Culver, D. C., and T. Pipan. 2009. *The Biology of Caves and Other Subterranean Habitats*. Oxford: Oxford University Press.

Curio, E. 1988. Cultural transmission of enemy recognition by birds. In: *Social Learning: Psychological and Biological Perspectives*, ed. J.R.T.R. Zentall and B. G. Galef, 75–97. Hillsdale, NJ: Lawrence Erlbaum.

D'Aguillo, M., C. Hazelwood, B. Quarles, and K. Donohue. 2021. Genetic consequences of biologically altered environments. *Journal of Heredity* 113:26–36.

Dall, S.R.X., J. M. McNamara, and O. Leimar. 2015. Genes as cues: Phenotypic integration of genetic and epigenetic information from a Darwinian perspective. *Trends in Ecology and Evolution* 30:327–333.

Dalle Nogare, D., and A. B. Chitnis. 2017. A framework for understanding morphogenesis and migration of the zebrafish posterior lateral line primordium. *Mechanisms of Development* 148:69–78.

Damer, B. F. 2019. The hot spring hypothesis for the origin of life and the extended evolutionary synthesis. Extended Evolutionary Synthesis project website, May 8, 2019. http://extendedevolutionarysynthesis.com/the-hot-spring-hypothesis-for-the-origin-of-life-and-the-extended-evolutionary-synthesis/.

Damer, B. F., and D. Deamer. 2020. The hot spring hypothesis for an origin of life. *Astrobiology* 20:429–452.

Danchin, E., A. Charmantier, F. A. Champagne, A. Mesoudi, B. Pujol, and S. Blanchet. 2011. Beyond DNA: Integrating inclusive inheritance into an extended theory of evolution. *Nature Reviews Genetics* 12:475–486.

Danchin, E., S. Nöbel, A. Pocheville, A. Dagaeff, L. Demay, M. Alphand, S. Ranty-Roby, et al. 2018. Cultural flies: Conformist social learning in fruitflies predicts long-lasting mate-choice traditions. *Science* 362:1025–1030.

Danchin, E., A. Pocheville, O. Rey, B. Pujol, and S. Blanchet. 2019. Epigenetically facilitated mutational assimilation: Epigenetics as a hub within the inclusive evolutionary synthesis. *Biological Review* 94:259–282.

Darwin, C. R. (1859) 1968. *On the Origin of Species by Means of Natural Selection, or The Preservation of Favoured Races in the Struggle for Life*, reprint of 1st ed. London: Penguin Books.

———. 1868. *The Variation of Plants and Animals under Domestication*. London: John Murray.

———. (1871) 1981. *The Descent of Man and Selection in Relation to Sex*, reprint of 1st ed. Princeton, NJ: Princeton University Press.

———. 1876. *The Variation in Animals and Plants under Domestication*, 2nd ed. London: John Murray.

———. 1881. *The Formation of Vegetable Mold through the Action of Worms, with Observations on Their Habits*. London: John Murray.

Davidson, E. H., and D. H. Erwin. 2006. Gene regulatory networks and the evolution of animal body plans. *Science* 311:796–800.

Davies, A. D., Z. Lewis, and L. R. Dougherty. 2020. A meta-analysis of factors influencing the strength of mate-choice copying in animals. *Behavioral Ecology* 31:1279–1290.

Davies, N. B., and Justin A. Welbergen. 2009. Social transmission of a host defense against cuckoo parasitism. *Science* 324:1318–1320.

Dawkins, R. 1976. *The Selfish Gene.* Oxford: Oxford University Press.

———. 1982. *The Extended Phenotype.* Oxford: Oxford University Press.

Day, R. L., T. MacDonald, C. Brown, K. N. Laland, and S. M. Reader. 2001. Interactions between shoal size and conformity in guppy social foraging. *Animal Behaviour* 62:917–925.

Day, T., 2016. Interpreting phenotypic antibiotic tolerance and persister cells as evolution via epigenetic inheritance. *Molecular Ecology* 25:1869–1882.

Day, T., and R. Bonduriansky. 2011. A unified approach to the evolutionary consequences of genetic and nongenetic inheritance. *American Naturalist* 178:E18–E36.

Deacon, T. W. 1990. Rethinking mammalian brain evolution. *American Zoologist* 30:629–705.

———. 1997. *The Symbolic Species: The Coevolution of Language and the Brain.* New York: W. W. Norton.

Deamer, D. 2019. *Assembling Life.* New York: Oxford University Press.

Dearing, M. D., W. J. Foley, and S. McLean. 2005. The influence of plant secondary metabolites on the nutritional ecology of herbivorous terrestrial vertebrates. *Annual Review of Ecology, Evolution, and Systematics* 36:169–189.

Dearing, M. D., M. Kaltenpoth, and J. Gershenzon. 2022. Demonstrating the role of symbionts in mediating detoxification in herbivores. *Symbiosis* 87:59–66.

De Bakker, M.A.G., D. A. Fowler, K. den Oude, E. M. Dondorp, M.C.G. Navas, J. O. Horbanczuk, J.-Y. Sire, et al. 2013. Digit loss in archosaur evolution and the interplay between selection and constraints. *Nature* 500:445–448.

DeCasien, A. R., S. A. Williams, and J. P. Higham. 2017. Primate brain size is predicted by diet but not sociality. *Nature Ecology and Evolution* 1:0112.

Degnan, P. H., A. E. Pusey, E. V. Lonsdorf, J. Goodall, E. E. Wroblewski, M. L. Wilson, R. S. Rudicell, et al. 2012. Factors associated with the diversification of the gut microbial communities within chimpanzees from Gombe National Park. *Proceedings of the National Academy of Sciences of the USA* 109:13034–13039.

Delagnes, A., and H. Roche. 2005. Late Pliocene hominid knapping skills: The case of Lokalalei 2C, West Turkana, Kenya. *Journal of Human Evolution* 48:435–472.

Del Negro, C. A., G. D. Funk, and J. L. Feldman. 2018. Breathing matters. *Nature Reviews Neuroscience* 19:351–367.

DeMenocal, P. B. 2011. Climate and human evolution. *Science* 331:540–542.

Dempsey, M., M. Fisk, J. Yavitt, T. Fahey, and T. Balser. 2013. Exotic earthworms alter soil microbial community composition and function. *Soil Biology and Biochemistry* 67:263–270.

Dennett, D. 1995. *Darwin's Dangerous Idea: Evolution and the Meanings of Life*. London: Penguin.

Dennis, M. Y., X. Nuttle, P. H. Sudmant, F. Antonacci, T. A. Graves, M. Nefedov, J. A. Rosenfeld, et al. 2012. Evolution of human-specific neural *SRGAP2* genes by incomplete segmental duplication. *Cell* 149:912–922.

De Regt, H. W. 2017. *Understanding Scientific Understanding*. Oxford Studies in Philosophy of Science. Oxford: Oxford University Press.

Derex, M., J.-F. Bonnefon, R. Boyd, and A. Mesoudi. 2019. Causal understanding is not necessary for the improvement of culturally evolving technology. *Nature Human Behaviour* 3:446–452.

De Rooij, S. R., R. C. Painter, F. Holleman, P.M.M. Bossuyt, and T. J. Roseboom. 2007. The metabolic syndrome in adults prenatally exposed to the Dutch famine. *American Journal of Clinical Nutrition* 86:1219–1224.

De Rooij, S. R., R. C. Painter, T. J. Roseboom, D. Phillips, C. Osmond, D. Barker, M. Tanck, et al. 2006. Glucose tolerance at age 58 and the decline of glucose tolerance in comparison with age 50 in people prenatally exposed to the Dutch famine. *Diabetologia* 49:637–643.

Derryberry, E. P., J. N. Phillips, G. E. Derryberry, M. J. Blum, and D. Luther. 2020. Singing in a silent spring: Birds respond to a half-century soundscape reversion during the COVID-19 shutdown. *Science* 370:575–579.

Deshmukh, S. V., and N. N. Pathak. 1989. Voluntary intake and dry matter digestibility of green fodders and tree leaves in New Zealand white rabbits. *Cheiron* 18:223–225.

DeSilva, J. 2021. *First Steps: How Walking Upright Made Us Human*. London: Collins.

DeSilva, J., and J. Browne. 2021. *A Most Interesting Problem: What Darwin's Descent of Man Got Right and Wrong about Human Evolution*. Princeton, NJ: Princeton University Press.

Dey, S., S. R. Proulx, and H. Teotónio. 2016. Adaptation to temporally fluctuating environments by the evolution of maternal effects. *PLOS Biology* 14:e1002388.

Dias, B. G., and K. J. Ressler. 2014. Parental olfactory experience influences behavior and neural structure in subsequent generations. *Nature Neuroscience* 17:89–96.

Dickins, T. E., and B.J.A. Dickins. 2023. *Evolutionary Biology: Contemporary and Historical Reflections upon Core Theory*. Berlin: Springer.

Dickins, T. E., and Q. Rahman. 2012. The extended evolutionary synthesis and the role of soft inheritance in evolution. *Proceedings of the Royal Society B* 279: 2913–2921.

Ding, Y., J. L. Lillvis, J. Cande, G. J. Berman, B. J. Arthur, X. Long, M. Xu, et al. 2019. Neural evolution of context-dependent fly song. *Current Biology* 29:1089–1099.e7.

Dingle, E. H., M. J. Creed, H. D. Sinclair, D. Gautam, N. Gourmelen, A.G.L. Borthwick, and M. Attal. 2020. Dynamic flood topographies in the Terai region of Nepal. *Earth Surface Processes and Landforms* 45:3092–3102.

Dingle, K., G. V. Pérez, and A. A. Louis. 2020. Generic predictions of output probability based on complexities of inputs and outputs. *Scientific Reports* 10:4415.

Dingle, K., S. Schaper, and A. A. Louis. 2015. The structure of the genotype–phenotype map strongly constrains the evolution of non-coding RNA. *Interface Focus* 5:20150053.

Diogo, R. 2017. *Evolution Driven by Organismal Behavior*. Cham, Switzerland: Springer.

Diogo, R., D. Razmadze, N. Siomava, N. Douglas, J.S.M. Fuentes, and A. Duerinckx. 2019. Musculoskeletal study of cebocephalic and cyclopic lamb heads illuminates links between normal and abnormal development, evolution and human pathologies. *Scientific Reports* 9:991.

Distin, M. R. 2023. Genetic evolvability: Using a restricted pluralism to tidy up the evolvability concept. In: *Evolutionary Biology: Contemporary and Historical Reflections upon Core Theory*, ed. T. E. Dickins and B.J.A Dickins, 587–609. Berlin: Springer.

Dixon, M. L., A. De La Vega, C. Mills, J. Andrews-Hanna, R. N. Spreng, M. W. Cole, and K. Christoff. 2018. Heterogeneity within the frontoparietal control network and its relationship to the default and dorsal attention networks. *Proceedings of the National Academy of Sciences of the USA* 115:E1598–E1607.

Dobzhansky, T. 1951. *Genetics and the Origin of Species*, 3rd ed. New York: Columbia University Press.

———. 1973. Nothing in biology makes sense except in the light of evolution. *American Biology Teacher* 35:125–129.

Dobzhansky, T., F. J. Ayala, G. L. Stebbins, and J. W. Valentine. 1977. *Evolution*. San Francisco: Freeman.

Dodd, D.M.B. 1989. Reproductive isolation as a consequence of adaptive divergence in *Drosophila pseudoobscura*. *Evolution* 43:1308–1311.

Dong, X. 2022. Evolution of plant niche construction traits in biogeomorphic landscapes. *American Naturalist* 199:758–775.

Donohue, K. 2014. Why ontogeny matters during adaptation: Developmental niche construction and pleiotropy across the life cycle in *Arabidopsis thaliana*. *Evolution* 68:32–47.

Doolittle, W. F., and S. A. Inkpen. 2018. Processes and patterns of interaction as units of selection: An introduction to ITSNTS thinking. *Proceedings of the National Academy of Sciences of the USA* 115:4006–4014.

Dornhaus, A., and L. Chittka. 1999. Evolutionary origins of bee dances. *Nature* 401:38.

Dorninger, C., L. P. Menendez, and G. Caniglia. 2023. Social-ecological niche construction for sustainability: Understanding destructive processes and exploring regenerative potentials. *Philosophical Transactions of the Royal Society B* 379:20220431.

Douglas, A. E., and J. H. Werren. 2016. Holes in the hologenome: Why host-microbe symbioses are not holobionts. *mBio* 7:e02099.

Draghi, J. A., and G. P. Wagner. 2008. Evolution of evolvability in a developmental model. *Evolution* 62:301–315.

———. 2009. The evolutionary dynamics of evolvability in a gene network model. *Journal of Evolutionary Biology* 22:599–611.

Draghi, J. A., and M. C. Whitlock. 2012. Phenotypic plasticity facilitates mutational variance, genetic variance, and evolvability along the major axis of environmental variation. *Evolution* 66:2891–2902.

D'Souza, D., and A. Karmiloff-Smith. 2011. When modularization fails to occur: A developmental perspective. *Cognitive Neuropsychology* 28:276–287.

Dubin, M. J., P. Zhang, D. Meng, M. S. Remigereau, E. J. Osborne, F. Paolo Casale, P. Drewe, et al. 2015. DNA methylation in *Arabidopsis* has a genetic basis and shows evidence of local adaptation. *eLife* 4:e05255.

Duckworth, R. A. 2008. Adaptive dispersal strategies and the dynamics of a range expansion. *American Naturalist* 172:S4–S17.

———. 2009. The role of behavior in evolution: A search for mechanism. *Evolutionary Ecology* 23:513–531.

———. 2018. Reconciling the tension between behavioral change and stability. Chapter 17 in *Evolution Science and Contextual Behavioral Science: A Reintegration*, ed. D. S. Wilson and S. C. Hayes. Oakland, CA: New Harbinger.

———. 2019. Biological dynamics and evolutionary causation. In: *Evolutionary Causation, Biological and Philosophical Reflection*, ed. T. Uller and K. N. Laland, 153–171. Cambridge, MA: MIT Press.

Duckworth, R. A., V. Belloni, and S. R. Anderson. 2015. Cycles of species replacement emerge from locally induced maternal effects on offspring behavior in a passerine bird. *Science* 347:875–877.

Duckworth, R. A., A. L. Potticary, and A. V. Badyaev. 2018. On the origins of adaptive behavioral complexity: Developmental channeling of structural trade-offs. In: *Advances in the Study of Behavior*, ed. M. Naguib, L. Barrett, S. D. Healy, J. Podos, L. W. Simmons, and M. Zuk, 1–36. Oxford: Academic.

Dugatkin, L. A. 1992. Sexual selection and imitation: Females copy the mate choice of others. *American Naturalist* 139:1384–1389.

Dugatkin, L. A., and J. J. Godin. 1992. Reversal of female mate choice by copying in the guppy *Poecilia reticulata*. *Proceedings of the Royal Society B* 249:179–184.

———. 1993. Female mate copying in the guppy (*Poecilia reticulata*): Age-dependent effects. *Behavioral Ecology* 4:289–292.

Dunbar, R.I.M., and S. Shultz. 2017. Why are there so many explanations for primate brain evolution? *Philosophical Transactions of the Royal Society B* 372:20160244.

Dunbar, R.I.M., R. Baron, A. Frangou, E. Pearce, E.J.C. van Leeuwen, J. Stow, G. Partridge, et al. 2012. Social laughter is correlated with an elevated pain threshold. *Proceedings of the Royal Society B* 279:1161–1167.

Dunsworth, H. M. 2020. Expanding the evolutionary explanations for sex differences in the human skeleton. *Evolutionary Anthropology* 29:108–116.

Dunsworth, H. M., A. G. Warrener, T. Deacon, P. T. Ellison, and H. Pontzer. 2012. Metabolic hypothesis for human altriciality. *Proceedings of the National Academy of Sciences of the USA* 109:15212–15216.

Durham, W. H. 1991. *Coevolution: Genes, Culture and Human Diversity*. Stanford, CA: Stanford University Press.

Dury, G. J., and M. J. Wade. 2020. When mother knows best: A population genetic model of transgenerational versus intragenerational plasticity. *Journal of Evolutionary Biology* 33:127–137.

Dury, G. J., A. P. Moczek, and D. B. Schwab. 2020. Maternal and larval niche construction interact to shape development, survival, and population divergence in the dung beetle *Onthophagus taurus*. *Evolution and Development* 22:358–369.

Dutton, C. L., A. L. Subalusky, A. Sanchez, S. Estrela, N. Lu, S. K. Hamilton, L. Njoroge, et al. 2021. The meta-gut: Community coalescence of animal gut and environmental microbiomes. *Scientific Reports* 11:23117.

Dyall, S. D., M. T. Brown, and P. J. Johnson. 2004. Ancient invasions: From endosymbionts to organelles. *Science* 304:253–257.

Dyble, M., G. D. Salali, N. Chaudhary, A. Page, D. Smith, J. Thompson, L. Vinicius, et al. 2015. Sex equality can explain the unique social structure of hunter-gatherer bands. *Science* 348:796–798.

Echavarri, F. 2020. 2020 was the deadliest year on record for migrants crossing the Arizona Desert. *MotherJones*, December 22, 2020. https://www.motherjones.com/politics/2020/12/record-deaths-migrants-arizona-desert/.

Edelaar, P., and D. I. Bolnick. 2019. Appreciating the multiple processes increasing individual or population fitness. *Trends in Ecology and Evolution* 34:435–446.

Edelaar, P., R. Jovani, and I. Gomez-Mestre. 2017. Should I change or should I go? Phenotypic plasticity and matching habitat choice in the adaptation to environmental heterogeneity. *American Naturalist* 190:506–520.

Edelaar, P., J. Otsuka, and V. J. Luque. 2023. A generalised approach to the study and understanding of adaptive evolution. *Biological Reviews* 98:352–375.

Edelman, G. M. 1987. *Neural Darwinism: The Theory of Neuronal Group Selection*. New York: Basic Books.

Egan, A.J.F., J. Errington, and W. Vollmer. 2020. Regulation of peptidoglycan synthesis and remodelling. *Nature Reviews Microbiology* 18:446–460.

Ehrenreich, I. M., and D. W. Pfennig. 2016. Genetic assimilation: A review of its potential proximate causes and evolutionary consequences. *Annals of Botany* 117:769–779.

Ehrlich, M., K. F. Norris, R. Y. Wang, K. C. Kuo, and C. W. Gehrke. 1986. DNA cytosine methylation and heat-induced deamination. *Bioscience Reports* 6:387–393.

Eldredge, N. 1985. *Unfinished Synthesis: Biological Hierarchies and Modern Evolutionary Thought*. Oxford: Oxford University Press.

Elena, S. F., and R. E. Lenski. 2003. Evolution experiments with microorganisms: The dynamics and genetic bases of adaptation. *Nature Reviews Genetics* 4:457–469.

Ellis, E. C. 2015. Ecology in an anthropogenic biosphere. *Ecological Monographs* 85:287–331.

———. 2018. *Anthropocene: A Very Short Introduction*. Oxford: Oxford University Press.

Emery, N. J. 2016. *Bird Brain: An Exploration of Avian Intelligence*. Lewes, UK: Ivy.

Emery, N. J., and N. S. Clayton. 2004. The mentality of crows: Convergent evolution of intelligence in corvids and apes. *Science* 306:1903–1907.

Emlen, D. J., J. Hunt, and L. W. Simmons. 2005. Evolution of sexual dimorphism and male dimorphism in the expression of beetle horns: Phylogenetic evidence for modularity, evolutionary lability, and constraint. *American Naturalist* 166:S42–S68.

ENCODE Project Consortium. 2012. An integrated encyclopedia of DNA elements in the human genome. *Nature*. 489:57–74.

Endler, J. 1986. *Natural Selection in the Wild*. Monographs in Population Biology 21. Princeton, NJ: Princeton University Press.

English, S., I. Pen, N. Shea, and T. Uller. 2015. The information value of non-genetic inheritance in plants and animals. *PLOS One* 10:e0116996.

Enquist, M., K. Eriksson, and S. Ghirlanda. 2007. Critical social learning: A solution to Rogers's paradox of nonadaptive culture. *American Anthropologist* 109:727–734.

Eriksson, J. G., T. Forsén, J. Tuomilehto, C. Osmond, and D.J.P. Barker. 2001. Early growth and coronary heart disease in later life: Longitudinal study. *BMJ* 322:949.

Erwin, D. H. 2005. Seeds of diversity. *Science* 308:1752–1753.

———. 2008. Macroevolution of ecosystem engineering, niche construction and diversity. *Trends in Ecology and Evolution* 23:304–310.

———. 2017. The topology of evolutionary novelty and innovation in macroevolution. *Philosophical Transactions of the Royal Society B* 372:20160422.

———. 2020. The origin of animal body plans: A view from fossil evidence and the regulatory genome. *Development* 147(4):dev182899.

Erwin, D. H., M. Laflamme, S. M. Tweedt, E. A. Sperling, D. Pisani, and K. J. Peterson. 2011. The Cambrian conundrum: Early divergence and later ecological success in the early history of animals. *Science* 334:1091–1097.

Espinas, N. A., H. Saze, and Y. Saijo. 2016. Epigenetic control of defense signaling and priming in plants. *Frontiers in Plant Science* 7. https://doi.org/10.3389/fpls.2016.01201.

Espinosa-Soto, C., O. C. Martin, and A. Wagner. 2011. Phenotypic robustness can increase phenotypic variability after nongenetic perturbations in gene regulatory circuits. *Journal of Evolutionary Biology* 24:1284–1297.

Essock-Burns, T., B. D. Bennett, D. Arencibia, S. Moriano-Gutierrez, M. Medeiros, M. J. McFall-Ngai, and E. G. Ruby. 2021. Bacterial quorum-sensing regulation induces morphological change in a key host tissue during the *Euprymna scolopes–Vibrio fischeri* symbiosis. *mBio* 12:e0240221.

Evans, A. R., E. S. Daly, K. K. Catlett, K. S. Paul, S. J. King, M. M. Skinner, H. P. Nesse, et al. 2016. A simple rule governs the evolution and development of hominin tooth size. *Nature* 530:477–480.

Everaert, M.B.H., M.A.C. Huybregts, N. Chomsky, R. C. Berwick, and J. J. Bolhuis. 2015. Structures, not strings: Linguistics as part of the cognitive sciences. *Trends in Cognitive Sciences* 19:729–743.

Evershed, R. P., G. D. Smith, M. Roffet-Salque, A. Timpson, Y. Diekmann, M. S. Lyon, L. J. E. Cramp, et al. 2022. Dairying, diseases and the evolution of lactase persistence in Europe. *Nature* 608:336–345.

Eyal, S., E. Blitz, Y. Shwartz, H. Akiyama, R. Schweitzer, and E. Zelzer. 2015. On the development of the patella. *Development* 142:1831–1839.

Ezenwa, V. O., L. S. Ekernas, and S. Creel. 2012. Unravelling complex associations between testosterone and parasite infection in the wild. *Functional Ecology* 26:123–133.

Fabris, F. 2021. Conrad Hal Waddington (1905–1975). In: *Evolutionary Developmental Biology: A Reference Guide*, ed. L. Nuno de la Rosa and G. B. Müller. Cham, Switzerland: Springer. https://doi.org/10.1007/978-3-319-33038-9_30-1.

Falconer, D. S., and T.F.C. Mackay. 1996. *Introduction to Quantitative Genetics*. Harlow, UK: Prentice Hall.

Falk, D. 2004. Prelinguistic evolution in early hominins: Whence motherese? *Behavioral and Brain Sciences* 27:491–503.

Favé, M. J., R. A. Johnson, S. Cover, S. Handschuh, B. D. Metscher, G. B. Müller, and S. Gopalan. 2015. Past climate change on Sky Islands drives novelty in a core developmental gene network and its phenotype. *BMC Evolutionary Biology* 15:183.

Feiner, N., M. Brun-Usan, and T. Uller. 2021. Evolvability and evolutionary rescue. *Evolution and Development* 23:308–319.

Feiner, N., I.S.C. Jackson, K. L. Munch, R. Radersma, and T. Uller. 2020. Plasticity and evolutionary convergence in the locomotor skeleton of greater antillean anolis lizards. *eLife* 9:e57468.

Feiner, N., R. Radersma, L. Vasquez, M. Ringnér, B. Nystedt, A. Raine, E. W. Tobi, et al. 2022. Environmentally induced DNA methylation is inherited across generations in an aquatic keystone species. *iScience* 25(5):104303.

Feldman, M. W., and L. L. Cavalli-Sforza. 1976. Cultural and biological evolutionary processes, selection for a trait under complex transmission. *Theoretical Population Biology* 9:238–259.

———. 1981. Further remarks on Darwinian selection and "altruism." *Theoretical Population Biology* 19:251–260.

———. 1989. On the theory of evolution under genetic and cultural transmission, with application to the lactose absorption problem. In: *Mathematical Evolutionary Theory*, ed. Feldman, 145–173. Princeton, NJ: Princeton University Press.

Feldman, M. W., and K. N. Laland. 1996. Gene-culture coevolutionary theory. *Trends in Ecology and Evolution* 11:453–457.

Feldman, M. W., and U. Liberman. 1986. An evolutionary reduction principle for genetic modifiers. *Proceedings of the National Academy of Sciences of the USA* 83:4824–4827.

Feldman, M. W., and S. Ramachandran. 2018. Missing compared to what? Revisiting heritability, genes and culture. *Philosophical Transactions of the Royal Society B* 373:20170064.

Feldman, M. W., and J. Riskin. 2022. Why biology is not destiny. Review of *The Genetic Lottery: Why DNA Matters for Social Equality*, ed. K. P. Harden (Princeton University Press). *New York Review of Books*, April 21, 2022.

Felice, R. N., M. Randau, and A. Goswami. 2018. A fly in a tube: Macroevolutionary expectations for integrated phenotypes. *Evolution* 72:2580–2594.

Feller, A. F., M. P. Haesler, C. L. Peichel, and O. Seehausen. 2020. Genetic architecture of a key reproductive isolation trait differs between sympatric and non-sympatric sister species of Lake Victoria cichlids. *Proceedings of the Royal Society B* 287:20200270.

Felsenstein, J. 2004. *Inferring Phylogenies*. Sunderland, MA: Sinauer Associates.

Fernandes, H.B.F., M. A. Woodley, and J. te Nijenhuis. 2014. Differences in cognitive abilities among primates are concentrated on G: Phenotypic and phylogenetic comparisons with two meta-analytical databases. *Intelligence* 46:311–322.

Field, D. J., J. A. Gauthier, B. L. King, D. Pisani, T. R. Lyson, and K. J. Peterson. 2014. Toward consilience in reptile phylogeny: miRNAs support an archosaur, not lepidosaur, affinity for turtles. *Evolution and Development* 16:189–196.

Fierst, J. L. 2011. A history of phenotypic plasticity accelerates adaptation to a new environment. *Journal of Evolutionary Biology* 24:1992–2001.

Finlay, B. L., and R. B. Darlington. 1995. Linked regularities in the development and evolution of mammalian brains. *Science* 268:1578–1584.

Finlay, B. L., and R. Uchiyama. 2015. Developmental mechanisms channeling cortical evolution. *Trends in Neurosciences* 38:69–76.

———. 2017. The timing of brain maturation, early experience, and the human social niche. In: *Evolution of Nervous Systems*, vol. 3, ed. J. H. Kaas, 123–148. Amsterdam: Elsevier.

Fischer, B., and P. Mitteroecker. 2015. Covariation between human pelvis shape, stature, and head size alleviates the obstetric dilemma. *Proceedings of the National Academy of Sciences of the USA* 112:5655–5660.

Fischer, B., N.D.S. Grunstra, E. Zaffarini, and P. Mitteroecker. 2021. Sex differences in the pelvis did not evolve de novo in modern humans. *Nature Ecology and Evolution* 5:625–630.

Fischer. J. J., G. Mikhael, J. B. Tenenbaum, and N. Kanwisher. 2016. Functional neuroanatomy of intuitive physical inference. *Proceedings of the National Academy of Sciences of the USA* 113:E5072–E5081.

Fisher, J. B., and R. A. Hinde. 1949. Opening of milk bottles by birds. *British Birds* 42:347–357.

Fisher, R. A. 1918. The correlation between relatives on the supposition of Mendelian inheritance. *Transaction of the Royal Society of Edinburgh* 52:399–433.

———. 1930. *The Genetical Theory of Natural Selection*. Oxford: Clarendon.

———. 1958. *The Genetical Theory of Natural Selection*, 2nd ed. New York: Dover.

Fitch, W. T. 2004. Kin selection and "mother tongues": A neglected component in language evolution. In: *Evolution of Communication Systems: A Comparative Approach*, ed. D. K. Oller and U. Griebel, 275–296. Cambridge, MA: MIT Press.

———. 2005. The evolution of language: A comparative review. *Biology and Philosophy* 20:193–230.

———. 2006. The biology and evolution of music: A comparative perspective. *Cognition* 100:173–215.

———. 2010. *The Evolution of Language*. Cambridge: Cambridge University Press.

———. 2013. Rhythmic cognition in humans and animals: Distinguishing meter and pulse perception. *Frontiers Systems Neuroscience* 7:1–16.

Flack, J. C., M. Girvan, F.B.M. de Waal, and D. C. Krakauer. 2006. Policing stabilizes construction of social niches in primates. *Nature* 439:426–429.

Flint, H. J., E. A. Bayer, M. T. Rincon, R. Lamed, and B. A. White. 2008. Polysaccharide utilization by gut bacteria: Potential for new insights from genomic analysis. *Nature Reviews Microbiology* 6:121–131.

Flohr, R.C.E., C. J. Blom, P. B. Rainey, and H.J.E. Beaumont. 2013. Founder niche constrains evolutionary adaptive radiation. *Proceedings of the National Academy of Sciences of the USA* 110:20663–20668.

Fogarty, L., and M. J. Wade. 2022. Niche construction in quantitative traits: Heritability and response to selection. *Proceedings of the Royal Society B* 289:20220401.

Fontana, W. 2002. Modelling "evo-devo" with RNA. *BioEssays* 24:1164–1177.

Foote, A. D., N. Vijay, M. C. Avila-Arcos, R. W. Baird, J. W. Durban, M. Fumagalli, R. A. Gibbs, et al. 2016. Genome-culture coevolution promotes rapid divergence of killer whale ecotypes. *Nature Communications* 7:11693.

Forgacs, G., and S. A. Newman. 2005. *Biological Physics of the Developing Embryo*. Cambridge: Cambridge University Press.

Forsman, J. T., and J. T. Seppänen. 2011. Learning what (not) to do: Testing rejection and copying of simulated heterospecific behavioural traits. *Animal Behaviour* 81:879–883.

Forss, S.I.F., E. Willems, J. Call, and C. P. van Schaik. 2016. Cognitive differences between orang-utan species: A test of the cultural intelligence hypothesis. *Scientific Reports* 6:30516.

Foust, C. M., V. Preite, A. W. Schrey, M. Alvarez, M. H. Robertson, K.J.F. Verhoeven, and C. L. Richards. 2016. Genetic and epigenetic differences associated with environmental gradients in replicate populations of two salt marsh perennials. *Molecular Ecology* 25:1639–1652.

Fragaszy, D. M. 2011. Community resources for learning: How capuchin monkeys construct technical traditions. *Biological Theory* 6:231–240.

Fragaszy, D. M., and S. Perry. 2003. *The Biology of Traditions: Models and Evidence*. Cambridge: Cambridge University Press.

Francis, R. C. 2011. *Epigenetics: How Environment Shapes Our Genes*. New York: W. W. Norton.

Frank, S. A. 2009. Natural selection maximizes Fisher information. *Journal of Evolutionary Biology* 22: 231–244.

———. 2011. Natural selection, II: Developmental variability and evolutionary rate. *Journal of Evolutionary Biology* 24:2310–2320.

Franzenburg, S., J. Walter, S. Künzel, J. Wang, J. F. Baines, T.C.G. Bosch, and S. Fraune. 2013. Distinct antimicrobial peptide expression determines host species-specific bacterial associations. *Proceedings of the National Academy of Sciences of the USA* 110:E3730–E3738.

Franz-Odendaal, T. A., and B. K. Hall. 2006a. Modularity and sense organs in the blind cavefish, *Astyanax mexicanus*. *Evolution and Development* 8:94–100.

———. 2006b. Skeletal elements within teleost eyes and a discussion of their homology. *Journal of Morphology* 267:1326–1337.

Friston K., D. A. Friedman, A. Constant, V. Bleu Knight., C. Fields, T. Parr, and J. O. Campbell. 2023. A variational synthesis of evolutionary and developmental dynamics. *Entropy* 25:964.

Fritz, J. A., J. Brancale, M. Tokita, K. J. Burns, M. B. Hawkins, A. Abzhanov, and M. P. Brenner. 2014. Shared developmental programme strongly constrains beak shape diversity in songbirds. *Nature Communications* 5:3700.

Fu, S., Y. Wang, E. Bin, H. Huang, F. Wang, and N. Tang. 2023. c-JUN-mediated transcriptional responses in lymphatic endothelial cells are required for lung fluid clearance at birth. *Proceedings of the National Academy of Sciences of the USA* 120:e2215449120.

Fu, Z. Q., and X. Dong. 2013. Systemic acquired resistance: Turning local infection into global defense. *Annual Review of Plant Biology* 64:839–863.

Fuentes, A. 2009. *Evolution of Human Behavior*. Oxford: Oxford University Press.

————. 2017. *The Creative Spark*. New York: Penguin.

Fuentes, A., M. Kissel, P. Spikins, K. Molopyane, J. Hawks, and L. R. Berger. 2023. Burials and engravings in a small-brained hominin, *Homo naledi*, from the late Pleistocene: Contexts and evolutionary implications. Preprint. BioRxiv. https://doi.org/10.1101/2023.06.01 .543135.

Funkhouser, L. J., and S. R. Bordenstein. 2013. Mom knows best: The universality of maternal microbial transmission. *PLoS Biology* 11:e1001631.

Furness, A. I., K. Lee, and D. N. Reznick. 2015. Adaptation in a variable environment: Phenotypic plasticity and bet-hedging during egg diapause and hatching in an annual killifish. *Evolution* 69:1461–1475.

Furrow, R. E., and M. W. Feldman. 2014. Genetic variation and the evolution of epigenetic regulation. *Evolution* 68:673–683.

Futuyma, D. J. 1998. *Evolutionary Biology*. Sunderland, MA: Sinauer Associates.

————. 2013 *Evolution*. Sunderland, MA: Sinauer Associates.

————. 2015. Can modern evolutionary theory explain macroevolution? In: *Macroevolution: Interdisciplinary Evolution Research*, vol. 2, ed. E. Serrelli and N. Gontier, 29–85. Cham, Switzerland: Springer.

————. 2017. Evolutionary biology today and the call for an extended synthesis. *Interface Focus* 7:20160145.

Gabora, L. 2006. The fate of evolutionary archaeology: Survival or extinction? *World Archaeology* 38:690–696.

Galef, B. G., Jr. 1988. Imitation in animals: History, definition, and interpretation of data from the psychological laboratory. In: *Social learning: Psychological and Biological Perspectives*, ed. B. G. Galef Jr and T. R. Zentall, 3–28. Hillsdale, NJ: Erlbaum.

————. 1995. Why behavior patterns that animals learn socially are locally adaptive. *Animal Behavior* 49: 1325–1334.

————. 2003. Traditional behaviours of brown and black rats. In: *The Biology of Traditions: Models and Evidence*, ed. S. Perry and D. Fragaszy, 159–186. Chicago: University of Chicago Press.

————. 2009. Culture in animals? In: *The Question of Animal Culture*, ed. K. N. Laland and B. G. Galef, 222–246. Cambridge, MA: Harvard University Press.

Galef, B. G., Jr., and M. Beck. 1985. Aversive and attractive marking of toxic and safe foods by Norway rats. *Behavioral and Neural Biology* 43:298–310.

Galef, B. G., Jr., and L. L Buckley. 1996. Use of foraging trails by Norway rats. *Animal Behaviour* 51:765–771.

Galef, B. G., Jr., and M. M. Clark. 1971a. Parent-offspring interactions determine time and place of first ingestion of solid food by wild rats. *Psychonomic Science* 25:15–16.

————. 1971b. Social factors in the poison avoidance and feeding behavior of wild and domesticated rat pups. *Journal of Comparative Physiology and Psychology* 78:341–357.

Galef, B. G., Jr., and L Heiber. 1976. Role of residual olfactory cues in the determination of feeding site selection and exploration patterns of domestic rats. *Journal of Comparative and Physiological Psychology* 90:727–739.

Galis, F. 1999. Why do almost all mammals have seven cervical vertebrae? Developmental constraints, Hox genes, and cancer. *Journal of Experimental Zoology* 285:19–26.

Galis, F., T. J. van Dooren, J. D. Feuth, J. A. Metz, A. Witkam, S. Ruinard, M. J. Steigenga, et al. 2006. Extreme selection in humans against homeotic transformations of cervical vertebrae. *Evolution* 60:2643–2654.

Gardner, R. A., and B. T. Gardner. 1969. Teaching sign language to a chimpanzee. *Science* 165:664–672.

Gehring, W. J. 2005. New perspectives on eye development and the evolution of eyes and photoreceptors. *Journal of Heredity* 96:171–184.

Geoghegan, J. L., and H. G. Spencer. 2012. Population-epigenetic models of selection. *Theoretical Population Biology* 81:232–242.

———. 2013. The adaptive invasion of epialleles in a heterogeneous environment. *Theoretical Population Biology* 88:1–8.

Gerbault, P., A. Liebert, Y. Itan, A. Powell, M. Currat, J. Burger, D. M. Swallow, et al. 2011. Evolution of lactase persistence: An example of human niche construction. *Philosophical Transactions of the Royal Society B* 366:863–877.

Gergely, G., and G. Csibra. 2005. The social construction of the cultural mind: Imitative learning as a mechanism of human pedagogy. *Interaction Studies* 6:463–481.

Gergely, G., K. Egyed, and I. Kiraly. 2007. On pedagogy. *Developmental Science* 10:139–146.

Gerhart, J. C., and M. W. Kirschner. 1997. *Cells, Embryos and Evolution*. Hoboken, NJ: Wiley.

———. 2007. The theory of facilitated variation. *Proceedings of the National Academy of Sciences of the USA* 104:8582–8589.

Gettler, L. T., T. W. McDade, A. B. Feranil, and C. W. Kuzawa. 2012. Prolactin, fatherhood, and reproductive behavior in human males. *American Journal of Physical Anthropology* 148:362–370.

Ghalambor, C. K., K. L. Hoke, E. W. Ruell, E. K. Fischer, D. N. Reznick, and K. A. Hughes. 2015. Non-adaptive plasticity potentiates rapid adaptive evolution of gene expression in nature. *Nature* 525:372–375.

Ghalambor, C. K., J. K. McKay, S. P. Carroll, and D. N. Reznick. 2007. Adaptive versus non-adaptive phenotypic plasticity and the potential for contemporary adaptation in new environments. *Functional Ecology* 21:394–407.

Gibson, R. M., J. W. Bradbury, and S. L. Vehrencamp. 1991. Mate choice in lekking sage grouse revisited: The roles of vocal display, female site fidelity, and copying. *Behavioral Ecology* 2:165–180.

Gilbert, C., Y. Le Maho, M. Perret, and A. Ancel. 2007. Body temperature changes induced by huddling in breeding male emperor penguins. *American Journal of Physiology–Regulatory, Integrative and Comparative Physiology* 292:R176–R185.

Gilbert, S. F. 1978. The embryological origins of the gene theory. *Journal of the History of Biology* 11:307–351.

———. 1991. Cytoplasmic action in development. *Quarterly Review of Biology* 66:309–316.

———. 1998. Bearing crosses: The historiography of genetics and embryology. *American Journal of Medical Genetics* 76:168–182.

———. 2001. Ecological developmental biology: Developmental biology meets the real world. *Developmental Biology* 233:1–12.

———. 2003a. The morphogenesis of evolutionary developmental biology. *International Journal of Developmental Biology* 47:467–477.

———. 2003b. Opening Darwin's black box: Teaching evolution through developmental genetics. *Nature Reviews Genetics* 4:735–741.

———. 2011a. Commentary: "The Epigenotype" by C. H. Waddington. *International Journal of Epidemiology* 41:20–23.

———. 2011b. The decline of soft inheritance. In: *Transformations of Lamarckism: From Subtle Fluids to Molecular Biology*, ed. E. Jablonka and S. Gissis, 121–125. Cambridge, MA: MIT Press.

———. 2018. Achilles and the tortoise: Some caveats to mathematical modeling in biology. *Progress in Biophysics and Molecular Biology* 137:37–45.

———. 2019. Evolutionary transitions revisited: Holobiont evo-devo. *Journal of Experimental Zoology Part B* 332:307–314.

———. 2020a. Developmental symbiosis facilitates the multiple origins of herbivory. *Evolution and Development* 22:154–164.

———. 2020b. *Holobionts Can Evolve by Changing Their Symbionts and Hosts*. Feral Atlas Field Reports. Stanford, CA: Stanford University Press.

Gilbert, S. F., and J. Bard. 2014. Formalizing theories of development: A fugue on the orderliness of nature. In: *Towards a Theory of Development*, ed. A. Minelli and T. Pradeu, 129–143. Oxford: Oxford University Press.

Gilbert, S. F., and D. Epel. 2009. *Ecological Developmental Biology: Integrating Epigenetics, Medicine, and Evolution*. Sunderland, MA: Sinauer Associates.

———. 2015. *Ecological Developmental Biology*, 2nd ed. Sunderland, MA: Sinauer Associates.

Gilbert, S. F., and S. Sarkar. 2000. Embracing complexity: Organicism for the twenty-first century. *Developmental Dynamics* 219:1–9.

Gilbert, S. F., T. C. Bosch, and C. Ledon-Rettig. 2015. Eco-evo-devo: Developmental symbiosis and developmental plasticity as evolutionary agents. *Nature Reviews Genetics* 16:611–622.

Gilbert, S. F., J. M. Opitz, and R. A. Raff. 1996. Resynthesizing evolutionary and developmental biology. *Development Biology* 173:357–372.

Gilbert, S. F., J. Sapp, and A. I. Tauber. 2012. A symbiotic view of life: We have never been individuals. *Quarterly Review of Biology* 87:325–341.

Gintis, H. 2009. *Game Theory Evolving: A Problem-Centered Introduction to Modelling Strategic Interaction*, 2nd ed. Princeton, NJ: Princeton University Press.

Gleeson, B. T., and L.A.B Wilson. 2023. Shared reproductive disruption, not neural crest or tameness, explains the domestication syndrome. *Proceedings of the Royal Society B* 290:20222464.

Glover, J. D., Z. R. Sudderick, B. Bo-Ju Shih, C. Batho-Samblas, L. Charlton, A. L. Krause, C. Anderson, et al. 2023. The developmental basis of fingerprint pattern formation and variation. *Cell* 186:940–956.e20.

Gluckman, P. D., and M. A. Hanson. 2006. The conceptual basis for the developmental origins of health and disease. In *Developmental Origins of Health and Disease*, ed. P. D. Gluckman and M. A. Hanson, 33–50. Cambridge: Cambridge University Press.

———. 2007. *Mismatch: Why Our World No Longer Fits Our Bodies*. Oxford: Oxford University Press.

Godfrey-Smith, P. 1996. *Complexity and the Function of Mind in Nature*. Cambridge: Cambridge University Press.

———. 2009. *Darwinian Populations and Natural Selection*. New York: Oxford University Press.

Godin J., E. Herman, and L. A. Dugatkin. 2005. Social influences on female mate choice in the guppy, *Poecilia reticulata*: Generalized and repeatable trait-copying behaviour. *Animal Behaviour* 69:999–1005.

Goldschmitdt, R. B. 1940. *The Material Basis of Evolution*. New Haven, CT: Yale University Press.

González-Forero, M. 2023. How development affects evolution. *Evolution* 77:562–579.

———. 2024. A mathematical framework for evo-devo dynamics. *Theoretical Population Biology* 155:24–50.

González-Forero, M. In Press. Evo-devo dynamics of hominin brain size. *Nature Human Behaviour*. https://www.biorxiv.org/content/10.1101/2023.03.20.533421v3.

González-Forero, M., and A. Gardner. 2018. Inference of ecological and social drivers of human brain-size evolution. *Nature* 557:554–557.

Goodall, J. 1986. *The Chimpanzees of Gombe: Patterns of Behavior*. Cambridge, MA: Harvard University Press.

Goodwin, B. 2001. *How the Leopard Changed Its Spots: The Evolution of Complexity*. Princeton, NJ: Princeton University Press.

Gopnik, A., and A. N. Meltzoff. 1997. *Words, Thoughts, and Theories*. Cambridge, MA: MIT Press.

Gopnik, A., S. O'Grady, C. G. Lucas, T. L. Griffiths, A. Wente, S. Bridgers, R. Aboody, et al. 2017. Changes in cognitive flexibility and hypothesis search across human life history from childhood to adolescence to adulthood. *Proceedings of the National Academy of Sciences of the USA* 114:7892–7899.

Gordon, D. M. 2023. *The Ecology of Collective Behavior*. Princeton, NJ: Princeton University Press.

Gore, A. V., K. A. Tomins, J. Iben, L. Ma, D. Castranova, A. E. Davis, A. Parkhurst, et, al. 2018. An epigenetic mechanism for cavefish eye degeneration. *Nature Ecology and Evolution* 2:1155–1160.

Gorelick, R., and S. M. Bertram. 2003. Maintaining heritable variation via sex-limited temporally fluctuating selection: A phenotypic model accommodating non-Mendelian epigenetic effects. *Theory in Biosciences* 122:321–338.

Goriely, A., and A.O.M. Wilkie. 2012. Paternal age effect mutations and selfish spermatogonial selection: Causes and consequences for human disease. *American Journal of Human Genetics* 90:175–200.

Gottlieb, G. 2002. *Individual Development and Evolution*. New York:Taylor and Francis.

Gottschling, D. E., O. M. Aparicio, B. L. Billington, and V. A. Zakian. 1990. Position effect at *S.cerevisiae* telomeres: Reversible repression of pol II transcription. *Cell* 63:751–762.

Gould, P. D., M. Domijan, M. Greenwood, I. T. Tokuda, H. Rees, L. Kozma-Bognar, A.J.W. Hall, et al. 2018. Coordination of robust single cell rhythms in the *Arabidopsis* circadian clock via spatial waves of gene expression. *eLife* 7:e31700.

Gould, S. J. 1977. *Ontogeny and Phylogeny*. Cambridge, MA: Belknap Press of Harvard University Press.

———. 1980a. G. G. Simpson, paleontology and the modern synthesis. In: *The evolutionary Synthesis*, ed. E. Mayr and W. Provine, 153–172. Cambridge, MA: Harvard University Press.

———. 1980b. Is a new and general theory of evolution emerging? *Paleobiology* 6:119–130.

———. 1982. Foreword to *Genetics and the Origin of Species* by T. Dobzhansky, ed. N. Eldredge and S. Gould, xvii–xli. Columbia Classics in Evolution Series. New York: Columbia University Press.

———. 1983a. Darwinism and the expansion of evolutionary thought. *Science* 216:380–387.

———. 1983b. The hardening of the modern synthesis. In: *Dimensions of Darwinism*, ed. M. Greene, 71–93. Cambridge: Cambridge University Press.

———. 1983c. Unorthodoxies in the first formulation of natural selection. *Evolution* 37:856–858.

———. 1989. *Wonderful Life: The Burgess Shale and the Nature of History*. New York: W. W. Norton.

———. 1991. Exaptation: A crucial tool for an evolutionary psychology. *Journal of Social Issues* 47:43–65.

———. 2002. *The Structure of Evolutionary Theory*. Cambridge, MA: Belknap Press of Harvard University Press.

Gould, S. J., and R. Lewontin. 1979. The spandrels of San Marco and the Panglossian paradigm: A critique of the adaptationist programme. *Proceedings of the Royal Society B* 205:581–598.

Gould, S. J., and E. Vrba. 1982. Exaptation: A missing term in the science of form. *Palaeobiology* 8:4–15.

Graeve, A., I. Ioannidou, J. Reinhard, D. M. Görl, A. Faissner, and L. C. Weiss. 2021a. Brain volume increase and neuronal plasticity underlie predator-induced morphological defense expression in *Daphnia longicephala*. *Scientific Reports* 11:12612.

Graeve, A., M. Janßen, M. Villalba de la Pena, R. Tollrian, and L. C. Weiss. 2021b. Higher, faster, better: Maternal effects shorten time lags and increase morphological defenses in *Daphnia lumholtzi* offspring generations. *Frontiers in Ecology and Evolution* 9. https://doi.org/10.3389/fevo.2021.637421.

Grant, P. R., and B. R. Grant. 2002. Unpredictable evolution in a 30-year study of Darwin's finches. *Science* 296:707–711.

Green, R. C., and G. J. Annas. 2008. The genetic privacy of presidential candidates. *New England Journal of Medicine* 359:2192–2193.

Greer, A. E. 1987. Limb reduction in the lizard genus *Lerista*, 1: Variation in the number of phalanges and presacral vertebrae. *Journal of Herpetology* 21:267–276.

———. 1990. Limb reduction in the scincid lizard genus *Lerista*, 2: Variation in the bone complements of the front and rear limbs and the number of postsacral vertebrae. *Journal of Herpetology* 24:142–150.

Griesemer, J. 2021. Levels, perspectives and thickets: Toward an ontology of complex scaffolded living systems. In: *Levels of Organization in the Biological Sciences*, ed. D. S. Brooks, J. DiFrisco, and W. C. Wimsatt, 89–110. Cambridge, MA: MIT Press.

Griffiths, P. E., and R. D. Gray. 1994. Developmental systems and evolutionary explanation. *Journal of Philosophy* 91:277–304.

Griffiths, P. E., and K. Stotz. 2013. *Genetics and Philosophy: An Introduction*. Cambridge: Cambridge University Press.

Gromko, M. H. 1995. Unpredictability of correlated response to selection: Pleiotropy and sampling interact. *Evolution* 49:685–693.

Groothuis, T.G.G., and H. Schwabl. 2008. Hormone-mediated maternal effects in birds: Mechanisms matter but what do we know of them? *Philosophical Transactions of the Royal Society B* 363:1647–1661.

Gross, J. B. 2012. The complex origin of astyanax cavefish. *BMC Evolutionary Biology* 12:105.

Groussin, M., F. Mazel, J. G. Sanders, C. S. Smillie, S. Lavergne, W. Thuiller, and E. J. Alm. 2017. Unraveling the processes shaping mammalian gut microbiomes over evolutionary time. *Nature Communications* 8:14319.

Grun, P. 1976. *Cytoplasmic Inheritance and Evolution*. New York: Columbia University Press.

Grunstra, N.D.S., F. E. Zachos, A. N. Herdina, B. Fischer, M. Pavličev, and P. Mitteroecker. 2019. Humans as inverted bats: A comparative approach to the obstetric conundrum. *American Journal of Human Biology* 31:e23227.

Guillaume, F., and M. C. Whitlock. 2007. Effects of migration on the genetic covariance matrix. *Evolution* 61:2398–2409.

Gunst, N., S. Boinski, and D. Fragaszy. 2008. Acquisition of foraging competence in wild brown capuchins (*Cebus apella*), with special reference to conspecifics' foraging artefacts as an indirect social influence. *Behaviour* 145:195.

———. 2010. Development of skilled detection and extraction of embedded prey by wild brown capuchin monkeys (*Cebus apella apella*). *Journal of Comparative Psychology* 124:194–204.

Haas, R., J. Watson, T. Buonasera, J. Southon, J. C. Chen, S. Noe, K. Smith, et al. 2020. Female hunters of the early Americas. *Science Advances* 6:eabd0310.

Haig, D. 2013. Proximate and ultimate causes: How come? And what for? *Biology and Philosophy* 28:781–786.

Hajheidari, M., C. Koncz, and M. Bucher. 2019. Chromatin evolution: Key innovations underpinning morphological complexity. *Frontiers in Plant Science* 10:454.

Halder, G., P. Callaerts, and W. J. Gehring. 1995. Induction of ectopic eyes by targeted expression of the eyeless gene in *Drosophila*. *Science* 267:1788–1792.

Hall, B. K. 1999. *Evolutionary Developmental Biology*. Berlin: Springer.

———. 2007. *Fins into limbs: Evolution, Development, Transformation*. Chicago: University of Chicago Press.

———. 2015. *Bones and Cartilage: Developmental and Evolutionary Skeletal Biology*, 2nd ed. San Diego, CA: Academic.

Hall, B. K., and B. Hallgrímsson. 2007. *Strickberger's Evolution*, 4th ed. Jones and Bartlett.

Hallgrímsson, B., J. D. Aponte, M. Vidal-Garcíal, H. Richbourg, R. Green, N. M. Young, J. M. Cheverud, et al. 2023. The developmental basis for evolvability. In: *Evolvability: A Unifying Concept in Evolutionary Biology?*, ed. T. F. Hansen, D. Houle, M. Pavlicev, and C. Pélabon, 171–197. Vienna Series in Theoretical Biology. Cambridge, MA: MIT Press.

Hallgrímsson, B., H. Jamniczky, N. M. Young, C. Rolian, T. E. Parsons, J. C. Boughner, and R. S. Marcucio. 2009. Deciphering the palimpsest: Studying the relationship between morphological integration and phenotypic covariation. *Evolutionary Biology* 36:355–376.

Hallgrímsson, B., H. Jamniczky, N. M. Young, C. Rolian, U. Schmidt-Ott, and Ralph S. Marcucio. 2012. The generation of variation and the developmental basis for evolutionary novelty. *Journal of Experimental Zoology Part B* 318:501–517.

Hamburger, V. 1980. Embryology and the modern synthesis in evolutionary theory. In: *The Evolutionary Synthesis*, ed. E. Mayr and W. Provine, 97–112. Cambridge, MA: Harvard University Press.

Hanna, L., and E. Abouheif. 2021. The origin of wing polyphenism in ants: An eco-evo-devo perspective. *Current Topics in Developmental Biology* 141:279–336.

Hansell, M. H. 1984. *Animal Architecture and Building Behaviour*. London: Longman.

———. 1993. The ecological impact of animal nests and burrows. *Functional Ecology* 7:5–12.

———. 2000. *Birds Nests and Construction Behaviour*. Cambridge: Cambridge University Press.

Hansen, T. F. 2006. The evolution of genetic architecture. *Annual Review of Ecology, Evolution, and Systematics* 37:123–157.

———. 2011. Epigenetics: Adaptation or contingency? In: *Epigenetics: Linking Genotype and Phenotype in Development and Evolution*, ed. B. Hallgrímsson and B. K. Hall, 357–376. Berkeley: University of California Press.

———. 2015. Evolutionary constraints. In: *Oxford Bibliographies in Evolutionary Biology*, ed. J. Losos. Oxford: Oxford University Press. Last modified January 15, 2015; last reviewed April 12, 2023. http://www.oxfordbibliographies.com/view/document/obo-9780199941 728/obo-9780199941728-0061.xml.

———. 2021. Epistasis. In: *Evolutionary Developmental Biology: A Reference Guide*, ed. L. Nuño de la Rosa and G. B. Müller, 1097–1110. Cham, Switzerland: Springer International.

Hansen, T. F., and D. Houle. 2008. Measuring and comparing evolvability and constraint in multivariate characters. *Journal of Evolutionary Biology* 21:1201–1219.

Hansen, T. F., and C. Pélabon. 2021. Evolvability: A quantitative-genetics perspective. *Annual Review of Ecology, Evolution, and Systematics* 52:153–175.

Hansen, T. F., D. Houle, M. Pavličev, and C. Pélabon, eds. 2023. *Evolvability: A Unifying Concept in Evolutionary Biology?* Vienna Series in Theoretical Biology. Cambridge, MA: MIT Press.

Harjunmaa, E., A. Kallonen, M. Voutilainen, K. Hämäläinen, M. L. Mikkola, and J. Jernvall. 2012. On the difficulty of increasing dental complexity. *Nature* 483:324–327.

Harjunmaa, E., K. Seidel, T. Häkkinen, E. Renvoisé, I. J. Corfe, A. Kallonen, Z. Zhang, et al. 2014. Replaying evolutionary transitions from the dental fossil record. *Nature* 512:44–48.

Harrison, N. M., and M. J. Whitehouse. 2011. Mixed-species flocks: An example of niche construction? *Animal Behaviour* 81:675–682.

Hartl, D. L., and C. H. Taubes. 1998. Towards a theory of evolutionary adaptation. *Genetica* 102–103:525–533.

Hasselquist, D., and J. Nilsson. 2009. Maternal transfer of antibodies in vertebrates: Transgenerational effects on offspring immunity. *Philosophical Transactions of the Royal Society B* 364:51–60.

Hauser, M. D. 1996. *The Evolution of Communication*. Cambridge, MA: MIT Press.

Hauser, M. D., N. Chomsky, and W. T. Fitch. 2002. The faculty of language: What is it, who has it, and how did it evolve? *Science* 298:1569–1579.

Hawkes, K. 2003. Grandmothers and the evolution of human longevity. *American Journal of Human Biology* 15:380–400.

Hawkes, K., J. F. O'Connell, N. G. Blurton Jones, H. Alvarez, and E. L. Charnov. 1998. Grandmothering, menopause, and the evolution of human life histories. *Proceedings of the National Academy of Sciences of the USA* 95:1336–1339.

Hawkes, L. A., A. C. Broderick, M. H. Godfrey, and B. J. Godley. 2009. Climate change and marine turtles. *Endangered Species Research* 7:137–154.

Hawks, J., E. T. Wang, G. M. Cochran, H. C. Harpending, and R. K. Moyzis. 2007. Recent acceleration of human adaptive evolution. *Proceedings of the National Academy of Sciences of the USA* 104:20753–20758.

Hayden, L., K. Lochovska, M. Sémon, S. Renaud, M.-L. Delignette-Muller, M. Vilcot, R. Peterkova, et al. 2020. Developmental variability channels mouse molar evolution. *eLife* 9:e50103.

He, X., D. Tillo, J. Vierstra, K. S. Syed, C. Deng, G. J. Ray, J. Stamatoyannopoulos, et al. 2015. Methylated cytosines mutate to transcription factor binding sites that drive tetrapod evolution. *Genome Biology and Evolution* 7:3155–3169.

He, Y., and Z. Li. 2018. Epigenetic environmental memories in plants: Establishment, maintenance, and reprogramming. *Trends in Genetics* 34:856–866.

Healy, S. D., and J. R. Krebs. 1996. Food storing and the hippocampus in Paridae. *Brain, Behavior and Evolution* 47:195–199.

Heard, E., and R. A Martienssen. 2014. Transgenerational epigenetic inheritance: Myths and mechanisms. *Cell* 157:95–109.

Hecht, E. E., D. A. Gutman, N. Khreisheh, S. V. Taylor, J. Kilner, A. A. Faisal, B. A. Bradley, et al. 2015. Acquisition of paleolithic toolmaking abilities involves structural remodeling to inferior frontoparietal regions. *Brain Structure and Function* 220:2315–2331.

Heckley, A. M., A. P. Pearce, K. M. Gotanda, A. P. Hendry, and K. B. Oke. 2022. Compiling forty years of guppy research to investigate the factors contributing to (non)parallel evolution. *Journal of Evolutionary Biology* 35:1414–1431.

Heckwolf, M. J., B. S. Meyer, R. Häsler, M. P. Höppner, C. Eizaguirre, and T.B.H. Reusch. 2020. Two different epigenetic information channels in wild three-spined sticklebacks are involved in salinity adaptation. *Science Advances* 6:eaaz1138.

Hedenstierna-Jonson, C., A. Kjellström, T. Zachrisson, M. Krzewińska, V. Sobrado, N. Price, T. Günther, et al. 2017. A female Viking warrior confirmed by genomics. *American Journal of Physical Anthropology* 164:853–860.

Heffner, R. S., and R. B. Masterton. 1975. Variation in the form of the pyramidal tract and its relationship to digital dexterity. *Brain, Behavior and Evolution* 12:161–200.

———. 1983. The role of the corticospinal tract in the evolution of human digital dexterity. *Brain, Behavior and Evolution* 23:165–183.

Hegrenes, S. 2001. Diet-induced phenotypic plasticity of feeding morphology in the orangespotted sunfish, *Lepomis humilis*. *Ecology of Freshwater Fish* 10:35–42.

Hehemann, J. H., G. Correc, T. Barbeyron, W. Helbert, M. Czjzek, and G. Michel. 2010. Transfer of carbohydrate-active enzymes from marine bacteria to Japanese gut microbiota. *Nature* 464:908–912.

Heijmans, B. T., W. T. Elmar, D. S. Aryeh, H. Putter, J. B. Gerard, S. S. Ezra, P. E. Slagboom, et al. 2008. Persistent epigenetic differences associated with prenatal exposure to famine in humans. *Proceedings of the National Academy of Sciences of the USA* 105:17046–17049.

Helanterä, H., and T. Uller. 2019. The causes of a major transition: How social insects traverse Darwinian space. In: *Evolutionary Causation, Biological and Philosophical Reflections*, ed. T. Uller and K. N. Laland, 173–196. Cambridge, MA: MIT Press.

Held, L. 2017. *Deep Homology? Uncanny Similarities of Humans and Flies Uncovered by Evo-Devo*. Cambridge: Cambridge University Press.

Heldstab, S. A., K. Isler, S. M. Graber, C. Schuppli, and C. P. van Schaik. 2022. The economics of brain size evolution in vertebrates. *Current Biology* 32:R697–R708.

Helfman, G. S., and E. T. Schultz. 1984. Social transmission of behavioural traditions in a coral reef fish. *Animal Behaviour* 32:379–384.

Helms, J. A., D. Cordero, and M. D. Tapadia. 2005. New insights into craniofacial morphogenesis. *Development* 132:851–861.

Hendrikse, J. L., T. E. Parsons, and B. Hallgrímsson. 2007. Evolvability as the proper focus of evolutionary developmental biology. *Evolution and Development* 9:393–401.

Hendry, A. P. 2020. *Eco-Evolutionary Dynamics*. Princeton, NJ: Princeton University Press.

Henrich, J. 2016. *The Secret of Our Success: How Culture Is Driving Human Evolution, Domesticating Our Species, and Making Us Smarter*. Princeton, NJ: Princeton University Press.

Henrich, J., and R. McElreath. 2003. The evolution of cultural evolution. *Evolutionary Anthropology* 12:123–135.

Henrich, J., and M. Muthukrishna. 2021. The origins and psychology of human cooperation. *Annual Review of Psychology* 72:207–240.

Henry, L. P., M. Bruijning, S.K.G. Forsberg, and J. F. Ayroles. 2021. The microbiome extends host evolutionary potential. *Nature Communications* 12:5141.

Henshaw, J. M., M. B. Morrissey, and A. G. Jones. 2020. Quantifying the causal pathways contributing to natural selection. *Evolution* 74:2560–2574.

Herculano-Houzel, S. 2020. What modern mammals teach about the cellular composition of early brains and mechanisms of brain evolution. In: *Evolutionary Neuroscience*, 2nd ed., ed. J. H. Kaas, 349–375. London: Academic.

———. 2021. Remarkable but not extraordinary: The evolution of the human brain. In *A Most Interesting Problem*, ed. J. DeSilva and J. Browne, 46–62. Princeton, NJ: Princeton University Press.

Herman, J. J., and S. E. Sultan. 2016. DNA methylation mediates genetic variation for adaptive transgenerational plasticity. *Proceedings of the Royal Society B* 283:20160988.

Herman, J. J., H. G. Spencer, K. Donohou, and S. E. Sultan. 2014. How stable "should" epigenetic modifications be? Insights from adaptive plasticity and bet hedging. *Evolution* 68:632–643.

Herman, L. M., D. G. Richards, and J. P. Wolz. 1984. Comprehension of sentences by bottlenose dolphins. *Cognition* 16:129–219.

Herring, S. W. 1993. Formation of the vertebrate face: Epigenetic and functional influences. *American Zoologist* 33:472–483.

Herrmann, E., J. Call, M. V. Hernandez-Lloreda, B. Hare, and M. Tomasello. 2007. Humans have evolved specialized skills of social cognition: The cultural intelligence hypothesis. *Science* 317:1360–1366.

Herrmann, E., M. V. Hernandez-Lloreda, J. Call, B. Hare, and M. Tomasello. 2010. The structure of individual differences in the cognitive abilities of children and chimpanzees. *Psychological Science* 21:102–110.

Herron, J. C., and S. Freeman. 2014. *Evolutionary Analysis*, 5th ed. New York: Pearson Education.

Heyes, C. M. 2018. *Cognitive Gadgets: The Cultural Evolution of Thinking*. Cambridge, MA: Belknap Press of Harvard University Press.

Hill, K., M. Barton, and A. M. Hurtado. 2009. The emergence of human uniqueness: Characters underlying behavioral modernity. *Evolutionary Anthropology* 18:174–187.

Hill, W. G., M. E. Goddard, and P. M. Visscher. 2008. Data and theory point to mainly additive genetic variance for complex traits. *PLoS Genetics* 4: e1000008.

Hinton, G. E., and S. J. Nowlan. 1987. How learning can guide evolution. *Complex Systems* 1:495–502.

Ho, M.-W., and P. T. Saunders. 1984. *Beyond Neo-Darwinism: An Introduction to the New Evolutionary Paradigm*. New York: Academic.

Ho, W.-C., and J. Zhang. 2018. Evolutionary adaptations to new environments generally reverse plastic phenotypic changes. *Nature Communications* 9:350.

Hobaiter, C., and R. W. Byrne. 2011. The gestural repertoire of the wild chimpanzee. *Animal Cognition* 14:745–767.

Hochberg, M. E., P. A. Marquet, R. Boyd, and A. Wagner. 2017. Innovation: An emerging focus from cells to societies. *Philosophical Transactions of the Royal Society B* 372:20160414.

Hoehl, S., S. Keupp, H. Schleihauf, N. McGuigan, D. Buttelmann, and A. Whiten. 2019. Overimitation: A review and appraisal of a decade of research. *Developmental Review* 51:90–108.

Hoekstra, H. E., R. J. Hirschmann, R. A. Bundey, P. A. Insel, and J. P. Crossland. 2006. A single amino acid mutation contributes to adaptive beach mouse color pattern. *Science* 313:101–104.

Hoekstra, H. E., J. M. Hoekstra, D. Berrigan, S. N. Vignieri, A. Hoang, C. E. Hill, P. Beerli, and J. G. Kingsolver. 2001. Strength and tempo of directional selection in the wild. *Proceedings of the National Academy of Sciences of the USA* 98:9157–9160.

Hoelzel, A. R., and A. E. Moura. 2016. Killer whales differentiating in geographic sympatry facilitated by divergent behavioural traditions. *Heredity* 117:481–482.

Hofmann, H. A., M. E. Benson, and R. D. Fernald. 1999. Social status regulates growth rate: Consequences for life-history strategies. *Proceedings of the National Academy of Sciences of the USA* 96:14171–14176.

Holden, C., and R. Mace. 1997. Phylogenetic analysis of the evolution of lactose digestion in adults. *Human Biology* 69:605–628.

Holliday, R., and J. E. Pugh. 1975. DNA modification mechanisms and gene activity during development. *Science* 187:226–232.

Hollwey, E., A. Briffa, M. Howard, and D. Zilberman. 2023. Concepts, mechanisms and implications of long-term epigenetic inheritance. *Current Opinion in Genetics and Development* 81:102087.

Holmes, S. G., and J. R. Broach. 1996. Silencers are required for inheritance of the repressed state in yeast. *Genes and Development* 10:1021–1032.

Holoch, D., and D. Moazed. 2015. RNA-mediated epigenetic regulation of gene expression. *Nature Reviews Genetics* 16:71–84.

Hongoh, Y., P. Deevong, T. Inoue, S. Moriya, S. Trakulnaleamsai, M. Ohkuma, and C. Vongkaluang. 2005. Intra- and interspecific comparisons of bacterial diversity and community structure support coevolution of gut microbiota and termite host. *Applied and Environmental Microbiology* 71:6590–6599.

Hoppitt, W.J.E., and K. N. Laland. 2013. *Social Learning: An Introduction to Mechanisms, Methods, and Models*. Princeton, NJ: Princeton University Press.

Horner, V., and A. Whiten. 2005. Causal knowledge and imitation/emulation switching in chimpanzees (*Pan troglodytes*) and children (*Homo sapiens*). *Animal Cognition* 8:164–181.

Houle, D. 1991. Genetic covariance of fitness correlates: What genetic correlations are made of and why it matters. *Evolution* 45:630–648.

———. 1992. Comparing evolvability and variability of quantitative traits. *Genetics* 130:195–204.

Houle, D., and J. Fierst. 2013. Properties of spontaneous mutational variance and covariance for wing size and shape in *Drosophila melanogaster*. *Evolution* 67:1116–1130.

Houle, D., G. H. Bolstad, K. van der Linde, and T. F. Hansen. 2017. Mutation predicts 40 million years of fly wing evolution. *Nature* 548:447–450.

Houri-Ze'evi, L., and O. Rechavi. 2017. A matter of time: Small RNAs regulate the duration of epigenetic inheritance. *Trends in Genetics* 33:46–57.

Houri-Ze'evi, L., Y. Korem, H. Sheftel, L. Faigenbloom, I. A. Toker, Y. Dagan, L. Awad, et al. 2016. A tunable mechanism determines the duration of the transgenerational small RNA inheritance in *C. elegans*. *Cell* 165:88–99.

Houri-Ze'evi, L., G. Teichman, H. Gingold, and O. Rechavi. 2021. Stress resets ancestral heritable small RNA responses. *eLife* 10:e65797.

Hovers, E. 2012. Invention, reinvention and innovation: Makings of Oldowan lithic technology. In: *Origins of Human Innovation and Creativity*, ed. S. Elias, 51–68. Developments in Quaternary Science 16. Maryland Heights, MO: Elsevier.

Hovland, A. S., M. Rothstein, and M. Simoes-Costa. 2020. Network architecture and regulatory logic in neural crest development. *WIREs Systems Biology and Medicine* 12(2):e1468.

Howson, C. 2000. *Hume's Problem: Induction and the Justification of Belief*. Oxford: Oxford University Press.

Hrdy, S. B. 2009. *Mothers and Others: The Evolutionary Origins of Mutual Understanding*. Cambridge, MA: Harvard University Press.

Hu, Y., and R. C. Albertson. 2017. Baby fish working out: An epigenetic source of adaptive variation in the cichlid jaw. *Proceedings of the Royal Society B* 284:20171018.

Hu, Y., D. M. Linz, and A. P. Moczek. 2019. Beetle horns evolved from wing serial homologs. *Science* 366:1004–1007.

Hu, Y., D. M. Linz, E. S. Parker, D. B. Schwab, S. Casasa, A.L.M. Macagno, and A. P. Moczek. 2020. Developmental bias in horned dung beetles and its contributions to innovation, adaptation, and resilience. *Evolution and Development* 22:165–180.

Hubel, D. H., and T. N. Weisel. 2005. *Brain and Visual Perception: The Story of a 25-Year Collaboration*. Oxford: Oxford University Press.

Huey, R. B., P. E. Hertz, and B. Sinervo. 2003. Behavioral drive versus behavioral inertia in evolution: A null model approach. *American Naturalist* 161:357–366.

Hughes, C. L., and T. C. Kaufman. 2002. Hox genes and the evolution of the arthropod body plan. *Evolution and Development* 4:459–499.

Hughes, M., S. Gerber, and M. A. Wills. 2013. Clades reach highest morphological disparity early in their evolution. *Proceedings of the National Academy of Sciences of the USA* 110:13875–13879.

Hull, D. L., R. E. Langman, and S. S. Glenn. 2001. A general account of selection: Biology, immunology, and behavior. *Behavioral and Brain Sciences* 24:511–528.

Hume, D. (1748) 1910. *An Enquiry Concerning Human Understanding*. New York: P. F. Collier.

Huneman, P. 2019. How the modern synthesis came to ecology. *Journal of the History of Biology* 52:635–686.

Huneman, P., and D. M. Walsh. 2017. *Challenging the Modern Synthesis: Adaptation, Development and Inheritance*. Oxford: Oxford University Press.

Hunt, G. R., and R. D. Gray. 2003. Diversification and cumulative evolution in New Caledonian crow tool manufacture. *Proceedings of the Royal Society B* 270:867–874.

Hunt, J. H. 2012. A conceptual model for the origin of worker behaviour and adaptation of eusociality. *Journal of Evolutionary Biology* 25:1–19.

Hutchinson, G. E. 1957. Concluding remarks. *Cold Spring Harbor Symposia on Quantitative Biology* 22:415–427.

Huxley, T. H. 1870. Excerpts from "Biogenesis and Abiogenesis," presidential address to the British Association for the Advancement of Science, Liverpool, September 14, 1870. Reproduced in *Nature* 2:400–406.

Ilardo, M. A., I. Moltke, T. S. Korneliussen, J. Cheng, A. J. Stern, F. Racimo, P. de Barros Damgaard, et al. 2018. Physiological and genetic adaptations to diving in sea nomads. *Cell* 173:569–580.e15.

Immler, S. 2018. The sperm factor: Paternal impact beyond genes. *Heredity* 121:239–247.

Ingram, P. J., M.P.H. Stumpf, and J. Stark. 2006. Network motifs: Structure does not determine function. *BMC Genomics* 7:108.

Inquiry into the History of Eugenics at UCL—Final Report. 2020. University College London, Office of the President and Provost. https://www.ucl.ac.uk/provost/sites/provost/files/ucl_history_of_eugenics_inquiry_report.pdf.

International Human Genome Sequencing Consortium. 2001. Initial sequencing and analysis of the human genome. *Nature* 409:860–921.

Izar, P., L. Peternelli-dos-Santos, J. M. Rothman, D. Raubenheimer, A. Presotto, G. Gort, E. M. Visalberghi, et al. 2022. Stone tools improve diet quality in wild monkeys. *Current Biology* 32: 4088–4092.e3.

Jablonka, E. 2017. The evolutionary implications of epigenetic inheritance. *Interface Focus* 7:20160135.

Jablonka, E., and M. J. Lamb. 1995. *Epigenetic Inheritance and Evolution: The Lamarckian Dimension*. Oxford: Oxford University Press.

———. 2005. *Evolution in Four Dimensions*. Cambridge, MA: MIT Press.

———. 2006. The evolution of information in the major transitions. *Journal of Theoretical Biology* 239:236–246.

———. 2014. *Evolution in Four Dimensions*, rev. ed. Cambridge, MA: MIT Press.

Jablonka, E., and G. Raz. 2009. Transgenerational epigenetic inheritance: Prevalence, mechanisms, and implications for the study of heredity and evolution. *Quarterly Review of Biology* 84:131–176.

Jablonski, D. 2017. Approaches to macroevolution, 1: General concepts and origin of variation. *Evolutionary Biology* 44:427–450.

———. 2020. Developmental bias, macroevolution, and the fossil record. *Evolution and Development* 22:103–125.

Jablonski, N. G., and G. Chaplin. 2000. The evolution of human skin coloration. *Journal of Human Evolution* 39:57–106.

Jacob, F. 1974. *The Logic of Life: A History of Heredity*, transl. B. E. Spillmann. New York: Pantheon.

Jaeger, J., and A. Crombach. 2012. Life's attractors: Understanding developmental systems through reverse engineering and in silico evolution. In: *Evolutionary Systems Biology*, ed. O. S. Soyer, 93–119. New York: Springer.

Jaeger, J., and N. Monk. 2014. Bioattractors: Dynamical systems theory and the evolution of regulatory processes. *Journal of Physiology* 592:2267–2281.

Jain, R., M. C. Rivera, J. E. Moore, and J. A. Lake. 2003. Horizontal gene transfer accelerates genome innovation and evolution. *Molecular Biology and Evolution* 20:1598–1602.

Jeffery, W. R. 2008. Emerging model systems in evo-devo: Cavefish and microevolution of development. *Evolution and Development* 10:265–272.

Jennings, H. S. 1937. Formation, inheritance and variation of the teeth in *Difflugia corona*: A study of the morphogenic activities of rhizopod protoplasm. *Journal of Experimental Zoology* 77: 287–336.

Jesmer, B. R., J. A. Merkle, J. R. Goheen, E. O. Aikens, J. L. Beck, A. B. Courtemanch, M. A. Hurley, et al. 2018. Is ungulate migration culturally transmitted? Evidence of social learning from translocated animals. *Science* 361:1023–1025.

Jiménez, A., J. Cotterell, A. Munteanu, and J. Sharpe. 2015. Dynamics of gene circuits shapes evolvability. *Proceedings of the National Academy of Sciences of the USA* 112:2103–2108.

Jin, X., F. Tecuapetla, and R. M. Costa. 2014. Basal ganglia subcircuits distinctively encode the parsing and concatenation of action sequences. *Nature Neuroscience* 17:423–430.

Jirtle, R. L., and M. K. Skinner. 2007. Environmental epigenomics and disease susceptibility. *Nature Reviews Genetics* 8:253–262.

Johannsen, W. 1911. The genotype conception of heredity. *American Naturalist* 45:129–159.

Jones, A.G., S. J. Arnold, and R. Bürger. 2007. The mutation matrix and the evolution of evolvability. *Evolution* 61:727–745.

Jones, A. G., R. Bürger, and S. J. Arnold. 2014. Epistasis and natural selection shape the mutational architecture of complex traits. *Nature Communications* 5:3709.

Jones, B. C., and E. H. DuVal. 2019. Mechanisms of social influence: A meta-analysis of the effects of social information on female mate choice decisions. *Frontiers in Ecology and Evolution* 7. https://doi.org/10.3389/fevo.2019.00390.

Jones, C. G., J. H. Lawton, and M. Shachak. 1994. Organisms as ecosystem engineers. Oikos. 69(3): 373–86.

Juan, C., M. T. Guzik, D. Jaume, and S.J.B. Cooper. 2010. Evolution in caves: Darwin's "wrecks of ancient life" in the molecular era. *Molecular Ecology* 19:3865–3880.

Judson, O. P. 2017. The energy expansions of evolution. *Nature Ecology and Evolution* 1:0138.

Just, E.E. 1933. Cortical cytoplasm and evolution. *American Naturalist* 67:20–29.

Kaati, G., L. O. Bygren, and S. Edvinsson. 2002. Cardiovascular and diabetes mortality determined by nutrition during parents' and grandparents' slow growth period. *European Journal of Human Genetics* 10:682–688.

Kaati, G., L. O. Bygren, M. Pembrey, and M. Sjöström. 2007. Transgenerational response to nutrition, early life circumstances and longevity. *European Journal of Human Genetics* 15:784–790.

Kahn, S., and Ehrlich, P. R. 2018. *Jaws: The Story of a Hidden Epidemic*. Stanford, CA: Stanford University Press.

Kamra, D. N. 2005. Rumen microbial ecosystem. *Current Science* 89:124–135.

Kangas, A. T., A. R. Evans, I. Thesleff, and J. Jernvall. 2004. Nonindependence of mammalian dental characters. *Nature* 432:211–214.

Kapheim, K. M., B. M. Jones, H. Pan, C. Li, B. A. Harpur, C. F. Kent, A. Zayed, et al. 2020. Developmental plasticity shapes social traits and selection in a facultatively eusocial bee. *Proceedings of the National Academy of Sciences of the USA* 117:13615–13625.

Kaplan, H. S., and J. A. Robson. 2002. The emergence of humans: The coevolution of intelligence and longevity with intergenerational transfers. *Proceedings of the National Academy of Sciences of the USA* 99:10221–10226.

Kaplan, H. S., K. Hill, J. Lancaster, and A. Magdalena Hurtado. 2000. A theory of human life history evolution: Diet, intelligence, and longevity. *Evolutionary Anthropology* 9:156–185.

Kardon, G. 1998. Muscle and tendon morphogenesis in the avian hind limb. *Development* 125:4019–4032.

Karmiloff-Smith, A. 1995. *Beyond Modularity: A Developmental Perspective on Cognitive Science*. Cambridge, MA: MIT Press.

Kashtan, N., E. Noor, and Uri Alon. 2007. Varying environments can speed up evolution. *Proceedings of the National Academy of Sciences of the USA* 104:13711–13716.

Kauffmann, S. A. 1983. Developmental constraints: Internal factors in evolution. In: *Development and Evolution*, ed. B. C. Goodwin, N. Holder, and C. G. Wylie, 183–199. Cambridge: Cambridge University Press.

Kauffman, S. A., and A. Roli. 2021. The world is not a theorem. *Entropy* 23:1467.

Kavanagh, K. D. 2020. Developmental plasticity associated with early structural integration and evolutionary patterns: Examples of developmental bias and developmental facilitation in the skeletal system. *Evolution and Development* 22:196–204.

Kavanagh, K. D., A. R. Evans, and J. Jernvall. 2007. Predicting evolutionary patterns of mammalian teeth from development. *Nature* 449:427–432.

Kavanagh, K. D., O. Shoval, B. B. Winslow, U. Alon, B. P. Leary, A. Kan, and C. J. Tabin. 2013. Developmental bias in the evolution of phalanges. *Proceedings of the National Academy of Sciences of the USA* 110:18190–18195.

Kawakatsu, T., S. C. Huang, F. Jupe, E. Sasaki, R. J. Schmitz, R. J. Urich, M. A. Castanon, et al. 2016. Epigenomic diversity in a global collection of *Arabidopsis thaliana* accessions. *Cell* 166:492–505.

Keller, E. F. 2010. *The Mirage of a Space between Nature and Nurture*. Durham, NC: Duke University Press.

———. 2014. From gene action to reaction genomes. *Journal of Physiology* 592:2423–2429.

Keller, R. A., C. Peeters, and P. Beldade. 2014. Evolution of thorax architecture in ant castes highlights trade-off between flight and ground behaviors. *eLife* 3:e01539.

Kelley, J. L., and A. E. Magurran. 2003. Learned predator recognition and antipredator responses in fishes. *Fish and Fisheries* 4:216–226.

Kemp, D. J., D. N. Reznick, J. Arendt, C. van den Berg, and J. A. Endler. 2023. How to generate and test hypotheses about colour: Insights from half a century of guppy research. *Proceedings of the Royal Society B* 290:20222492.

Kemp, M. E., A. M. Mychajliw, J. Wadman, and A. Goldberg. 2020. 7000 years of turnover: Historical contingency and human niche construction shape the Caribbean's Anthropocene biota. *Proceedings of the Royal Society B* 287:20200447.

Kendal, J., J. J. Tehrani, and F. J. Odling-Smee. 2011. Human niche construction in interdisciplinary focus. *Philosophical Transactions of the Royal Society B* 366:785–792.

Kendal, R. L., N. J. Boogert, L. Rendell, K. N. Laland, M. Webster, and P. L. Jones. 2018. Social learning strategies: Bridge-building between fields. *Trends in Cognitive Sciences* 22:651–665.

Kendal, R. L., I. Coolen, and K. N. Laland. 2004. The role of conformity in foraging when personal and social information conflict. *Behavioral Ecology* 15:269–277.

Kerr, M., H. Stattin, and K. Trost. 1999. To know you is to trust you: Parents' trust is rooted in child disclosure of information. *Journal of Adolescence* 22:737–752.

Kikuchi, Y., M. Hayatsu, T. Hosokawa, A. Nagayama, K. Tago, and T. Fukatsu. 2012. Symbiont-mediated insecticide resistance. *Proceedings of the National Academy of Sciences of the USA* 109:8618–8622.

Kimura, I., J. Miyamoto, R. Ohue-Kitano, K. Watanabe, T. Yamada, M. Onuki, R. Aoki, et al. 2020. Maternal gut microbiota in pregnancy influences offspring metabolic phenotype in mice. *Science* 367:eaaw8429.

King, A. J., and C. Sueur. 2011. Where next? Group coordination and collective decision making by primates. *International Journal of Primatology* 32:1245–1267.

Kingsolver, J. G., S. E. Diamond, A. M. Siepielski, and S. M. Carlson. 2012. Synthetic analyses of phenotypic selection in natural populations: Lessons, limitations and future directions. *Evolutionary Ecology* 26:1101–1118.

Kingsolver, J. G., H. E. Hoekstra, J. M. Hoekstra, D. Berrigan, S. N. Vignieri, C. E. Hill, A. Hoang, et al. 2001. The strength of phenotypic selection in natural populations. *American Naturalist* 157:245–261.

Kiontke, K., A. Barriere, I. Kolotuev, B. Podbilewicz, R. Sommer, D.H.A. Fitch, and M. A. Felix. 2007. Trends, stasis, and drift in the evolution of nematode vulva development. *Current Biology* 17:1925–1937.

Kirby, S., H. Cornish, and K. Smith. 2008. Cumulative cultural evolution in the laboratory: An experimental approach to the origins of structure in human language. *Proceedings of the National Academy of Sciences of the USA* 105:10681–10686.

Kirby, S., M. Dowman, and T. L. Grifths. 2007. Innateness and culture in the evolution of language. *Proceedings of the National Academy of Sciences of the USA* 104:5241–5245.

Kirby, S., M. Tamariz, H. Cornish, and K. Smith. 2015. Compression and communication in the cultural evolution of linguistic structure. *Cognition* 141:87–102.

Kirkpatrick, M. 2009. Patterns of quantitative genetic variation in multiple dimensions. *Genetica* 136:271–284.

Kirkpatrick, M., and L. A. Dugatkin. 1994. Sexual selection and the evolutionary effects of copying mate choice. *Behavioral Ecology and Sociobiology* 34:443–449.

Kirkpatrick, M., and R. Lande. 1989. The evolution of maternal characters. *Evolution* 43:485–503.

Kirsch, R., L. Gramzow, G. Theißen, B. D. Siegfried, R. H. Ffrench-Constant, D. G. Heckel, and Y. Pauchet. 2014. Horizontal gene transfer and functional diversification of plant cell wall degrading polygalacturonases: Key events in the evolution of herbivory in beetles. *Insect Biochemistry and Molecular Biology* 52:33–50.

Kirschner, M. W. 2015. The road to facilitated variation. Chapter 9 in *Conceptual Change in Biology: Scientific and Philosophical Perspectives on Evolution and Development*, ed. A.C. Love. Dordrecht: Springer.

Kirschner, M. W., and J. C. Gerhart. 1998. Evolvability. *Proceedings of the Royal Society B* 95:8420–8427.

———. 2005. *The Plausibility of Life: Resolving Darwin's Dilemma*. New Haven, CT: Yale University Press.

Kirschner, S., and M. Tomasello. 2010. Joint music making promotes prosocial behavior in 4-year-old children. *Evolution and Human Behavior* 31:354–364.

Kitcher, P. 1989. Explanatory unification and the causal structure of the world. In: *Scientific Explanation*, ed. P. Kitcher and W. Salmon, Minnesota Studies in the Philosophy of Science 13, 410–505. Minneapolis: University of Minnesota Press.

Kittelmann, S., and A. P. McGregor. 2019. Modulation and evolution of animal development through microRNA regulation of gene expression. *Genes* 10:321.

Klaus, S., J.C.E. Mendoza, J. H. Liew, M. Plath, R. Meier, and D.C.J. Yeo. 2013. Rapid evolution of troglomorphic characters suggests selection rather than neutral mutation as a driver of eye reduction in cave crabs. *Biology Letters* 9:20121098.

Klenerman, P. 2017. *The Immune System: A Very Short Introduction*. Oxford: Oxford University Press.

Klingenberg, C. P. 2008. Morphological integration and developmental modularity. *Annual Review of Ecology, Evolution, and Systematics* 39:115–132.

Klingenberg, C. P., and S. D. Zaklan. 2000. Morphological integration between developmental compartments in the *Drosophila* wing. *Evolution* 54:1273–1285.

Klosin, A., and B. Lehner. 2016 Mechanisms, timescales and principles of trans-generational epigenetic inheritance in animals. *Current Opinion in Genetics and Development* 36:41–49.

Koch, H., and P. Schmid-Hempel. 2011. Socially transmitted gut microbiota protect bumble bees against an intestinal parasite. *Proceedings of the National Academy of Sciences of the USA* 108:19288–19292.

Kohl, K. D., A. Stengel, and M. D. Dearing. 2016. Innoculation of tannin-degrading bacteria into novel hosts increases performance on tannin-rich diets. *Environmental Microbiology* 18:1720–1729.

Kohl, K. D., R. B Weiss, J. Cox, C. Dale, and M. D. Dearing. 2014. Gut microbes of mammalian herbivores facilitate intake of plant toxins. *Ecology Letters* 17:1238–1246.

Kolodny, O., and S. Edelman. 2018. The evolution of the capacity for language: The ecological context and adaptive value of a process of cognitive hijacking. *Philosophical Transactions of the Royal Society B* 373:20170052.

Kolodny, O., and M. W. Feldman. 2017. A parsimonious neutral model suggests Neanderthal replacement was determined by migration and random species drift. *Nature Communications* 8:1040.

Kolodny, O., M. Weinberg, L. Reshef, L. Harten, A. Hefetz, U. Gophna, M. W. Feldman, and Y. Yovel 2019. Coordinated change at the colony level in fruit bat fur microbiomes through time. *Nature Ecology and Evolution* 3:116–124.

Kondo, S., M. Watanabe, and S. Miyazawa. 2021. Studies of Turing pattern formation in zebrafish skin. *Philosophical Transactions of the Royal Society B* 379:1–14.

Koonin, E. V. 2016. Horizontal gene transfer: Essentiality and evolvability in prokaryotes, and roles in evolutionary transitions. *F1000 Research* 5:1805.

Kosiol, C., T. Vinař, R. R. da Fonseca, M. J. Hubisz, C. D. Bustamante, R. Nielsen, and A. Siepel. 2008. Patterns of positive selection in six mammalian genomes. *PLOS Genetics* 4:e1000144.

Kounios, L., J. Clune, K. Kouvaris, G. Wagner, M. Pavličev, D. Weinreich, and R. Watson. 2016. Resolving the paradox of evolvability with learning theory: How evolution learns to improve evolvability on rugged fitness landscapes. *arXiv*:1612.05955v1.

Kouvaris, K., J. Clune, L. Kounios, M. Brede, and R. A. Watson. 2017. How evolution learns to generalise: Using the principles of learning theory to understand the evolution of developmental organisation. *PLOS Computational Biology* 13:e1005358.

Kowalko, J. 2020. Utilizing the blind cavefish *Astyanax mexicanus* to understand the genetic basis of behavioral evolution. *Journal of Experimental Biology* 223(suppl. 1):jeb208835.

Kowalko, J., T. A. Franz-Odendaal, and N. Rohner. 2020. Introduction to the special issue— cavefish—adaptation to the dark. In Cavefish special issue, ed. Kowalko, Franz-Odendaal, and Rohner, *Journal of Experimental Zoology Part B* 334(7–8):393–396.

Koyama, T., C. Mendes, and C. Mirth. 2013. Mechanisms regulating nutrition-dependent developmental plasticity through organ-specific effects in insects. *Frontiers in Physiology* 4. https://doi.org/10.3389/fphys.2013.00263.

Kozmik, Z. 2005. Pax genes in eye development and evolution. *Current Opinion in Genetics and Development* 15:430–438.

Kozmik, Z., J. Ruzickova, K. Jonasova, Y. Matsumoto, P. Vopalensky, I. Kozmikova, H. Strnad, et al. 2008. Assembly of the cnidarian camera-type eye from vertebrate-like components. *Proceedings of the National Academy of Sciences of the USA* 105:8989–8993.

Krakauer, D. C., K. M. Page, and D. H. Erwin. 2009. Diversity, dilemmas, and monopolies of niche construction. *American Naturalist* 173:26–40.

Krause, J., G. D. Ruxton, and S. Krause. 2010. Swarm intelligence in animals and humans. *Trends in Ecology and Evolution* 25:28–34.

Krebs, J. E., S. T. Kilpatrick, and E. S. Goldstein. 2008. *Lewin's Genes X*. Burlington, MA: Jones and Bartlett.

Kronholm, I., A. Bassett, D. Baulcombe, and S. Collins. 2017. Epigenetic and genetic contributions to adaptation in *Chlamydomonas*. *Molecular Biology and Evolution* 34:2285–2306.

Krubitzer, L., and D. S. Stolzenberg. 2014. The evolutionary masquerade: Genetic and epigenetic contributions to the neocortex. *Current Opinion in Neurobiology* 24:157–165.

Kuijper, B., and R. B. Hoyle. 2015. When to rely on maternal effects and when on phenotypic plasticity? *Evolution* 69:950–968.

Kuratani, S. 2021. Evo-devo studies of cyclostomes and the origin and evolution of jawed vertebrates. *Current Topics in Developmental Biology* 141:207–239.

Kurzban, R. 2011. *Why Everyone (Else) Is a Hypocrite: Evolution and the Modular Mind*. Princeton, NJ: Princeton University Press.

Kuzawa, C. W., H. T. Chugani, L. I. Grossman, L. Lipovich, O. Muzik, P. R. Hof, D. E. Wildman, et al. 2014. Metabolic costs and evolutionary implications of human brain development. *Proceedings of the National Academy of Sciences of the USA* 111:13010–13015.

Kylafis, G., and M. Loreau. 2008. Ecological and evolutionary consequences of niche construction for its agent. *Ecology Letters* 11:1072–1081.

———. 2011. Niche construction in the light of niche theory. *Ecology Letters* 14:82–90.

Lacal, I., and R. Ventura. 2018. Epigenetic inheritance: Concepts, mechanisms and perspectives. *Frontiers in Molecular Neuroscience* 11.

Lachlan, R. F., and P.J.B. Slater. 1999. The maintenance of vocal learning by gene culture interaction: The cultural trap hypothesis. *Proceedings of the Royal Society B* 266:701–706.

Lachmann, M., and E. Jablonka. 1996. The inheritance of phenotypes: An adaptation to fluctuating environments. *Journal of Theoretical Biology* 181:1–9.

Laidre, M. E. 2012. Niche construction drives social dependence in hermit crabs. *Current Biology* 22:R861–R863.

Lakatos, I. 1978. *Philosophical Papers*, vol. 1, *The Methodology of Scientific Research Programmes*. Cambridge: Cambridge University Press.

Lala, K., and M. J. O'Brien. 2023. The cultural contribution to evolvability. *PaleoAnthropology* 2023, no. 2. https://doi.org/10.48738/2023.iss2.115.

Laland, K. N. 1990. A theoretical investigation of the role of social transmission in evolution. PhD dissertation, University College London.

———. 1994a. On the evolutionary consequences of sexual imprinting. *Evolution* 48:477–489.

———. 1994b. Sexual selection with a culturally transmitted mating preference. *Theoretical Population Biology* 45:1–15.

———. 2004. Social learning strategies. *Learning and Behavior* 32:4–14.

———. 2017a. *Darwin's Unfinished Symphony: How Culture Made the Human Mind*. Princeton, NJ: Princeton University Press.

———. 2017b. The origins of language in teaching. *Psychonomic Bulletin and Review* 24:225–231.

Laland, K. N., and G. R. Brown. 2006. Niche construction, human behavior, and the adaptive-lag hypothesis. *Evolutionary Anthropology* 15:95–104.

Laland, K. N., and M. J. O'Brien. 2010. Niche construction theory and archaeology. *Journal of Archaeological Method and Theory* 17:303–322.

Laland, K. N., and H. C. Plotkin. 1991. Excretory deposits surrounding food sites facilitate social learning of food preferences in Norway rats. *Animal Behaviour* 41:997–1005.

———. 1993. Social transmission of food preferences amongst Norway rats by marking of food sites, and by gustatory contact. *Animal Learning and Behavior* 21:35–41.

Laland, K. N., and A. Seed. 2021. Understanding human cognitive uniqueness. *Annual Review of Psychology* 72:689–716.

Laland, K. N., and K. Williams. 1997. Shoaling generates social learning of foraging information in guppies. *Animal Behaviour* 53:1161–1169.

———. 1998. Social transmission of maladaptive information in the guppy. *Behavioral Ecology* 9:493–499.

Laland, K. N., F. J. Odling-Smee, and J. Endler. 2017. Niche construction, sources of selection and trait coevolution. *Interface Focus* 7:20160147.

Laland, K. N., F. J. Odling-Smee, and M. W. Feldman. 1996. The evolutionary consequences of niche construction: A theoretical investigation using two-locus theory. *Journal of Evolutionary Biology* 9:293–316.

———. 1999. Evolutionary consequences of niche construction and their implications for ecology. *Proceedings of the National Academy of Sciences of the USA* 96:10242.

———. 2000. Niche construction, biological evolution and cultural change. *Behavioral and Brain Sciences* 23:131–146.

———. 2001. Cultural niche construction and human evolution. *Journal of Evolutionary Biology* 14:22–33.

———. 2019. Understanding niche construction as an evolutionary process. In: *Evolutionary Causation: Biological and Philosophical Reflections*, ed. T. Uller and K. N. Laland, 127–152. Cambridge, MA: MIT Press.

Laland, K. N., F. J. Odling-Smee, and S. F. Gilbert. 2008. Evo-devo and niche construction: Building bridges. *Journal of Experimental Zoology Part B* 310:549–566.

Laland K. N., F. J. Odling-Smee, W. Hoppitt, and T. Uller. 2012. More on how and why: Cause and effect in biology revisited. *Biology and Philosophy* 28:719–745.

———. 2013. More on how and why: A response to commentaries. *Biology and Philosophy* 28:793–810.

Laland, K. N., F. J. Odling-Smee, and S. Myles. 2010. How culture shaped the human genome: Bringing genetics and the human sciences together. *Nature Reviews Genetics* 11:137–148.

Laland, K. N., F. J. Odling-Smee, and J. S. Turner. 2014a. The role of internal and external constructive processes in evolution. *Journal of Physiology* 592:2413–2422.

Laland, K. N., T. Oudman, and W. Toyokawa. 2022. How learning affects evolution. In: *Evolution of Learning and Memory Mechanisms*, ed. M. A Krause, K. L. Hollis, and M. R. Papini, 265–282. Cambridge: Cambridge University Press.

Laland, K. N., K. Sterelny, F. J. Odling-Smee, W. Hoppitt, and T. Uller. 2011. Cause and effect in biology revisited: Is Mayr's proximate-ultimate dichotomy still useful? *Science* 334:1512–1516.

Laland, K. N., W. Toyokawa, and T. Oudman. 2020. Animal learning as a source of developmental bias. *Evolution and Development* 22:126–142.

Laland, K. N., T. Uller, M. W. Feldman, K. Sterelny, G. B. Müller, A. Moczek, E. Jablonka, et al. 2014b. Does evolutionary theory need a rethink? Yes. *Nature* 514:161–164.

———. 2015. The extended evolutionary synthesis: Its structure, assumptions and predictions. *Proceedings of the Royal Society B* 282:20151019.

Laland, K. N., C. Wilkins, and N. S. Clayton. 2016. The evolution of dance. *Current Biology* 26:R1–R21.

Lande, R. 1979. Quantitative genetic analysis of multivariate evolution, applied to brain:body size allometry. *Evolution* 33:402–416.

———. 1980. Sexual dimorphism, sexual selection, and adaptation in polygenic characters. *Evolution* 34:292–305.

———. 2009. Adaptation to an extraordinary environment by evolution of phenotypic plasticity and genetic assimilation. *Journal of Evolutionary Biology* 22:1435–1446.

Lande. R., and S. J. Arnold. 1983. The measurement of selection on correlated characters. *Evolution* 37: 1210–1226.

Lange, A. 2020. *Evolutionstheorie im Wandel: Ist Darwin überholt?* Berlin: Springer.

———. 2023. *Extending the Evolutionary Synthesis: Darwin's Legacy Redesigned.* Boca Raton, FL: CRC.

Lange, A., and G. B. Müller. 2017. Polydactyly in development, inheritance, and evolution. *The Quarterly Review of Biology* 92:1–38.

Lange, A., H. L. Nemeschkal, and G. B. Müller. 2014. Biased polyphenism in polydactylous cats carrying a single point mutation: The Hemingway model for digit novelty. *Evolutionary Biology* 41:262–275.

———. 2018. A threshold model for polydactyly. *Progress in Biophysics and Molecular Biology* 137:1–11.

Lansing, J. S., and K. M. Fox. 2011. Niche construction on Bali: The gods of the countryside. *Philosophical Transactions of the Royal Society B* 366:927–934.

Lansing, J. S., M. P. Cox, S. S. Downey, M. A. Janssen, and J. W. Schoenfelder. 2009. A robust budding model of Balinese water temple networks. *World Archaeology* 41:112–133.

La Reau, A. J., and G. Suen. 2018. The Ruminococci: Key symbionts of the gut ecosystem. *Journal of Microbiology* 56:199–208.

Lassig. M., V. Mustonen, and A. M. Walczak. 2017. Predicting evolution. *Nature, Ecology and Evolution* 1:77.

Laubichler, M. D., and J. Renn. 2015. Extended evolution: A conceptual framework for integrating regulatory networks and niche construction. *Journal of Experimental Zoology Part B* 324:565–577.

Lawson, R. 2006. The science of cycology: Failures to understand how everyday objects work. *Memory and Cognition* 34:1667–1675.

Lax, S., D. P. Smith, J. Hampton-Marcell, S. M. Owens, K. M. Handley, N. M. Scott, S. M. Gibbons, et al. 2014. Longitudinal analysis of microbial interaction between humans and the indoor environment. *Science* 345:1048–1052.

Le, Q.V., L. A. Isbell, J. Matsumoto, M. Nguyen, E. Hori, R. S. Maior, C. Tomaz, et al. 2013. Pulivar neurons reveal neurobiological evidence of past selection for rapid detection of snakes. *Proceedings of the National Academy of Sciences of the USA* 11:19000–19005.

Leadbeater, E., and L. Chittka. 2007. Social learning in insects: From miniature brains to consensus building. *Current Biology* 17: R703–R713.

Leadbeater, E., and E. H. Dawson. 2017. A social insect perspective on the evolution of social learning mechanisms. *Proceedings of the National Academy of Sciences of the USA* 114:7838.

Ledón-Rettig, C. C., D. W. Pfennig, and E. J. Crespi. 2009. Stress hormones and the fitness consequences associated with the transition to a novel diet in larval amphibians. *Journal of Experimental Biology* 212:3743–3750.

———. 2010. Diet and hormonal manipulation reveal cryptic genetic variation: Implications for the evolution of novel feeding strategies. *Proceedings of the Royal Society B* 277: 3569–3578.

Ledón-Rettig, C. C., D. W. Pfennig, and N. Nascone-Yoder. 2008. Ancestral variation and the potential for genetic accommodation in larval amphibians: Implications for the evolution of novel feeding strategies. *Evolution and Development* 10:316–325.

Lee, K. 1985. *Earthworms: Their Ecology and Relation with Soil and Land Use.* London: Academic.

Lee, T. M., and I. Zucker. 1988. Vole infant development is influenced perinatally by maternal photoperiodic history. *American Journal of Physiology* 255:R831–R838.

Lefebvre, L. 1995. The opening of milk-bottles by birds: Evidence for accelerating learning rates, but against the wave-of-advance model of cultural transmission. *Behavioral Processes* 34:43–53.

Lefebvre, L., S. M. Reader, and D. Sol. 2004. Brains, innovations and evolution in birds and primates. *Brain, Behavior and Evolution* 63:233–246.

Lefebvre, L., P. Whittle, E. Lascaris, and A. Finkelstein. 1997. Feeding innovations and forebrain size in birds. *Animal Behaviour* 53:549–560.

Lehmann, L. 2007. The evolution of trans-generational altruism: Kin selection meets niche construction. *Journal of Evolutionary Biology* 20:181–189.

———. 2008. The adaptive dynamics of niche constructing traits in spatially subdivided populations: Evolving posthumous extended phenotypes. *Evolution* 62:549–566.

Leigh, S. R. 2004. Brain growth, life history, and cognition in primate and human evolution. *American Journal of Primatology* 62:139–164.

Leimar, O., and J. M. McNamara. 2015. The evolution of transgenerational integration of information in heterogeneous environments. *American Naturalist* 185:E55–E69.

Leimar, O., P. Hammerstein, and T.J.M. van Dooren. 2006. A new perspective on developmental plasticity and the principles of adaptive morph determination. *American Naturalist* 167:367–376.

Lettice, L. A., A. E, Hill, P. S. Devenney, and R. E. Hill. 2008. Point mutations in a distant sonic hedgehog cis-regulator generate a variable regulatory output responsible for preaxial polydactyly. *Human Molecular Genetics* 17:978–985.

Levin, M. 2020. The biophysics of regenerative repair suggests new perspectives on biological causation. *BioEssays* 42:1900146.

Levins, R., and R. C. Lewontin. 1985. *The Dialectical Biologist*. Cambridge, MA: Harvard University Press.

Levis, N. A., and D. W. Pfennig. 2016. Evaluating "plasticity-first" evolution in nature: Key criteria and empirical approaches. *Trends in Ecology and Evolution* 31:563–574.

———. 2020. Plasticity-led evolution: A survey of developmental mechanisms and empirical tests. *Evolution and Development* 22:71–87.

Levis, N. A., A. J. Isdaner, and D. W. Pfennig. 2018. Morphological novelty emerges from preexisting phenotypic plasticity. *Nature Ecology and Evolution* 2:1289–1297.

Levis, N. A., E.M.X. Reed, D. W. Pfennig, and M. O. Burford Reiskind. 2020. Identification of candidate loci for adaptive phenotypic plasticity in natural populations of spadefoot toads. *Ecology and Evolution* 10:8976–8988.

Lev-Yadun, S., G. Katzir, and G. Ne'eman. 2009. *Rheum palaestinum* (desert rhubarb), a self-irrigating desert plant. *Naturwissenschaften* 96:393–397.

Lewens, T. 2019. The extended evolutionary synthesis: What is the debate about, and what might success for the extenders look like? *Biological Journal of the Linnean Society* 127:707–721.

Lewontin, R. C. 1955. The effects of population density and composition on viability in *Drosophila melanogaster*. *Evolution* 9:27–41.

———. 1970. The units of selection. *Annual Review of Ecology and Systematics* 1:1–18.

———. 1982. Organism and environment. In: *Learning, Development and Culture*, ed. H. C. Plotkin, 151–170. Chichester, UK: Wiley.

———. 1983. Gene, organism, and environment. In: *Evolution from Molecules to Men*, ed. D. S. Bendall, 273–285. Cambridge: Cambridge University Press.

———. 2000. The problems of population genetics. In: *Evolutionary Genetics*, ed. R. S. Krimbas, 5–23. Cambridge: Cambridge University Press.

———. 2002. *The Triple Helix*. Cambridge, MA: Harvard University Press.

Lewontin, R. C., and Y. Matsuo. 1963. Interaction of genotypes determining viability in *Drosophila busckii*. *Proceedings of the National Academy of Sciences of the USA* 49:270–278.

Ley, R. E., C. A. Lozupone, M. Hamady, R. Knight, and J. I. Gordon. 2008. Worlds within worlds: Evolution of the vertebrate gut microbiota. *Nature Reviews Microbiology* 6:776–788.

Li, C., X. C. Wu, O. Rieppel, L. T. Wang, and L. J. Zhao. 2008. An ancestral turtle from the late Triassic of southwestern China. *Nature* 456:497–501.

Libbrecht, K., and P. Rasmussen. 2004. The snowflake: Winter's secret beauty. *American Journal of Physics* 72:1134–1135.

Lieberman, D. E., D. M. Bramble, D. A. Raichlen, and J. J. Shea. 2009. Brains, brawn, and the evolution of human endurance running capabilities. Chapter 8 in *The First Humans: Origin and Early Evolution of the Genus* Homo, ed. F. E. Grine, J. G. Fleagle, and R. E. Leakey. Stony Brook, NY: Stony Brook University.

Liebman, S. W., and Y. O. Chernoff. 2012. Prions in yeast. *Genetics* 191:1041–1072.

Lind, M. I., M. K. Zwoinska, J. Andersson, H. Carlsson, T. Krieg, T. Larva, and A. A. Maklakov. 2020. Environmental variation mediates the evolution of anticipatory parental effects. *Evolution Letters* 4:371–381.

Lister, R., R. C. O'Malley, J. Tonti-Filippini, B. D. Gregory, C. C. Berry, A. H. Millar, and J. R. Ecker. 2008. Highly integrated single-base resolution maps of the epigenome in *Arabidopsis*. *Cell* 133:523–536.

Little, A. C., R. P. Burriss, B. C. Jones, L. M. DeBruine, and C. Caldwell. 2008. Social influence in human face preference: Men and women are influenced more for long-term than short-term attractiveness decisions. *Evolution and Human Behavior* 29:140–146.

Liu, X., Y. I. Li, and J. K. Pritchard. 2019. Trans effects on gene expression can drive omnigenic inheritance. *Cell* 177:1022–1034.e6.

Livnat, A., and D. Melamed. 2023. Evolutionary honing in and mutational replacement: How long-term directed mutational responses to specific environmental pressures are possible. *Theory in Biosciences* 142:87–105.

Lloyd Morgan, C. 1896. *Habit and Instinct*. London: Arnold.

Lombardo, M. P. 2008. Access to mutualistic endosymbiotic microbes: An underappreciated benefit of group living. *Behavioral Ecology and Sociobiology* 62:479–497.

Lombardo, U., J. Iriarte, L. Hilbert, J. Ruiz-Perez, J. M. Campril, and H. Veit. 2020. Early Holocene crop cultivation and landscape modification in Amazonia. *Nature* 581:190–193.

Longcamp, A., and J. Draghi. 2023. Evolutionary rescue via niche construction: Infrequent niche construction can prevent post-invasion extinction. *Theoretical Population Biology* 153:37–49.

Lord, K. A., G. Larson, R. P. Coppinger, and E.K.K. Karlsson. 2019. The history of fox farm foxes undermines the animal domestication syndrome. *Trends in Ecology and Evolution* 35:125–136.

Loreau, M. 2010. *From Populations to Ecosystems*. Princeton, NJ: Princeton University Press.

Losos, J. B. 2014. *The Princeton Guide to Evolution*. Princeton, NJ: Princeton University Press.

Losos, J. B., D. A. Creer, D. Glossip, R. Goellner, A. Hampton, G. Roberts, N. Haskell, et al. 2000. Evolutionary implications of phenotypic plasticity in the hindlimb of the lizard *Anolis sagrei*. *Evolution* 54:301–305.

Losos, J. B., T. W. Schoener, and D. A. Spiller. 2004. Predator-induced behaviour shifts and natural selection in field-experimental lizard populations. *Nature* 432:505–508.

Louçã, F. 2009. Emancipation through interaction: How eugenics and statistics converged and diverged. *Journal of the History of Biology* 42:649–684.

Love, A. C. 2003. Evolvability, dispositions, and intrinsicality. *Philosophy of Science* 70:1015–1027.

———. 2008. Explaining evolutionary innovations and novelties: Criteria of explanatory adequacy and epistemological prerequisites. *Philosophy of Science* 75:874–886.

———. 2015. Conceptual change and evolutionary developmental biology. In: *Conceptual Change in Biology: Scientific and Philosophical Perspectives on Evolution and Development*, ed. Love, 1–53. Dordrecht: Springer.

Love, A. C., and G. P. Wagner. 2022. Co-option of stress mechanisms in the origin of evolutionary novelties. *Evolution* 76:394–413.

Lu, C. P., L. Polak, B. E. Keyes, and E. Fuchs. 2016. Spatiotemporal antagonism in mesenchymal-epithelial signaling in sweat versus hair fate decision. *Science* 35:aah6102.

Lumsden, C. J., and E. O. Wilson. 1981. *Genes, Mind and Culture*. Cambridge, MA: Harvard University Press.

Luna, E., T.J.A. Bruce, M. R. Roberts, V. Flors, and J. Ton. 2012. Next-generation systemic acquired resistance. *Plant Physiology* 158:844–853.

Lyko, F., S. Foret, R. Kucharski, S. Wolf, C. Falckenhayn, and R. Maleszka. 2010. The honey bee epigenomes: Differential methylation of brain DNA in queens and workers. *PLOS Biology* 8:e1000506.

Lynch, M. 2007. The frailty of adaptive hypotheses for the origins of organismal complexity. *Proceedings of the National Academy of Sciences of the USA* 104:8597–8604.

Lynch, M., and B. Walsh. 1998. *Genetics and Analysis of Quantitative Traits*. Sunderland, MA: Sinauer Associates.

Lynch, V. J., R. D. Leclerc, G. May, and G. P. Wagner. 2011. Transposon-mediated rewiring of gene regulatory networks contributed to the evolution of pregnancy in mammals. *Nature Genetics* 43:1154–1159.

Lyons, S. L. 1995. The origins of T. H. Huxley's saltationism: History in Darwin's shadow. *Journal of the History of Biology* 28:463–494.

Lys, J. A., and R. H. Leuthold. 1994. Forces affecting water imbibition in *Macrotermes* workers (Termitidae, Isoptera). *Insectes sociaux* 41:79–84.

Lyson, T. R., and G. S. Bever. 2020. Origin and evolution of the turtle body plan. *Annual Review of Ecology, Evolution, and Systematics* 51:143–166.

Lyson, T. R., G. S. Bever, T. M. Scheyer, A. Y. Hsiang, and J. A. Gauthier. 2013. Evolutionary origin of the turtle shell. *Current Biology* 23:1113–1119.

Lyson, T. R., B. S. Rubidge, T. M. Scheyer, K. de Queiroz, E. R. Schachner, R.M.H. Smith, J. Botha-Brink, et al. 2016. Fossorial origin of the turtle shell. *Current Biology* 26:1887–1894.

Ma, L., M. Ng, J. Shi, A. V. Gore, D. Castranova, B. M. Weinstein, and W. R. Jeffery. 2021. Maternal control of visceral asymmetry evolution in *Astyanax* cavefish. *Scientific Reports* 11:10312.

Macagno, A.L.M., A. P. Moczek, and A. Pizzo. 2016. Rapid divergence of nesting depth and digging appendages among tunneling dung beetle populations and species. *American Naturalist* 187:E143–E151.

Macagno, A.L.M., E. E. Zattara, O. Ezeakudo, A. P. Moczek, and C. C. Ledón-Rettig. 2018. Adaptive maternal behavioral plasticity and developmental programming mitigate the transgenerational effects of temperature in dung beetles. *Oikos* 127:1319–1329.

MacArthur, R. 1962. Growth and regulation of animal population. *Ecology* 43:185–215.

MacCord, K. 2024. *How Does Germline Regenerate?* Chicago: University of Chicago Press.

Maden, M. 2009. Axolotl/newt. In: *Molecular Embryology: Methods and Protocols*, ed. P. T. Sharpe and I. Mason, 467–480. Totowa, NJ: Humana.

———. 2018. The evolution of regeneration: Where does that leave mammals? *International Journal of Developmental Biology* 62:369–372.

Maestripieri, D., and J. M. Mateo. 2009. *Maternal Effects in Mammals.* Chicago: University of Chicago Press.

Mallarino, R., P. R. Grant, B. R. Grant, A. Herrel, W. P. Kuo, and A. Abzhanov. 2011. Two developmental modules establish 3D beak-shape variation in Darwin's finches. *Proceedings of the National Academy of Sciences of the USA* 108:4057–4062.

Mangan, S., and U. Alon. 2003. Structure and function of the feed-forward loop network motif. *Proceedings of the National Academy of Sciences of the USA* 100:11980–11985.

Manrubia, S., J. A. Cuesta, J. Aguirre, S. E. Ahnert, L. Altenberg, A. V. Cano, P. Catalán, et al. 2021. From genotypes to organisms: State-of-the-art and perspectives of a cornerstone in evolutionary dynamics. *Physics of Life Reviews* 38:55–106.

Marler, P., and M. Tamura. 1964. Culturally transmitted patterns of vocal behavior in sparrows. *Science* 146:1483.

Marroig, G., and J. M. Cheverud. 2001. A comparison of phenotypic variation and covariation patterns and the role of phylogeny, ecology, and ontogeny during cranial evolution of new world monkeys. *Evolution* 55:2576–2600.

Marshall, D., and T. Uller. 2007. When is a maternal effect adaptive? *Oikos* 116:1957–1963.

Martin, S. A., B. H. Alhajeri, and S. J. Steppan. 2016. Dietary adaptations in the teeth of murine rodents (Muridae): A test of biomechanical predictions. *Biological Journal of the Linnean Society* 119:766–784.

Martin, S. J., R. R. Funch, P. R. Hanson, and E.-H. Yoo. 2018. A vast 4,000-year-old spatial pattern of termite mounds. *Current Biology* 28:R1292–R1293.

Martin-DeLeon, P. A. 2016. Uterosomes: Exosomal cargo during the estrus cycle and interaction with sperm. *Frontiers in Bioscience (Scholar Edition)* 8:115–122.

Martino, F., A. R. Perestrelo, V. Vinarský, S. Pagliari, and G. Forte. 2018. Cellular mechanotransduction: From tension to function. *Frontiers in Physiology* 9. https://doi.org/10.3389/fphys.2018.00824.

Masel, J. 2004. Genetic assimilation can occur in the absence of selection for the assimilating phenotype, suggesting a role for the canalization heuristic. *Journal of Evolutionary Biology* 17:1106–1110.

Mason, J. R., and R. F. Reidinger. 1982. Observational learning of food aversions in red-winged blackbirds (*Agelaius phoeniceus*). *Auk* 99:548–554.

Matthews, B., L. de Meester, C. G. Jones, B. W. Ibelings, T. J. Bouma, V. Nuutinen, J. van de Koppel, and J. Odling-Smee. 2014. Under niche construction: An operational bridge between ecology, evolution, and ecosystem science. *Ecological Monographs* 84:245–263.

Maynard Smith, J. 1982. *Evolution and the Theory of Games*. Cambridge: Cambridge University Press.

Maynard Smith, J., and E. Szathmáry. 1995. *The Major Transitions in Evolution*. Oxford: Oxford University Press.

Maynard Smith, J., R. Burian, S. Kauffman, P. Alberch, J. Campbell, B. Goodwin, R. Lande, et al. 1985. Developmental constraints and evolution: A perspective from the Mountain Lake Conference on Development and Evolution. *Quarterly Review of Biology* 60:265–287.

Mayr, E. 1942. *Systematics and the Origin of Species*. New York: Columbia University Press.

———. 1961. Cause and effect in biology. *Science* 134:1501–1506.

———. 1980. Some thoughts on the history of the evolutionary synthesis. In *The Evolutionary Synthesis*, ed. E. Mayr and W. Provine, 1–48. Cambridge, MA: Harvard University Press.

———. 1982. *The Growth of Biological Thought: Diversity, Evolution, and Inheritance*. Cambridge, MA: Harvard University Press.

———. 1984. The triumph of the evolutionary synthesis. *Times Literary Supplement*, November 2, 1984, 1261–1262.

Mayr, E., and W. Provine. 1980. *The Evolutionary Synthesis*. Cambridge, MA: Harvard University Press.

McCabe, C., S. Reader, and C. Nunn. 2015. Infectious disease, behavioural flexibility and the evolution of culture in primates. *Proceedings of the Royal Society B* 282:20140862.

McCleary, D. F., and J. Rine. 2017. Nutritional control of chronological aging and heterochromatin in *Saccharomyces cerevisiae*. *Genetics* 205:1179–1193.

McClure, S. B. 2015. The pastoral effect. *Current Anthropology* 56: 901–910.

McDonnell, M. J., and A. K. Hahs. 2015. Adaptation and adaptedness of organisms to urban environments. *Annual Review of Ecology, Evolution, and Systematics* 46:261–280.

McElreath, R. 2018. Sizing up human brain evolution. *Nature* 557:496–497.

McElreath, R., and R. Boyd. 2007. *Mathematical Models of Social Evolution: A Guide for the Perplexed*. Chicago: University of Chicago Press.

McFall-Ngai, M. J., and E. G. Ruby. 1991. Symbiont recognition and subsequent morphogenesis as early events in an animal-bacterial mutualism. *Science* 254:1491–1494.

McGaugh, S. E., S. Weaver, E. N. Gilbertson, B. Garrett, M. L. Rudeen, S. Grieb, J. Roberts, et al. 2019. Evidence for rapid phenotypic and behavioural shifts in a recently established cavefish population. *Biological Journal of the Linnean Society* 129:143–161.

McGhee, G. 2007. *The Geometry of Evolution: Adaptive Landscapes and Theoretical Morphospaces*. Cambridge: Cambridge University Press.

McGlothlin, J. W., M. E. Kobiela, H. V. Wright, D. L. Mahler, J. J. Kolbe, J. B. Losos, and E. D. Brodie III. 2018. Adaptive radiation along a deeply conserved genetic line of least resistance in *Anolis* lizards. *Evolution Letters* 2:310–322.

McGuigan, N., and A. Whiten. 2009. Emulation and "overemulation" in the social learning of causally opaque versus causally transparent tool use by 23- and 30-month-olds. *Journal of Experimental Child Psychology* 104:367–381.

McLean, C. Y., P. L. Reno, A. A. Pollen, A. I. Bassan, T. D. Capellini, C. Guenther, and V. B. Indjeian. 2011. Human-specific loss of regulatory DNA and the evolution of human-specific traits. *Nature* 471:216–219.

McLeod, D. V., G. Wild, and F. Úbeda. 2021. Epigenetic memories and the evolution of infectious diseases. *Nature Communications* 12:4273.

McNamara, J. M., and S.R.X. Dall. 2010. Information is a fitness enhancing resource. *Oikos* 119:231–236.

McNamara, J. M., and O. Leimar. 2020. *Game Theory in Biology: Concepts and Frontiers.* Oxford: Oxford University Press.

McNamara, J. M., S.R.X. Dall, P. Hammerstein, and O. Leimar. 2016. Detection vs. selection: Integration of genetic, epigenetic and environmental cues in fluctuating environments. *Ecology Letters* 19:1267–1276.

McShea, D. W. 1994. Mechanisms of large-scale evolutionary trends. *Evolution* 48:1747–1763.

Meaney, M. J., and M. Szyf. 2005. Environmental programming of stress responses through DNA methylation: Life at the interface between a dynamic environment and a fixed genome. *Dialogues in Clinical Neuroscience* 7:103–123.

Medawar, P. B. 1952. *An Unsolved Problem of Biology.* London: H. K. Lewis.

Mercader, J., H. Barton, J. Gillespie, J. Harris, S. Kuhn, R. Tyler, and Christophe Boesch. 2007. 4,300-year-old chimpanzee sites and the origins of percussive stone technology. *Proceedings of the National Academy of Sciences of the USA* 104:3043–3048.

Mercader, N., E. Leonardo, N. Azpiazu, A. Serrano, G. Morata, C. Martínez, and M. Torres. 1999. Conserved regulation of proximodistal limb axis development by Meis1/Hth. *Nature* 402:425–429.

Mery, F., S. Varela, E. Danchin, S. Blanchet, D. Parejo, I. Coolen, and R. Wagner. 2009. Public versus personal information for mate copying in an invertebrate. *Current Biology* 19: 730–734.

Mesoudi, A. 2011. *Cultural Evolution: How Darwinian Evolutionary Theory Can Explain Human Culture and Synthesize the Social Sciences.* Chicago: University of Chicago Press.

Mesoudi, A., K. N. Laland, R. Boyd, B. Buchanan, E. Flynn, R. N. McCauley, J. Renn, et al. 2013. The cultural evolution of technology and science. In *Cultural Evolution: Society, Technology, Language, and Religion,* ed. P. J. Richerson and M. H. Christiansen, 193–218. Cambridge, MA: MIT Press.

Metz, J.A.J.H. 2011. Thoughts on the geometry of meso-evolution: Collecting mathematical elements for a postmodern synthesis. In: *The Mathematics of Darwin's Legacy: Mathematics and Biosciences in Interaction,* ed. F. Chalub and J. Rodrigues, 193–232. Basel, Switzerland: Springer.

Meyer, A., T. D. Kocher, P. Basasibwaki, and A. C. Wilson. 1990. Monophyletic origin of Lake Victoria cichlid fishes suggested by mitochondrial DNA sequences. *Nature* 347:550–553.

Mezey, J. G., and D. Houle. 2005. The dimensionality of genetic variation for wing shape in *Drosophila melanogaster. Evolution* 59:1027–1038.

Miller, A. W., K. F. Oakeson, C. Dale, and M. D. Dearing. 2016. Effect of dietary oxalate on the gut microbiota of the mammalian herbivore *Neotoma albigula. Applied and Environmental Microbiology* 82:2669–2675.

Mills, R., and R. A. Watson. 2006. On crossing fitness valleys with the Baldwin effect. In: *Proceedings of the Tenth International Conference on the Simulation and Synthesis of Living Systems,* ed. L. M. Rocha, 493–499. Cambridge, MA: MIT Press.

Milocco, L., and I. Salazar-Ciudad. 2020. Is evolution predictable? Quantitative genetics under complex genotype-phenotype maps. *Evolution* 74:230–244.

Mineka, S., and M. Cook. 1988. Social learning and the acquisition of snake fear in monkeys. In: *Social Learning: Psychological and Biological Perspectives,* ed. B. G. Galef Jr. and T. R. Zentall, 51–73. Hillsdale, NJ: Erlbaum.

Minelli, A., and T. Pradeu. 2014. *Towards a Theory of Development.* Oxford: Oxford University Press.

Mitteroecker, P., S. M. Huttegger, B. Fischer, and M. Pavlicev. 2016. Cliff-edge model of obstetric selection in humans. *Proceedings of the National Academy of Sciences of the USA* 113: 14680–14685.

Mivart, St G. J. 1871. *On the Genesis of Species*. New York: D. Appleton.

Mizrahi, I., and E. Jami. 2021. A method to the madness. *EMBO Reports* 22:e52269.

Moczek, A. P. 2005. The evolution and development of novel traits, or how beetles got their horns. *BioScience* 55:937–951.

———. 2008. On the origins of novelty in development and evolution. *BioEssays* 30:432–447.

———. 2012. The nature of nurture and the future of evodevo: Toward a theory of developmental evolution. *Integrative and Comparative Biology* 52:108–119.

———. 2015. Re-evaluating the environment in developmental evolution. *Frontiers in Ecology and Evolution* 3. https://doi.org/10.3389/fevo.2015.00007.

———. 2017. What evolutionary developmental biology (evo devo) brings to evolutionary biology. *Extended Evolutionary Synthesis* (blog), January 9, 2017. https://extendedevolutionary synthesis.com/what-evo-devo-brings-to-evolutionary-biology/.

———. 2020. Biases in the study of developmental bias. *Evolution and Development* 22:3–6.

Moczek, A. P., and D. J. Emlen. 1999. Proximate determination of male horn dimorphism in the beetle *Onthophagus taurus* (Coleoptera: Scaravaeidae). *Journal of Evolutionary Biology* 12:27–37.

———. 2000. Male horn dimorphism in the scarab beetle, *Onthophagus taurus*: Do alternative reproductive tactics favour alternative phenotypes? *Animal Behaviour* 59:459–466.

Moczek, A. P., S. Sultan, S. Foster, C. Ledón-Rettig, I. Dworkin, H. F. Nijhout, E. Abouheif, et al. 2011. The role of developmental plasticity in evolutionary innovation. *Proceedings of the Royal Society B* 278:2705–2713.

Moeller, A. H., A. Caro-Quintero, D. Mjungu, A. V. Georgiev, E. V. Lonsdorf, M. N. Muller, A. E. Pusey, et al. 2016. Cospeciation of gut microbiota with hominids. *Science* 353:380–382.

Møller, A. P., and M. D. Jennions. 2002. How much variance can be explained by ecologists and evolutionary biologists? *Oecologia* 132:492–500.

Monks, N., and P. Palmer. 2002. *Ammonites*. Washington, DC: Smithsonian Books.

Montgomery, M. K., and M. McFall-Ngai. 1994. Bacterial symbionts induce host organ morphogenesis during early postembryonic development of the squid *Euprymna scolopes*. *Development* 120:1719–1729.

Montllor, C. B., A. Maxmen, and Alexander H. Purcell. 2002. Facultative bacterial endosymbionts benefit pea aphids *Acyrthosiphon pisum* under heat stress. *Ecological Entomology* 27:189–195.

Moore, R. S., R. Kaletsky, and C. T. Murphy. 2019. Piwi/PRG-1 argonaute and TGF-β mediate transgenerational learned pathogenic avoidance. *Cell* 177:1827–1841.e12.

Moran, N. A. 2007. Symbiosis as an adaptive process and source of phenotypic complexity. *Proceedings of the National Academy of Sciences of the USA* 104:8627–8633.

Moran, N. A., and D. B. Sloan. 2015. The hologenome concept: Helpful or hollow? *PLOS Biology* 13:e1002311.

Moran, N. A., and Y. Yun. 2015. Experimental replacement of an obligate insect symbiont. *Proceedings of the National Academy of Sciences of the USA* 112:2093–2096.

Moran, N. A., H. Ochman, and T. J. Hammer. 2019. Evolutionary and ecological consequences of gut microbial communities. *Annual Review of Ecology, Evolution, and Systematics* 50:451–475.

Moran, P.A.P. 1964. On the nonexistence of adaptive topographies. *Annals of Human Genetics* 27:383–393.

Morand-Ferron, J., and J. L. Quinn. 2011. Larger groups of passerines are more efficient problem solvers in the wild. *Proceedings of the National Academy of Sciences of the USA* 108: 15898–15903.

Morgan, T. H. 1910. Chromosomes and heredity. *American Naturalist* 44:449–496.

———. 1919. *The Physical Basis of Heredity*. Philadelphia, PA: Lippincott.

———. 1926. *The Theory of the Gene*. New Haven, CT: Yale University Press,.

Morgan, T.J.H., N. Uomini, L. E. Rendell, L. Chouinard- Thuly, S. E. Street, H. M. Lewis, C. P. Cross, et al. 2015. Experimental evidence for the co-evolution of hominin tool- making, teaching and language. *Nature Communications* 6:6029.

Morin, P. A., F. I. Archer, A. D. Foote, J. Vilstrup, E. E. Allen, P. Wade, J. Durban, et al. 2010. Complete mitochondrial genome analysis of killer whales (*Orcinus orca*) indicates multiple species. *Genome Research* 20:908–916.

Morita, W., N. Morimoto, and J. Jernvall. 2020. Mapping molar shapes on signaling pathways. *PLOS Computational Biology* 16:e1008436.

Moritz, C. 2002. Strategies to protect biological diversity and the evolutionary processes that sustain it. *Systematic Biology* 51:238–254.

Morton, A. 2002. *The Importance of Being Understood: Folk Psychology as Ethics*. London: Routledge.

Moss, L. 2003. *What Genes Can't Do*. Cambridge, MA: MIT Press.

Moura, A. E., J. G. Kenny, R. R. Chaudhuri, M. A. Hughes, R. R. Reisinger, P. J. de Bruyn, M. E. Dahlheim, et al. 2015. Phylogenomics of the killer whale indicates ecotype divergence in sympatry. *Heredity Edinburgh* 114:48–55.

Moura, A. E., J. G. Kenny, R. R. Chaudhuri, M. A. Hughes, A. J. Welch, R. R. Reisinger, P. J. Nico, et al. 2014. Population genomics of the killer whale indicates ecotype evolution in sympatry involving both selection and drift. *Molecular Ecology* 23:5179–5192.

Mousseau, T. A., and C. W. Fox. 1998a. The adaptive significance of maternal effects. *Trends in Ecology and Evolution* 13:403–407.

———. 1998b. *Maternal Effects as Adaptations*. New York: Oxford University Press.

Muegge, B. D., J. Kuczynski, D. Knights, J. C. Clemente, A. González, L. Fontana, B. Henrissat, et al. 2011. Diet drives convergence in gut microbiome functions across mammalian phylogeny and within humans. *Science* 332:970–974.

Müller, G. B. 2003a. Embryonic motility: Environmental influences and evolutionary innovation. *Evolution and Development* 5:56–60.

———. 2003b. Homology: The evolution of morphological organization. In: *Origination of Organismal Form*, ed. G. B. Müller and S. A. Newman, 51–69. Cambridge, MA: MIT Press.

———. 2007. Evo-devo: Extending the evolutionary synthesis. *Nature Reviews Genetics* 8:943–949.

———. 2010. Epigenetic innovation. In: *Evolution: The Extended Synthesis*, ed. M. Pigliucci and G. B. Müller, 307–331. Cambridge, MA: MIT Press.

———. 2017. Why an extended evolutionary synthesis is necessary. *Interface Focus* 7:20170015.

———. 2021. Evo-devo's contributions to the extended evolutionary synthesis. In: *Evolutionary Developmental Biology, a Reference Guide*, ed. L. Nuño de la Rosa and G. B. Müller, 1127–1138. Cham, Switzerland: Springer International.

Müller, G. B., and P. Alberch. 1990. Ontogeny of the limb skeleton in *Alligator mississippiensis*: Developmental invariance and change in the evolution of archosaur limbs. *Journal of Morphology* 203:151–164.

Müller, G. B., and G. P. Wagner. 1991. Novelty in evolution: Restructuring the concept. *Annual Review of Ecology and Systematics* 22:229–256.

Muller, H. 1929. The gene as the basis of life. In: *Proceedings of the International Congress of Plant Sciences, Ithaca, NY* 1:897–921.

Müller, F. M. 1861. *Lectures on the Science of Language, Delivered at the Royal Institution of Great Britain in April, May, and June, 1861*. Vol. 1. London: Longman, Green, Longman and Roberts.

Muñoz, M. M., and J. B. Losos. 2018. Thermoregulatory behavior simultaneously promotes and forestalls evolution in a tropical lizard. *American Naturalist* 191:E15–E26.

Munteanu, A., and R. V. Solé. 2008. Neutrality and robustness in evo-devo: Emergence of lateral inhibition. *PLOS Computational Biology* 4:e1000226.

Muschick, M., M. Barluenga, W. Salzburger, and A. Meyer. 2011. Adaptive phenotypic plasticity in the Midas cichlid fish pharyngeal jaw and its relevance in adaptive radiation. *BMC Evolutionary Biology* 11:116.

Muschick, M., A. Indermaur, and W. Saltzburger. 2012. Convergent evolution within an adaptive radiation of cichlid. *Current Biology* 22:2362–2368.

Musgrave, S., D. Morgan, E. Lonsdorf, R. Mundry, and C. Sanz. 2016. Tool transfers are a form of teaching among chimpanzees. *Scientific Reports* 6:34783.

Muthukrishna, M. 2023. *A Theory of Everyone*. Cambridge, MA: MIT Press.

Muthukrishna, M., and J. Henrich. 2016. Innovation in the collective brain. *Philosophical Transactions of the Royal Society B* 371:20150192.

Muthukrishna, M., M. Doebeli, M. Chudek, and J. Henrich. 2018. The cultural brain hypothesis: How culture drives brain expansion, sociality and life history. *PLOS Computational Biology*. https://doi.org/10.1371/journal.pcbi.1006504.

Nagashima, H., F. Sugahara, M. Takechi, R. Ericsson, Y. Kawashima-Ohya, Y. Narita, and S. Kuratani. 2009. Evolution of the turtle body plan by the folding and creation of new muscle connections. *Science* 325:193–196.

Nakahara, H., K. Doya, and O. Hikosaka. 2001. Parallel cortico-basal ganglia mechanisms for acquisition and execution of visuomotor sequences: A computational approach. *Journal of Cognitive Neuroscience* 13:626–647.

Naqvi, S., Y. Sleyp, H. Hoskens, K. Indencleef, J. P. Spence, R. Bruffaerts, A. Radwan, et al. 2021. Shared heritability of human face and brain shape. *Nature Genetics* 53:830–839.

Navalón, G., J. Marugán-Lobón, J. A. Bright, C. R. Cooney, and E. J. Rayfield. 2020. The consequences of craniofacial integration for the adaptive radiations of Darwin's finches and Hawaiian honeycreepers. *Nature Ecology and Evolution* 4:270–278.

Navarrete, A. F., S. M. Reader, S. E. Street, A. Whalen, and K. N. Laland. 2016. The coevolution of innovation and technical intelligence in primates. *Philosophical Transactions of the Royal Society B* 371:20150186.

Navarro, N., and A. Murat Maga. 2018. Genetic mapping of molar size relations identifies inhibitory locus for third molars in mice. *Heredity Edinburgh* 121:1–11.

Navon, D., I. Male, E. R. Tetrault, B. Aaronson, R. O. Karlstrom, and R. Craig Albertson. 2020. Hedgehog signaling is necessary and sufficient to mediate craniofacial plasticity in teleosts. *Proceedings of the National Academy of Sciences of the USA* 117:19321–19327.

Nayfach, S., B. Rodriguez-Mueller, N. Garud, and K. S. Pollard. 2016. An integrated metagenomics pipeline for strain profiling reveals novel patterns of bacterial transmission and biogeography. *Genome Research* 26:1612–1625.

Neeman, N., N. J. Robinson, F. V. Paladino, J. R. Spotila, and M. P. O'Connor. 2015. Phenology shifts in leatherback turtles (*Dermochelys coriacea*) due to changes in sea surface temperature. *Journal of Experimental Marine Biology and Ecology* 462:113–120.

Nei, M. 2013. *Mutation-Driven Evolution*. Oxford: Oxford University Press.

Newman, S. A. 2007. The Turing mechanism in vertebrate limb patterning. *Nature Reviews Molecular Cell Biology* 8:1.

———. 2010. Dynamic patterning modules. In: *Evolution: The Extended Synthesis*, ed. M. Pigliucci and G. B. Müller, 281–305. Cambridge, MA: MIT Press.

———. 2019. Inherency and homomorphy in the evolution of development. *Current Opinion in Genetics and Development* 57:1–8.

———. 2023. Inherency and agency in the origin and evolution of biological functions. *Biological Journal of the Linnean Society* 139:487–502.

Newman, S. A., and R. Bhat. 2009. Dynamical patterning modules: A "pattern language" for development and evolution of multicellular form. *International Journal of Developmental Biology* 53:693–705.

Newman, S. A., and H. L. Frisch. 1979. Dynamics of skeletal pattern formation in developing chick limb. *Science* 205:662–668.

Newman, S. A., and G. B. Müller. 2000. Epigenetic mechanisms of character origination. *Journal of Experimental Zoology* 288:304–317.

Newman, S. A., T. Glimm, and R. Bhat. 2018. The vertebrate limb: An evolving complex of self-organizing systems. *Progress in Biophysics and Molecular Biology* 137:12–24.

Newson, L., and P. Richerson. 2021. *A Story of Us: A New Look at Human Evolution*. New York: Oxford Academic.

Nicholson, D. 2014. The return of the organism as a fundamental explanatory concept in biology. *Philosophy Compass* 9:347–359.

Nicholson, D. J., and R. Gawne. 2015. Neither logical empiricism nor vitalism, but organicism: What the philosophy of biology was. *History and Philosophy of the Life Sciences* 37:345–381.

Nickles, T. 1989. Heuristic appraisal: A proposal. *Social Epistemology* 3:175–188.

Nicolakakis, N., D. Sol, and L. Lefebvre. 2003. Behavioural flexibility predicts species richness in birds, but not extinction risk. *Animal Behaviour* 65:445–452.

Nielsen, R., I. Hellmann, M. Hubisz, C. Bustamante, and A. G. Clark. 2007. Recent and ongoing selection in the human genome. *Nature Reviews Genetics* 8:857–868.

Niemiller, M. L., B. M. Fitzpatrick, and B. T. Miller. 2008. Recent divergence with gene flow in Tennessee cave salamanders (Plethodontidae: *Gyrinophilus*) inferred from gene genealogies. *Molecular Ecology* 17:2258–2275.

Nijhout, H. F. 1999. Control mechanisms of polyphenic development in insects. *BioScience* 49:181–192.

———. 2003. Development and evolution of adaptive polyphenisms. *Evolution and Development* 5:9–18.

———. 2019. The multistep morphing of beetle horns. *Science* 366:946–947.

Nijhout, H. F., and M.L.C. Reed. 2014. Homeostasis and dynamic stability of the phenotype link robustness and plasticity. *Integrative and Comparative Biology* 54:264–275.

Nijhout, H. F., A. M. Kudla, and C. C. Hazelwood. 2021. Genetic assimilation and accommodation: Models and mechanisms. *Current Topics in Developmental Biology* 141:337–369.

Nishida, H., and K. Sawada. 2001. Macho-1 encodes a localized mRNA in ascidian eggs that specifies muscle fate during embryogenesis. *Nature* 409:724–729.

Noble, D.W.A., R. Radersma, and T. Uller. 2019. Plastic responses to novel environments are biased towards phenotype dimensions with high additive genetic variation. *Proceedings of the National Academy of Sciences of the USA* 116:13452–13461.

Nowak, M., and R. Highfield. 2011. *SuperCooperators: Altruism, Evolution, and Why We Need Each Other to Succeed*. New York: Free Press.

Nuño de la Rosa, L. 2017. Computing the extended synthesis: Mapping the dynamics and conceptual structure of the evolvability research front. *Journal of Experimental Zoology Part B* 328:395–411.

Nyirenda, M. J., R. S. Lindsay, C. J. Kenyon, A. Burchell, and J. R. Seckl. 1998. Glucocorticoid exposure in late gestation permanently programs rat hepatic phosphoenolpyruvate carboxykinase and glucocorticoid receptor expression and causes glucose intolerance in adult offspring. *Journal of Clinical Investigation* 101:2174–2181.

O'Brien, M. J., and K. N. Laland. 2012. Genes, culture and agriculture: An example of human niche construction. *Current Anthropology* 53:434–470.

O'Dea, R. E., D.W.A. Noble, S. L. Johnson, D. Hesselson, and S. Nakagawa. 2016. The role of non-genetic inheritance in evolutionary rescue: Epigenetic buffering, heritable bet hedging and epigenetic traps. *Environmental Epigenetics* 2:dvv014.

Odling-Smee, F. J. 1983. Multiple levels in evolution: An approach to the nature-nurture issue via "applied epistemology." In: *Animal Models of Human Behaviour*, ed. G. Davey, 135–158. Chichester, UK: Wiley.

———. 1988. Niche-constructing phenotypes. In: *The Role of Behavior in Evolution*, ed. H. C. Plotkin, 73–132. Cambridge, MA: MIT Press,.

———. 2024. *Niche Construction: How Life Contributes to Its Own Evolution*. Cambridge, MA: MIT Press.

Odling-Smee, F. J., K. N. Laland, and M. W. Feldman. 2003. *Niche Construction: The Neglected Process in Evolution*. Princeton, NJ: Princeton University Press.

Oh, K. P., and A. V. Badyaev. 2010. Structure of social networks in a passerine bird: Consequences for sexual selection and the evolution of mating strategies. *American Naturalist* 176:E80–E89.

Oliver, K. M., J. A. Russell, N. A. Moran, and M. S. Hunter. 2003. Facultative bacterial symbionts in aphids confer resistance to parasitic wasps. *Proceedings of the National Academy of Sciences of the USA* 100:1803–1807.

Olson, E. N., 2006. Gene regulatory networks in the evolution and development of the heart. *Science* 313:1922–1927.

Olsson, A., and E. A. Phelps. 2007. Social learning of fear. *Nature Neuroscience* 10:1095–1102.

Orenstein, R. 2012. *Turtles, Tortoises and Terrapins: A Natural History*. Richmond Hill, ON: Firefly Books.

Ornelas-García, C. P., and C. Pedraza-Lara. 2016. Phylogeny and evolutionary history of *Astyanax mexicanus*. In: *Biology and Evolution of the Mexican Cavefish*, ed. A. Keene, M. Yoshizawa, and S. McGaugh, 77–90. San Diego, CA: Academic.

Osborn, H. F. 1896. Ontogenetic and phylogenetic variation. *Science* 4:786–789.

Oskin, B. 2013. Antarctic hills haven't seen water in 14 million years. LiveScience, November 1, 2013. https://www.livescience.com/40890-antarctic-dry-valleys-no-water.html.

Osmanovic, D., D. A. Kessler, Y. Rabin, and Y. Soen. 2018. Darwinian selection of host and bacteria supports emergence of Lamarckian-like adaptation of the system as a whole. *Biology Direct* 13:24.

Otsuka, J. 2016. Causal foundations of evolutionary genetics. *British Journal for the Philosophy of Science* 67:247–269.

———. 2019. Ontology, causality, and methodology of evolutionary research programs. In: *Evolutionary Causation: Biological and Philosophical Reflections*, ed. T. Uller and K. N. Laland, 247–264. Cambridge, MA: MIT Press.

Otto, S. P. 2018. Adaptation, speciation and extinction in the Anthropocene. *Proceedings of the Royal Society B* 285:20182047.

Otto, S. P., and T. Day. 2007. *A Biologist's Guide to Mathematical Modeling in Ecology and Evolution*. Princeton, NJ: Princeton University Press.

Oudman, T., K. N. Laland, G. Ruxton, I. Tombre, P. Shimmings, and J. Prop. 2020. Young birds switch but old birds lead: How barnacle geese adjust migratory habits to environmental change. *Frontiers in Ecology and Evolution* 7. https://doi.org/10.3389/fevo.2019.00502.

Overington, S. E., J. Morand-Ferron, N. J. Boogert, and L. Lefebvre. 2009. Technical innovations drive the relationship between innovativeness and residual brain size in birds. *Animal Behaviour* 78:1001–1010.

Owren, M. J., and D. Rendall. 1997. An affect-conditioning model of nonhuman primate vocal signaling. In *Perspectives in Ethology*, vol. 12, *Communication*, ed. D. H. Owings, M. D. Beecher, and N. S. Thompson, 299–346. New York: Plenum.

Oyama, S. 1985. *The Ontogeny of Information: Developmental Systems and Evolution*, 2nd ed. Durham, NC: Duke University Press.

———. 2000. *The Ontogeny of Information: Developmental Systems and Evolution*, 2nd ed., revised and expanded. Durham, NC: Duke University Press.

Oyama, S., P. E. Griffiths, and R. D. Gray. 2001. *Cycles of Contingency: Developmental Systems and Evolution*. Cambridge, MA: MIT Press.

Oyston, J. W., M. Hughes, S.Gerber, and M. A. Wills. 2016. Why should we investigate the morphological disparity of plant clades? *Annals of Botany* 117:859–879.

Paenke, I., B. Sendhoff, and T. J. Kawecki. 2007. Influence of plasticity and learning on evolution under directional selection. *American Naturalist* 170:E47–E58.

Page, K. M., and R. Perez-Carrasco. 2018. Degradation rate uniformity determines success of oscillations in repressive feedback regulatory networks. *Journal of the Royal Society Interface* 15:20180157.

Pagel, M. 2012. *Wired for Culture: Origins of the Human Social Mind*. New York: W. W. Norton.

Painter, K. J., P. K. Maini, and H. G. Othmer. 1999. Stripe formation in juvenile *Pomacanthus* explained by a generalized Turing mechanism with chemotaxis. *Proceedings of the National Academy of Sciences of the USA* 96:5549–5554.

Painter, R. C., S. R. de Rooij, P.M.M. Bossuyt, C. Osmond, D.J.P. Barker, O. P. Bleker, and T. J. Roseboom. 2006a. A possible link between prenatal exposure to famine and breast cancer: A preliminary study. *American Journal of Human Biology* 18:853–856.

Painter, R. C., S. R. de Rooij, P. M. Bossuyt, T. A. Simmers, C. Osmond, D. J. Barker, O. P. Bleker, et al. 2006b. Early onset of coronary artery disease after prenatal exposure to the Dutch famine. *American Journal of Clinical Nutrition* 84:322–327.

Painter, R. C., C. Osmond, P. Gluckman, M. Hanson, D.I.W. Phillips, and T. J. Roseboom. 2008. Transgenerational effects of prenatal exposure to the Dutch famine on neonatal adiposity and health in later life. *BJOG: An International Journal of Obstetrics and Gynaecology* 115:1243–1249.

Pal, C. 1998. Plasticity, memory and the adaptive landscape of the genotype. *Proceedings of the Royal Society B* 265:1319–23.

Pang, J. C., K. M. Aquino, M. Oldehinkel, P. A. Robinson, B. D. Fulcher, M. Breakspear, and A. Fornito. 2023. Geometric constraints on human brain function. *Nature* 618:566–574.

Panganiban, G., S. M. Irvine, C. Lowe, H. Roehl, L. S. Corley, B. Sherbon, J. K. Grenier, et al. 1997. The origin and evolution of animal appendages. *Proceedings of the National Academy of Sciences of the USA* 94:5162–5166.

Parsons, K. J., and R. C. Albertson. 2009. Roles for Bmp4 and CaM1 in shaping the jaw: Evodevo and beyond. *Annual Review of Genetics* 43:369–388.

Parsons, K. J., M. Concannon, D. Navon, J. Wang, I. Ea, K. Groveas, C. Campbell, et al. 2016. Foraging environment determines the genetic architecture and evolutionary potential of trophic morphology in cichlid fishes. *Molecular Ecology* 25:6012–6023.

Parsons, K. J., K. McWhinnie, N. Pilakouta, and L. Walker. 2019. Does phenotypic plasticity initiate developmental bias? *Evolution and Development* 22:56–70.

Parsons, K. J., A. Rigg, A. J. Conith, A. C. Kitchener, S. Harris, and H. Zhu. 2020. Skull morphology diverges between urban and rural populations of red foxes mirroring patterns of domestication and macroevolution. *Proceedings of the Royal Society B* 287:20200763.

Parsons, K. J., H. D. Sheets, S. Skúlason, and M. M. Ferguson. 2011. Phenotypic plasticity, heterochrony and ontogenetic repatterning during juvenile development of divergent Arctic charr (*Salvelinus alpinus*). *Journal of Evolutionary Biology* 24:1640–1652.

Parsons, K. J., A. T. Taylor, K. E. Powder, and R. C. Albertson. 2014. Wnt signalling underlies the evolution of new phenotypes and craniofacial variability in Lake Malawi cichlids. *Nature Communications* 5:3629.

Parter, M., N. Kashtan, and U. Alon. 2008. Facilitated variation: How evolution learns from past environments to generalize to new environments. *PLOS Computational Biology* 4:e1000206.

Patel, A. D. 2006. Musical rhythm, linguistic rhythm, and human evolution. *Music Perception* 24:99–104.

Patel, A. D., J. R. Iversen, M. R. Bregman, and I. Schulz. 2009. Experimental evidence for synchronization to a musical beat in a nonhuman animal. *Current Biology* 19:827–830.

Patterson, L. B., and D. M. Parichy. 2019. Zebrafish pigment pattern formation: Insights into the development and evolution of adult form. *Annual Review of Genetics* 53:505–530.

Pauchet, Y., R. Kirsch, S. Giraud, H. Vogel, and D. G. Heckel. 2014. Identification and characterization of plant cell wall degrading enzymes from three glycoside hydrolase families in the cerambycid beetle *Apriona japonica*. *Insect Biochemistry and Molecular Biology* 49:1–13.

Paul, D. B. 1988. The selection of the "survival of the fittest." *Journal of the History of Biology* 21:411–424.

Paulsen, S. M. 1996. Evidence against the constancy of G: Quantitative genetics of the wing color pattern in the buckeye butterfly (*Precis coenia* and *Precis evarete*). *Evolution* 50:1585–1597.

Pavličev, M., and G. P. Wagner. 2012. A model of developmental evolution: Selection, pleiotropy and compensation. *Trends in Ecology and Evolution* 27:316–322.

Pavličev, M., S. Bourg, and A. Le Rouzic. 2023. The genotype-phenotype map structure and its role in evolvability. Chapter 8 in *Evolvability: A Unifying Concept in Evolutionary Biology?*, ed. T. F. Hansen, D. Houle, M. Pavličev, and C. Pélabon, Vienna Series in Theoretical Biology. Cambridge, MA: MIT Press.

Pavličev, M., E. A. Norgard, G. L. Fawcett, and J. M. Cheverud. 2011. Evolution of pleiotropy: Epistatic interaction pattern supports a mechanistic model underlying variation in genotype–phenotype map. *Journal of Experimental Zoology Part B* 316:371–385.

Pavličev, M., R. Romero, and P. Mitteroecker. 2020. Evolution of the human pelvis and obstructed labor: New explanations of an old obstetrical dilemma. *American Journal of Obstetrics and Gynecology* 222:3–16.

Payne, J. L., and A. Wagner. 2015. Mechanisms of mutational robustness in transcriptional regulation. *Frontiers in Genetics* 27:322.

———. 2019. The causes of evolvability and their evolution. *Nature Reviews Genetics* 20:24–38.

Pearl, J. 2009. *Causality: Models, Reasoning, and Inference*, 2nd ed. New York: Cambridge University Press.

Pélabon, C., M. B. Morrissey, J. M. Reid, and J. Sztepanacz. 2023. Can we explain variation in evolvability on ecological timescales? Chapter 13 in *Evolvability: A Unifying Concept in Evolutionary Biology?*, ed. T. F. Hansen, D. Houle, M. Pavličev, and Pélabon, Vienna Series in Theoretical Biology. Cambridge, MA: MIT Press.

Pelegri, F. 2003. Maternal factors in zebrafish development. *Developmental Dynamics* 228:535–554.

Pelletier, F., D. Garant, and A. P. Hendry. 2009. Eco-evolutionary dynamics. *Philosophical Transactions of the Royal Society B* 364:1483–1489.

Peluffo, A. E. 2015. The "genetic program": Behind the genesis of an influential metaphor. *Genetics* 200:685–696.

Pembrey, M. E., L. O. Bygren, G. Kaati, S. Edvinsson, K. Northstone, M. Sjöström, J. Golding, and the Alspac Study Team. 2006. Sex-specific, male-line transgenerational responses in humans. *European Journal of Human Genetics* 14:159–166.

Pendleton, A. L., F. Shen, A. M. Taravella, S. Emery, K. R. Veeramah, A. R. Boyko, and J. M. Kidd. 2018. Comparison of village dog and wolf genomes highlights the role of the neural crest in dog domestication. *BMC Biology* 16:64.

Pepperberg, I. M. 1999. *The Alex Studies*. Cambridge, MA: Harvard University Press.

———. 2017. Animal language studies: What happened? *Psychonomic Bulletin and Review* 24:181–185.

Perez-Carrasco, R., C. P. Barnes, Y. Schaerli, M. Isalan, J. Briscoe, and K. M. Page. 2018. Combining a toggle switch and a repressilator within the ac-dc circuit generates distinct dynamical behaviors. *Cell Systems* 6:521–530.e3.

Perkins, S. 2013. Giant camels roamed Arctic realms. *Science*, March 5, 2013. https://www.science.org/content/article/giant-camels-roamed-arctic-realms.

Perofsky, A. C., R. J. Lewis, L. A. Abondano, A. Di Fiore, and L. A. Meyers. 2017. Hierarchical social networks shape gut microbial composition in wild Verreaux's sifaka. *Proceedings of the Royal Society B* 284:20172274.

Perry, G. H., N. J. Dominy, K. G. Claw, A. S. Lee, H. Fiegler, R. Redon, J. Werner, et al. 2007. Diet and the evolution of human amylase gene copy number variation. *Nature Genetics* 39:1256–1260.

Perry, S., M. Panger, L. M. Rose, M. Baker, J. Gros-Louis, K. Jack, K. C. Mackinnon, et al. 2003. Traditions in wild white-faced capuchin monkeys. In *The Biology of Traditions: Models and Evidence*, ed. D. M. Fragaszy and S. Perry, 391–425. Cambridge: Cambridge University Press.

Pértille, F., V. H. da Silva, A. M. Johansson, T. Lindström, D. Wright, L. L. Coutinho, P. Jensen, et al. 2019. Mutation dynamics of CpG dinucleotides during a recent event of vertebrate diversification. *Epigenetics* 14:685–707.

Peter, I., and E. H. Davidson. 2015. *Genomic Control Process: Development and Evolution*. San Diego, CA: Academic.

Peterson, E. 2016. *The Life Organic. The Theoretical Biology Club and the Roots of Epigenetics*. Pittsburg, PA: University of Pittsburg Press.

Peterson, T., and G. B. Müller. 2016. Phenotypic novelty in evodevo: The distinction between continuous and discontinuous variation and its importance in evolutionary theory. *Evolutionary Biology* 43:314–335.

———. 2018. Developmental finite element analysis of cichlid pharyngeal jaws: Quantifying the generation of a key innovation. *PLOS One* 13:e0189985.

Petroski, H. 1992. *The Evolution of Useful Things*. New York: Vintage Books.

Pfennig, D. W. 1990. The adaptive significance of an environmentally-cued developmental switch in an anuran tadpole. *Oecologia* 85:101–107.

———. 2021. *Phenotypic Plasticity and Evolution: Causes, Consequences, and Controversies*. New York: Academic.

Pfennig, D. W., and M. McGee. 2010. Resource polyphenism increases species richness: A test of the hypothesis. *Philosophical Transactions of the Royal Society B* 365:577–591.

Pfennig, D. W., and P. J. Murphy. 2000. Character displacement in polyphenic tadpoles. *Evolution* 54:1738–1749.

Pfennig, D. W., M. A. Wund, E C. Snell-Rood, T. Cruickshank, C. D. Schlichting, and A. P. Moczek. 2010. Phenotypic plasticity's impacts on diversification and speciation. *Trends in Ecology and Evolution* 25:459–467.

Pfenning, A. R., E. Hara, O. Whitney, M. V. Rivas, R. Wang, P. L. Roulhac, J. T. Howard, et al. 2014. Convergent transcriptional specializations in the brains of humans and song-learning birds. *Science* 346:1256846.

Phillips, P. C., M. C. Whitlock, and K. Fowler. 2001. Inbreeding changes the shape of the genetic covariance matrix in *Drosophila melanogaster*. *Genetics* 158:1137–1145.

Piersma, T., and J. A. van Gils. 2011. *The Flexible Phenotype*. Oxford: Oxford University Press.

Pigliucci, M. 2008. Is evolvability evolvable? *Nature Reviews Genetics* 9:75–82.

———. 2019. Causality and the role of philosophy of science. In: *Evolutionary Causation: Biological and Philosophical Reflections*, ed. T. Uller and K. N. Laland, 13–28. Cambridge, MA: MIT Press.

Pigliucci, M., and G. B. Müller. 2010. *Evolution, the Extended Synthesis*. Cambridge, MA: MIT Press.

Pigliucci, M., C. J. Murren, and C. D. Schlichting. 2006. Phenotypic plasticity and evolution by genetic assimilation. *Journal of Experimental Biology* 209:2362–2367.

Pika, S., K. Liebal, J. Call, and M. Tomasello. 2005. The gestural communication of apes. *Gesture* 5:41–56.

Pilakouta, N., J. L. Humble, I.D.C. Hill, J. Arthur, A.P.B. Costa, B. A. Smith, B. K. Kristjánsson, et al. 2023. Testing the predictability of morphological evolution in contrasting thermal environments. *Evolution* 77:239–253.

Pineda-Munoz, S., I. A. Lazagabaster, J. Alroy, and A. R. Evans. 2017. Inferring diet from dental morphology in terrestrial mammals. *Methods in Ecology and Evolution* 8:481–491.

Pinker, S. 1995. *The Language Instinct*. New York: Penguin.

———. 2010. The cognitive niche: Coevolution of intelligence, sociality, and language. *Proceedings of the National Academy of Sciences of the USA* 107:8993–8999.

Piperno, D. R. 2017. Assessing elements of an extended evolutionary synthesis for plant domestication and agricultural origin research. *Proceedings of the National Academy of Sciences of the USA* 114:6429–6437.

Plotkin, H. C. 1988. *The Role of Behavior in Evolution*. Cambridge, MA: MIT Press.

———. 1994. *Darwin Machines and the Nature of Knowledge*. London: Penguin.

———. 2010. *Evolutionary Worlds without End*. Oxford: Oxford University Press.

Plotkin, H. C., and F. J. Odling-Smee. 1981. A multiple-level model of evolution and its implications for sociobiology. *Behavioral and Brain Sciences* 4:225–235.

Pocheville, A. 2019. A Darwinian dream and a few whimsical nightmares: On time, levels, and processes in evolution. In: *Evolutionary Causation: Biological and Philosophical Reflections*, ed. T. Uller and K. N. Laland, 265–298. Cambridge, MA: MIT Press.

Pohle, A.-K., A. Zalewski, M. Muturi, C. Dullin, L. Farková, L. Keicher, and D.K.N. Dechmann. 2023. Domestication effect of reduced brain size is reverted when mink become feral. *Royal Society Open Science* 10:230463.

Pollard, A. S., B. G. Charlton, J. R. Hutchinson, T. Gustafsson, I. M. McGonnell, J. A. Timmons, and A. A. Pitsillides. 2017. Limb proportions show developmental plasticity in response to embryo movement. *Scientific Reports* 7:41926.

Pollard, K. S., S. R. Salama, N. Lambert, M. A. Lambot, S. Coppens, J. S. Pedersen, S. Katzman, et al. 2006. An RNA gene expressed during cortical development evolved rapidly in humans. *Nature* 443:167–172.

Pontarotti, G., M. Mossio, and A. Pocheville. 2022. The genotype-phenotype distinction: From Mendelian genetics to 21st century biology. *Genetica* 150:223–234.

Pontzer, H., M. H. Brown, D. A. Raichlen, H. Dunsworth, B.Hare, K. Walker, A. Luke, et al. 2016. Metabolic acceleration and the evolution of human brain size and life history. *Nature* 533:390–392

Post, D. M., and E. P. Palkovacs. 2009. Eco-evolutionary feedbacks in community and ecosystem ecology: Interactions between the ecological theatre and the evolutionary play. *Philosophical Transactions of the Royal Society B* 364:1629–1640.

Potochnik, A. 2017. *Idealization and the Aims of Science*. Chicago: University of Chicago Press.

Potts, R., and J. T. Faith. 2015. Alternating high and low climate variability: The context of natural selection and speciation in Plio-Pleistocene hominin evolution. *Journal of Human Evolution* 87:5–20.

Povinelli, D. 2000. *Folk Physics for Apes*. New York: Oxford University Press.

Powder, K. E., and R. C. Albertson. 2016. Cichlid fishes as a model to understand normal and clinical craniofacial variation. *Developmental Biology* 415:338–346.

Powers, A. K., D. J. Berning, and J. B. Gross. 2020a. Parallel evolution of regressive and constructive craniofacial traits across distinct populations of *Astyanax mexicanus* cavefish. *Journal of Experimental Zoology Part B* 334:450–462.

Powers, A. K., T. E. Boggs, and J. B. Gross. 2020b. An asymmetric genetic signal associated with mechanosensory expansion in cave-adapted fish. *Symmetry* 12:1951.

Powers, A. K., S. A. Kaplan, T. E. Boggs, and J. B. Gross. 2018. Facial bone fragmentation in blind cavefish arises through two unusual ossification processes. *Scientific Reports* 8:7015.

Prabhakar, S., J. P. Noonan, S. Pääbo, and E. M. Rubin. 2006. Accelerated evolution of conserved noncoding sequences in humans. *Science* 314:786.

Pray, L. A. 2004. Epigenetics: Genome, meet your environment. *Scientist (Philadelphia, PA)* 18(13):14–20.

Price, T. D. 2006. Phenotypic plasticity, sexual selection and the evolution of colour patterns. *Journal of Experimental Biology* 209:2368–2376.

Price, T. D., A. Qvarnström, and D. E. Irwin. 2003. The role of phenotypic plasticity in driving genetic evolution. *Proceedings of the Royal Society B* 270:1433–1440.

Proulx, S. R., S. Dey, T. Guzella, and H. Teotónio. 2019. How differing modes of non-genetic inheritance affect population viability in fluctuating environments. *Ecology Letters* 22:1767–1775.

Provine, W. 1971. *The Origins of Theoretical Population Genetics*. Chicago: University of Chicago Press.

Pujol, B., S. Blanchet, A. Charmantier, E. Danchin, B. Facon, P. Marrot. F. Roux, et al. 2018. The missing response to selection in the wild. *Trends in Ecology and Evolution* 33:337–346.

Pulliam, H. R., and C. Dunford. 1980. *Programmed to Learn: An Essay on the Evolution of Culture*. New York: Columbia University Press.

Purves, D. 1988. *Body and Brain: A Trophic Theory of Neural Connections*. Cambridge, MA: Harvard University Press.

Purves, D., and J. W. Lichtman. 1985. *Principles of Neural Development*. Sunderland, MA: Sinauer Associates.

Quiñones, A. E., and I. Pen. 2017. A unified model of hymenopteran preadaptations that trigger the evolutionary transition to eusociality. *Nature Communications* 8:15920.

Quintus, S., and E. E. Cochrane. 2018. The prevalence and importance of niche construction in agricultural development in Polynesia. *Journal of Anthropological Archaeology* 51:173–186.

Rachinsky, A., and K. Hartfelder. 1990. Corpora allata activity, a prime regulating element for caste-specific juvenile hormone titre in honey bee larvae (Apis mellifera carnica). *Journal of Insect Physiology* 36:189–194.

Radersma, R., D.W.A. Noble, and T. Uller. 2020. Plasticity leaves a phenotypic signature during local adaptation. *Evolution Letters* 4:360–370.

Radford, E. J., M. Ito, H. Shi, J. A. Corish, K. Yamazawa, E. Isganaitis, S. Seisenberger, et al. 2014. In utero undernourishment perturbs the adult sperm methylome and intergenerational metabolism. *Science* 345:1255903.

Radick, G. 2007. *The Simian Tongue: The Long Debate about Animal Language*. Chicago: University of Chicago Press.

Raff, R. A. 1996. *The Shape of Life: Genes, Development, and the Evolution of Animal Form*. Chicago: University of Chicago Press.

Raghanti, M. A., M. K. Edler, A. R. Stephenson, E. L. Munger, B. Jacobs, P. R. Hof, C. C. Sherwood, et al. 2018. A neurochemical hypothesis for the origin of hominids. *Proceedings of the National Academy of Sciences of the USA* 115:E1108–E1116.

Rago, A., K. Kouvaris, T. Uller, and R. A. Watson. 2019. How adaptive plasticity evolves when selected against. *PLOS Computational Biology* 15:e1006260.

Rajakumar, R., D. San Mauro, M. B. Dijkstra, M. H. Huang, D. E. Wheeler, F. Hiou-Tim, A. Khila, et al. 2012. Ancestral developmental potential facilitates parallel evolution in ants. *Science* 335:79–82.

Rakic, P. 1986. Mechanisms of ocular dominance segregation in the lateral geniculate nucleus: Competitive elimination hypothesis. *Trends in Neuroscience* 9:11–15.

Ram, Y., U. Liberman, and M. W. Feldman. 2019. Vertical and oblique cultural transmission fluctuating in time and in space. *Theoretical Population Biology* 125:11–19.

Ramakers, J., P. Gienapp, and M. Visser. 2019. Phenological mismatch drives selection on elevation, but not on slope, of breeding time plasticity in a wild songbird. *Evolution* 73:175–187.

Rampelli, S., S. Turroni, F. Debandi, A. Alberdi, S. L. Schnorr, C. A. Hofman, A. Taddia, et al. 2021. The gut microbiome buffers dietary adaptation in Bronze Age domesticated dogs. *iScience* 24:102816.

Raspopovic, J., L. Marcon, L. Russo, and J. Sharpe. 2014. Digit patterning is controlled by a Bmp-Sox9-Wnt Turing network modulated by morphogen gradients. *Science* 345:566–570.

Raup, D. M. 1966. Geometric analysis of shell coiling: General problems. *Journal of Paleontology* 40:1178–1190.

Reader, S. M., and K. N. Laland. 2002. Social intelligence, innovation and enhanced brain size in primates. *Proceedings of the National Academy of Sciences of the USA* 99:4436–4441.

———. 2003. *Animal Innovation*. New York: Oxford University Press.

Reader, S. M., Y. Hager, and K. N. Laland. 2011. The evolution of primate general and cultural intelligence. *Philosophical Transactions of the Royal Society B* 366:1017–1027.

Reader, S. M., J. R. Kendal, and K. N. Laland. 2003. Social learning of foraging sites and escape routes in wild Trinidadian guppies. *Animal Behaviour* 66:729–739.

Reader, S. M., J. Morand-Ferron, and E. Flynn. 2016. Animal and human innovation: Novel problems and novel solutions. *Philosophical Transactions of the Royal Society B* 371:20150182.

Redman, R. S., K. B. Sheehan, R. G. Stout, R. J. Rodriguez, and J. M. Henson. 2002. Thermotolerance generated by plant/fungal symbiosis. *Science* 298:1581.

Reinhold, K. 2000. Maintenance of a genetic polymorphism by fluctuating selection on sex-limited traits. *Journal of Evolutionary Biology* 13:1009–1014.

Rembold, H. 1987. Caste specific modulation of juvenile hormone titres in Apis mellifera. *Insect Biochemistry* 17:1003–1006

Ren, X., X. Lemoine, D. Mo, T. R. Kidder, Y. Guo, Z. Qin, and X. Liu. 2016. Foothills and inter-mountain basins: Does China's fertile arc have "hilly flanks"? *Quaternary International* 426:86–96.

Rendell, L., and H. Whitehead. 2001. Culture in whales and dolphins. *Behavioral and Brain Sciences* 24:309–324.

Rendell, L., R. Boyd, D. Cownden, M. Enquist, K. Eriksson, M. W. Feldman, L. Fogarty, et al. 2010. Why copy others? Insights from the social learning strategies tournament. *Science* 328: 5975.

Rendell, L., L. Fogarty, W.J.E. Hoppitt, T.J.H. Morgan, M. M. Webster, and K. N. Laland. 2011. Cognitive culture: Theoretical and empirical insights into social learning strategies. *Trends in Cognitive Science* 15:68–76.

Reznick, D., L. Nunney, and A. Tessier. 2000. Big houses, big cars, superfleas and the costs of reproduction. *Trends in Ecology and Evolution* 15:421–425.

Rheinberger, H. J., and S. Müller-Wille. 2017. *The Gene: From Genetics to Postgenomics*. Chicago: University of Chicago Press.

Riahi, I. A. 2023. Macroevolutionary origins of comparative development. *Economic Journal* 2023:uead095.

Riber, L., and L. H. Hansen. 2021. Epigenetic memories: The hidden drivers of bacterial persistence? *Trends in Microbiology* 29:190–194.

Rice, R., A. Kallonen, J. Cebra-Thomas, and S. F. Gilbert. 2016. Development of the turtle plastron, the order-defining skeletal structure. *Proceedings of the National Academy of Sciences of the USA* 113:5317.

Rice, S. H. 2004. *Evolutionary Theory: Mathematical and Conceptual Foundations*. Sunderland, MA: Sinauer Associates.

———. 2008a. A stochastic version of the Price equation reveals the interplay of deterministic and stochastic processes in evolution. *BMC Evolutionary Biology* 8:262.

———. 2008b. Theoretical approaches to the evolution of development and genetic architecture. *Annals of the New York Academy of Sciences* 1133:67–86.

Richards, C. L., A. W. Schrey, and M. Pigliucci. 2012. Invasion of diverse habitats by few Japanese knotweed genotypes is correlated with epigenetic differentiation. *Ecology Letters* 15:1016–1025.

Richards, M. P., R. J. Schulting, and R.E.M. Hedges. 2003. Archaeology: Sharp shift in diet at onset of Neolithic. *Nature* 425:366.

Richerson, P. J. 2019. An integrated Bayesian theory of phenotypic flexibility. *Behavioural Processes* 161:54–64.

Richerson, P. J., and R. Boyd. 2005. *Not by Genes Alone: How Culture Transformed Human Evolution*. Chicago: University of Chicago Press.

———. 2013. Rethinking paleoanthropology: A world queerer than we supposed. In: *Evolution of Mind*, ed. G. Hatfield and H. Pittman, 263–302. Philadelphia: University of Pennsylvania Museum of Archaeology and Anthropology.

———. 2020. The human life history is adapted to exploit the adaptive advantages of culture. *Philosophical Transactions of the Royal Society B* 375:20190498.

———. 2022. What sort of mind/brain is compatible with cultural adaptation? *Journal of Cognition and Culture* 22:390–405.

Richerson, P. J., R. Baldini, A. V. Bell, K. Demps, K. Frost, V. Hillis, S. Mathew, et al. 2016. Cultural group selection plays an essential role in explaining human cooperation: A sketch of the evidence. *Behavioral and Brain Sciences* 39:e30.

Richerson, P. J., R. Boyd, and J. Henrich. 2010. Gene-culture coevolution in the age of genomics. *Proceedings of the National Academy of Sciences of the USA* 107:8985–8992.

Ridley, M. 2004. *Evolution*, 3rd ed. San Diego, CA: Blackwell.

Riederer, J. M., S. Tiso, T.J.B. van Eldijk, and F. J. Weissing. 2022. Capturing the facets of evolvability in a mechanistic framework. *Trends in Ecology and Evolution* 37:430–439.

Riedl, R. 1978. *Order in Living Systems: A Systems Analysis of Evolution*. New York: John Wiley.

Rieppel, O. 2001. Turtles as hopeful monsters. *BioEssays* 23:987–991.

———. 2009. Hennig's enkaptic system. *Cladistics* 25:311–317.

Riesch, R., L. Barrett-Lennard, G. Ellis, J. Ford, and V. Deecke. 2012. Cultural traditions and the evolution of reproductive isolation: Ecological speciation in killer whales? *Biological Journal of the Linnean Society* 106:1–17.

Riggs, A. D. 1975. X inactivation, differentiation, and DNA methylation. *Cytogenetics and Cell Genetics* 14:9–25.

Rine, J., and I. Herskowitz. 1987. Four genes responsible for a position effect on expression from HML and HMR in *Saccharomyces cerevisiae*. *Genetics* 116:9–22.

Rivoire, O., and S. Leibler. 2014. A model for the generation and transmission of variations in evolution. *Proceedings of the National Academy of Sciences of the USA* 111:E1940–E1949.

Roach, D. A., and R. D. Wulff. 1987. Maternal effects in plants. *Annual Review of Ecology and Systematics* 18:209–235.

Roberts, A. 2021. The fetus, the fish heart and the fruit fly. In *A Most Interesting Problem*, ed. J. DeSilva and J. Browne, 24–45. Princeton, NJ: Princeton University Press.

Robinson, S. R., and V. Méndez-Gallardo. 2011. Amniotic fluid as an extended milieu intérieur. In: *Developmental Science, Behaviour, and Genetics*, ed. K. E. Hood, C. T. Halpern, G. Greenberg, and R. M. Lerner, 234–284. Malden, MA: Wiley-Blackwell.

Roche, H., A. Delagnes, J.-P. Brugal, C. Feibel, M. Kibunjia, V. Mourre, and P.-J. Texier. 1999. Early hominid stone tool production and technical skill 2.34 myr ago in West Turkana, Kenya. *Nature* 399:57–60.

Rodriguez, R. J., J. Henson, E. van Volkenburgh, M. Hoy, L. Wright, F. Beckwith, Y.-O. Kim, and R. S. Redman. 2008. Stress tolerance in plants via habitat-adapted symbiosis. *International Society for Microbial Ecology Journal* 2:404–416.

Rodriguez-Guzman, M., J. A. Montero, E. Santesteban, Y. Ganan, D. Macias, and J. M. Hurle. 2007. Tendon-muscle crosstalk controls muscle bellies morphogenesis, which is mediated by cell death and retinoic acid signaling. *Developmental Biology* 302:267–280.

Rohner, N., D. F. Jarosz, J. E. Kowalko, M. Yoshizawa, W. R. Jeffery, R. L. Borowsky, S. Lindquist, et al. 2013. Cryptic variation in morphological evolution: HSP90 as a capacitor for loss of eyes in cavefish. *Science* 342:1372–1375.

Rohner, P. T., and D. Berger. 2023. Developmental bias predicts 60 million years of wing shape evolution. *Proceedings of the National Academy of Sciences of the USA* 120:e2211210120.

Rohner, P. T., and A. P. Moczek. 2020. Rapid differentiation of plasticity in life history and morphology during invasive range expansion and concurrent local adaptation in the horned beetle *Onthophagus taurus*. *Evolution* 74:2059–2072.

———. 2023. Vertically inherited microbiota and environment modifying behaviors indirectly shape the exaggeration of secondary sexual traits in the gazelle dung beetle. *Ecology and Evolution* 13:e10666.

———. 2024. Vertically inherited microbiota and environment modifying behaviors conceal genetic variation in dung beetle life history. *Proceedings Royal Society B. bioRxiv* 2024.01.17.576108.

Rohner, P. T., Y. Hu, and A. P. Moczek. 2022. Developmental bias in the evolution and plasticity of beetle horn shape. *Proceedings of the Royal Society B* 289:20221441.

Rohner, P. T., I. J. Jones, A. P. Moczek. 2024. Plasticity, symbionts, and niche construction interact in shaping dung beetle development and evolution. *Journal of Experimental Biology* 227:jeb245976.

Rohner, P. T., D. M. Linz, and A. P. Moczek. 2021. Doublesex mediates species, sex, environment and trait specific exaggeration of size and shape. *Proceedings of the Royal Society B* 288:20210241.

Rose, S. 1998. *Lifelines: Biology beyond Determinism*. Oxford: Oxford University Press.

Rose, S., R. C. Lewontin, and L. J. Kamin. 1984. *Not in Our Genes: Biology, Ideology and Human Nature*. London: Penguin.

Roseboom, T. J., S. R. de Rooij, and R. C. Painter. 2006. The Dutch famine and its long-term consequences for adult health. *Early Human Development* 82:485–491.

Roseboom, T. J., R C. Painter, A.F.M. van Abeelen, M.V.E. Veenendaal, and S. R. de Rooij. 2011. Hungry in the womb: What are the consequences? Lessons from the Dutch famine. *Maturitas* 70:141–145.

Roseman, C. C. 2020. Exerting an influence on evolution. *eLife* 9:e55952.

Roseman, C. C., and L. K. Delezene. 2019. The inhibitory cascade model is not a good predictor of molar size covariation. *Evolutionary Biology* 46:229–238.

Rosenberg, E., and I. Zilber-Rosenberg. 2016. Microbes drive evolution of animals and plants: The hologenome concept. *mBio* 7:e01395.

Rosenberg, K. R. 2023. The evolution of human infancy: Why it helps to be helpless. *Annual Review of Anthropology* 50:423–440.

Rosenblatt, J. S. 2010. Behavioral development during the mother-young interaction in placental mammals. In: *Developmental Science, Behaviour, and Genetics*, ed. K. E. Hood, C. T. Halpern, G. Greenberg, and R. M. Lerner, 205–233. Malden, MA: Wiley-Blackwell.

Rosenzweig, M. L. 1981. A theory of habitat selection. *Ecology* 62:327–335.

Roth, G., and D. B. Wake. 1985. Trends in the functional morphology and sensorimotor control of feeding behavior in salamanders: An example of the role of internal dynamics in evolution. *Acta biotheoretica* 34:175–191.

Roth, O., A. Beemelmanns, S. M. Barribeau, and B. M. Sadd. 2018. Recent advances in vertebrate and invertebrate transgenerational immunity in the light of ecology and evolution. *Heredity* 121:225–238.

Roughgarden, J. 1995. *Theory of Population Genetics and Evolutionary Ecology: An Introduction.* Hoboken, NJ: Prentice-Hall.

———. 2020. Holobiont evolution: Mathematical model with vertical vs. horizontal microbiome transmission. *Philosophy, Theory, and Practice in Biology* 12. https://doi.org/10.3998/ptpbio.16039257.0012.002.

Roughgarden, J., S. F. Gilbert, E. Rosenberg, I. Zilber-Rosenberg, and E. A. Lloyd. 2018. Holobionts as units of selection and a model of their population dynamics and evolution. *Biological Theory* 13:44–65.

Rountree, D. B., and H. F. Nijhout. 1995. Hormonal control of a seasonal polyphenism in *Precis coenia* (Lepidoptera: Nymphalidae). *Journal of Insect Physiology* 41:987–992.

Rouse, J. 2023. *Social Practices as Biological Niche Construction.* Chicago: Chicago University Press.

Royle, N. J., P. T. Smiseth, and M. Kolliker. 2012. *The Evolution of Parental Care.* Chicago: University of Chicago Press.

Rubio, A. O., and K. Summers. 2022. Neural crest cell genes and the domestication syndrome: A comparative analysis of selection. *PLoS One* 17: e0263830.

Ruiz, A. M., and L. R. Santos. 2013. Understanding differences in the way human and non-human primates represent tools. In *Tool Use in Animals*, ed. C. M. Sanz, J. Call, and C. Boesch, 119–133. Cambridge: Cambridge University Press.

Rumpho, M. E., K. N. Pelletreau, A. Moustafa, and D. Bhattacharya. 2011. The making of a photosynthetic animal. *Journal of Experimental Biology* 214:303–311.

Russell, J. A., C. S. Moreau, B. Goldman-Huertas, M. Fujiwara, D. J. Lohman, and N. E. Pierce. 2009. Bacterial gut symbionts are tightly linked with the evolution of herbivory in ants. *Proceedings of the National Academy of Sciences of the USA* 106:21236–21241.

Russon, A. E. 2003. Innovation and creativity in forest-living rehabilitant orangutans. In: *Animal Innovation*, ed. S. M. Reader and K. N. Laland, 279–306. Oxford: Oxford University Press.

Rutz, C., L. A. Bluff, N. Reed, J. Troscianko, J. Newton, R. Inger, A. Kacelnik, and S. Bearhop. 2010. The ecological significance of tool use in New Caledonian crows. *Science* 329: 1523–1526.

Ruxton, G. D., and D. M. Wilkinson. 2011. Avoidance of overheating and selection for both hair loss and bipedality in hominins. *Proceedings of the National Academy of Sciences of the USA* 108:20965–20969.

Ryan, F. P. 2004. Human endogenous retroviruses in health and disease: a symbiotic perspective. *Journal of the Royal Society of Medicine* 97:560–565.

Saastamoinen, M., D. van der Sterren, N. Vastenhout, B. J. Zwaan, and P. M. Brakefield. 2010. Predictive adaptive responses: Condition-dependent impact of adult nutrition and flight in the tropical butterfly *Bicyclus anynana*. *American Naturalist* 176:686–698.

Sabater, B. 2018. Evolution and function of the chloroplast: Current investigations and perspectives. *International Journal of Molecular Science* 19:3095

Sabeti, P. C., S. F. Schaffner, B. Fry, J. Lohmueller, P. Varilly, O. Shamovsky, A. Palma, et al. 2006. Positive natural selection in the human lineage. *Science* 312:1614–1620.

Sabeti, P. C., P. Varilly, B. Fry, J. Lohmueller, E. Hostetter, C. Cotsapas, X. Xie, et al. 2007. Genome-wide detection and characterization of positive selection in human populations. *Nature* 449:913–918.

Sagan, L. 1967. On the origin of mitosing cells. *Journal of Theoretical Biology* 14:225–274.

Sagner, A., and J. Briscoe. 2017. Morphogen interpretation: Concentration, time, competence, and signaling dynamics. *WIREs Developmental Biology* 6:e271.

Saini, A. 2023. *The Patriarchs: How Men Came to Rule*. London: Fourth Estate.

Salazar-Ciudad, I. 2006. Developmental constraints vs. variational properties: How pattern formation can help to understand evolution and development. *Journal of Experimental Zoology Part B* 306:107–125.

——. 2021. Why call it developmental bias when it is just development? *Biology Direct* 16:3.

Salazar-Ciudad, I., and H. Cano-Fernandez. 2023. Evo-devo beyond development: Generalizing evo-devo to all levels of the phenotypic evolution. *BioEssays* 45:2200205.

Salazar-Ciudad, I., and J. Jernvall. 2002. A gene network model accounting for development and evolution of mammalian teeth. *Proceedings of the National Academy of Sciences of the USA* 99:8116–8120.

——. 2005. Graduality and innovation in the evolution of complex phenotypes: Insights from development. *Journal of Experimental Zoology Part B* 30:619–631.

——. 2010. A computational model of teeth and the developmental origins of morphological variation. *Nature* 464:583–586.

Saltz, J. B., and B. R. Foley. 2011. Natural genetic variation in social niche construction: Social effects of aggression drive disruptive sexual selection in *Drosophila melanogaster*. *American Naturalist* 177:645–654.

Saltz, J. B., and S. V. Nuzhdin. 2014. Genetic variation in niche construction: Implications for development and evolutionary genetics. *Trends in Ecology and Evolution* 29:8–14.

Saltzburger, W. 2018. Understanding explosive diversification through cichlid fish genomics. *Nature Reviews Genetics* 19:705–717.

Salvini-Plawen, L., and E. Mayr. 1961. On the evolution of photoreceptors and eyes. *Evolutionary Biology* 10:207–263.

Sánchez-Villagra, M. R., and C. P. van Schaik. 2019. Evaluating the self-domestication hypothesis of human evolution. *Evolutionary Anthropology* 28:133–143.

Sánchez-Villagra, M. R., M. Geiger, and R. A. Schneider. 2016. The taming of the neural crest: A developmental perspective on the origins of morphological covariation in domesticated mammals. *Royal Society Open Science* 3:160107.

Sander, E. G., R. G. Warner, H. N. Harrison, and J. K. Loosli. 1959. The stimulatory effect of sodium butyrate and sodium propionate on the development of rumen mucosa in the young calf. *Journal of Dairy Science* 42:1600–1605.

Sandgathe, D. M., H. L. Dibble, P. Goldberg, S. P. McPherron, A. Turq, L. Niven, and J. Hodgkins. 2011. Timing of the appearance of habitual fire use. *Proceedings of the National Academy of Sciences of the USA* 108:E298.

San Roman, M., and A. Wagner. 2018. An enormous potential for niche construction through bacterial cross-feeding in a homogeneous environment. *PLOS Computational Biology* 14:e1006340.

Sapp, J. 1983. The struggle for authority in the field of heredity, 1900–1932: New perspectives on the rise of genetics. *Journal of the History of Biology* 16:311–342.

——. 1987. *Beyond the Gene: Cytoplasmic Inheritance and the Struggle for Authority in Genetics*. New York: Oxford University Press.

——. 2003. *Genesis: The Evolution of Biology*. New York: Oxford University Press.

Sarapas, C., G. Cai, L. M. Bierer, J. A. Golier, S. Galea, M. Ising, T. Rein, et al. 2011. Genetic markers for PTSD risk and resilience among survivors of the World Trade Center attacks. *Disease Markers* 30:328054.

Sarin, S., and R. Dukas. 2009. Social learning about egg-laying substrates in fruitflies. *Proceedings of the Royal Society B* 276:4323–4328.

Sarto-Jackson, I. 2022. *The Making and Breaking of Minds.* Wilmington, DE: Vernon.

Sarto-Jackson, I., D. O. Larson, and W. Callebaut. 2017. Culture, neurobiology, and human behavior: New perspectives in anthropology. *Biology and Philosophy* 32:729–748.

Sasaki, T., and D. Biro. 2017. Cumulative culture can emerge from collective intelligence in animal groups. *Nature Communications* 8:15049.

Schacter D. L., D. R. Addis, and R. L. Buckner. 2007. Remembering the past to imagine the future: The prospective brain. *Nature Reviews Neuroscience* 8:657–661.

Schaerli, Y., A. Jiménez, J. M. Duarte, L. Mihajlovic, J. Renggli, M. Isalan, J. Sharpe, at al. 2018. Synthetic circuits reveal how mechanisms of gene regulatory networks constrain evolution. *Molecular Systems Biology* 14:e8102.

Scharloo, W. 1970. Stabilizing and disruptive selection on a mutant character in *Drosophila*, II: Polymorphism caused by a genetical switch mechanism. *Genetics* 65:681–691.

Schick, K., and N. Toth, eds. 2006. *Oldowan Case Studies into Earliest Stone Age.* Gosport, UK: Stone Age Institute.

Schlichting, C. D., and M Pigliucci. 1998. *Phenotypic Evolution: A Reaction Norm Perspective.* Sunderland, MA: Sinauer Associates.

Schlichting, C. D., and M. A. Wund. 2014. Phenotypic plasticity and epigenetic marking: An assessment of evidence for genetic accommodation. *Evolution* 68:656–672.

Schlosser, G., and G. P. Wagner. 2004. *Modularity in Development and Evolution.* Chicago: University of Chicago Press.

Schluter, D. 1996. Adaptive radiation along genetic lines of least resistance. *Evolution* 50:1766–1774.

Schmalhausen, I. I. 1949. *Factors of Evolution: The Theory of Stabilizing Selection.* New York: Blakiston.

Schmid, M. W., C. Heichinger, D. C. Schmid, D. Guthörl, V. Gagliardini, R. Bruggmann, S. Aluri, et al. 2018. Contribution of epigenetic variation to adaptation in *Arabidopsis. Nature Communications* 9:4446.

Schneider, R. F., and A. Meyer. 2017. How plasticity, genetic assimilation and cryptic genetic variation may contribute to adaptive radiations. *Molecular Ecology* 26:330–350.

Schoch, R. R., and H. D. Sues. 2015. A middle Triassic stem-turtle and the evolution of the turtle body plan. *Nature* 523:584–587.

Schoener, T. W. 2011. The newest synthesis: Understanding the interplay of evolutionary and ecological dynamics. *Science* 331:426–429.

Schuster-Böckler, B., and B. Lehner. 2012. Chromatin organization is a major influence on regional mutation rates in human cancer cells. *Nature* 488:504–507.

Schwab, D. B., S. Casasa, and A. P. Moczek. 2017. Evidence of developmental niche construction in dung beetles: Effects on growth, scaling and reproductive success. *Ecology Letters* 20:1353–1363.

Schwab, D. B., H. E. Riggs, I.L.G. Newton, and Armin P. Moczek. 2016. Developmental and ecological benefits of the maternally transmitted microbiota in a dung beetle. *American Naturalist* 188:679–692.

Schwenk, K., and G. P. Wagner. 2004. The relativism of constraints on phenotypic evolution. In: *Phenotypic Integration: Studying the Ecology and Evolution of Complex Phenotypes*, ed. M. Pigliucci and K. Preston, 391–408. Oxford: Oxford University Press.

Scott-Phillips, T. C., T. E. Dickins, and S. A. West. 2011. Evolutionary theory and the ultimate–proximate distinction in the human behavioral sciences. *Perspectives on Psychological Science* 6:38–47.

Scott-Phillips, T. C., K. N. Laland, D. M. Shuker, T. E. Dickins, and S. A. West. 2014. The niche construction perspective: A critical appraisal. *Evolution* 68:1231–1243.

Scoville, A. G., and M. E. Pfrender. 2010. Phenotypic plasticity facilitates recurrent rapid adaptation to introduced predators. *Proceedings of the National Academy of Sciences of the USA* 107:4260–4263.

Sear, R. 2016. Beyond the nuclear family: An evolutionary perspective on parenting. *Current Opinion in Psychology* 7:98–103.

Seeley, T. D. 1995. *The Wisdom of the Hive*. Cambridge, MA: Harvard University Press.

Seely, M. K., and W. J. Hamilton. 1976. Fog catchment sand trenches constructed by tenebrionid beetles, *Lepidochora*, from the Namib Desert. *Science* 193:484–486.

Seger, J., and H. Brockmann. 1987. What is bet-hedging? In: *Oxford Surveys in Evolutionary Biology*, vol. 4, ed. H. Partridge and L. Partridge, 182–211. Oxford: Oxford University Press.

Senaldi, L., and M. Smith-Raska. 2020. Evidence for germline non-genetic inheritance of human phenotypes and diseases. *Clinical Epigenetics* 12:136.

Sender, R., S. Fuchs, and R. Milo. 2016. Revised estimates for the number of human and bacteria cells in the body. *PLOS Biology* 14:e1002533.

Seong, K., D. Li, H. Shimizu, R. Nakamura, and S. Ishii. 2011. Inheritance of stress-induced, ATF-2-dependent epigenetic change. *Cell* 145:1049–1061.

Sepkoski, D. 2012. *Re-reading the Fossil Record: The Growth of Paleobiology as an Evolutionary Discipline*. Chicago: University of Chicago Press.

Seppänen, J. T., J. T. Forsman, M. Mönkkönen, I. Krams, and T. Salmi. 2011. New behavioural trait adopted or rejected by observing heterospecific tutor fitness. *Proceedings of the Royal Society B* 278:1736–1741.

Seyfarth, R. M., and D. L. Cheney. 2000. Social awareness in the monkey. *American Zoologist* 40:902–909.

Seyfarth, R. M., D. L. Cheney, and P. Marler. 1980. Vervet monkey alarm calls: Semantic communication in a free-ranging primate. *Animal Behaviour* 28:1070–1094.

Shachak, M., and Y. Steinberger. 1980. An algae–desert snail food chain: Energy flow and soil turnover. *Oecologia* 46:402–411.

Shapiro, M. D., M. E. Marks, C. L. Peichel, B. K. Blackman, K. S. Nereng, B. Jónsson, D. Schluter, et al. 2004. Genetic and developmental basis of evolutionary pelvic reduction in threespine sticklebacks. *Nature* 428:717–723.

Sharma, A. 2014. Bioinformatic analysis revealing association of exosomal mRNAs and proteins in epigenetic inheritance. *Journal of Theoretical Biology* 357:143–149.

Sharma, U., F. Sun, C. C. Conine, B. Reichholf, S. Kukreja, V. A. Herzog, S. L. Ameres, et al. 2018. Small RNAs are trafficked from the epididymis to developing mammalian sperm. *Developmental Cell* 46:481–494.e6.

Sharon, G., D. Segal, J. M. Ringo, A. Hefetz, I. Zilber-Rosenberg, and E. Rosenberg. 2010. Commensal bacteria play a role in mating preference of *Drosophila melanogaster*. *Proceedings of the National Academy of Sciences of the USA* 107:20051–20056.

Shea, N., I. Pen, and T. Uller. 2011. Three epigenetic information channels and their different roles in evolution. *Journal of Evolutionary Biology* 24:1178–1187.

Shen, J., W. M. Rideout III, and Peter A. Jones. 1994. The rate of hydrolytic deamination of 5-methylcytosine in double-stranded DNA. *Nucleic Acids Research* 22:972–976.

Sherry, D., and B. Galef. 1984. Cultural transmission without imitation: Milk bottle opening by birds. *Animal Behaviour* 32:937–938.

Sherwood, C. C. 2018. Are we wired differently? *Scientific American* 319:60–63.

Sherwood, C. C., and A. Gómez-Robles. 2017. Brain plasticity and human evolution. *Annual Review of Anthropology* 46:399–419.

Sheth, R., L. Marcon, M. F. Bastida, M. Junco, L. Quintana, R. Dahn, M. Kmita, et al. 2012. Hox genes regulate digit patterning by controlling the wavelength of a Turing-type mechanism. *Science* 338:1476–1480.

Shettleworth, S. J. 2010. *Cognition, Evolution, and Behavior*, 2nd ed. New York: Oxford University Press.

Shi, Y., D. S. Stornetta, R. J. Reklow, A. Sahu, Y. Wabara, A. Nguyen, K. Li, et al. (2021). A brainstem peptide system activated at birth protects postnatal breathing. *Nature* 589:426–430.

Shinde, D. N., D. P. Elmer, P. Calabrese, J. Boulanger, N. Arnheim, and I. Tiemann-Boege. 2013. New evidence for positive selection helps explain the paternal age effect observed in achondroplasia. *Human Molecular Genetics* 22:4117–4126.

Shipley, B. 2016. *Cause and Correlation in Biology: A User's Guide to Path Analysis, Structural Equations and Causal Inference with R*, 2nd ed. Cambridge: Cambridge University Press.

Shubin, N., C. Tabin, and S. Carroll. 1997. Fossils, genes and the evolution of animal limbs. *Nature* 388:639–648.

———. 2009. Deep homology and the origins of evolutionary novelty. *Nature* 457:818–823.

Shultz, A. J., and T. B. Sackton. 2019. Immune genes are hotspots of shared positive selection across birds and mammals. *eLife* 8:e41815.

Shyer, A. E., A. R. Rodrigues, G. G. Schroeder, E. Kassianidou, S. Kumar, and R. M. Harland. 2017. Emergent cellular self-organization and mechanosensation initiate follicle pattern in the avian skin. *Science* 357:811–815.

Sibly, R. M., and R. N. Curnow. 2022. Sexual imprinting leads to speciation in locally adapted populations. *Ecology and Evolution* 12:e9479.

Sieber, R., and E. D. Kokwaro. 1982. Water intake by the termite *Macrotermes michaelseni*. *Entomologia experimentalis et applicata* 37:147–153.

Siegal, M. L., D.E.L. Promislow, and A. Bergman. 2007. Functional and evolutionary inference in gene networks: Does topology matter? *Genetica* 129:83–103.

Siepielski, A. M., J. D. DiBattista, and S. M. Carlson. 2009. It's about time: The temporal dynamics of phenotypic selection in the wild. *Ecology Letters* 12:1261–1276.

Siepielski, A. M., K. M. Gotanda, M. B. Morrissey, S. E. Diamond, J. D. DiBattista, and S. M. Carlson. 2013. The spatial patterns of directional phenotypic selection. *Ecology Letters* 16:1382–1392.

Siepielski, A. M., M. B. Morrissey, M. Buoro, S. M. Carlson, C. M. Caruso, S. M. Clegg, T. Coulson, et al. 2017. Precipitation drives global variation in natural selection. *Science* 355:959–962.

Sifuentes-Romero, I., E.F.S. Thakur, L. A. Laboissonniere, M. Solomon, C. L. Smith, A. C. Keene, J. M. Trimarchi, et al. 2020. Repeated evolution of eye loss in Mexican cavefish: Evidence of similar developmental mechanisms in independently evolved populations. *Journal of Experimental Zoology Part B* 334:423–437.

Silver, M., and E. Di Paolo. 2006. Spatial effects favour the evolution of niche construction. *Theoretical Population Biology* 70:387–400.

Simons, A. M. 2011. Modes of response to environmental change and the elusive empirical evidence for bed hedging. *Proceedings of the Royal Society B* 278:1601–1609.

———. 2014. Playing smart vs. playing safe: The joint expression of phenotypic plasticity and potential bet hedging across and within thermal environments. *Journal of Evolutionary Biology* 27:1047–1056.

Simpson, G. G. 1953. The Baldwin effect. *Evolution* 2:110–117.

Simpson, S. J., G. A. Sword, and N. Lo. 2011. Polyphenism in Insects. *Current Biology* 21: R738–R749.

Sinervo, B. 1999. Mechanistic analysis of natural selection and a refinement of Lack's and Williams's principles. *American Naturalist* 154: S26–S42.

Skeide, M. A., U. Kumar, R. K. Mishra, V. N. Tripathi, A. Guleria, J. P. Singh, F. Eisner, et al. 2017. Learning to read alters cortico-subcortical cross-talk in the visual system of illiterates. *Science Advances* 3:e1602612.

Skinner, B. F. 1938. *The Behavior of Organisms*. New York: D. Appleton-Century.

Slagsvold, T., and K. L. Wiebe. 2007. Learning the ecological niche. *Proceedings of the Royal Society B* 274:19–23.

———. 2011. Social learning in birds and its role in shaping a foraging niche. *Philosophical Transactions of the Royal Society B* 366:969–977.

Slagsvold, T., K. Wigdahl Kleiven, A. Eriksen, and L. E. Johannessen. 2013. Vertical and horizontal transmission of nest site preferences in titmice. *Animal Behaviour* 85:323–328.

Slater, P.J.B. 1986. The cultural transmission of bird song. *Trends in Ecology and Evolution* 1:94–97.

Slater, P.J.B., S. A. Ince, and P. W. Colgan. 1980. Chaffinch song types: Their frequencies in the population and distribution between repertoires of different individuals. *Behaviour* 75:207–218.

Slatkin, M. 2009. Epigenetic inheritance and the missing heritability problem. *Genetics* 182:845–850.

Slobodkin, L. 1961. *Growth and Regulation of Animal Populations.* New York: Holt, Rinehart and Winston.

Slobodkin, L. B., and A. Rapoport. 1974. An optimal strategy of evolution. *Quarterly Review of Biology* 49:181–200.

Smith, B. D. 2007a. Niche construction and the behavioral context of plant and animal domestication. *Evolutionary Anthropology* 16:188–199.

———. 2007b. The ultimate ecosystem engineers. *Science* 315:1797–1798.

———. 2011. A cultural niche construction theory of initial domestication. *Biological Theory* 6: 260–271.

———. 2016. Neo-Darwinism, niche construction theory, and the initial domestication of plants and animals. *Evolutionary Ecology* 30:307–324.

Smith, K., and S. Kirby. 2008. Cultural evolution: Implications for understanding the human language faculty and its evolution. *Philosophical Transactions of the Royal Society B* 363: 3591–3603.

Smocovitis, V. B. 1996. *Unifying Biology.* Princeton, NJ: Princeton University Press.

———. 2000. The 1959 Darwin centennial celebration in America. *Osiris* 14:274–323.

Snell-Rood, E. C. 2012. Selective processes in development: Implications for the costs and benefits of phenotypic plasticity. *Integrative and Comparative Biology* 52:31–42.

Snell-Rood, E. C., M. E. Kobiela, K. L. Sikkink, and A. M. Shephard. 2018. Mechanisms of plastic rescue in novel environments. *Annual Review of Ecology, Evolution, and Systematics* 49:331–354.

Sober, E. 1984. *The Nature of Selection: Evolutionary Theory in Philosophical Focus.* Cambridge, MA: MIT Press.

Sober, E., and D. S. Wilson. 1999. *Unto Others: The Evolution and Psychology of Unselfish Behavior.* Cambridge, MA: Harvard University Press.

Soen, Y., M. Knafo, and M. Elgart. 2015. A principle of organization which facilitates broad Lamarckian-like adaptations by improvisation. *Biology Direct* 10:68.

Sol, D. 2003. Behavioral flexibility: A neglected issue in the ecological and evolutionary literature? In: *Animal Innovation,* ed. S. M. Reader and K. N. Laland, 63–82. Oxford: Oxford University Press.

Sol, D., and L. Lefebvre. 2000. Behavioural flexibility predicts invasion success in birds introduced to New Zealand. *Oikos* 90:599–605.

Sol, D., R. P. Duncan, T. M. Blackburn, P. Cassey, and L. Lefebvre. 2005a. Big brains, enhanced cognition, and response of birds to novel environments. *Proceedings of the National Academy of Sciences of the USA* 102:5460–5465.

Sol, D., M. Elie, M. Marcoux, E. Chrostovsky, C. Porcher, and L. Lefebvre. 2005b. Ecological mechanisms of a resource polymorphism in zenaida doves of Barbados. *Ecology* 86:2397–2407.

Sol, D., O. Lapiedra, and C. González-Lagos. 2013. Behavioural adjustments for a life in the city. *Animal Behaviour* 85:1101–1112.

Sol, D., L. Lefebvre, and J. D. Rodríguez-Teijeiro. 2005c. Brain size, innovative propensity and migratory behavior in temperate Palaearctic birds. *Proceedings of the Royal Society B* 272:1433–1441.

Sol, D., D. G. Stirling, and L. Lefebvre. 2005d. Behavioral drive or behavioral inhibition in evolution: Subspecific diversification in Holarctic passerines. *Evolution* 59:2669–2677.

Sol, D., S. Timmermans, and L. Lefebvre. 2002. Behavioural flexibility and invasion success in birds. *Animal Behaviour* 63:495–502.

Somel, M., X. Liu, and P. Khaitovich. 2013. Human brain evolution: Transcripts, metabolites and their regulators. *Nature Reviews Neuroscience* 14:112–127.

Somel, M., R. Rohlfs, and X. Liu. 2014. Transcriptomic insights into human brain evolution: Acceleration, neutrality, heterochrony. *Current Opinion in Genetics and Development* 29:110–119.

Sommer, F., M. Ståhlman, O. Ilkayeva, J. M. Arnemo, J. Kindberg, J. Josefsson, C. B. Newgard, et al. 2016. The gut microbiota modulates energy metabolism in the hibernating brown bear *Ursus arctos. Cell Reports* 14:1655–1661.

Soto, A. M., and C. Sonnenschein. 2018. Reductionism, organicism, and causality in the biomedical sciences: A critique. *Perspectives in Biology and Medicine* 61:489–502.

Soubry, A., C. Hoyo, R. L. Jirtle, and S. K. Murphy. 2014. A paternal environmental legacy: Evidence for epigenetic inheritance through the male germ line. *BioEssays* 36:359–371.

Soucy, S. M., J. Huang, and J. P. Gogarten. 2015. Horizontal gene transfer: Building the web of life. *Nature Reviews Genetics* 16:472–482.

Sousa, A.M.M., K. A. Meyer, G. Santpere, F. O. Gulden, and N. Sestan. 2017. Evolution of the human nervous system function, structure, and development. *Cell* 170:226–247.

Spalding, D. A. 1873. Instinct: With original observations on young animals. *Macmillan's Magazine* 27:282–293.

Spelke, E. S. 2009. Forum. In *Why We Cooperate*, ed. M. Tomasello, 149–172. Cambridge, MA: MIT Press.

Staddon, J.E.R. 2016. *Adaptive Behavior and Learning*, 2nd ed. Cambridge: Cambridge University Press.

Stajic, D., and L.E.T. Jansen. 2021. Empirical evidence for epigenetic inheritance driving evolutionary adaptation. *Philosophical Transactions of the Royal Society B* 376:20200121.

Stajic, D., L. Perfeito, and L.E.T. Jansen. 2019. Epigenetic gene silencing alters the mechanisms and rate of evolutionary adaptation. *Nature Ecology and Evolution* 3:491–498.

Stamatoyannopoulos, J. A. 2012. What does our genome encode? *Genome Research* 22:1602–1611.

Stamps, J. A., V. V. Krishnan, and M. L. Reid. 2005. Search costs and habitat selection by dispersers. *Ecology* 86:510–518.

Standen, E. M., T. Y. Du, and H.C.E. Larsson. 2014. Developmental plasticity and the origin of tetrapods. *Nature* 513:54–58.

Stanhope, M. J., V. G. Waddell, O. Madsen, W. de Jong, S. B. Hedges, G. C. Cleven, and D. Kao. 1998. Molecular evidence for multiple origins of Insectivora and for a new order of endemic African insectivore mammals. *Proceedings of the National Academy of Sciences of the USA* 95:9967–9972.

Stanley, E. L., R. L. Kendal, J. R. Kendal, S. Grounds, and K. N. Laland. 2008. The effects of group size, rate of turnover and disruption to demonstration on the stability of foraging traditions in fish. *Animal Behaviour* 75:565–572.

Staszewski, V., J. Gasparini, K. D. Mccoy, T. Tveraa, and T. Boulinier. 2007. Evidence of an interannual effect of maternal immunization on the immune response of juveniles in a long-lived colonial bird. *Journal of Animal Ecology* 76:1215–1223.

Stearns S. C. 1992. *The Evolution of Life Histories*. Oxford: Oxford University Press.

Stearns, S. C., and R. F. Hoekstra. 2005. *Evolution: An Introduction*, 2nd ed. Oxford: Oxford University Press.

Stebbins, G. L. 1959. The synthetic approach to problems of organic evolution. *Cold Spring Harbor Symposia on Quantitative Biology* 24:305–311.

Stedman, H. H., B. W. Kosyak, A. Nelson, D. M. Thesier, L. T. Su, D. W. Low, C. R. Bridges, et al. 2004. Myosin gene mutation correlates with anatomical changes in the human lineage. *Nature* 428:415–418.

Steele, J., P. F. Ferrari, and L. Fogassi. 2012. From action to language: Comparative perspectives on primate tool use, gesture and the evolution of human language. *Philosophical Transactions of the Royal Society B* 367:4–9.

Steinberg, M. S., and M. Takeichi. 1994. Experimental specification of cell sorting, tissue spreading, and specific spatial patterning by quantitative differences in cadherin expression. *Proceedings of the National Academy of Sciences of the USA* 91:206–209.

Steinhart, Z., and S. Angers. 2018. Wnt signaling in development and tissue homeostasis. *Development* 145:dev146589.

Stellatelli, O. A., A. Villalba, C. Block, L. E. Vega, J. E. Dajil, and F. B. Cruz. 2018. Seasonal shifts in the thermal biology of the lizard *Liolaemus tandiliensis* (Squamata, Liolaemidae). *Journal of Thermal Biology* 73:61–70.

Stephenson, G. 1967. Cultural acquisition of a specific learned response among rhesus monkeys. In *Progress in Primatology*, ed. R. Schneider and H.K.D. Starck, 279–288. Stuttgart: Gustav Fisher Verlag.

Stern, D. L. 2010. *Evolution, Development, and the Predictable Genome*. Greenwood Village, CO: Roberts.

Stoltzfus, A. 2019. Understanding bias in the introduction of variation as an evolutionary cause. In: *Evolutionary Causation: Biological and Philosophical Reflections*, ed. T. Uller and K. N. Laland, 29–62. Cambridge, MA: MIT Press.

———. 2021. *Mutation, Randomness and Evolution*. Oxford: Oxford University Press.

Stoltzfus, A., and D. M. McCandlish. 2017. Mutational biases influence parallel adaptation. *Molecular Biology and Evolution* 34:2163–2172.

Stoltzfus, A., and R. W. Norris. 2016. On the causes of evolutionary transition:transversion bias. *Molecular Biology and Evolution* 33:595–602.

Stotz, K. C. 2019. Biological information in developmental and evolutionary systems. In: *Evolutionary Causation: Biological and Philosophical Reflections*, ed. T. Uller and K. N. Laland, 323–344. Cambridge, MA: MIT Press.

Stotz, K. C., A. Bostanci, and P. E. Griffiths. 2006. Tracking the shift to "postgenomics." *Community Genetics* 9:190–196.

Stout, D. 2011. Stone toolmaking and the evolution of human culture and cognition. *Philosophical Transactions of the Royal Society B* 366:1050–1059.

Stout, D., and T. Chaminade. 2012. Stone tools, language and the brain in human evolution. *Philosophical Transactions of the Royal Society B* 367:75–87.

Street, S. E., A. F. Navarrete, S. M. Reader, and K. N. Laland. 2017. Coevolution of cultural intelligence, extended life history, sociality, and brain size in primates. *Proceedings of the National Academy of Sciences of the USA* 114:7908.

Striedter, G. F. 2005. *Principles of Brain Evolution*. Sunderland, MA: Sinauer Associates.

Stringer, C., and P. Andrews. 2012. *The Complete World of Human Evolution*, 2nd ed. London: Thames and Hudson.

Suboski, M. D., S. Bain, A. E. Carty, L. M. McQuoid, M. I. Seelen, and M. Seifert. 1990. Alarm reaction in acquisition and social transmission of simulated-predator recognition by zebra danio fish (*Brachydanio rerio*). *Journal of Comparative Psychology* 104:101–112.

Sugita, Y. 1980. Imitative choice behavior in guppies. *Japan Psychological Research* 22:7–12.

Sultan, S. E. 2015. *Organism and Environment: Ecological Development, Niche Construction and Adaptation*. Oxford: Oxford University Press.

———. 2019. Genotype-environment interaction and the unscripted reaction norm. In: *Cause and Process in Evolution*, ed. T. Uller and K. N. Laland, 109–126. Cambridge, MA: MIT Press.

Sumpter, D.J.T. 2010. *Collective Animal Behavior*. Princeton, NJ: Princeton University Press.

Sun, J., M. Lu, N. E. Gillette, and M. J. Wingfield. 2013. Red turpentine beetle: Innocuous native becomes invasive tree killer in China. *Annual Review of Entomology* 58:293–311.

Sutherland, W. J. 1998. Evidence for flexibility and constraint in migration systems. *Journal of Avian Biology* 29:441–446.

Suzuki, T. A., J. L. Fitzstevens, V. T. Schmidt, H. Enav, K. E. Huus, M. M. Ngwese, A. Grießhammer, et al. 2022. Codiversification of gut microbiota with humans. *Science* 377:1328–1332.

Suzuki, Y., and H. F. Nijhout. 2006. Evolution of a polyphenism by genetic accommodation. *Science* 311:650–652.

Svensson, E. I. 2023. Phenotypic selection in natural populations: What have we learned in 40 years? *Evolution* 77:1493–1504.

Svensson, E. I., and R. Calsbeek. 2012. *The Adaptive Landscape in Evolutionary Biology*. Oxford: Oxford University Press.

Swaddle, J. P., M. G. Cathey, M. Correll, and B. P. Hodkinson. 2005. Socially transmitted mate preferences in a monogamous bird: A non-genetic mechanism of sexual selection. *Proceedings of the Royal Society B* 272:1053–1058.

Tadros, W., and H. D. Lipshitz. 2009. The maternal-to-zygotic transition: A play in two acts. *Development* 136:3033–3042.

Taerum, S. J., T. A. Duong, Z. W. de Beer, N. Gillette, J.-H. Sun, D. R. Owen, and M. J. Wingfield. 2013. Large shift in symbiont assemblage in the invasive red turpentine beetle. *PLOS One* 8:e78126.

Tallinen, T., J. Y. Chung, F. Rousseau, N. Girard, J. Lefèvre, and L. Mahadevan. 2016. On the growth and form of cortical convolutions. *Nature Physics* 12:588–593.

Tan, C.M.J., and A. J. Lewandowski. 2020. The transitional heart: From early embryonic and fetal development to neonatal life. *Fetal Diagnosis and Therapy* 47:373–386.

Tanaka, M. M., P. Godfrey-Smith, and B. Kerr. 2020. The dual landscape model of adaptation and niche construction. *Philosophy of Science* 87:478–498.

Tarr, B., J. Launay, and R.I.M. Dunbar. 2014. Music and social bonding: "self-other" merging and neurohormonal mechanisms. *Frontiers in Psychology* 5:1096.

Tebbich, S., K. Sterelny, and I. Teschke. 2010. The tale of the finch: Adaptive radiation and behavioural flexibility. *Philosophical Transactions of the Royal Society B* 365:1099–1109.

Tehranifar, P., H.-C. Wu, J. A. McDonald, F. Jasmine, R. M. Santella, I. Gurvich, J. D. Flom, et al. 2018. Maternal cigarette smoking during pregnancy and offspring DNA methylation in midlife. *Epigenetics* 13:129–134.

Ten Cate, C., and C. Rowe. 2007. Biases in signal evolution: Learning makes a difference. *Trends in Ecology and Evolution* 22:380–387.

Terkel, J. 1996. Cultural transmission of feeding behavior in the black rat (*Rattus rattus*). In *Social Learning in Animals: The Roots of Culture*, ed. C. M. Heyes and B. G. Galef Jr., 17–47. San Diego, CA: Academic.

Terrace, H. S. 1979. *How Nim Chimpsky Changed My Mind*. San Francisco, CA: Ziff-Davis.

Theis, K. R., T. M. Schmidt, and K. E. Holekamp. 2012. Evidence for a bacterial mechanism for group-specific social odors among hyenas. *Scientific Reports* 2:615.

Theofanopoulou, C., S. Gastaldon, T. O'Rourke, B. D. Samuels, A. Messner, P. T. Martins, F. Delogu, et al. 2017. Self-domestication in *Homo sapiens*: Insights from comparative genomics. *PLOS One* 12:e0185306.

Thom, R. 1989. *Mutation Theory: Ideas and Applications*. Shanghai: Shanghai Translation.

Thompson, D. W. 1917. *On Growth and Form*. Cambridge: Cambridge University Press.

Thompson, J. C., D. K. Wright, and S. J. Ivory. 2021. The emergence and intensification of early hunter-gatherer niche construction. *Evolutionary Anthropology* 30:17–27.

Thorogood, R., and N. B. Davies. 2012. Cuckoos combat socially transmitted defenses of reed warbler hosts with a plumage polymorphism. *Science* 337:578.

Thorson, J.L.M., M. Smithson, D. Beck, I. Sadler-Riggleman, E. Nilsson, M. Dybdahl, and M. K. Skinner. 2017. Epigenetics and adaptive phenotypic variation between habitats in an asexual snail. *Scientific Reports* 7:14139.

Tobi, E. W., R. C. Slieker, R. Luijk, K. F. Dekkers, A. D. Stein, K. M. Xu, P. E. Slagboom, et al. 2018. DNA methylation as a mediator of the association between prenatal adversity and risk factors for metabolic disease in adulthood. *Science Advances* 4:eaao4364.

Todd, P.M.G. 1991. Exploring adaptive agency II: Simulating the evolution of associative learning. In: *From Animals to Animals: Proceedings of the First International Conference on Simulation of Adaptive Behavior*, ed. J.M.S. Wilson, 306–315. Cambridge, MA: MIT Press.

Toler, P. D. 2019. *Women Warriors: An Unexpected History*. Boston, MA: Beacon.

Tollefsbol, T. 2019. *Transgenerational Epigenetics*, 2nd ed. London: Academic.

Tomasello, M. 2008. *Origins of Human Communication*. Cambridge, MA: MIT Press.

———. 2009. *Why We Cooperate*. Cambridge, MA: MIT Press.

———. 2011. Human culture in evolutionary perspective. In: *Advances in Culture and Psychology*, ed. M. J. Gelfand, C. Chui, and Y. Hong, 5–51. Oxford: Oxford University Press.

———. 2018. How children come to understand false beliefs: A shared intentionality account. *Proceedings of the National Academy of Sciences of the USA* 115:8491–8498.

Tomasello, M., B. Hare, H. Lehmann, and J. Call. 2007. Reliance on head versus eyes in the gaze following of great apes and human infants: The cooperative eye hypothesis. *Journal of Human Evolution* 52:314–320.

Tomillo, P. S., V. S. Saba, G. S. Blanco, C. A. Stock, F. V. Paladino, and J. R. Spotila. 2012. Climate driven egg and hatchling mortality threatens survival of Eastern Pacific leatherback turtles. *PLOS One* 7:e37602.

Tomkova, M., and B. Schuster-Böckler. 2018. DNA modifications: Naturally more error prone? *Trends in Genetics* 34:627–638.

Tomlinson, G. 2018. *Culture and the Course of Human Evolution*. Chicago: University of Chicago Press.

Tomoyasu, Y., Y. Arakane, K. J. Kramer, and R. E. Denell. 2009. Repeated co-options of exoskeleton formation during wing-to-elytron evolution in beetles. *Current Biology* 19:2057–2065.

Torday, J. S. 2016. The cell as the first niche construction. *Biology* 5:19.

Torres-Garcia, S., I. Yaseen, M. Shukla, P.N.C.B. Audergon, S. A. White, A. L. Pidoux, and R. C. Allshire. 2020. Epigenetic gene silencing by heterochromatin primes fungal resistance. *Nature* 585:453–458.

Toth, N. 1987. Behavioral inferences from early stone artifact assemblages: An experimental model. *Journal of Human Evolution* 16:763–787.

Towler, J,. and J. Bramall. 1986. *Midwives in History and Society*. London: Routledge.

Townes, P. L., and J. Holtfreter. 1955. Directed movements and selective adhesion of embryonic amphibian cells. *Journal of Experimental Zoology* 128:53–120.

Toyokawa, W., A. Whalen, and K. N. Laland. 2019. Social learning strategies regulate the wisdom and madness of interactive crowds. *Nature Human Behaviour* 3:183–193.

Trevathan, W. R. 2011. *Human Birth: An Evolutionary Perspective*. Abingdon, UK: Routledge.

Trut L. N., A. V. Kharlamova, and Y. E. Herbeck. 2020. Belyaev's and PEI's foxes: A far cry. *Trends in Ecology and Evolution* 35:649–651.

Trut, L. N., I. Oskina, and A. Kharlamova. 2009. Animal evolution during domestication: The domesticated fox as a model. *BioEssays* 31:349–360.

Tsuchida, T., R. Koga, and T. Fukatsu. 2004. Host plant specialization governed by facultative symbiont. *Science* 303:1989.

Tung, A., and M. Levin. 2020. Extra-genomic instructive influences in morphogenesis: A review of external signals that regulate growth and form. *Developmental Biology* 461:1–12.

Turing, A. M. 1952. The chemical basis of morphogenesis. *Philosophical Transactions of the Royal Society B* 237:37–72.

Turner, J. S. 2000. *The Extended Organism: The Physiology of Animal-Built Structures.* Cambridge, MA: Harvard University Press.

———. 2005. Extended physiology of an insect-built structure. *American Entomologist* 51:36–38.

Turner, J. S., E. Marais, M. Vinte, A. Mudengi, and W. Park. 2006. Termites, water and soils. *Agricola* 16.

Uchiyama, R., R. Spicer, and M. Muthukrishna. 2022. Cultural evolution of genetic heritability. *Behavioral and Brain Sciences* 45:e152.

Uddin, M., D. E. Wildman, G. Liu, W. Xu, R. M. Johnson, P. R. Hof, G. Kapatos, et al. 2004. Sister grouping of chimpanzees and humans as revealed by genome-wide phylogenetic analysis of brain gene expression profiles. *Proceedings of the National Academy of Sciences of the USA* 101:2957–2962.

Uller, T. 2008. Developmental plasticity and the evolution of parental effects. *Trends in Ecology and Evolution* 2:432–438.

———. 2012. Parental effects in development and evolution. In *Evolution of Parental Care*, ed. P. Smiseth, N. J. Royle, and M. Kölliker, 247–266. Oxford: Oxford University Press.

———. 2013. Non-genetic inheritance and evolution. In *The Philosophy of Biology: A Companion for Educators*, ed. K. Kampourakis, 267–287. Dordrecht: Springer Netherlands.

———. 2019. Evolutionary perspectives on transgenerational epigenetics. In: *Transgenerational Epigenetics*, 2nd ed., ed. T. O. Tollefsbol, 333–350. London: Academic.

Uller, T., and H. Helanterä. 2017. Heredity and evolutionary theory. In: *Challenging the Modern Synthesis: Adaptation, Development and Inheritance*, ed. P. Huneman and D. Walsh, 280–316. Oxford: Oxford University Press.

———. 2019. Niche construction and conceptual change in evolutionary biology. *British Journal for the Philosophy of Science* 70:351–375.

Uller, T., and K. N. Laland. 2019. *Evolutionary Causation: Biological and Philosophical Reflections.* Cambridge, MA: MIT Press.

Uller, T., S. English, and I. Pen. 2015. When is incomplete epigenetic resetting in germ cells favoured by natural selection? *Proceedings of the Royal Society B* 282:20150682.

Uller, T., N. Feiner, R. Radersma, I.S.C. Jackson, and A. Rago. 2020. Developmental plasticity and evolutionary explanations. *Evolution and Development* 22:47–55.

Uller, T., L. Milocco, J. Isanta-Navarro, C. K. Cornwallis, and N. Feiner. 2024. Twenty years on from *Developmental Plasticity and Evolution:* middle-range theories and how to test them. *J Exp Biol* 7, 227 (Suppl_1): jeb246375.

Uller, T., A. P. Moczek, R. A. Watson, P. M. Brakefield, and K. N. Laland. 2018. Developmental bias and evolution: A regulatory network perspective. *Genetics* 209:949.

Underhill, P. A., G. Passarino, A. A. Lin, P. Shen, M. Mirazon Lahr, R. A. Foley, P. J. Oefner, et al. 2001. The phylogeography of Y chromosome binary haplotypes and the origins of modern human populations. *Annals of Human Genetics* 65:43–62.

Uomini, N. T., and G. F. Meyer. 2013. Shared brain lateralization patterns in language and Acheulean stone tool production: A functional transcranial Doppler ultrasound study. *PLOS One* 8:e72693.

Vågerö, D., P. R. Pinger, V. Aronsson, and G. J. van den Berg. 2018. Paternal grandfather's access to food predicts all-cause and cancer mortality in grandsons. *Nature Communications* 9:5124.

Valentine, J. W., D. Jablonski, and D. H. Erwin. 1999. Fossils, molecules and embryos: New perspectives on the Cambrian explosion. *Development* 126:851–859.

Vanadzina, K., S. E. Street, S. D. Healy, K. N. Laland, and C. Sheard. 2023. Global drivers of variation in cup nest size in passerine birds. *Journal of Animal Ecology* 92:338–351.

Van der Graaff, J., S. Branje, M. de Wied, S. Hawk, P. van Lier, and W. Meeus. 2014. Perspective taking and empathic concern in adolescence: Gender differences in developmental changes. *Developmental Psychology* 50:881–888.

Van Dyken, J. D., and M. J. Wade. 2012. Origins of altruism diversity, II: Runaway coevolution of altruistic strategies via "reciprocal niche construction." *Evolution* 66:2498–2513.

Van Gestel, J., and F. J. Weissing. 2016. Regulatory mechanisms link phenotypic plasticity to evolvability. *Scientific Reports* 6:24524.

Van Schaik, C. P., M. Ancrenaz, G. Borgen, B. Galdikas, C. D. Knott, I. Singleton, A. Suzuki, et al. 2003. Orangutan cultures and the evolution of material culture. *Science* 299:102–105.

Van Valen, L. 1973. Festschrift. *Science* 180:488.

Vasquez Kuntz, K. L., S. A. Kitchen, T. L. Conn, S. A. Vohsen, A. N. Chan, M.J.A. Vermeij, C. Page, et al. 2022. Inheritance of somatic mutations by animal offspring. *Science Advances* 8:eabn0707.

Verd, B., N.A.M. Monk, and J. Jaeger. 2019. Modularity, criticality, and evolvability of a developmental gene regulatory network. *eLife* 8:e42832.

Vermeij, G. J., and D. R. Lindberg. 2000. Delayed herbivory and the assembly of marine benthic ecosystems. *Paleobiology* 26:419–430.

Verzijden, M. N., and C. ten Cate. 2007. Early learning influences species assortative mating preferences in Lake Victoria cichlid fish. *Biology Letters* 3:134–136.

Verzijden, M. N., C. ten Cate, M. R. Servedio, G. M. Kozak, J. W. Boughman, and E. I. Svensson. 2012. The impact of learning on sexual selection and speciation. *Trends in Ecology and Evolution* 27:511–519.

Via, S. 1999. Reproductive isolation between sympatric races of pea aphids, i: Gene flow restriction and habitat choice. *Evolution* 53:1446–1457.

Via, S., R. Gomulkiewicz, G. de Jong, S. M. Scheiner, C. D. Schlichting, and P. H. van Tienderen. 1995. Adaptive phenotypic plasticity: Consensus and controversy. *Trends in Ecology and Evolution* 10:212–217.

Villegas C., A. C. Love, L. Nuño de la Rosa, I. Brigandt, and G. P.Wagner. 2023. Conceptual roles of evolvability across evolutionary biology: Between diversity and unification. Chapter 3 in *Evolvability: A Unifying Concept in Evolutionary Biology?*, ed. T. F. Hansen, D. Houle, M. Pavličev, and C. Pélabon, Vienna Series in Theoretical Biology. Cambridge, MA: MIT Press.

Vincent, T. L., and J. S. Brown. 2005. *Evolutionary Game Theory, Natural Selection and Darwinian Dynamics*. New York: Cambridge University Press.

Visalberghi, E., G. Sabbatini, A. H. Taylor, and G. R. Hunt. 2017. Cognitive insights from tool use in nonhuman animals. In: *APA Handbook of Comparative Psychology*, vol. 2, *Perception, Learning, and Cognition*, ed. J. Call, G. M. Burghardt, I. M. Pepperberg, C. T. Snowdon, and T. Zentall, 673–701. Washington, DC: American Psychological Association.

Vogt, G. 2015. Stochastic developmental variation, an epigenetic source of phenotypic diversity with far-reaching biological consequences. *Journal of Biosciences* 40:159–204.

———. 2021. Epigenetic variation in animal populations: Sources, extent, phenotypic implications, and ecological and evolutionary relevance. *Journal of Biosciences* 46:24.

Voight, B. F., S. Kudaravalli, X. Wen, and J. K. Pritchard. 2006. A map of recent positive selection in the human genome. *PLOS Biology* 4:e72.

Volpe, T. A., C. Kidner, I. M. Hall, G. Teng, S. I. Grewal, and R. A. Martienssen. 2002. Regulation of heterochromatic silencing and histone H3 lysine-9 methylation by RNAi. *Science* 297:1833–1837.

Von Steiniger, F. 1950. Beiträge zur Soziologie und sonstigen Biologie der Wanderratte. *Zeitschrift fur Tierpsychologie* 7:356–379.

Vuong, H. E., G. N. Pronovost, D. W. Williams, E.J.L. Coley, E. L. Siegler, A. Qiu, M. Kazantsev, et al. 2020. The maternal microbiome modulates fetal neurodevelopment in mice. *Nature* 586:281–286.

Waddington, C. H. 1939. *An Introduction to Modern Genetics*. London: Allen and Unwin

———. 1940. *Organisers and Genes*. Cambridge: Cambridge University Press.

———. (1942) 2012. The epigenotype. *International Journal of Epidemiology* 41:10–13. Originally published in *Endeavor* 1:18–20.

———. 1953. Genetic assimilation of an acquired character. *Evolution* 7:118–126.

———. 1957. *The Strategy of the Genes: A Discussion of Some Aspects of Theoretical Biology*. London: Allen and Unwin.

———. 1959. Evolutionary systems: Animal and human. *Nature* 183:1634–1638.

———, ed. 1968. *The Origin of Life*. Toward a Theoretical Biology 1. Chicago: Aldine.

———, ed. 1969a. *Sketching Theoretical Biology*. Toward a Theoretical Biology 2. Chicago: Aldine.

———. 1969b. Paradigm for an evolutionary process. In: *Sketching Theoretical Biology*, ed. Waddington, 106–128, Toward a Theoretical Biology 2. Chicago: Aldine.

———, ed. 1970. *Organization Stability and Process*. Toward a Theoretical Biology 3. Chicago: Aldine.

———, ed. 1972. *Biological Processes in Living Systems*. Toward a Theoretical Biology 4. Chicago: Aldine.

Wade, M. J. 1998. The evolutionary genetics of maternal effects. In: *Maternal Effects as Adaptations*, ed. T. A. Mousseau and C. W. Fox, 5–21. New York: Oxford University Press.

Wade, M. J., and S. Kalisz. 1990. The causes of natural selection. *Evolution* 44:1947–1955.

Wade, M. J., and S. E. Sultan. 2024. Niche construction and the environmental term of the Price equation: How natural selection changes when organisms alter their environments. *Evolution and Development* 25:451–469.

Wagner, A. 2008. Robustness and evolvability: A paradox resolved. *Proceedings of the Royal Society B* 275:91–100.

———. 2011a. The molecular origins of evolutionary innovations. *Trends in Genetics* 27:397–410.

———. 2011b. *The Origins of Evolutionary Innovations*. Oxford: Oxford University Press.

———. 2013. *Robustness and Evolvability in Living Systems*. Princeton, NJ: Princeton University Press.

———. 2014. Mutational robustness accelerates the origin of novel RNA phenotypes through phenotypic plasticity. *Biophysical Journal* 106:955–965.

Wagner, G. P. 1989a. The biological homology concept. *Annual Review of Ecology and Systematics* 20:51–69.

———. 1989b. The origin of morphological characters and the biological basis of homology. *Evolution* 43:1157–1171.

———. 1996. Homologues, natural kinds and the evolution of modularity. *American Zoologist* 36:36–43.

———. 2014. *Homology, Genes, and Evolutionary Innovation*. Princeton, NJ: Princeton University Press.

Wagner G. P., and L. Altenberg. 1996. Perspective: Complex adaptations and the evolution of evolvability. *Evolution* 50:967–976.

Wagner, G. P., and J. Draghi. 2010. Evolution of evolvability. In *Evolution: The Extended Synthesis*, ed. M. Pigliucci and G. B. Müller, 379–399. Cambridge, MA: MIT Press.

Wagner, G. P., C.-H. Chiu, and M. Laubichler. 2000. Developmental evolution as a mechanistic science: The inference from developmental mechanisms to evolutionary processes. *American Zoologist* 40:819–831.

Wagner, G. P., E. M. Erkenbrack, and A. C. Love. 2019. Stress-induced evolutionary innovation: A mechanism for the origin of cell types. *BioEssays* 41:1800188.

Wahlbuhl, M., S. Reiprich, M. R. Vogl, M. R. Bösl, and M. Wegner. 2012. Transcription factor Sox10 orchestrates activity of a neural crest-specific enhancer in the vicinity of its gene. *Nucleic Acids Research* 40:88–101.

Wakano, J. W., K. Aoki, and M. W. Feldman. 2004. Evolution of social learning: A mathematical analysis. *Theoretical Population Biology* 66:249–258.

Wall, M. E., M. J. Dunlop, and W. S. Hlavacek. 2005. Multiple functions of a feed-forward-loop gene circuit. *Journal of Molecular Biology* 349:501–514.

Walsh, B., and M. W. Blows. 2009. Abundant genetic variation + strong selection = multivariate genetic constraints: A geometric view of adaptation. *Annual Review of Ecology, Evolution, and Systematics* 40:41–59.

Walsh, B., and M. Lynch. 2018. *Evolution and Selection of Quantitative Traits*. Oxford: Oxford University Press.

Walsh, D. M. 2015. *Organism, Agency, and Evolution*. New York: Cambridge University Press.

———. 2019. The paradox of population thinking: First order causes and higher order effects. In: *Evolutionary Causation: Biological and Philosophical Reflections*, ed. T Uller and K. N. Laland, 227–246. Cambridge, MA: MIT Press.

Walsh, D. M., and G. Rupik. 2023. The agential perspective: Countermapping the modern synthesis. In Agency in living systems, ed. A. Moczek and S. Sultan, special issue, *Evolution and Development* 25(6):335–352.

Wang, E. T., G. Kodama, P. Baldi, and R. K. Moyzis. 2006. Global landscape of recent inferred Darwinian selection for *Homo sapiens*. *Proceedings of the National Academy of Sciences of the USA* 103:135–140.

Wang, J., Y. Wurm, M. Nipitwattanaphon, O. Riba-Grognuz, Y. Huang, D. Shoemaker, and L. Keller. 2013. A Y-like social chromosome causes alternative colony organization in fire ants. *Nature* 493:664–668.

Wang, X., L. Pipes, L. N. Trut, Y. Herbeck, A. V. Vladimirova, R. G. Gulevich, A. V. Kharlamova, et al. 2018. Genomic responses to selection for tame/aggressive behaviors in the silver fox (*Vulpes vulpes*). *Proceedings of the National Academy of Sciences of the USA* 115:10398–10403.

Wang, Y., Z. P. Chen, H. Hu, J. Lei, Z. Zhou, B. Yao, L. Chen, et al. 2021. Sperm microRNAs confer depression susceptibility to offspring. *Science Advances* 7:eabd7605.

Ward, D. 2016. *The Biology of Deserts*, 2nd ed. Oxford: Oxford University Press.

Waring, T. M., and Z. T. Wood. 2021. Long-term gene–culture coevolution and the human evolutionary transition. *Proceedings of the Royal Society B* 288:20210538.

Warner, R. R. 1988. Traditionality of mating-site preferences in a coral reef fish. *Nature* 335:719–721.

———. 1990. Male versus female influences on mating-site determination in a coral-reef fish. *Animal Behaviour* 39:540–548.

Washburn, S. L. 1960. Tools and human evolution. *Scientific American* 203:63–75.

Watson, J. D., and F.H.C. Crick. 1953. Molecular structure of nucleic acids: A structure for deoxyribose nucleic acid. *Nature* 171:737–738.

Watson, R. A. 2020. Evolvability. In *Evolutionary Developmental Biology: A Reference Guide*, ed. L. Nuño de la Rosa and G. Müller. 1–16. Cham, Switzerland: Springer International.

Watson, R. A., and E. Szathmary. 2016. How can evolution learn? *Trends in Ecology and Evolution* 31:147–157.

Watson, R. A., and C. Thies. 2019. Are developmental plasticity, niche construction and extended inheritance necessary for evolution by natural selection? The role of active phenotypes in the minimal criteria for Darwinian individuality. In: *Evolutionary Causation: Biological and Philosophical Reflections*, ed. T. Uller and K. N. Laland, 197–226. Cambridge, MA: MIT Press.

Watson, R. A., R. Mills, C. L. Buckley, K. Kouvaris, A. Jackson, S. T. Powers, C. Cox, et al. 2016. Evolutionary connectionism: Algorithmic principles underlying the evolution of biological organisation in evo-devo, evo-eco and evolutionary transitions. *Evolutionary Biology* 43:553–581.

Watson, R. A., G. P. Wagner, M. Pavličev, D. M. Weinreich, and R. Mills. 2014. The evolution of phenotypic correlations and "developmental memory." *Evolution* 68:1124–1138.

Watt, W. B. 2013. Causal mechanisms of evolution and the capacity for niche construction. *Biology and Philosophy* 28:757–766.

Wcislo, W. T. 1989. Behavioral environments and evolutionary change. *Annual Review of Ecology and Systematics* 20:137–169.

Weber, C., Y. Zhou, J. G. Lee, L. L. Looger, G. Qian, C. Ge and B. Capel. 2020. Temperature-dependent sex determination is mediated by pSTAT3 repression of Kdm6b. *Science* 368:303–306.

Webster, M. M., and K. N. Laland. 2011. Reproductive state affects reliance on public information in sticklebacks. *Proceedings of the Royal Society B* 278: 619–627.

Webster, M. M., L. Chouinard-Thuly, G. Herczeg, J. Kitano, R. Riley, S. Rogers, M. D. Shapiro, et al. 2019. A four-questions perspective on public information use in sticklebacks (Gasterosteidae). *Royal Society Open Science* 6:181735.

Webster, N. S., and T. Reusch. 2017. Microbial contributions to the persistence of coral reefs. *ISME Journal* 11:2167–2174.

Wells, J.C.K., and J. T. Stock. 2020. Life history transitions at the origins of agriculture: A model for understanding how niche construction impacts human growth, demography and health. *Frontiers in Endocrinology* 11. https://doi.org/10.3389/fendo.2020.00325.

West, M. J., and A. P. King. 1987. Settling nature and nurture into an ontogenetic niche. *Developmental Psychobiology* 20:549–562.

West, S. A., C. El Mouden, and A. Gardner. 2011. Sixteen common misconceptions about the evolution of cooperation in humans. *Evolution and Human Behavior* 32:231–262.

West, S. A., A. S. Griffin, and A. Gardner. 2007. Social semantics: Altruism, cooperation, mutualism, strong reciprocity and group selection. *Journal of Evolutionary Biology* 20:415–432.

West, W. F. 1970. The Bulawayo Symposium papers, no 2: Termite prospecting. *Chamber of Mines Journal* 30:32–35.

West-Eberhard, M. J. 1989. Phenotypic plasticity and the origins of diversity. *Annual Review of Ecology and Systematics* 20:249–278.

———. 2003. *Developmental Plasticity and Evolution*. New York: Oxford University Press.

———. 2005. Phenotypic accommodation: Adaptive innovation due to developmental plasticity. *Journal of Experimental Zoology Part B* 304:610–618.

———. 2021. Foreword: A perspective on plasticity. In: *Phenotypic Plasticity and Evolution: Causes, Consequences, Controversies*, ed. D. W. Pfennig, 4–21. New York: CRC.

Whalen, A., D. Cownden, and K. N. Laland. 2015. The learning of action sequences through social transmission. *Animal Cognition* 18:1093–1103.

Wheeler, B. C., and J. Fischer. 2012. Functionally referential signals: A promising paradigm whose time has passed. *Evolutionary Anthropology* 21:195–205.

———. 2015. The blurred boundaries of functional reference: A response to Scarantion and Clay. *Animal Behaviour* 100:e9–e13.

Wheeler, D. A., M. Srinivasan, M. Egholm, Y. Shen, L. Chen, A. McGuire, W. He, et al. 2008. The complete genome of an individual by massively parallel DNA sequencing. *Nature* 452:872–876.

White, D. J. 2004. Influences of social learning on mate-choice decisions. *Learning and Behavior* 32:105–113.

White, D. J., and B. G. Galef. 2000. "Culture" in quail: Social influences on mate choices of female *Coturnix japonica*. *Animal Behaviour* 59:975–979.

Whitehead, H., and L. Rendell. 2015. *The Cultural Lives of Whales and Dolphins*. Chicago: University of Chicago Press.

Whitehead, H., K. N. Laland, L. Rendell, R. Thorogood, and A. Whiten. 2019. The reach of gene–culture coevolution in animals. *Nature Communications* 10:2405.

Whiten, A. 2017. A second inheritance system: The extension of biology through culture. *Interface Focus* 7:20160142.

———. 2019. Cultural evolution in animals. *Annual Review of Ecology, Evolution, and Systematics* 50:27–48

———. 2021. The burgeoning reach of animal culture. *Science* 372:eabe6514.

Whiten, A., and D. Erdal. 2012. The human socio-cognitive niche and its evolutionary origins. *Philosophical Transactions of the Royal Society B* 367:2119–2129.

Whiten, A., and C. P. van Schaik. 2007. The evolution of animal "cultures" and social intelligence. *Philosophical Transactions of the Royal Society B* 362:603–620.

Whiten, A., F. J. Ayala, M.W. Feldman, and K. N. Laland. 2017. The extension of biology through culture. *Proceedings of the National Academy of Sciences of the USA* 114:7775–7781.

Whiten, A., J. Goodall, W. C. McGrew, T. Nishida, V. Reynolds, Y. Sugiyama, C. E. Tutin, et al. 1999. Cultures in chimpanzees. *Nature* 399:682–685.

———. 2001. Charting cultural variation in chimpanzees. *Behaviour* 138:1481–1516.

Whittle, C. A., S. P. Otto, M. O. Johnston, and J. E. Krochko. 2009. Adaptive epigenetic memory of ancestral temperature regime in *Arabidopsis thaliana*. *Botany* 87:650–657.

Wilke, C. O., and C. Adami. 2003. Evolution of mutational robustness. *Mutation Research/ Fundamental and Molecular Mechanisms of Mutagenesis* 522:3–11.

Wilkin, D. J., J. K. Szabo, R. Cameron, S. Henderson, G. A. Bellus, M. L. Mack, I. Kaitila, et al. 1998. Mutations in fibroblast growth-factor receptor 3 in sporadic cases of achondroplasia occur exclusively on the paternally derived chromosome. *American Journal of Human Genetics* 63:711–716.

Wilkins, A. S. 2017. Revisiting two hypotheses on the "domestication syndrome" in light of genomic data. *Vavilov Journal of Genetics and Breeding* 21:435–442.

———. 2020. A striking example of developmental bias in an evolutionary process: The "domestication syndrome." *Evolution and Development* 22:143–153.

Wilkins, A. S., R. W. Wrangham, and W. T. Fitch. 2014. The domestication syndrome in mammals: A unified explanation based on neural crest cell behavior and genetics. *Genetics* 197:795–808.

———. 2021. The neural crest/domestication syndrome hypothesis, explained: Reply to Johnsson, Henriksen, and Wright. *Genetics* 219:iyab098.

Wilkinson, G. S. 1992. Information transfer at evening bat colonies. *Animal Behaviour* 44:501–518.

Williams, G. C. 1966. *Adaptation and Natural Selection*. Princeton, NJ: Princeton University Press.

Williams, J. B. 2001. Energy expenditure and water flux of free-living dune larks in the Namib: A test of the reallocation hypothesis on a desert bird. *Functional Ecology* 15:175–185.

Wilson, D. S. 2010. Multilevel selection and major transitions. In: *Evolution: The Extended Synthesis*, ed. M. Pigliucci and G. B. Müller, 81–94. Cambridge, MA: MIT Press.

Wilson, D. S., M. van Vugt, and R. O'Gorman. 2008. Multilevel selection theory and major evolutionary transitions: Implications for psychological science. *Current Directions in Psychological Science* 17:6–9.

Wimsatt, W. C. 2013. The role of generative entrenchment and robustness in the evolution of complexity. In *Complexity and the Arrow of Time*, ed. C. H. Lineweaver, P.C.W. Davies, and M. Ruse, 308–331. Cambridge: Cambridge University Press.

———. 2021. Levels, robustness, emergence and heterogeneous dynamics: Finding partial organization in causal thickets. In: *Levels of Organization in the Biological Sciences*, ed. D. S. Brooks, J. DiFrisco, and W. C. Wimsatt, 21–38. Cambridge, MA: MIT Press.

Witte, K., and R. Massmann. 2003. Female sailfin mollies (*Poecilia latipinna*) remember males and copy the choice of others after 1 day. *Animal Behaviour* 65:1151–1159.

Wolf, J. B., and M. J. Wade. 2009. What are maternal effects (and what are they not)? *Philosophical Transactions of the Royal Society B* 364:1107–1115.

Wolf, L. V., Y. Yang, J. Wang, Q. Xie, B. Braunger, E. R. Tamm, and J. Zavadil. 2009. Identification of Pax6-dependent gene regulatory networks in the mouse lens. *PLOS One* 4:e4159.

Wolpert, L. 1969. Positional information and the spatial pattern of cellular differentiation. *Journal of Theoretical Biology* 25:1–47.

———. 2011. *Developmental Biology: A Very Short Introduction*. Oxford: Oxford University Press.

Woodward, J. 2006. Sensitive and insensitive causation. *Philosophical Review* 115:1–50.

Workman, A. D., C. J. Charvet, B. Clancy, R. B. Darlington, and B. L. Finlay. 2013. Modeling transformations of neurodevelopmental sequences across mammalian species. *Journal of Neuroscience* 33:7368–7383.

Woznica, A., J. P. Gerdt, R. E. Hulett, J. Clardy, and N. King. 2017. Mating in the closest living relatives of animals is induced by a bacterial chondroitinase. *Cell* 170:1175–1183.e11.

Wragg Sykes, R. 2020. *Kindred: Neanderthal Life, Love, Death and Art*. London: Bloomsbury.

Wrangham, R. 2009. *Catching Fire: How Cooking Made Us Human*. London: Profile Books.

Wray, G. A., D. A. Futuyma, R. E. Lenski, T.F.C. MacKay, D. Schluter, J. E. Strassman, and H. E. Hoekstra. 2014. Does evolutionary biology need a rethink? Counterpoint: no, all is well. *Nature* 5:161–164.

Wright D., R. Henriksen, and M. Johnsson. 2020. Defining the domestication syndrome: Comment on Lord et al 2020. *Trends in Ecology and Evolution* 35:1059–1060.

Wright, D. F. 2017. Phenotypic innovation and adaptive constraints in the evolutionary radiation of Paleozoic crinoids. *Scientific Reports* 7:13745.

Wright, S. 1988. Surfaces of selective value revisited. *American Naturalist* 131:115–123.

Wu, H., X. Guang, M. B. Al-Fageeh, J. Cao, S. Pan, H. Zhou, L. Zhang, et al. 2014. Camelid genomes reveal evolution and adaptation to desert environments. *Nature Communications* 5:5188.

Wu, H., R. Hauser, S. A. Krawetz, and J. R. Pilsner. 2015. Environmental susceptibility of the sperm epigenome during windows of male germ cell development. *Current Environmental Health Reports* 2:356–366.

Wund, M. A., J. A. Baker, B. Clancy, J. L. Golub, and S. A. Foster. 2008. A test of the "flexible stem" model of evolution: Ancestral plasticity, genetic accommodation, and morphological divergence in the threespine stickleback radiation. *American Naturalist* 172:449–462.

Wyles, J. S., J. G. Kunkel, and A. C. Wilson. 1983. Birds, behavior, and anatomical evolution. *Proceedings of the National Academy of Sciences of the USA* 80:4394–4397.

Wylie, C. D. 2019. The plurality of assumptions about fossils and time. *History and Philosophy of the Life Sciences* 41:21.

Xavier-Neto, J., R. A. Castro, A. C. Sampaio, A. P. Azambuja, H. A. Castillo, R. M. Cravo, and M. S. Simões-Costa. 2007. Cardiovascular development: Towards biomedical applicability. *Cellular and Molecular Life Sciences* 64:719.

Xia, B., W. Zhang, G. Zhao, X. Zhang, J. Bai, R. Brosh, A. Wudzinska, et al. 2024. On the genetic basis of tail-loss evolution in humans and apes. *Nature* 626:1042–1048.

Xiong, X., S. L. Loo, and M. M. Tanaka. 2022. Gut mutualists can persist in host populations despite low fidelity of vertical transmission. *Evolutionary Human Sciences* 4:e41.

Xue, C., Z. Liu, and N. Goldenfeld. 2020. Scale-invariant topology and bursty branching of evolutionary trees emerge from niche construction. *Proceedings of the National Academy of Sciences of the USA* 117:7879–7887.

Yamamoto, Y., M. S. Byerly, W. R. Jackman, and W. R. Jeffery. 2009. Pleiotropic functions of embryonic sonic hedgehog expression link jaw and taste bud amplification with eye loss during cavefish evolution. *Developmental Biology* 330:200–211.

Yamamoto, Y., D. W. Stock, and W. R. Jeffery. 2004. Hedgehog signalling controls eye degeneration in blind cavefish. *Nature* 431:844–847.

Yampolsky, L. Y., and A. Stoltzfus. 2001. Bias in the introduction of variation as an orienting factor in evolution. *Evolution and Development* 3:73–83.

Yan, Z., J. Sun, O. Don, and Z. Zhang. 2005. The red turpentine beetle, *Dendroctonus valens leconte* (Scolytidae): An exotic invasive pest of pine in China. *Biodiversity and Conservation* 14:1735–1760.

Yehuda, R., L. M. Bierer, J. Schmeidler, D. H. Aferiat, I. Breslau, and S. Dolan. 2000. Low cortisol and risk for PTSD in adult offspring of Holocaust survivors. *American Journal of Psychiatry* 157:1252–1259.

Yehuda, R., G. Cai, J. A. Golier, C. Sarapas, S. Galea, M. Ising, T. Rein, et al. 2009. Gene expression patterns associated with posttraumatic stress disorder following exposure to the World Trade Center attacks. *Biological Psychiatry* 66:708–711.

Yehuda, R., S. M. Engel, S. R. Brand, J. Seckl, S. M. Marcus, and G. S. Berkowitz. 2005. Transgenerational effects of posttraumatic stress disorder in babies of mothers exposed to the World Trade Center attacks during pregnancy. *Journal of Clinical Endocrinology and Metabolism* 90:4115–4118.

Ylikoski, P., and J. Kuorikoski. 2010. Dissecting explanatory power. *Philosophical Studies* 148: 201–219.

Yoffe, M., K. Patel, E. Palia, S. Kolawole, A. Streets, G. Haspel, and Daphne Soares. 2020. Morphological malleability of the lateral line allows for surface fish (*Astyanax mexicanus*) adaptation to cave environments. *Journal of Experimental Zoology Part B* 334:511–517.

Yong, E. 2016. Why turtles evolved shells: It wasn't for protection. *Atlantic*, July 14, 2016. https://www.theatlantic.com/science/archive/2016/07/the-turtle-shell-first-evolved-for-digging-not-defence/491087/.

Young, J. W., D. Fernández, and J. G. Fleagle. 2010. Ontogeny of long bone geometry in capuchin monkeys *Cebus albifrons* and *Cebus apella*: Implications for locomotor development and life history. *Biology Letters* 6:197–200.

Young, N. M., and B. Hallgrímsson. 2005. Serial homology and the evolution of mammalian limb covariation structure. *Evolution* 59:2691–2704.

Young, N. M., B. Winslow, S. Takkellapati, and K. Kavanagh. 2015. Shared rules of development predict patterns of evolution in vertebrate segmentation. *Nature Communications* 6:6690.

Zardoya, R., and A. Meyer. 1998. Complete mitochondrial genome suggests diapsid affinities of turtles. *Proceedings of the National Academy of Sciences of the USA* 95:14226–14231.

Zattara, E. E., H. A. Busey, D. M. Linz, Y. Tomoyasu, and A. P. Moczek. 2016. Neofunctionalization of embryonic head patterning genes facilitates the positioning of novel traits on the dorsal head of adult beetles. *Proceedings of the Royal Society B* 283:20160824.

Zattara, E. E., A.L.M. Macagno, H. A. Busey, and A. P. Moczek. 2017. Development of functional ectopic compound eyes in scarabaeid beetles by knockdown of orthodenticle. *Proceedings of the National Academy of Sciences of the USA* 114:12021–12026.

Zeder, M. A. 2015. Core questions in domestication research. *Proceedings of the National Academy of Sciences of the USA* 112:3191–3198.

———. 2017. Domestication as a model system for the extended evolutionary synthesis. *Interface Focus* 7:20160133.

———. 2018. Why evolutionary biology needs anthropology: Evaluating core assumptions of the extended evolutionary synthesis. *Evolutionary Anthropology* 27:267–284.

———. 2020. Straw foxes: Domestication syndrome evaluation comes up short. *Trends in Ecology and Evolution* 35:647–649.

Zeder, M. A., and B. D. Smith. 2009. A conversation on agricultural origins: Talking past each other in a crowded room. *Current Anthropology* 50:681–690.

Zeeman, E. C. 1977. *Catastrophe Theory: Selected papers, 1972–1977.* Reading, MA: Addison-Wesley.

Zhang, J. X., and W. P. Maddison. 2013. Molecular phylogeny, divergence times and biogeography of spiders of the subfamily Euophryinae (Araneae: Salticidae). *Molecular Phylogenetics and Evolution* 68:81–92.

Zhu, J., Y.-T. Zhang, M. S. Alber, and S. A. Newman. 2010. Bare bones pattern formation: A core regulatory network in varying geometries reproduces major features of vertebrate limb development and evolution. *PLOS One* 5:e10892.

Zimmer, C. 2018. *She Has Her Mother's Laugh.* New York: Dutton.

Zimmer, C., and D. J. Emlen. 2013. *Evolution: Making Sense of Life.* Greenwood Village, CO: Roberts.

Zogbaum, L., P. G. Friend, and R. C. Albertson. 2021. Plasticity and genetic basis of cichlid gill arch anatomy reveal novel roles for hedgehog signaling. *Molecular Ecology* 30:761–774.

INDEX

Page numbers in *italics* refer to figures and tables

GPSR Authorized Representative: Easy Access System Europe - Mustamäe tee 50, 10621 Tallinn, Estonia, gpsr.requests@easproject.com

www.ingramcontent.com/pod-product-compliance
Ingram Content Group UK Ltd.
Pitfield, Milton Keynes, MK11 3LW, UK
UKHW042251300325
456823UK00003B/3/J